EXAMINING THE BIG BANG
AND DIFFUSE BACKGROUND RADIATIONS

T0183517

INTERNATIONAL ASTRONOMICAL UNION

UNION ASTRONOMIQUE INTERNATIONALE

EXAMINING THE BIG BANG AND DIFFUSE BACKGROUND RADIATIONS

PROCEEDINGS OF THE 168TH SYMPOSIUM OF THE
INTERNATIONAL ASTRONOMICAL UNION,
HELD IN THE HAGUE, THE NETHERLANDS,
AUGUST 23–26, 1994

EDITED BY

MENAS KAFATOS

*Center for Earth Observing and Space Research,
Institute for Computational Sciences & Informatics,
George Mason University, Fairfax, VA, U.S.A.*

and

YOJI KONDO

*NASA/Goddard Space Flight Center,
Greenbelt, MD, U.S.A.*

KLUWER ACADEMIC PUBLISHERS

DORDRECHT / BOSTON / LONDON

Library of Congress Cataloging-in-Publication Data

Examining the big bang and diffuse background radiations / edited by
  Menas Kafatos and Yoji Kondo.
      p.   cm.
    "International Astromonical Union Symposia. Volume 168"--T.p.
  verso.
    Includes index.
    ISBN 978-0-7923-3815-4 (hardcover : alk. paper)
    1. Cosmic background radiation--Congresses. 2. Big bang theory-
  -Congresses. 3. Galaxies--Congresses. I. Kafatos, Minas C.
  II. Kondo, Yoji.
  QB991.C64E93 1996
  523.1--dc20   95-40169

ISBN-13: 978-0-7923-3815-4   e-ISBN-13:978-94-009-0145-2
DOI: 10.1007/978-94-009-0145-2

Published on behalf of
the International Astronomical Union
by
Kluwer Academic Publishers, P.O. Box 17, 3300 AA Dordrecht, The Netherlands.

Kluwer Academic Publishers incorporates
the publishing programmes of
D. Reidel, Martinus Nijhoff, Dr W. Junk and MTP Press.

Sold and distributed in the U.S.A. and Canada
by Kluwer Academic Publishers,
101 Philip Drive, Norwell, MA 02061, U.S.A.

In all other countries, sold and distributed
by Kluwer Academic Publishers Group,
P.O. Box 322, 3300 AH Dordrecht, The Netherlands.

Printed on acid-free paper

TABLE OF CONTENTS

PREFACE ix

SCIENTIFIC ORGANIZING COMMITTEE xi

PART 1: INVITED REVIEWS 1

L. WOLTJER: Introductory Remarks 3

VIRGINIA TRIMBLE: Backgrounds and the Big Bang: Some Extracts
   from Their History 9

JOHN MATTER: Measurement and Implications of the Cosmic Microwave
   Background Spectrum 17

G.F. SMOOT: Of Cosmic Background Anisotropies 31

I. STRUKOV, et al: Status of the Relict-2 Mission and Our Future Plans ... 45

R. B. PARTRIDGE: Fluctuations in the Microwave Sky 59

A.K. SINGAL: Radio Galaxies and Quasars as Cosmological Probes 71

JAMES S. DUNLOP: High Redshift Radio Galaxies 79

HUBB ROTTGERING: Cold Material in Distant Radio Galaxies 89

P.N. WILKINSON, et al: A Search for Gravitational Milli-Lenses 95

MICHAEL G. HAUSER: The COBE DIRBE Search for the Cosmic
   Infrared Background 99

C.J. CESARSKY and D. ELBAZ: Galaxy Formation -- ISOCAM Counts
   109

H. OKUDA: Observations of Diffuse IR Background Radiation by IRTS
   and IRIS 117

P.M. LUBIN: Degree Scale Anisotropy -- Current Status 125

JOHN P. HUCHRA: Galaxy Motions in the Nearby Universe 143

SIDNEY VAN DEN BERGH: The Hubble Parameter - A Status Report at
   EPOCH 1994.5 157

G.A. TAMMANN and ALLAN SANDAGE: The Local Velocity Field and
   the Hubble Constant 163

D.S. MATHEWSON and V.L. FORD: Large-Scale Flows in the Local
   Universe 175

RICCARDO GIOVANELLI, et al: Spiral Galaxies and the Peculiar
   Velocity Field 183

J. EINASTO: Formation of the Supercluster-Void Network 193

DAVID C. KOO: Faint Field Galaxy Counts, Colors, and Redshifts 201

PETER SCHNEIDER: Cosmological Applications of Gravitational Lensing
   209

R.E. GRIFFITHS, et al: The HST Medium Deep Survey -- Galaxy
   Morphology at High Redshift 219

PETER JAKOBSEN: The Far-Ultraviolet Background 229

JILL BECHTOLD: Evolution of the EUV Background from Quasar
Absorption Line Studies 237

G. HASINGER: The X-Ray Background 245

GIANCARLO SETTI and ANDREA COMASTRI: The Sources of the
Hard X-Ray Background 263

P. LAURENT: The Cosmological Diffuse γ-Ray Background: Myth or
Reality? 271

P. SREEKUMAR and D.A. KNIFFEN: Diffuse Gamma Rays of Galactic
and Extragalactic Origin: Preliminary Results from EGRET 279

I. NOVIKOV: Big Bang Scenario and Nature of Dark Matter 289

MICHAEL S. TURNER: The Hot Big Bang and Beyond 301

J.A. FRIEMAN: Inflation, Microwave Background Anisotropy, and Open
Universe Models 321

F. HOYLE, G. BURBIDGE and J.V. NARLIKAR: The Quasi-Steady State
Cosmology 329

H. ARP: X-Ray Observations of Galaxy-Quasar Associations 369

D.W. SCIAMA: The Present Status of the Decaying Neutrino Theory . . . 381

MARTIN J. REES: Background Radiation -- Probes and Future Tests . . . 389

YOJI KONDO: A Panel Discussion of "Major Unsolved Problems of
Cosmology" 399

H. ARP: Fundamental Observational Problems 401

G. BURBIDGE: Redshifts of Unknown Origin 407

JOHN C. MATHER: Future Cosmic Microwave and Cosmic Infrared
Background Measurements 419

PHILIP MORRISON: Three Cosmological Remarks 423

R.B. PARTRIDGE: Panel Contribution -- IAU Symposium 168 427

MENAS KAFATOS: Knowledge Limits in Cosmology 431

PART 2: CONTRIBUTED PAPERS 439

S.N. DUTTA and G. EFSTATHIOU: Temperature Fluctuations of the
Microwave Background in Primeval Isocurvature Baryon Models
. . . . 441

E. MARTINEZ-GONZALEZ and J.L. SANZ: CMB -- Anisotropies Due
to Non-Linear Clustering 445

J.L. PUGET, et al: Planning Future Space Measurements of the CMB . . . 447

R.D. DAVIES, et al: New Results on CMB Structure from the Tenerife
Experiments 453

M. MELEK: On the Use of COBE Results 461

E.M. DE GOUVEIA DAL PINO, et al: Is the Early Universe Fractal . . . 465

P. DRIESSEN: A Fractal Model of the Universe 467

R.A. DALY: Powerful Extended Radio Sources -- A Goldmine for
Cosmology 469

I.B. VAVILOVA: On the Use of Fractal Concepts in Analysis of
Distributions of Galaxies 473

L.I. GURVITS: Milliarcsecond Radio Structure of AGN as a Cosmological
Probe 477
J.V. WALL, C.R. BENN and A.J. LOAN: Mapping Large-Scale Structure
with Radio Sources 481
J.L. JONAS: The Rhodes/Hartrao 2300 Mhz Horn Telescope 487
J.C. JACKSON and MARINA DODGSON: Vacuum-Energy and the
Angular-Size/Redshift Diagram for Milliarcsecond Radio Sources
. . . . 489
RENDONG NAN and ZHENGDONG CAI: A Hint to Possible Anisotropy
in Radio Universe 491
E. MARTINEZ-GONZALEZ, et al: Projected Clustering Around $1 < Z < 2$
Radiogalaxies 493
E. MARTINEZ-GONZALEZ, et al: Upper Limits on the $LY\alpha$ Emission at
$Z = 3.4$ 497
J.L. SANZ, et al: Intermediate Resolution Spectroscopy of the Radio
Galaxy B2 $0902 + 34$ at $Z \approx 3.4$ 499
J. VON LINDE, et al: Foreground Galaxies around Luminous Quasars . . 501
MIRTA MOSCONI, PATRICIA TISSERA and DIEGO GARCIA
LAMBAS: Evolution of Galaxy Luminosity in the CDM Model . . 503
DORU MARIAN SURAN: Cosmological Parameters Determinations from
Deep Sky Redshift Surveys 505
CARLOS A. VALOTTO and DIEGO GARCIA LAMBAS: A Statistical
Study of Environment Effects on Galaxy Properties 509
JOANNA ANOSOVA and LUDMILA KISELEVA: Chance and Non-
Chance Clustering in the Universe and Problems of High Redshift
Galaxies in Compact Groups 511
S.P. BHAVSAR and D.A. LAUER: Analysis of the CFA "Great Wall"
Using the Minimal Spanning Tree 517
I. KUNEVA and M. KALINKOV: Superclusters and Supervoids 521
J. ANOSOVA, S. IYER and R.K. VARMA: Clusters and Voids in General
Galaxy Field 523
LUDMILA KISELEVA and JOANNA ANOSOVA: Typical Characteristics
of Chance and Non-Chance Compact Groups of Galaxies 525
U. BORGEEST, K.-J. SCHRAMM and J. VON LINDE: A Dedicated
Quasar Monitoring Telescope 527
JEFFREY LINSKY: Accurate Measurements of the Local Deuterium
Abundance from HST Spectra 529
A.V. MANDZHOS: Mutual Interference and Structural Properties of
Object Images in the Vicinity of the Gravitational Lens CUSP
Point 533
T. SCHRAMM: Complex Theory of Gravitational Lenses Part I 535
V.L. AFANASIEV, et al: Field Spectroscopy Observations the
Gravitational Lenses $H1413 + 117$ and $Q2237 + 030$, Preliminary
Results 537
J. PELT, et al: The Time Delay Between QSO $0957 + 561$ A, B 539

viii

S.V. KHMIL: Gravitational Macrolensing and Quasar Spectra 541
JONATHAN E. GRINDLAY and EYAL MAOZ: Halo/Thick Disk CVS
    and the Cosmic X-Ray Background 543
I.M. GIOIA, et al: Arcs in X-Ray Selected Clusters 549
NOAH BROSCH: Tauvex and the Nature of the Cosmological UV
    Background 553
D. DUARI, P. DASGUPTA and J.V. NARLIKAR: Peaks and Periodicities
    in the Redshift Distribution of Quasi-Stellar Object 555
YUVAL NEEMAN: Inflationary Cosmogony, Copernican Relevelling and
    Extended Reality 559
M. ROOS, et al: Do Massive Neutrinos Ionize Intergalactic HI? 563
ALEXANDER GUSEV: Phase Transition at the Metric Elastic Universe . . 569
ALEXANDER GUSEV: Non-Singular Metric Elastic Universe 571
M.I. WANAS and M.A. BAKRY: Stability of Cosmological Models 573
M.I. WANAS: Cosmological Models in AP-Spaces 575
MARIO G. ABADI, DIEGO G. LAMBAS and PATRICIA B. TISSERA:
    Cosmological Simulations with Smoothed Particle Hydrodynamics
    577

AUTHOR INDEX 579
SUBJECT INDEX 581
OBJECT INDEX 585

PREFACE

IAU Symposium No. 168, "Examining the Big Bang and Diffuse Background Radiations", took place on 1994 August 23-26 at the XXIInd IAU General Assembly in the Hague, Netherlands.

The meeting attracted a large number -- over 250 -- of astronomers, reflecting the strong interest engendered by the great advances in cosmology made in recent years. There still remains a multitude of unresolved problems in modern cosmology and the symposium offered a salubrious occasion to examine them objectively -- if not altogether dispassionately -- at a place where many leading workers in related fields gathered together.

After the introduction by IAU President L. Woltjer and the historical background by Vice President Virginia Trimble, this volume begins with reviews of the cosmic microwave radiation from COBE (Cosmic Background Explorer). Reviews of recent observations then extend from radio to infrared, visible light, ultraviolet, X-rays and gamma-rays. It is followed by theoretical models for the Big Bang and Inflation, and alternative views to the Big Bang. Following a discourse on Probes and Future Tests, the meeting ended with Panel Discussion on "Major Unsolved Problems of Cosmology". Some forty-four contributed papers -- both oral and poster reports -- are included after the invited talks and panel discussions. Regrettably, manuscripts of some invited reviews and panel talks have not reached us as we went to press in 1995.

Following the lucid reviews on the Hubble parameters, the session chair took a survey of the audience. Apparently, a similar poll had been taken a few years earlier at another meeting on the subject. The results were entirely different this time. The majority voted for the view that we do not know if the Hubble parameter is greater or smaller than 75 km/sec/Mpc. Of the remaining 30-40 percent, most voted for a value greater than 75. Of course, scientific progresses are not made by consensus, but it demonstrated an age-old (but often neglected) wisdom that persuasive and impassioned lectures can greatly influence the views of a body of scientists.

With that caveat in mind, we will be pleased indeed if this volume should assist the readers in forming their own views based on what is currently known and unknown. If their conclusion should turn out the be that we do not yet have answers to a number of cosmological questions, that too will be a satisfactory outcome for this symposium.

We would like to extend our appreciation to the members of the SOC for their contributions, help and support. We would also like to thank the General Assembly LOC for their competent assistance. Mr. John Zhu provided invaluable service in the editing of the present volume.

Menas Kafatos and Yoji Kondo
Co-editors

PART I

Invited Reviews

Introductory Remarks

L. Woltjer *

Those of us who were around a few decades ago, frequently became initiated into cosmology by a very clear and concise small book, written by Herman Bondi, entitled "Cosmology" (1952/1960). It is interesting to see the list of chapter headings:

    Physics and Cosmology

    The Cosmological Principle

Observational Evidence :

The background light of the sky	(Olbers' paradox)
The problem of inertia	(Mach's principle)
Observations of distant nebulae	(isotropy, $m < 19$, $z < 0.2$)
Astrophysical and geophysical data	(ages, abundances)
Microphysics and Cosmology	(large numbers)

Theories :

Newtonian Cosmology	
Relativistic Cosmology	(Friedmann/Lemaitre models)
Steady State Cosmology	
Milne/Eddington/Dirac/Jordan.	

Today we still believe in the cosmological principle, the background light is the main topic of this symposium, Mach's principle still has an uncertain status, and distant nebulae now are found at $B = 26$ and fainter and at $z = 4$ or more. A particular problem at the time concerned ages, with the earth at 4 Gyr being about twice as old as the Universe. Today we worry about the same problem with the globular clusters at some 15 Gyr perhaps older than the Universe, at

---

* Observatoire de Haute Provence, F-04870 Saint Michel l'Observatoire, France

M. Kafatos and Y. Kondo (eds.), Examining the Big Bang and Diffuse Background Radiations, 3–7.
© 1996 IAU.

least if the short distance scale were correct. The "large numbers" ("radius" of Universe / "radius" of electron, etc.) nowadays tend to be used in support of "anthropic" ideas according to which there would be nobody to observe the Universe if these numbers were very different. If one believes that it is legitimate to consider ensembles of universes this would appear to make sense. The relativistic cosmologies are still with us with gradually more physics incorporated; the strict Steady State being at variance with the evidence for evolution everywhere in the Universe has been replaced with a steady state on average with more active episodes. The theories of the "heretics" from Bondi's book have been forgotten, but not surprisingly others have taken their place. On the whole one can say that the main stream models of the Universe have not changed all that much, but the current problems are more connected to the very early phases (inflation, etc.), the formation of galaxies, the origin of some chemical elements and other aspects of the physical content of the Universe.

Sociologically cosmologists have become a more abundant species, which is no surprise since the subjects mentioned offer ample opportunity for a wide variety of quantitative researches. Whereas the Astrophysical Journal 25 years ago contained about 1 % of cosmology, today the figure is around 6 % in a very much enlarged journal.

Perhaps the greatest progress has been made in the observation and analysis of the extragalactic background light and its implications for the physics of the Universe. This light has been observed in two spectral domains: first in X-rays in 1962 by Giacconi, Gursky, Paolini and Rossi (1962), and a year later in the microwave region by Penzias and Wilson (1965). The former appears to be due largely to discrete sources at modest redshifts, while the latter is more likely to be truly diffuse, mainly reflecting conditions at redshifts of 1000.

The first satisfactory calculation of the likely background due to sources was made by Loys de Cheseaux in 1744, long before Olbers (1823) published his account of the "paradox". It is interesting to compare the two accounts. Whereas Olbers gives a long philosophical introduction, de Cheseaux gives the calculation in a few lines which are very clear and in a form in which we can use it to calculate the X-ray background. Starting out from shells of equal thickness containing stars at a fixed density, he shows that each shell contributes the same amount to the background (if the inverse square law holds) and then calculates at which radius of his "universe" the surface brightness becomes equal to that of the sun. Obviously the night sky is much less bright than that, and he therefore concluded that either the volume in which the fixed stars are contained has a much smaller cutoff radius or absorption invalidates the inverse square law; since not much absorption per light year would be needed, he concluded that the latter is the more probable reason why the sky is so dark.

Exactly this same reasoning may be applied to the X-ray background. Early suggestions (Setti and Woltjer 1973) that this background could be due to the integrated effect of sources have been amply confirmed. In fact, as reported at this symposium, at least 2/3 of the background is due to unresolved sources, mainly AGN. Since intergalactic space should be transparent to X-rays it follows that the limited surface brightness of the background implies a maximum radius for the volume in which the sources are contained - at least if a Newtonian calculation is valid. Of course, the maximum radius is globally the radius where redshift effects terminate that validity. The implication of this is that the X-ray sources responsible for most of the background must be at a cosmological redshift of order unity as, in fact, they are observed to be. However, if we were to believe that the observed AGN have intrinsic redshifts and that they are much closer by then the background due to more distant sources would become too large. This simple argument has never been dealt with by the proponents of "local" quasars.

In a way it is unfortunate that the source contribution to the X-ray background is so large. It makes it very difficult, if not impossible, to determine if there is also a truly diffuse component, due to hot gas in the Universe.

Of a much more fundamental importance is the microwave background which will be extensively discussed at this symposium. Gamov predicted this background as a result of an expanding initially very hot universe and made the connection with nucleogenesis and with the decoupling of radiation and matter in the recombination epoch. Alpher, Gamov and Herman in various papers, published between 1948 and 1956, predicted values for the temperature of the black body radiation in the order of 5 K (see Alpher and Herman 1988). Because the nucleogenesis proposed turned out to be not very satisfactory , these predictions were forgotten remarkably soon. In 1957 Denisse, Lequeux and Le Roux not only determined that T_B at 0.9 GHz was less than 3 K, they also determined that the fluctuations on a 40 deg^2 beam were less than 0.5 K, not surprising in view of the fact that COBE found values about 10^5 lower. It remains remarkable that no connection was made to the then recent results of Gamov et al.

The 3 K microwave background finally was discovered by Penzias and Wilson in 1963. Small improvements followed thereafter until in 1992 the COBE satellite determined the spectrum and angular distribution with unprecedented accuracy. The perfection of the black body fit to the spectrum places interesting limits on the energy input following recombination. The main aspect of the angular distribution corresponds to a dipole, generally associated with the motion of the sun with respect to the local rest frame, while the fluctuations around the mean dipole contain interesting, but as yet inconclusive, information on galaxy formation. A major question that remains is why the solar motion with respect to the galaxies within 100 Mpc is so different from that inferred for the dipole (Mathewson et al. 1992).

Cosmology has always been a subject where strong opinions are generally far ahead of the facts and where the tolerance for diverging opinions is low. It is true that some of the diverging opinions show a remarkable ease in forgetting certain facts. Nevertheless, the publication problems of some papers which presented many solid new facts, but hinted at the possibility that the canonical view could be incomplete, is somewhat worrisome. It would after all be remarkable if the present consensus picture of the Universe turned out to be definitive.

References

Alpher, R.A. and Herman, R., 1988, Phys. Today 41 (8), 24

Denisse, J.F., Lequeux, J. and Le Roux, E., 1957, CR Ac. Sci. 244, 3030

Giacconi, R., Gursky, H., Paolini, F.R. and Rossi, B., 1962, Phys. Rev. Lett. 9, 439

Loys de Cheseaux, 1744, Traité de la comète qui a paru en 1743 ... (Lausanne, Bousquet)

Mathewson, D. et al., 1992, ApJ 389, L5

Olbers, W., 1823, Berliner Astronomisches Jahrbuch für das Jahr 1826

Penzias, A.A. and Wilson, R.W., 1965, ApJ 142, 419

Setti, G. and Woltjer, L., 1973, IAU Symp. 55, 208

BACKGROUNDS AND THE BIG BANG:
SOME EXTRACTS FROM THEIR HISTORY

VIRGINIA TRIMBLE

Astronomy Department, University of Maryland,
College Park, MD 20742
& Department of Physics, University of California,
Irvine CA 92717, USA

ABSTRACT.   The first astronomical background firmly established was that of cosmic rays. Photons at various wavelengths came later; and in some bands we have not year clearly peeled away all the sources to see a true background, if one even exists. All known backgrounds are astrophysically important and at least several cosmologically so. The path by which the standard hot big bang came to be generally regarded as standard is littered with the detritus of mistaken impressions, misunderstandings, and missed opportunities.

1. Introduction

The present symposium had its origins in two initially separate and rather different sorts of proposals. The first, put forward by M. Hanner and S. Bowyer, was for a discussion of all known diffuse backgrounds, many of which, like zodiacal light, have very little to do with cosmology. The other was for a meeting on the robustness of the standard hot big bang and alternatives to it, proposed by M. Kafatos and Y. Kondo. These entered into the mysterious interior of the IAU executive committee and came forth as a single symposium, bearing some resemblance to the mythical animals of Dr. Suess.

The remaining sections deal with (2) the discovery and significance of the backgrounds, (3) the standard HBB along the lines of "how we know that" it is standard, (4) items within that picture still to be determined, (5) the cosmic tolerance quotient for deviations from it, and (6) a compromise answer to whether the big bang is important.

2. Known and Expected Backgrounds

The table shows all of these I could think of, electromagnetic and other. Henry (1991) displays the photon backgrounds in an insightful graph. The cautious reader will note that the observed infrared consists of inter-stellar cirrus and thermal and scattered (zodiacal light) contributions from interplanetary dust, not emission from hypothetical primeval

9

M. Kafatos and Y. Kondo (eds.), Examining the Big Bang and Diffuse Background Radiations, 9–16.
© 1996 *IAU.*

galaxies, while detection and significance of the ultraviolet and EUV ones are in considerable disorder (hence the accompanying alternative review by Bowyer, 1991). There is no space to discuss all the wavebands, and I focus here on the gamma ray background (the $\sim$ 100 MeV part of which, due to pion decay, was predicted by Hayakawa etc al. 1958) and the relict microwaves.

KNOWN AND EXPECTED ASTRONOMICAL BACKGROUNDS			
TYPE	D/S	WHO/WHEN	Cosm Sign?
Particles (cosmic rays)	D	Hess, Bothe, Kohlhörster 1911-29	½

Neutrinos
  - 1.9 K
  - high energy
	D
	S
		To be discovered	Yes
			½

Gravitational radiation
  - early universe
  - core collapses etc.
	D
	S
		To be discovered	Yes
			½

Gamma rays
  $\sim$100 MeV (pions)
  $\sim$1 MeV
	D
	S?
		Kraushaar, Clark, Garmire 1961-68
		Arnold et al. 1962
			No
			½

X-rays
  - from hot IGM
  - from AGNs etc
	D
	S
		Known to be small
		Giacconi et al. 1962; Bowyer, Friedman 1963
			Yes
			½

Soft X-ray to EUV - $\sim$0.1 keV (local ISM)	D	Yentis, Novick, Vanden Bout 1972. Many later confirmations	No

1216 - 2500 A
  - dust-scattered starlight
  - extragalactic
	D
	S?
		See reviews, Henry (1991), Bowyer (1991)	No
			?

Optical
  - airglow + Zodiacal light
 	
  - stars and galaxies
  - residual
	D
	
	S
	D?
		- Known to ancients; first photoelectric
		  measurement van Rhijn 1921
		- Seares, van Rhijn et al. 1925; Roach 1960ff
		- To be discovered
			No
			
			½
			?

Infrared
  - 10-25μ (warm interplan dust)
  - 60-100μ (cold interst. dust)
  - 1-200μ from galaxy formation
	D
	D
	S
		IRAS confirmation
		IRAS confirmation
		DIRBE - last word is not yet in
			N
			N
			Y

Submillimeter excess (era of confusion)	X	Nagoya-Berkeley balloon to COBE	N!

Radio
  - mm to cm
 	
  - cm to decameter (galactic
    synchrotron and thermal)
	D
	
	D
		Penzias, Wilson 1964-65 (Ohm 1961 and
		others)
		Jansky 1933; Reber 1940
			Y
			
			N

Cosmic rays (indicated in the table as D, for truly diffuse, as opposed to S, for sum of courses) were initially thought to be very high energy photons. Their particulate nature was established from their extraordinary penetrating power, in a classic one-page paper by Bothe and Kohlhörster (1929). No significant knowledge of German is required to make sense of the change from "Gammastrahlung" in the first sentence to "Korpuskularstrahlen" in the last. The "Cosmological Significance" column for cosmic rays says neither Yes or No, but ½, meaning that not all sources (especially for the highest energies) have been identified, and there may or may not be information about the large scale structure and evolution of the universe to be learned from them.

No extra-solar-system gamma rays were seen until the 1960s, and the numbers were initially small. Kraushaar and Clark (1962), for instance, speak of "the remaining 22 events, which came from a variety of directions in space...", accounting for the rumor that gamma ray astronomy is the field where one photon is a discovery, two is a spectrum, and three is the Rossi Prize. Jim Arnold, who found the 1 MeV flux in Ranger 3 data early in his career, recently retired as director of Calspace, indicating the length of a generation in the filed. Additional early history is described in Fazio's 91967) review article -- written before the first non-solar source had even been firmly established.

The early history of the microwave background is closely coupled with that of big bang nucleosynthesis. McKellar's (1941) interpretation of CN absorption lines in a spectrum taken by W.S. Adams is widely known. But the real stinger is Herzberg's (1950) discussion of the result in his classic volume, Spectra of Diatomic Molecules, "...a rotational temperature of $2.3°$ K follows, which has of course only a very restricted meaning." the real (and highly unrestricted!) meaning just eluded at least three pre-Penzias-and-Wilson measurers of microwave sky temperatures. Woltjer (elsewhere in this volume) alludes to a French measurement, and a symposium participant mentioned a Russian one that may be earliest of all. I note here the work of Ohm (1961), because its misinterpretation led Zeldovich (1962, 1963) and at least some of his colleagues to confine their attention to a cold big bang for some time.

The problem was at least partly a verbal one. The measured sky temperature was very closely what had been predicted, without any cosmological component. But as Ohm (1994) has explained, this is exactly as it should be -- the phrase "sky temperature" meant the radiation attributable to thermal emission from the earth's atmosphere (a cosecant θ component). the total coming into the whole system is the "system temperature", and, as Ohm's Table II shows, this was about $3.3°$ K larger than the sum of all the contributions he could think of. The excess was only about 1σ and was, correctly, reported as an upper limit.

Better calibration gave much higher significance to the "measurement of excess antenna temperature at 4080 Mc/S" by Penzias and Wilson (1965). A radio astronomer participant explained that it is not just a coincidence that their observing frequency was exactly 10 times the 408 Mc/s frequency of some Cambridge source surveys, but rather a result of the way radio frequencies are assigned for various purposes.

The "Alpha Beta Gamma" (Alpher et al. 1948) paper is a real one, though very short, published on April 1st. I own an authentic reprint, "deaccessioned" by Sir. Fred Hoyle in about 1970. This was part of an extensive office clearing effort and probably has no significance for the history of science.

The first relict radiation prediction (Alpher and Hermann, 1948) appeared later the same year. Like the Gamow et al. nucleosynthesis discussions, it presupposes a primordial soup of pure neutrons, as does their more detailed 1949 treatment of matter and radiation densities in the universe. The correct initial conditions, with protons and neutrons in thermal equilibrium, were first treated by Hayashi (1950). In this very little read (or cited) paper, the author expresses some hopes that the mix of protons, neutrons, deuterons, tritons, and helium nuclei may permit bridging the A = 5 and 8 gaps to form carbon and other heavy elements.

3. How the Standard Hot Big Bang Got That Way

The established facts of observational cosmology can still be counted with your shoes one. First, apparent brightness falls like (distance)$^{-2}$, as much have been noticed by the first paleolithic tribe to carry their campfires from place to place. Second, light travels at a finite speed, first measured by the Dane Ole Roemer (while in Paris), from timing of eclipses of the moons of Jupiter. He also built the first transit circle instrument, in case anyone is interested.

Third comes the large wavelength shifts of the spiral nebulae, first recorded for M31 by Slipher in 1912. His spectrogram, which took two December nights to record, is reproduced in volume 2 of the classic Russell, Dugan and Stewart text and shows, rather dimly, the F and G bands and the calcium H and K absorption lines (and no emission features). Theory then intrudes, with Einstein and general relativity followed in the same year (1916) by Friedmann's solutions, describing a homogeneous, isotropic universe. It could expand, contract, or sit still (but only unstably). If GR is the right (classical) description of gravity -- we know it is wrong in quantum mechanical limits -- then theoretical freedom is much restricted. Our confidence in GR derives from solar system (weak field) tests, but also from strong field effects in binary pulsars, which GR describes better than the available alternatives, even ones motivated by the desire for a quantum theory and unification of all the forces (Taylor et al. 1992).

The actual existence of galaxies outside the Milky Way was established only in 1924, when Hubble identified Cepheid variables in M31, in some sense our fourth important fact. Meanwhile, Slipher and then Hubble and Humason were busily adding to the body of measured wavelength shifts. Lundmark, Witz, Stromberg, and Robertson were among those early tempted to plot the shifts vs. distances to the galaxies, in about 1925. With no allowance for Malmquist bias and angular diameter as the common distance indicator, the resulting functional form was typically a quadratic, as predicted for an empty, de Sitters, universe. At least one contemporary writer on the subject continues to find his quadratic.

The fifth and most important pre-war fact of observational cosmology is the linearity of the redshift-distance relation, put forward in 1929 by Hubble. His original drawing, from the Proceedings of the National Academy of Sciences is widely reproduced in astronomy history books. The velocity units were, accidentally, km rather than km/sec, and the full range is only 1000. With maximum galaxy distances of 2 Mpc, Hubble arrived at H = 536 $\pm$ 25 km/s/Mpc. His fractional error bars are about the same size as those reported by van den Bergh and Tammann (elsewhere in this volume).

Hubble himself oscillated between universal expansion and tired light explanations of his correlation. In principle, the latter can be ruled out observationally. Tired light predicts that surface brightnesses will scale as $(1+z)^{-1}$ vs. $(1+z)^{-4}$ for expansion, and that time scales of variable phenomena will not be time dilated as they would be by true cosmic expansion. The surface brightness test (recently pursued by Sandage) remains mired in observational difficulties. Two (and only two) type Ia supernova at $z = 0.3$ and 0.5 seem to show time dilation, and it has arguably also been seen in gamma ray busters (but other explanations of time scale vs. flux correlations are possible). It is, however, fair to say that there is no conventional physical mechanism for tired light: the Feynman graphs for it sum to zero.

If (and perhaps only if) you accept both GR and expansion, then the universe has a hot, dense state in its past. The time scale for "past" in independently established by considerations of stellar evolution and nucleocosmochronology. It has been known at least crudely since 1905, when Rutherford showed that some earth rocks with radioactive content are at least 10^9 years old.

This hot, dense stage provides, of course, a simple explanation for both the 2.7 K background radiation and for the abundance of helium (etc.) in unevolved objects. If you try to produce the helium in galaxies over the characteristic 10^{10} year lifetime to their stars, you end up with a ratio of luminosity to baryon mass of about 10 in solar units, far larger than in the galaxies we see. The problem with accounting for the radiation in a non-evolving universe is not so much the energy required as the observed isotropy and black body spectrum. If your try to achieve these with discrete sources and reprocessing, your universe is likely to be opaque to microwaves at redshifts where we see sources unabsorbed. A separate new constraint comes from recent observations that the small fluctuations of background intensity in the sky also have a black body character. The importance of this in ruling out absorption and reemission or wavelength-dependent scattering (vs. electron scattering in the conventional hot big bang) as the mechanism for thermalization and isotropization is emphasized by Rees (elsewhere in this volume).

4. To Be Determined.

The hot big bang may be standard, but its parameters clearly are not. Factors of two or more still surround Hubble's H, q (deceleration), Ω and Ω_b (total and baryonic density), χ (cosmological constant), the age of the universe, and its radius of curvature (or k). That we have been asking for 40-some years has become something of an embarrassment to the astronomical community and is unfortunately sometimes used to support claims that cosmologists haven't a clue what they are talking about. Obviously most Symposium 169 participants disagree!

I would like, however, to put in a plea here for both observers and theorists to be clear about which parameter they are trying to measure. Geometry of the universe, but not about deceleration, unless you have assumed a value of χ, and conversely.

Other items not yet established include, notoriously, the nature of (at least much of) the ubiquitous dark matter and the extent and topology of the largest-scale structures and deviations from Hubble flow in the present universe and how those have changed with redshift.

5. The Limits of Cosmic Tolerance

How much deviation from the standard hot big bang can you take without getting nervous? I list these in order of increasing nervousness on my part, while reminding you that inflation (hence $\Omega = 1$, the Harrison-Zeldovich spectrum, and so forth) is not really part of the standard model (according to Rees), or at any rate not yet (according to Turner).

Deviations that would be interesting but not threatening include $\Omega \neq 1$ and $\chi \neq 0$. Fluctuations (either in the background radiation or assumed in the initial density field to make galaxies) that are non-Gaussian, not Harrison-Zeldovich, not adiabatic, etc. are also OK. Seeding by strings for galaxy information, tensor (gravitational radiation) contributions to irregularities in the background radiation, and inhomogeneous nucleosynthesis are even expected in some versions of the "standard" and are similarly non-distressing.

We have gradually got used to larger and larger cluster/void structures at least up to 100 h^{-1} Mpc and to streaming and coherent peculiar velocities over similar length scales up to 1000 km/sec or so. Varying levels of skepticism seem to have greeted Brent Tully's super-duper clusters of 10^{18} M_θ, the Broadhurst et al. indications of regularity at large scales, and the Lauer and Postman report of very extensive streaming. Much of the skepticism is probably justified by the limited data; some is probably nervousness, expressed in the usual way of conservative scientists. I would not myself be horrified to learn that part of all of our local 3K dipole is really anisotropic of the universe (but others would be).

The universe has non-zero baryon number (or we would not be here to talk about it). Standard big bang nucleosynthesis sets the lepton number equal to baryon number (or to zero, depending on exactly what you do with your neutrinos, and the difference does not matter). No current observations, however, exclude an unbalanced density of photons. Large lepton number in this sense makes a major difference to the amounts of H^2, He^4, Li^7 etc. coming out of the hot big bang, and this particular non-standard case may deserve investigation with a state-of-art nucleosynthesis code.

Red (or blue) shifts due neither to Hubble expansion, to ordinary physical motion, nor to strong gravitational fields cross the border into the intolerable for most cosmologist. I can imagine incorporating them into some rational model only if they carried a clear signature (apart from the red/blue shift itself). Non-constant l might be one thing to look for.

The standard model breaks down ever more completely if the early universe was cold (though dense), redshifts are quantized or not linear in distance, or there was no dense phase at any temperature.

6. Is the Big Bang Important?

Ask a silly question (my mother used to say) and you get a silly answer. Thus an analogy may be useful. If you believe (as many apparently do) that the universe was created in 4004 BC or thereabouts (12 noon on 29 October is an optional refinement), then you can have a successful career in medicine and physiology, laboratory physics or chemistry, engineering or mathematics. But you had better stay away from astronomy, the geosciences, evolutionary and ecological biology.

By the same token, if we are all wrong and there was never a hot dense stage (or, alternatively, if there was, but you don't believe it), then some parts of astronomy are still perfectly OK. You can work on the solar system, cosmic rays sources and acceleration (though watch out for the highest energy ones not confined by galactic magnetic fields), formation, structure and evolution of stars (at least within the Local Group), and physics and chemistry of the interstellar medium. The possible range of initial helium abundances begins to produce difficulties in studying stellar populations in other galaxies, even nearby. And if redshift is not an accurate guide to distance (hence luminosity and lookback time) then all bets are off for any investigation of galaxy (normal or active) formation and evolution, and even the interpretation of colors in terms of stellar populations at large redshift.

It is perhaps appropriate to end this section with quotes from two senior astronomers who have questioned the correctness of the standard hot big bang. Sir Fred Hoyle is supposed to have said (in a cosmological context), "I can see no reason to disbelieve something just because it is impossible." (to which Oort is said to have replied, "I can think of no better reason."). And I have heard Thomas Gold say (in another context), "If we are all going the same direction, it must be forward."

Acknowledgements

Over the years, I have been privileged to hear astronomers from Arp to Zwicky discuss these issues. Some of them (starting with G.R. Burbidge) would undoubtedly want to go on record as saying that they do not at all agree with how I have interpreted their remarks. A special thanks this time to Ed Ohm for copies of his classic papers and explanations of how the data should be thought about. Since maternal great-grandmother's father was an Ohm, I have hopes that we might be related.

16

References

Alpher, R.A., H. Bethe & G. Gamow 1948. Phys. Rev. 73, 803
Alpher, R.A. & R. Herman 1948. Nature 162, 775 1949. Phys. Rev. 75, 1092
Arnold, J.R. et al. 1962. JGR 67, 4878
Bothe, W. & W. Kohlhörster 1929. Phys. Zeit. 30, 516
Bowyer, S. 1991. ARA&A 29, 59
Fazio, G.G. 1967. ARA&A, 481
Hayakawa, S. et al. 1958. Prog. Theor. Phys. Sup. 6,1
Hayashi, C. 1950. Prog. Theor. Phys. Sup. 6, 1
Henry, R.C. 1991. ARA&A 29, 89
Herzberg, G. 1950. Spectra of Diatomic Molecules (Van Nostrand) p. 496
Jansky, K. 1933. Nature 132, 66
Kraushaar, W.L. & G.W. Clark 1962. Phys. Rev. Lett. 8, 106
McKellar, A. 1941. Publ. DAO 7, 251
Ohm, E.A. 1961. Bell System Tech. J. 60, 1068 1994. Personal communication
Penzias, A.A. & R.E. Wilson 1965. ApJ 142, 420
Reber, G. 1940. ApJ 91, 621; Proce. IRE 28, 681
Seares, F.H., P.J. van Rhign, Mary C. Joyner & Myrtle L. Richmond 1925. ApJ 62,320
Taylor, J.H. et al. 1992. Nature 355, 125
Yentis, D.K., R. Novick & P. Vanden Bout 1972. ApJ 177, 365 & 375
Zeldovich, Ya. B. 1962. JETP 43, 1561 1963. Usphekii 6, 475

MEASUREMENT AND IMPLICATIONS OF THE COSMIC MICROWAVE BACKGROUND SPECTRUM

JOHN C. MATHER

Code 685, Laboratory for Astronomy and Solar Physics
NASA Goddard Space Flight Center, Greenbelt, MD 20771
USA

Abstract.

The Cosmic Background Explorer (COBE) was developed by NASA Goddard Space Flight Center to measure the diffuse infrared and microwave radiation from the early universe. It also measured emission from nearby sources such as the stars, dust, molecules, atoms, ions, and electrons in the Milky Way, and dust and comets in the Solar System. It was launched 18 November 1989 on a Delta rocket, carrying one microwave instrument and two cryogenically cooled infrared instruments. The Far Infrared Absolute Spectrophotometer (FIRAS) mapped the sky at wavelengths from 0.01 to 1 cm, and compared the CMBR to a precise blackbody. The spectrum of the CMBR differs from a blackbody by less than 0.03%. The Differential Microwave Radiometers (DMR) measured the fluctuations in the CMBR originating in the Big Bang, with a total amplitude of 11 parts per million on a 10° scale. These fluctuations are consistent with scale-invariant primordial fluctuations. The Diffuse Infrared Background Experiment (DIRBE) spanned the wavelength range from 1.2 to 240 μm and mapped the sky at a wide range of solar elongation angles to distinguish foreground sources from a possible extragalactic Cosmic Infrared Background Radiation (CIBR). In this paper we summarize the COBE mission and describe the results from the FIRAS instrument. The results from the DMR and DIRBE were described by Smoot and Hauser at this Symposium.

M. Kafatos and Y. Kondo (eds.), Examining the Big Bang and Diffuse Background Radiations, 17–29.
© 1996 IAU.

1. COBE Concept

The three COBE instruments cover the entire wavelength range from 1.2 μm to 1 cm. The FIRAS (Far Infrared Absolute Spectrophotometer) spans 105 μm to 1 cm in two bands with a 7° beam. The DMR (Differential Microwave Radiometer) covers 31.5, 53, and 90 GHz (9.5, 5.6, and 3 mm), also with a 7° beam. The DIRBE (Diffuse Infrared Background Experiment) spans 1.2 to 240 μm with 10 bands, and also measures polarization in its three shorter bands. Previous works have described the COBE and early results (Mather et al. 1990; Mather et al. 1991a,b; Janssen and Gulkis 1991; Wright 1990, 1991; Hauser 1991a,b; Smoot et al. 1991; Smoot 1991; Bennett 1991; Boggess 1991; Mather 1982; Gulkis et al. 1990; Boggess 1992), and the engineering of the spacecraft has also been published (Barney 1991; Milam 1991; Hopkins and Castles 1985; Hopkins and Payne 1987; Volz and Ryschkewitsch 1990; Volz et al. 1990, 1991; Volz and DiPirro 1992; Mosier 1991; Coladonato et al. 1990; Sampler 1990; Bromberg and Croft 1985).

A liquid helium cryostat cooled the FIRAS and DIRBE to about 1.5 K for 10 months, and after the evaporation of the helium the temperature slowly rose to about 60 K. The DMR radiometers are radiatively cooled, with the 31.5 GHz receivers at 290 K and the others at 140 K. A conical shield protects them from the Sun and Earth, which do not illuminate the instruments directly. The orbit is a sun-synchronous polar orbit, at 900 km altitude and 99° inclination, with a 6 PM ascending node. The Earth's gravitational quadrupole precesses the orbit plane to maintain a 6 PM ascending node, so that the plane is always approximately perpendicular to the Sun. The spacecraft is oriented with its spin axis away from the Earth and about 94° to the Sun. It spins at 0.8 rpm and the DMR and DIRBE beams therefore scan the range from 64° to 124° from the Sun.

The COBE design evolved significantly since it was first proposed in 1974. The initial concept was proposed by a Goddard-MIT-Princeton team, while versions with only microwave radiometers were proposed by teams at JPL and Berkeley. In 1976 NASA formed a science team of 6 people, drawn from all three groups, and selected Goddard to support the mission. The science team eventually grew to include 20 members, and the Goddard effort required about 1500 civil servants and contractors. The initial concept called for launch on a Delta rocket, but in the late 1970's NASA decided to launch all payloads on the Shuttle. Following the Challenger explosion in 1986, the COBE was redesigned for launch on the Delta. To achieve this, its weight was cut in half, but no compromise in performance and few changes were needed in the scientific instruments. Only three years after the redesign was approved, the COBE was launched. Its liquid helium lasted for 10 months, but the DMR and the short wavelength channels of the

DIRBE continued operation until January 1994. The COBE now serves for satellite communication tests.

2. FIRAS

2.1. FIRAS INSTRUMENT

The COBE FIRAS (Far Infrared Absolute Spectrophotometer) was designed to compare the CMBR spectrum to a blackbody with great precision. The most recent results show that the CMBR deviates from blackbody form by less than 0.03% of the peak intensity over the wavelength range from 0.5 to 5 mm (Mather et al. 1994). Our interpretation of these results has also been given (Wright et al. 1994). The FIRAS also showed that the cosmic dipole has the expected shape of the derivative of the Planck function with respect to temperature (Fixsen et al. 1994a). The primary basis of the comparison is a full beam external blackbody calibrator that can be adjusted to match the temperature of the sky. This calibration is described in Fixsen et al. (1994b). The FIRAS is the first instrument to have this opportunity in the protected space environment. The instrument is designed with multiple modes of operation and multiple detectors to test for possible systematic errors.

The instrument is a polarizing Michelson interferometer used as a spectrometer (Mather, Shafer, and Fixsen 1994). It has a beamwidth of 7° , defined by a compound parabolic concentrator. It is symmetrical, with differential inputs and differential outputs. It is a two-beam interferometer, whose output signal (called an interferogram) depends on the path difference between the two beams. A monochromatic input produces an output proportional to the cosine of the phase difference between the two beams. A general input produces an interferogram which is the Fourier transform of the input spectrum. These interferograms are detected, amplified, filtered, digitized, filtered and averaged again, and transmitted to the computers on the ground. They are then Fourier transformed numerically and calibrated to determine the input spectra. The mirror mechanism can scan at two different scan speeds and over two different stroke lengths, to obtain low and high spectral resolution.

The FIRAS was used to map almost the entire sky. To derive cosmological results, two terms were subtracted from the spectrum of each pixel. A dipole term compensated for the Doppler shift of the spectrum, due to the motion of the Earth. The mean spectrum of the dust in the Galaxy was also derived from the map, and a fraction of this mean spectrum was fitted to each pixel spectrum and subtracted. The resulting sky map showed almost no trace of the strong Galactic emission, so that a single dust brightness map times a single characteristic spectrum is a very good representation of the far infrared Galaxy.

To quantify the distortion of the mean spectrum of the whole sky, it was fitted to a sum of additional terms. The first two terms are a Planck function for the mean temperature, and a temperature adjustment times the derivative of the Planck function with respect to temperature. The second term just allows for the possibility that the first may not be calculated for the correct mean temperature, since we do not know it a priori. The third term is a small number times the function $g(\nu)$ that represents the mean Galactic spectrum. This allows for the possibility that the initial subtraction done on each pixel might have been incomplete or biased. The fourth term represents the cosmic distortion. There are two interesting forms for this: a μ form and a y form (Zeldovich & Sunyaev, 1969, 1970). The μ form is a chemical potential distortion, that might arise between the redshifts of about 3×10^6 and 10^5. The second form is a Comptonization form that could arise after $z = 10^5$.

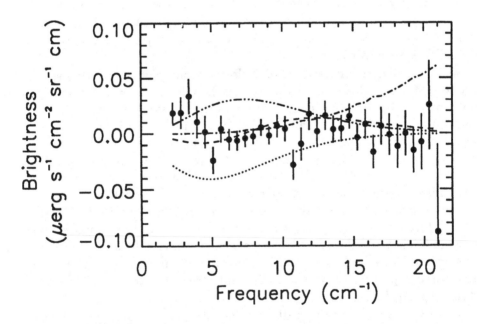

Figure 1. FIRAS measured CMBR residuals, $I_0 - B_\nu(T_0) - \Delta T(\partial B_\nu/\partial T) - G_0 g(\nu)$. Spectrum model components: the maximum allowed distortions (95% CL) $y = 2.5 \times 10^{-5}$ (— — —) and $|\mu| < 3.3 \times 10^{-4}$ (···); the Galaxy spectrum $g(\nu)$ scaled to one fourth the flux at the Galactic pole (· –), and the effects of a 200 μK temperature shift in T_0, $0.0002(\partial B_\nu(T)/\partial T)$, (··· —). (Mather *et al.* 1994)

Figure 1 (Mather *et al.* 1994) shows the results of these least squares fits. The residuals are for the case where $y = \mu = 0$ and show that the CMBR spectrum is the same as a blackbody within 0.03% of the peak intensity. The weighted rms residual is only 0.01% of the peak brightness. Nevertheless, these residuals are about twice as large as those expected from detector noise alone. We do not presently know the cause of this but have arbitrarily increased the size of the error bars throughout. The y and μ curves show the shapes of distortion that would be produced if these coefficients have the 95% confidence limit values of $|y| = 2.5 \times 10^{-5}$ and $|\mu| = 3.3 \times 10^{-4}$. These limits allow for the increased uncertainty noted above, and for variations of the derived y and μ as the boundary of the selected data set moves farther from the Galactic plane.

Other possible distortions can also be explored. As an example, if the CMBR spectrum is a graybody, then we can limit its emissivity to 1 ± 0.00041 (95% CL). If the true cosmic signal contains a distortion with the same shape as the Galactic signal, then we cannot determine it with this method alone. Similarly, a cosmic energy release that simply changed the CMBR temperature could not be recognized.

The temperature of the CMBR is also important, although few cosmological calculations require a more precise number. The measured temperature is primarily needed for comparisons with different experiments, such as ground-based measurements and interstellar CN measurements. The FIRAS has two ways to determine this number. First, the thermometers in the external calibrator are the primary scale. When the calibrator is set to match the sky brightness, its thermometers measure the sky temperature. This method gives 2.730 K. The second method uses the wavelength scale of the FIRAS to determine color temperatures. In effect, we measure the frequency of the peak brightness of the CMBR and use fundamental constants to compute a temperature. The wavelength scale was determined from FIRAS observations of interstellar [C II] and confirmed with other lines of [N II], [C I], and CO. The color temperature method yields a CMBR temperature of 2.722 K. These numbers differ by much more than the random errors, which are only a few μK. We have no obvious explanation for the difference and simply take the mean, obtaining T = 2.726 ± 0.010 K (95% confidence). Clearly this is not a statistical determination of the error bar, which is entirely systematic error.

Fortunately, there are other confirmations that this number is approximately correct. A sounding rocket experiment launched only weeks after the COBE by Gush, Halpern, and Wishnow (1990) with a similar instrument and greatly superior detectors obtained the result 2.736 ± 0.017 K. Interstellar CN results are also in agreement with these numbers (Roth, Meyer, & Hawkins 1993, and Kaiser & Wright 1990). We can also determine the temperature from the spectrum of the dipole anisotropy. This can be done from both DMR and FIRAS data. The FIRAS result is determined only from the color temperature, since the absolute value of the Earth's velocity is not known. The FIRAS yields 2.714 ± 0.022 K (95% CL). The DMR result (see below) is 2.76 ± 0.18 K, and is derived from the known variation of the Earth's velocity around the Sun. The spectrum of the FIRAS dipole was fit to an amplitude of 3.343 mK. The dipole residuals are very small and are close to the value expected from detector noise alone.

2.2. FIRAS INTERPRETATION

Our interpretation of these spectrum distortions is given by Wright *et al.* (1994) and summarized here. Large CMBR spectrum distortions are very difficult to produce in plausible versions of the hot Big Bang universe. After the annihilation of positrons, the CMBR energy density far exceeded even the rest mass energy density of the baryonic matter until quite recently. Consequently, there are few processes involving the baryonic matter that can liberate much energy and change the CMBR spectrum significantly. It is even more difficult to produce enough energy to create the whole CMBR radiation from anything except a hot Big Bang.

Therefore, the most immediate conclusion of the FIRAS spectrum distortion measurements is that the hot Big Bang is the only natural explanation for a nearly perfect blackbody. Carefully tailored models are required if one desires to produce the whole energy content of a blackbody spectrum by adding up non-blackbody contributions at different redshifts, as required by alternatives to the hot Big Bang. If one imagines that dust in intergalactic space can thermalize the radiation, then that dust must have substantial optical depth over an interval of cosmic history. That moment cannot be recent, or we would not be able to see distant galaxies at far infrared wavelengths. The IRAS galaxy at $z = 2.286$ demonstrates that one can see very far, and if the millimeter wave optical depth were large we would not see even such a spectacular object.

The next conclusion is that rather little of the energy of the CMBR was added to it after the first year of the expansion. The fraction of the CMBR energy added is approximately 0.71μ in the redshift range $3 \times 10^6 > z > 10^5$. For later redshifts, the fraction is $4y$. A more precise calculation gives the results shown in Fig. 2.

There are many possible sources of such energy conversions, including decay of primeval turbulence, elementary particles, cosmic strings, or black holes. The growth of black holes, quasars, galaxies, clusters, and superclusters might also convert energy from other forms. The FIRAS data, together with the DMR anisotropy measurements, provide a limit on the spectral index of primordial density fluctuations. Wright *et al.* (1994) found an upper

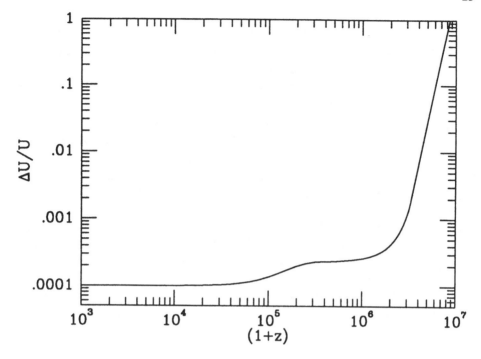

Figure 2. Limits on Energy Release Prior to Recombination (Wright *et al.* 1994)

limit of 1.9, while Hu, Scott, and Silk (1994) claim an upper limit of about 1.55. It is interesting that these calculations give tighter limits than existing direct measurements, even though the FIRAS numbers are a non-detection. These results are dependent on assuming that a power law is the correct form for the fluctuations over 7 orders of magnitude of scale. There is little possibility of observational evidence to confirm this assumption over such a wide range, since small scale fluctuations have long since been replaced by nonlinear phenomena.

Wright *et al.* (1994) also give limits on hydrogen burning following the decoupling. These results depend on using geometrical arguments (a csc |b| fit) to estimate the maximum amount of extragalactic energy that could have a spectrum similar to that of our own Galactic dust. We found a limit that is a factor of about 3 smaller than the polar brightness of the Milky Way. A better understanding of the Galactic dust would help produce a tighter limit on these extragalactic signals.

Consider first population III stars liberating energy that is converted by dust into far infrared light (using an optical depth of 0.02 per Hubble radius), and assume that $\Omega_b h^2 = 0.015$. In that case less than 0.6% of the hydrogen could have been burned after $z = 80$. As a second example, consider evolving infrared galaxies as observed by the IRAS. For reasonable assumptions, we found that less than 0.8% of the hydrogen could have been burned in evolving IR galaxies.

We also obtained limits on the heating and reionization of the intergalactic medium. It does not take very much energy to reionize the medium, relative to the CMBR energy, because there are so few baryons relative to CMBR photons. Even the strict FIRAS limits permit a single reionization event to occur as recently as $z = 5$. More detailed calculations by Durrer (1993) show that the energy required to keep the intergalactic medium ionized over long periods of time is much more substantial and quite strict limits can be obtained. If the FIRAS limits were about a factor of 5 more strict, then it would be possible to test the ionization state of the IGM all the way back to the decoupling.

If the IGM were hot and dense enough to emit the diffuse X-ray background light, it should distort the spectrum of the CMBR by inverse Compton scattering. This is a special case of the Comptonization process, with small optical depth and possibly relativistic particles. Calculations show that a smooth hot IGM could have produced less than 10^{-4} of the X-ray background, and that the electrons that do produce the X-ray background must also have a filling factor of less than 10^{-4}.

2.3. INFRARED ASTRONOMY WITH FIRAS

The FIRAS produced the first nearly all-sky far IR survey of Galactic emission at wavelengths greater than 120 μm (Wright et al. 1991). The general features seen in the maps are very similar to those found by the IRAS satellite at 60 and 100 μm . Dust is concentrated in the Galactic plane and is heated by starlight to a temperature ranging from 15 to 30 K, depending on the intensity of the starlight incident on the dust. This dust controls radiative transfer of energy through the Galaxy, and thus the heating and cooling of the interstellar gas. The dust has a mean spectrum which is approximated well by the form $\nu^{\alpha} B_{\nu}(T)$, where the index $\alpha = 1.65$ and $T = 23.3$ K. The mean dust spectrum of the Galaxy can also be well represented by the sum of two such terms with the index $\alpha = 2$, with cold dust with a temperature of about 4.8 K and warmer dust with a temperature of 20.4 K. In this representation, the optical depth of the cold dust is 6.7 times that of the warmer dust component. There are many possible explanations for this kind of dust spectrum. It is possible that there are multiple types of dust grains, so that some can have a very different equilibrium temperature than others. Dust grains could have unusual shapes, like needles (Wright 1982) or fractals (Wright 1987). Small grains with low heat capacities could have variable temperatures, as they are excited by discrete UV photons or cosmic rays. Some dust is protected within optically thick clouds and must reach much lower temperatures than that exposed to the full intensity of the interstellar radiation field, and other dust is very close to stars and must be much warmer than average. The total dust luminosity of the Galaxy is $(1.8 \pm 0.6) \times 10^{10} L_\odot$ (Wright et al. 1991).

The FIRAS also provided the first panoramic and unbiased spectral survey of the far IR line emissions from the interstellar medium, including CO, [C I], [C II], and [N II] lines. These results were reported by Wright et al. (1991) and by Bennett et al. (1994). The FIRAS map of the emission of [C II] at 158 μm shows it to be extremely widespread, both in and out of the Galactic plane. This line is the dominant cooling mechanism for most of the interstellar gas, and is the brightest emission line of the Galaxy, accounting for 0.3% of the dust luminosity. The strength of the line is so great that it can be seen by eye in the raw interferograms as an undamped sinusoid, and with more analysis the Galactic rotation curve can be seen in the line center frequency. The [C II] line is very strongly correlated with the dust emission, as expected from the theory of photodissociation regions. The heating of the gas is provided mostly by photoelectrons emitted from dust grains, and must be in balance with the cooling provided mostly by the [C II] line. Deviations from this correlation are expected where the dust properties change, where the interstellar radiation field changes color, or where other heating and cooling mechanisms become important at extremes of temperature or density. In particular, the ratio of [C II] to dust decreases to nearly zero at the Galactic poles.

The FIRAS also detected the emission of ionized nitrogen [N II] at 205 μm for the first time and measured its frequency well enough for further astronomical (Petuchowski and Bennett 1992) and laboratory (Brown et al. 1994) observations. This is the second brightest far IR line from the interstellar gas and probes a different phase of the interstellar medium because nitrogen ionizes less easily than hydrogen and carbon. The nitrogen and carbon line intensities are also strongly correlated but show the relation [N II] $\propto$ [C II]$^{3/2}$. The meaning of this correlation was explored by Petuchowski and Bennett (1993).

Carbon monoxide is the standard tracer for interstellar molecular gas. The FIRAS has mapped the brightness of the lines of CO from the 2-1 up to the 6-5 transitions, showing that the line ratios are highly dependent on location in the Galaxy. In particular, the CO lines are extremely bright relative to the others in the Galactic Center, where the gas temperature and density are very high and the UV field is shielded by dust. The two lines of neutral carbon [C I] are also seen and mapped by the FIRAS.

Weaker lines observed by the FIRAS include oxygen and water. The water line at 269 μm has never been observed before in astrophysical sources because of the extremely high optical depth of the terrestrial absorption line. The line is seen in absorption by FIRAS (Bennett *et al.* 1994) against the continuum emission of the Galactic Center, but at a low level of significance relative to the detector noise and possible systematic errors. The observed absorption is comparable to that expected from the concentration of interstellar water vapor.

3. Data Delivery

A set of initial data products from the Cosmic Background Explorer Satellite is now available for analysis, and further data releases are planned for summer of 1994. A COBE Guest Investigator Facility, which supports proposers funded under NASA's Astrophysics Data Program, has been established at the Goddard Space Flight Center. The data are distributed from the National Space Science Data Center (NSSDC). All of the initial data products are available via anonymous ftp and accessible on the World Wide Web; most of the datasets larger than about 10 MB will be distributed on a set of CD-ROMs.

The data now available, and the associated documentation, may be obtained by anonymous ftp to nssdca.gsfc.nasa.gov (equivalent IP address 128.183.36.23). Please provide your e-mail address as the password. Change to directory (cd) anon_dir:[000000.cobe] (" cd cobe" should work) and download the file called AAREADME.DOC for further instructions. Alternatively, the data and documentation may be obtained on tape by request to the Coordinated Request and User Support Office (CRUSO), NASA Goddard Space Flight Center, Code 633.4, Greenbelt, MD 20771. They may be requested by phone at 301-286-6695, or by e-mail from request@nssdca.gsfc.nasa.gov. For assistance in the use of COBE data, contact David Leisawitz by phone (301-286-0807), FAX (301-286-1771), e-mail (leisawitz@stars.gsfc.nasa.gov), or surface mail at the address given above. Please address requests for preprints or reprints of COBE Team publications to Susan Adams as follows: Ms. Susan Adams, NASA Goddard Space Flight Center, Code 685, Greenbelt, MD 20771, phone: 301-286-4257, or FAX: 301-286-1617, or e-mail: adams@stars.gsfc.nasa.gov. The COBE data are accessible on the World Wide Web at the following Uniform Resource Locator (URL) address:

```
http://www.gsfc.nasa.gov/astro/cobe_home.html
```

4. Summary and Conclusions

The COBE was designed to answer three questions: what is the spectrum of the CMBR, what are its anisotropies, and what is the brightness of the CIBR? The first two have been answered. The CMBR spectrum is very close to a blackbody of T = 2.726 K, and the weighted rms deviation from the blackbody form is less than 0.01%. The CMBR has anisotropies on all size scales larger than 7° , with a size spectrum consistent with the predicted scale-invariant primordial fluctuations. They have an amplitude of 11 parts per million on a 10° scale. The CIBR has not yet been found, but the data are in hand and analysis is in progress. All three instruments have given new views of the Galaxy, showing large scale features due to the interstellar medium and the general distribution of material. The Galactic bulge at the center is revealed at wavelengths where it is less obscured by dust absorption, and the Galaxy appears to have a central bar. Line emission from neutral and ionized carbon, ionized nitrogen, and carbon monoxide are widespread, and water may have been seen in absorption against the Galactic Center. The interplanetary dust (zodiacal light) has been observed more extensively than ever before, at a wide range of elongations from the Sun and throughout the year, enabling a new and more complete understanding of the dynamics and sources of the cloud.

5. Acknowledgements

The National Aeronautics and Space Administration/Goddard Space Flight Center (NASA /GSFC) is responsible for the design, development, and operation of the Cosmic Background Explorer (COBE). GSFC is also responsible for the software development and the final processing of the mission data. The COBE SWG is responsible for the definition, integrity, and delivery of the public data products, and includes the following members:

NAME	AFFILIATION	SPECIAL ROLE
J. C. Mather	GSFC	Project Scientist and FIRAS Principal Investigator
M. G. Hauser	GSFC	DIRBE Principal Investigator
G. F. Smoot	UC Berkeley	DMR Principal Investigator
C. L. Bennett	GSFC	DMR Deputy PI
N. W. Boggess	GSFC–ret	Deputy Proj. Scientist
E. S. Cheng	GSFC	Deputy Proj. Scientist
E. Dwek	GSFC	
S. Gulkis	JPL	

28

M. A. Janssen	JPL	
T. Kelsall	GSFC	DIRBE Deputy PI
P. M. Lubin	UCSB	
S. S. Meyer	MIT	
S. H. Moseley	GSFC	
T. L. Murdock	Gen.Res.Corp.	
R. A. Shafer	GSFC	FIRAS Deputy PI
R. E. Silverberg	GSFC	
J. Vrtilek	NASA HQ	Program Scientist
R. Weiss	MIT	Chairman of SWG
D. T. Wilkinson	Princeton	
E. L. Wright	UCLA	Data Team Leader

Many people have made essential contributions to the success of COBE in all its stages, from conception and approval through hardware and software development, launch, and flight operations. To all these people, in government agencies, universities, and industry, the science team and I extend my thanks and gratitude. In particular, I thank the large number of people at the GSFC who brought this challenging in-house project to fruition.

References

Barney, R. D. (1991) *Illuminating Eng. Soc. J.*, **34**, 34

Bennett, C. L. (1991) *IAU XXI Highlights Astron*, J. Bergeron, Ed., **9**, 335

Bennett, C. L. et al. (1992) *ApJ*, **391**, 466

Bennett, C. L. et al. (1994) *ApJ*, **434**, 587-598

Boggess, N. W. "The Cosmic Background Explorer (COBE): The Mission and Science Overview," (1991) *IAU XXI Highlights Astron*, J. Bergeron, Ed., **9**, 273

Boggess, N. W. et al. (1992) *ApJ*, **397**, 420

Bromberg, B. W. & Croft, J. (1985) *Adv. Astron. Sci.*, **57**, 217

Brown, J. M. et al. (1994) *ApJL*, **428**, L37

Coladonato, R. J. et al. (1990) *Proc. Third Air Force/NASA Symp. on Recent Advances in Multidisciplinary Analysis and Optimization*, 370, Anamet, Hayward, CA

Durrer, R. (1993) "Early Reionization in Cosmology," *Infrared Phys. Technol.*, **35**, 83-94

Fixsen, D. J. et al. (1994a) *ApJ*, **420**, 445-449

Fixsen, D. J. et al. (1994b) *ApJ*, **420**, 457-473

Gulkis, S., Lubin, P. M., Meyer, S. S., & Silverberg, R. F. (1990) *Sci. Amer.*, **262**, 132

Gush, H.P., Halpern, M., & Wishnow, E.H. (1990) *Phys. Rev. Lett.*, **65**, 537

Hauser, M. G. et al. (1991) *After the First Three Minutes*, AIP Conf. Proc, **222**, 161, eds. S. S. Holt, C. L. Bennett, & V. Trimble, New York

Hauser, M. G. (1991) *IAU XXI Highlights Astron*, J. Bergeron, Ed., **9**, 291

Hawkins, I., and Wright, E. (1988) *ApJ*, **324**, 46-59

Hopkins, R. A., & Castles, S. H. (1985) *Proc. SPIE*, **509**, 207

Hopkins, R. A., & Payne, D. A. (1987) *Adv. Cryogenic Engineering*, **33**, 925

Hu, W., Scott, D., and Silk, J. (1994), "Power Spectrum Constraints from Spectral Distortions in the Cosmic Microwave Background," *ApJ*, **430**, L5

Janssen, M. A. & Gulkis, S. (1991) *Proc. The Infrared and Submillimetre Sky After COBE, Les Houches*, 391, ed. M. Signore & C. Dupraz, Kluwer, Dordrecht

Kaiser, M. E., and Wright, E. L. (1990) *ApJ*, **356**, L1

Mather, J. C. (1982) *Opt. Eng.*, **21**, 769

Mather, J. C. et al. (1990) *IAU Colloq. 123, Observatories in Earth Orbit and Beyond, Proc.*, ed. Y. Kondo, 9, Kluwer, Boston

Mather, J. C. et al. (1991a) *AIP Conf. Proc. After the First Three Minutes*, **222**, 43, ed. S. S. Holt, C L. Bennett, & V. Trimble, AIP, New York

Mather, J. C. (1991b) *IAU XXI Highlights Astron*, J. Bergeron, Ed., **9**, 275

Mather, J. C., Shafer, R.A., and Fixsen, D. J. (1993) Proc. SPIE, **2019**, 146-157

Mather, J. C. et al. (1994) *ApJ*, **420**, 439-444

Milam, L. J. (1991) *Illuminating Eng. Soc. J.*, **34**, 27

Mosier, C. L. (1991) *AIAA 29th Aerospace Sciences Conference*, 91-361

Petuchowski, S. J. & Bennett, C. L. (1992) *ApJ*, **391**, 137-140

Petuchowski, S. J. & Bennett, C. L. (1993) *ApJ*, **405**, 595-598

Roth, K. C., Meyer, D., & Hawkins, I. (1993) *ApJL*, **413**, L67-L71

Sampler, H. P. (1990) *Proc. SPIE*, **1340**, 417

Smoot, G. F. (1991) *IAU XXI Highlights Astron*, J. Bergeron, Ed., **9**, 281

Smoot, G. F. et al. (1991) *ApJ*, **371**, L1

Volz, S. M. & DiPirro, M. J. (1992) *Cryogenics*, **32**, 77

Volz, S. M. & Ryschkewitsch, M. G. (1990) *Superfluid Helium Heat Transfer, HTD*, **134**, 23, ed. J. P. Kelly & W. J. Schneider AME, New York

Volz, S. M., Dipirro, M. J., Castles, S. H., Rhee, M. S., Ryschkewitsch, M. G., & Hopkins, R. (1990) *Proc. Internat. Symp. Optical and Opto-electronic Applied Sci. and Eng.*, 268, SPIE, San Diego

Volz, S. M., Dipirro, M. J., Castles, S. H., Ryschkewitsch, M. G., & Hopkins, R. (1991) *Adv. Cryogenic Engineering*, **37A**, 1183

Wright, E. L. (1982) *ApJ*, **255**, 401-407

Wright, E. L. (1991) *Proc. The Infrared and Submillimetre Sky After COBE, Les Houches*, 231, ed M. Signore & C. Dupraz, Kluwer, Dordrecht

Wright, E. L. (1987) *ApJ*, **320**, 818-824

Wright, E. L. (1990) *Ann. NY Acad. Sci., Proc. Texas-ESO-CERN Sym*, **647**, 190

Wright, E. L. et al. (1991) *ApJ*, **381**, 200

Wright, E. L. et al. (1994) *ApJ*, **420**, 450-456

Zeldovich, Ya. B., and Sunyaev, R.A. (1969) *Ap&SS*, **4**, 301

Zeldovich, Ya. B., and Sunyaev, R.A. (1970) *Ap&SS*, **7**, 20

OF COSMIC BACKGROUND ANISOTROPIES

G.F. SMOOT

Lawrence Berkeley Laboratory, Space Sciences Laboratory, Center for Particle Astrophysics, and Department of Physics, University of California, Berkeley, USA 94720

Abstract: Observations of the Cosmic Microwave Background (CMB) Radiation have put the standard model of cosmology, the Big Bang, on firm footing and provide tests of various ideas of large scale structure formation. CMB observations now let us test the role of gravity and General Relativity in cosmology including the geometry, topology, and dynamics of the Universe. Foreground galactic emissions, dust thermal emission and emission from energetic electrons, provide a serious limit to observations. Nevertheless, observations may determine if the evolution of the Universe can be understood from fundamental physical principles.

1. Introduction

With the discovery of the cosmic microwave background radiation the Big Bang has emerged as the standard model of cosmology. It is the goal of cosmologists to explain the origin and evolution of the Universe leading to what we observe in the present and then to predict what will occur in the future. An important example is the formation of large scale structure such as galaxies, clusters of galaxies, and even larger structures or voids. A major issue is the role of gravity. Modern cosmology uses the concepts of General Relativity to tie together the structure of space-time to the distribution of matter and energy as well as its dynamics - e.g. expansion. More recently quantum gravity has been considered in a role involving the creation of the Universe and its interesting contents. On a smaller scales we believe that gravity holds together all the structure that we observe. We assume it holds together stars, planets, the solar system, the stars in galaxies, clusters of galaxies, and so on. It is natural to assume that these structures form by gravitational instability. The otherwise smooth early Universe has seeds - density variations in an otherwise uniform and homogeneous medium.

31

M. Kafatos and Y. Kondo (eds.), Examining the Big Bang and Diffuse Background Radiations, 31–44.
© 1996 IAU.

Regions that are more dense than average attract the material around them which in turn attracts more material and so on until the over-density goes non-linear and a stable structure forms. Slightly less dense regions would lose material to the more dense medium surrounding them. The less dense regions would then evolve to voids.

Observations of large scale structure led Edward Harrison and Yakov Zeldovich to speculate in 1970 and 1972 respectively that a scale-invariant spectrum of primordial density perturbations would grow into the combination of large scale structures present today. As a result the scale-invariant spectrum is often called the Harrison-Zeldovich spectrum. They selected a scale-invariant primordial spectrum though the data actually indicated that the dependence only needed to be close to scale invariant - roughly within the physical scale to the one half power. At that time there was no known process for producing initial perturbations on any scale of astronomical interest so that scale-free and, particularly, scale-invariant were appropriate choices. Cosmologists now commonly consider two mechanisms that produce nearly scale-invariant spectra. These are inflation (quantum fluctuations expanded to astronomical scales by accelerating expansion) and spontaneous symmetry breaking (topological defect relics of a higher energy space vacuum state). Each of these has implications for the actual density fluctuation spectra.

Observations of the spectral and angular distributions of the CMB provide a most powerful means for investigating the early Universe. The observation that the CMB has to very high precision a blackbody spectrum not only constrains the possible thermal history of the Universe but also allows us to rule out many alternative cosmological models to the Big Bang. Figure 1 shows the current observations of the CMB spectrum.

The CMB's place as the relic radiation of the Big Bang makes CMB anisotropy observations a direct probe of the early Universe. The discovery of anisotropy by the COBE DMR (Smoot et al. 1992, Bennett et al. 1992, Wright et al. 1992, Kogut et al. 1992) marks the start of the epoch when the CMB observations began to fulfill their promise of revealing what was actually happening in the early Universe. It also inspired a veritable avalanche of activity in the field with nine groups now reporting anisotropy and new theoretical work appearing nearly every day. A consensus has formed on the observations to be made and the implications and understanding that can be derived from these observations. The goals of CMB anisotropy observation and theory include those shown in Table 1.

Wavelength [cm]

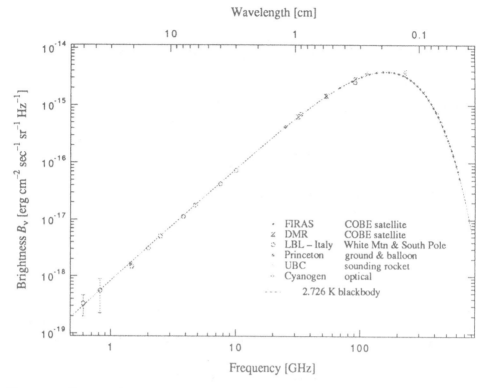

Figure 1. Current observations of the Cosmic Microwave Background spectrum: The measurements in the peak region are from COBE FIRAS (Mather et al. 1994) and the University of British Colombia rocket experiment (Gush et al. 1990) The Rayleigh-Jeans region data are from our LBL-Italy ground–based results from White Mountain and from the South Pole are shown together with a 2.73 K blackbody curve.

2. COBE DMR Observations

Measurements of anisotropy are clear and well-defined experimentally, even though technically difficult. One finds the sky brightness and brightness variations after calibrating the observations and removing the instrument signature and all foreground emissions. The results are interpreted in terms of our standard model. According to Big Bang theory the Universe is expanding from an original primeval fireball - a near thermal equilibrium state of very high density and temperature. As the Universe expanded it cooled, and when it reached a temperature of about 3000 K the primeval plasma coalesced to neutral atomic hydrogen and helium. At that time, about 300,000 years after the Big Bang, the cosmic background radiation was freed to move through the Universe with negligible scattering by free electrons. (This assumes a standard thermal history of the Universe, which implies no significant ionization of the Universe until redshifts less than

TABLE 1. CMB Anisotropy Observation Goals

I	**Initial Conditions for Large Scale Structure Formation** The formation of galaxies, clusters of galaxies, and larger scale structures as well as voids is a key issue in cosmology.
II	**Physics of the Early Universe** probes inflationary model predictions and/or quantum gravity. test of topological defects - monopoles, strings, domain walls, textures. probes the nature of the dark matter. probes of the baryonic content of the Universe.
III	**Geometry and Dynamics of the Universe** directly traces metric of space-time and isotropy of expansion and space-time. probe the curvature of space through the detection & location of Doppler peaks. measure the rotation and shear of the Universe.

about 100.) As a result the cosmic background photons are undisturbed, except for the universal expansion, since that epoch. If the primeval Universe was slightly inhomogeneous, the cosmic background radiation is slightly anisotropic.

The concept of a large angular scale anisotropy experiment is to look out from our local environment, past the solar system, past our Galaxy, past all galaxies and forming structure back to the cosmic photosphere. On large angular scales the dominant causes of anisotropy are gravitational redshift of the radiation (Sachs & Wolfe 1967) and the motion of the observer (Doppler shift) which produces a dipole anisotropy.

The COBE Differential Microwave Radiometer (DMR) experiment was designed to map the microwave sky and find fluctuations of cosmological origin. For the 7° angular scales observed by the DMR, structure is super-horizon (significantly larger than light could have traveled in the 300,000 years at last scattering) size so the features of the primordial perturbations are preserved unchanged since their primordial formation.

COBE was launched on 18 November 1989 into a 99° inclination, terminator (day-night) polar orbit. The DMR instrument observes the difference in power received by two horn antennas each pointing 30° from the space craft spin axis. The orbit and pointing of *COBE* result in a complete survey of the sky every six months while shielding the DMR from terrestial and solar radiation.

The DMR data were converted to maps that show the overall high level of uniformity of the CMB, the dipole anisotropy at about 1.2 parts per thousand (which is thought to be due to the Doppler effect caused by the motion of the solar system and Galaxy relative to the rest frame of the CMB), and fluctuations on all angular scales from the antenna resolution up to the quadrupole (90°) at levels of a few parts per million to about a part in 100,000. A key question is: which part of the signal is due to fluctuations in the CMB relating to the last scattering surface and which part might be foreground effects? The DMR maps the sky at frequencies of 31.5, 53, and 90 GHz (wavelengths of 9, 5.7, and 3.3 mm). Multiple frequencies were used to help separate the cosmic signal from possible foreground sources, particularly Galactic emission.

3. Separation of the Galactic and Cosmic Signals

The DMR anisotropy maps are sufficiently sensitive and free from systematic errors that our knowledge of Galactic emission is a limiting factor in interpreting the measurements of the DMR maps (also eventually other observations). The detected signals are nearly constant when expressed in terms of antenna temperature. When converted to thermodynamic temperature based on variations of a 2.73 K Planckian spectrum, they are consistent with a single constant amplitude. On the other hand the antenna temperatures of the Galactic emissions: from synchrotron, free-free, and dust have strong frequency dependences. If approximated by a power law, the power law index is roughly: -2.75 to -3.1, -2.1, and 1.5±0.5, respectively. The relative dependence and typical signal levels are shown in Figure 2.

Synchrotron emission arises when relativistic cosmic-ray electrons are accelerated by magnetic fields in the Galaxy. The energy spectrum of the cosmic ray electrons (a power-law with an index starting at about -2.75 and steepening with increasing energy) and the magnetic field strength determine the synchrotron emission effective power-law index. Free-free emission occurs when free electrons are accelerated by interactions with ions. Free-free emission is thermal bremsstrahlung radiation. It has nearly a flat intensity spectrum (index of -2 in antenna temperature) with a small modification accounted for by the Gaunt factor that brings the antenna temperature power-law index into the range -2.07 to -2.13 for reasonable interstellar plasma temperatures and densities.

The properties and factors that determine the emissivity of the dust at cm and mm wavelengths are a major topic of discussion. It is clear that dust is not an efficient radiator and absorber for wavelengths much larger than the size of the dust. Thus the antenna temperature of dust emission decreases at lower frequencies (larger wavelengths). The precise emissivity frequency dependence and temperature distribution of high galactic lati-

Galactic Foregrounds

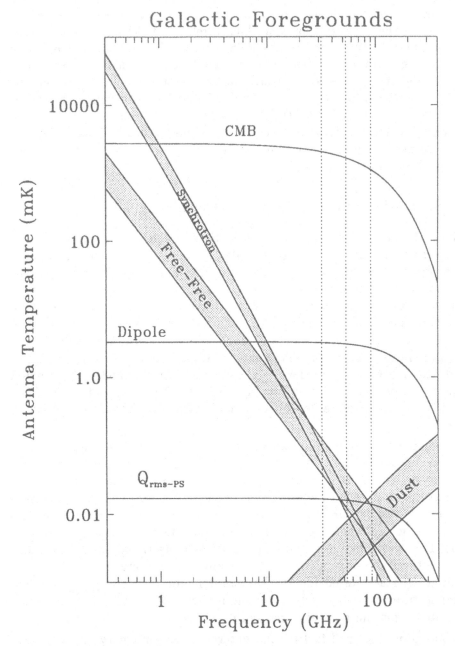

Figure 2. Galactic foregrounds compared with the Cosmic Microwave Background (CMB) signal all expressed in antenna temperature. (Antenna temperature is power per unit bandwidth per mode calibrated in Kelvin. It is defined by the relation: P= kT_aB, where P is the power, k is Boltzmann's constant, T_a is antenna temperature, and B is the bandwidth.) The three vertical dotted lines indicate the COBE DMR observing frequencies. The width of the Galactic emission bands indicate the typical range of signal level for Galactic latitudes between 20° and 70°.

tude dust remain to be determined.

The flat spectral index of the DMR anisotropy, without correction for Galactic emissions, is consistent with a cosmic origin and inconsistent with an origin from a single Galactic component. Ganga et al.(1993) have shown that the 'MIT' FIRS balloon-borne bolometer observations correlate well with the COBE DMR maps. The FIRS bolometer observations have an effective frequency of about 170 GHz. The Tenerife (Hancock et al. 1994) ground-based observations of a strip in the northern sky have also been compared to a simulation of their observations on the COBE DMR maps indicating correlations (Lineweaver et al. 1994) and extending the spectral coverage down to 15 GHz (2-cm wavelength). These additions make the flat spectral index argument stronger. However, from this fact alone we are unable to rule out a correlated superposition of dust, synchrotron, and free-free emission and thus more detailed Galactic emission models are required. The worst case scenario would be a dust component with an emissivity power law index that reached into the range 0 to about 1 in the mm wavelength range or pockets of very cold (< 4 K) dust – either interstellar or cosmic. This seems an unlikely scenario but must be investigated carefully. Using maps made at non-DMR frequencies as well as the DMR maps, a multicomponent fit to CMB anisotropies and the three Galactic emissions indicates that the signal is predominantly CMB anisotropies (Bennett et al. 1992).

4. COBE DMR Results

We proceed assuming that the Galactic foreground emissions are small ($\leq 10\%$) compared to the CMB fluctuations. The DMR maps are a full-sky panorama showing a whole spectrum of primordial perturbations in the early Universe. Though some spots have statistical significance (e.g. Cayon & Smoot 1994), in general the early DMR data signal to noise is roughly unity. To quantify the anisotropies, it is natural to expand the temperature fluctuations on the celestial sphere in spherical harmonics:

$$T(\theta, \phi) = \sum_{\ell m} a_{\ell m} Y_{\ell m}(\theta, \phi)$$

where ℓ is the Legendre polynomial number and $Y_{\ell m}$ are the spherical harmonics. A complete description of the CMB fluctuations would require observers to find the full set of coefficients $a_{\ell m}$. The DMR data can be fitted to determine the best estimated spherical harmonic coefficients.

At this stage we approach a somewhat simpler task: finding the mean power at each ℓ. We are making use of the idea of rotational invariance of the mean power in the anisotropy. That is that there are no preferred directions. The direction independent mean square power at ℓ is $T_\ell^2 = \sum_m a_{\ell m}^2$. Since the COBE anisotropies are consistent with being drawn from a random-phase, Gaussian distribution (Smoot et al. 1994), then in principle the power spectrum contains the information characterizing the parent distribution.

We compare the power observed by the DMR to a scale-invariant (or Harrison-Zeldovich) spectrum by plotting the ratio as a function of ℓ. Figure 3 shows a sample of the DMR measured anisotropy power spectrum. The data are clearly consistent with a scale-invariant (Harrison-Zeldovich) power spectrum. The question then becomes: what is the best estimate of the power spectrum and what limits can we place in an effort to exploit the CMB for learning cosmology as outlined in the table above? A careful analysis (Gorski et al. 1994) shows that the DMR data are very close to Harrison-Zeldovich with errors in power-law index of less than about 0.3 and that the amplitude normalized to the quadrupole ($\ell = 2$) is approximately 20 μK or about 7×10^{-6} of the CMB temperature. CMB anisotropy results are often ploted in terms of the mean power at some ℓ or band of ℓ's normalized to a Harrison-Zeldovich spectrum (often as the equivalent quadrupole for this flat spectrum or as a dimensionless anisotropy).

The CMB power spectrum can readily be interpreted for what it tells us about the primordial power spectrum of fluctuations and then applied either backwards to tell us about early-Universe, high-energy physics or viewed forward in its relation to large scale structure. The field, especially theoretically and rightly so, is focused on these issues. However, there are other things that we can learn about the Universe using these data. We can learn a lot about the geometry and dynamics of the Universe and set very tight limits on anisotropic Hubble expansion and on shear and rotation (vorticity) in the Universe. We also know that the geometry of the Universe is very near to the idealized Robertson-Walker metric with only small perturbations which must be on the same scale as the CMB anisotropies. In 1968 Ehlers, Geren, and Sachs proved the theorem: If a family of freely-falling observers measure self-gravitating background radiation to be *exactly* isotropic, then the Universe is *exactly* Friedmann-Lemaitre-Robertson-Walker. This has been interpreted/extrapolated by me and others that the observed high degree of observed CMB isotropy and the assumption that our location is not special imply that the Universe's metric is nearly Robertson-Walker with perturbations of the order of 10^{-5}. I have been quoting that result and encouraging the extension of the theorem. Now Stoeger, Maartens, and Ellis (1994) have shown that if all fundamental observers measure the cosmic microwave background radiation to be almost isotropic in an expanding Universe, then that Universe is almost spatially homogeneous and isotropic. This puts us on firmer footing and allows us to use the CMB isotropy ob-

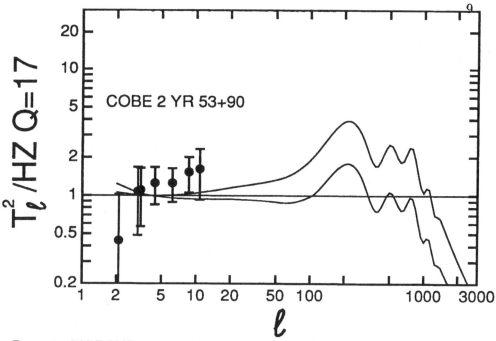

Figure 3. COBE DMR two-year data power spectrum: The horizontal line indicates the anisotropy level for a primordial scale-invariant model with rms quadrupole amplitude of $17\mu K$. The other two lines show the expected anisotropies with a scale-invariant and a tilted primordial spectrum as processed in a standard CDM model.

servations to set limits on the anisotropy, homogeneity, and dynamics of the Universe.

The observation of anisotropies opens one new cosmological test: probing the topology of the Universe. The simplest topology that we can imagine is an essentially isotropic and homogeneous Universe that is simply connected. However, we know of no constraints that actually require that the Universe be simply connected. It might in fact have the topology of a donut or many other possible objects. In recent times there have been reports of periodicity in the Universe both in terms of large scale structure of galaxies ($\sim$128 Mpc) and quasars (much larger scale). This has led a number of people to suggest that the Universe is small with opposite faces identified (or some other combination). Such universes would be periodic on the identified axes and thus could not have anisotropies with wavelengths longer than their symmetry axes. The existence of very large angular scale anisotropies, i.e. the quadrupole, octopole, and hexadecapole, put stringent limits on the size of the Universe. Various analyses of the DMR data set a limit on the smallness of the Universe at about 0.5 of the Hubble diameter (e.g. Jing & Fang 1994, and Costa & Smoot 1994). This is an illustration of the power of the CMB as a cosmological probe.

However, the main work of the field is to investigate the origin of the primordial perturbations and their tie to large scale structure. The best way to do this will be to combine the DMR data with the data from other experiments in an effort to map out the full sky and power spectrum.

5. Other Anisotropy Results

Since the DMR announced the discovery of anisotropy, nine groups have reported CMB anisotropies. Figure 4 shows the current status of the CMB anisotropy power spectrum observations.

The 'MIT' FIRS experiment (Meyer et al. 1991, Page et al. 1990) is the only other experiment to map a significant portion of the sky. The FIRS experiment has a $\sim 3°$ beam width and covered nearly a quarter of the sky with a single balloon flight. The FIRS data correlate well with the DMR data (Ganga et al. 1993) and show a similar power spectrum (Ganga et al. 1994) consistent with scale invariance.

Tenerife (Watson et al. 1991) is also a large angular scale experiment (beam width 5°) that covers a differenced (8°) strip scanned on the sky by the earth's rotation. The Tenerife experiment has pointed out bumps on the sky as specific locations of anisotropy (Hancock et al. 1994). The ULISSE experiment (de Bernardis et al. 1992) reported upper limits on 6° CMB anisotropy using balloon-borne bolometric observations. The Advanced Cosmic Microwave Explorer (ACME South Pole) (Gaier et al. 1992, Schuster et al. 1993) reported upper limits and detections of fluctuations operating with HEMT amplifiers. The Saskatoon "SK93" experiment (Wollack et al. 1993) used HEMT amplifiers to detect CMB anisotropy from Saskatoon, SK, Canada. Fluctuations were reported from South Pole observations by the Python experiment (Dragovan et al. 1993). The ARGO balloon-borne experiment (de Bernardis et al. 1994) observed a statistically significant signal with a 52' beam. The Italian Antarctic Base (IAB) experiment (Piccirillo & Calisse 1993) used bolometric techniques with a 50' Gaussian beam and reports anisotropy. The Millimeter-wave Anisotropy eXperiment (MAX) is a balloon-borne bolometric instrument with high sensitivity in the medium angular scale that has completed four flights detecting significant CMB fluctuations (Alsop et al. 1992, Meinhold et al. 1993, Devlin et al. 1994, Clapp et al. 1994). The Medium Scale Anisotropy Measurement (MSAM) balloon-borne experiment (Cheng et al. 1994) is a very similar balloon-borne medium-scale CMB anisotropy instrument but with a different chopping scheme that allows the results to be reported either as a difference or a triple difference, providing two effective window functions. Also from the South Pole the White Dish (WD) experiment (Tucker

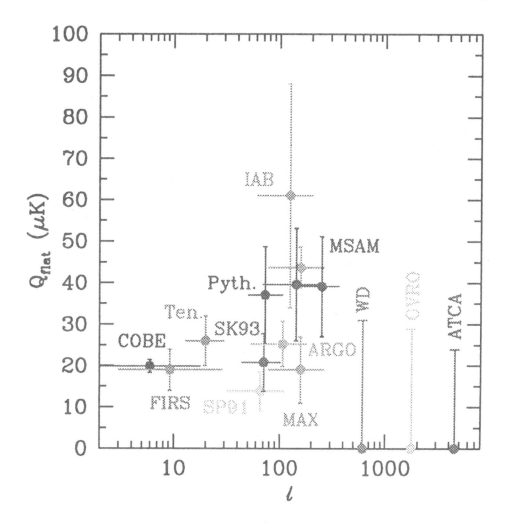

Figure 4. Current status of CMB anisotropy power spectrum observations adapted from Scott and White 1994. The amplitudes plotted are the quadrupole amplitudes for a flat (unprocessed scale-invariant spectrum of primordial perturbations, i.e. the horizontal line in Figure 3) anisotropy spectrum that would give the observed results for the experiment. Figure 3 gives an indication of the expected spectrum for a processed spectrum in a CDM model for comparison.

et al. 1993) reports an upper limit on CMB anisotropy. Arc-minute scale anisotropy limits came from the Owens Valley Radio Observatory (OVRO) (Myers et al. 1993). The Australia Telescope Compact Array (ATCA) was used to place upper limits on CMB anisotropy in a synthesized image (Subrahmayan et al. 1993).

It appears that the reported results, though scattered, are actually in rough agreement with each other and with many models. The current goal moment is to refine the results and our ability to distinguish between models. A hot topic is whether the data show evidence for a Doppler peak and if that peak is in the right location. The existence and location of a peak would go far in telling us not only whether we are on the right track with these models, and if there is a connection with large scale structure formation, but also about various cosmological parameters.

6. Interpretation, Future

In two short years the field of CMB anisotropy observations and theory has made great strides. Until April 1992 all plots of CMB anisotropy showed only upper limits, except for the $\ell = 1$ dipole. Now we are beginning to trace out the shape of the power spectrum and to make maps of the anisotropies. This promises to deliver a wealth of new information to cosmology and to connect to other fields. Already we are seeing plots showing the CMB anisotropy spectrum related to and overlaid on the primordial density perturbation power spectrum and attempts to reconstruct the inflaton potential. These are the first new steps in a new period of growth.

The COBE DMR has now released the first two years of its data and the full four-year data set is being processed and prepared for release in 1995. We can expect improved results from the DMR on the large angular scales but the scientific interest has moved to covering the full spectrum and learning what the medium and small angular scales will tell us. Already new experiments are underway. Nearly every group has new data under analysis and is also at work on developing new experiments. The first of these are the natural extensions of the ongoing experiments. Some groups are considering novel approaches. Real long-term progress depends on avoiding potential foregrounds: fluctuations of the atmosphere emission, a source of noise that that largely overwhelms recent advances in detector technology, and Galactic and extragalactic signals. This requires sufficient information and observing frequencies to separate the various components. It also means going above the varying atmosphere. Collaborations are working on long-duration ballooning instruments. Ultimately, as COBE has shown, going to space really allows one to overcome the atmospheric effects and to get data in a very stable and shielded environment. A number of groups are working on designs for new satellite experiments. The COBRAS/SAMBA mission (Mandolesi et al. 1994) leads the way in the multi-wavelength and benign orbit location. With the new data that are appearing and can be expected and ultimately with the COBRAS/SAMBA mission we can look forward to a very significant improvement in our knowledge of cosmology.

Acknowledgements
Thanks to Eric Gawiser and Alan Kogut for reading and reviewing this paper. We acknowledge the excellent work of those contributing to the COBE-DMR. COBE is supported by the office of Space Sciences of NASA Headquarters. Goddard Space Flight Center (GSFC), with the scientific guidance of the COBE Science Working Group, is responsible for the development and operation of COBE. This work is supported in part by the Director, Office of Energy Research, Office of High Energy and Nuclear Physics, Division of High Energy Physics of the U.S. Department of Energy under Contract No. DE-AC03-76SF00098.

References

Alsop, D. C., et al. 1992, ApJ, 395, 317
Bennett, C.L., et al. 1992, ApJ, 396, 5.
Bennett, C.L., et al. 1994 ApJ, 436, 423.
de Bernardis, P., et al., 1992, ApJ, 396, L57
de Bernardis, P., et al. 1994, ApJ, 422, L33
Cayon, L., & Smoot, G.F., 1994 submitted to ApJ
Cheng, E.S., et al., 1994, ApJ, 422, L37
Clapp, A.C., et al., 1994, ApJL, 433, L57
Costa, A,O., & Smoot, G.F., submitted to ApJ
Devlin, M.J., et al., 1994, ApJL, **430**, L1
Dragovan, M., et al., 1993, ApJ, 427, L67-70
Ehlers, J., Geren, P., Sachs, R.K., J. Math. Phys., 9, 1344.
Gaier, T., et al. 1992, ApJ, 398, L1
Ganga, K., Page, L., Cheng, E., & Meyer, S. 1993, ApJ., 410, L57.
Ganga, K., Page, L., Cheng, E., & Meyer, S. 1994, ApJ, 432, L15
Górski, K. M., Hinshaw, G., Banday, A. J., Bennett, C. L., Wright, E. L., Kogut, A., Smoot, G. F. & Lubin, P. 1994, ApJ, 430, L85
Gush, H., Halpern, M., & Winshow, E., 1990, Phys. Rev, Lett., 65, 537.
Hancock, S., et al., 1994, Nature, 367, 333
Harrison, E.R., 1970, *Phys. Rev. D*1, 2726
Jing, Y.P., & Fang, L.Z., 1994, PRL, 73 (14), 1882
Kogut et al., 1992, ApJ, 401, 1.
Lineweaver, C.L., Hancock, S., et al. 1994 submitted to ApJ
Mandolesi, N., et al. 1994 accepted Planetary and Space Sciences.

Mather, J. C., et al. 1994, *ApJL*, **420**, 439-44

Meinhold, P., & Lubin, P. 1991, ApJ, 370, L11

Meinhold, P., et al. 1993, ApJ, 409, L1

Meyer S. S., Page, L., & Cheng, E. S. 1991, ApJ, 371, L7

Myers, S. T., Readhead, A. C. S. & Lawrence, C. R. 1993, ApJ, 405, 8

Penzias, A.A., & Wilson, R.W. 1965, ApJ, 142, 419

Piccirillo. L., and Calisse, O. ApJ, 411, 529-533.

Sachs, R.K., & Wolfe, A.M., 1967, ApJ, 147, 73

Scott, D., & White, M., 1994, CWRU Workshop Proceedings.

Smoot, G.F., et al., 1992, Ap.J. 396, L1.

Smoot, G.F., et al., 1994, Ap.J. 437, 1.

Subrahmayan, R., Ekers, R., Sinclair, M. & Silk, J. 1993, MNRAS, 263, 416

Stoeger, W., Maartens, R., & Ellis, G.F.R., ApJ in press

Tucker, G.S., Griffin, G., Nguyen, H. & Peterson, J.B. 1993, ApJ, 419, L45

Watson, R.A. *et al.*, 1992 *Nature* **357**, 660

Wollack, E. J., et al., 1993, ApJ, 419, L49

Wright, E.L., et al., 1992, ApJ, 396, L13.

Wright, E.L., Smoot, G.F., Bennett, C., & Lubin, P., 1994, ApJ, 436, 443

Zel'dovich, Ya.B., 1972 *Mon. Not. R. astr. Soc.* **160**, 1.

STATUS OF THE RELICT-2 MISSION AND OUR FUTURE PLANS

STRUKOV I., SKULACHEV D., BUDILOVICH N., BRJUKHANOV A., NEMLIKHER YU., KOROGOD V. , KOSOV A. , RUKAVICIN A. , BOROVSKY R. AND NECHAEV V.
Space Research Institute of Russian Academy of Sciences
Profsojuznaja ul., 84/32, Moscow, 117810, Russia

Abstract. We review the main results obtained in the first cosmic experiment for study of the large scale anisotropy of CBR at 8 mm RELICT-1 and compare them with data of COBE mission.

1. Introduction

Investigation of the anisotropy of the cosmic background radiation is one of the most important directions for the more precise determination of such fundamental parameters as H, Ω, h.

In this paper we will discuss main results obtained in the first cosmic experiment for study of the large scale anisotropy of CBR at 8 mm RELICT-1.

Results of comparison of the experimental data of RELICT-1 and first year COBE data and low-frequency surveys are presented. The performed analyze showed that the accuracy of the COBE data is not enough for separation background anisotropy caused by galaxy emission from the CBR anisotropy. For such separation one need of many-frequency survey with more accuracy. Such accuracy may be achieved in the RELICT-2 project that has an unique radiometric equipment in the frequency band from 22 to 90 GHz with sensitivity more than 10 times better than COBE.

The experimental procedure of RELICT-2, its orbit and results of the engineering testing of the RELICT-2 equipment are discussed in the paper.

For more precise determination of the spectral index n it is useful to investigate CBR anisotropy with angular resolution about 3° .

M. Kafatos and Y. Kondo (eds.), Examining the Big Bang and Diffuse Background Radiations, 45–58.

2. Comparison of the RELICT-1 and COBE Data

RELICT-1 was our first space-borne experiment, Strukov and Skulachev (1984, 1987). It was performed by means of small satellite Prognoz-9 that was launched in 1983 to a high altitude orbit with apogee about 700 000 km as shown in Fig.1.

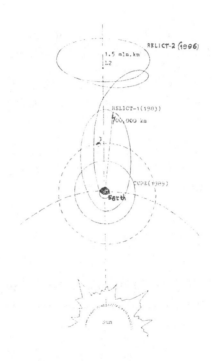

Figure 1.   Orbits

RELICT-1 was our first attempt to investigate the large-scale CMB anisotropy. We understood that one need of more sensitive receiver for such investigation which can be used at the frequency band from 22 to 100 GHz. So the next most important goal of the experiment was study of environment conditions during long term space experiment and choice of the type of receiver for such investigation.

The detectors of the receivers can be cooled to 80 K by means of radiative loss of heat into space. Degenerated paramp and Schottky mixers showed promise as sensitive devices for the conditions mentioned above. Degenerated paramp is the most sensitive type of detectors, but it is regenerative device and investigation of the stability of such devices under long-term use in deep space in terms of optimizing the parameters of the next generation of instrumentation was one of the most fundamental goals of the RELICT-1 mission.

The main goal of all investigations of the large scale anisotropy of CMB performed before the RELICT-1 mission was determination of the model independent magnitude of the quadrupole component, Cheng et al.(1979), Fabbri et al.(1980).

Choice of the strategy of data reduction was based on the next assumptions, Strukov and Skulachev (1987), Strukov (1991):

1) after averaging we have only one sample of noise;
2) we have only one Universe;
3) in theoretical models the population variance is used, but not sample one;
4) quadrupole has only 5 degrees of freedom:
5) in any case in comparison of experimental and theoretical data always the concrete model is used;
6) observational sample is not uniform: different sample units have different statistical weight. Weighting process causes the transformation of signal and noise response

$$S_{output}(l, m) = ||T(l, m)|| \cdot S_{input}(l, m)$$

$$||T(l, m)||^{-1} \quad - \text{is not exist}$$

Of course after some regularization matrix $||T(l, m)||$ is transformed to the form of square matrix.

In this case inversed transformation exist but regularization process brings additional uncertainies if signal/noise ratio is small.

7) The mapping strategy in experiment RELICT-1 allows to determine population variance of the instrumental noise.
8) The instrumental transform function and the statistical weight can be determined with high accuracy.
9) The number of degrees of freedom for sum of spherical harmonics is more than for the quadrupole one.

The mentioned above assumption gave us a possibility to abandon model independent estimate of quadrupole component and using Monte Carlo simulation we can determine the constraints on the power spectrum for any cosmological model and after that to receive the constraints on the quadrupole magnitude or on magnitude of an arbitrary spherical harmonic, Strukov et al.(1988).

After careful data reduction we have discovered anisotropy in the background radiation, Strukov et al.(1992, 1992, 1993).

Parameters of the instrumental noise and measured values for the different smoothing parameters are tabulated in Table 1.

Smoothing parameter φ_0, degree	Measured variance on the map μK^2	Variance of the noise μK^2	Degrees of freedom of noise	Sky variance
6	127^2	124^2	106	27.4^2
12	67^2	56^2	19	36.8^2
24	34.5^2	23.7^2	9	25.1^2

TABLE 1. *Parameters of noise and measured values*

It should be stress that the main part of the observed by RELICT-1 mission anisotropy is caused by decreasing of background temperature. We proposed that the discovered signal has cosmological nature, Strukov et al (1992).

It is necessary to stress that the direct comparison of massives of the RELICT-1 and COBE data is impossible. Poor signal/noise ratio in RELICT-1 data causes a necessity to use weighting, that dramatically changes a form of output signal. In this case the spherical harmonics lose their orthogonality and power of some input harmonics transfer in different output harmonics, moreover the transform functions are different for COBE and RELICT-1 radio maps.

We can correctly compare the results of two experiments only after receiving constraints for one or other type of cosmological model.

In case of Harrison-Zeldovich spectrum we have received constraints on the quadrupole component and on the amplitude of the metric fluctuations, Strukov et al. (1992)

$$17\,\mu K < \langle Q_2 \rangle < 95\,\mu K$$

and

$$5.2 \cdot 10^{-6} < \varepsilon_H < 2.9 \cdot 10^{-5}$$

The first results of COBE mission gave the next constraint on that parameters, Smoot et al.(1992)

$$2.6 \cdot 10^{-6} < \varepsilon_H < 1.1 \cdot 10^{-5}$$

and Tenerife experiment gave, Hancock et al. (1994)

$$5.8 \cdot 10^{-6} < \varepsilon_H < 9.7 \cdot 10^{-5}$$

Thus for Harrison-Zeldovich spectrum we have rather good agreement between estimates of RELICT-1 and Tenerife experiments. COBE experiment gave the less level two times more.

But after 2 years of COBE data reduction for Harrison-Zeldovich spectrum was obtained the next estimate for the quadrupole component, Gorski et al.(1994)

$$18.3 < \langle Q_2 \rangle < 21.5 mK$$

$$6 \cdot 10^{-6} < \varepsilon_H < 7 \cdot 10^{-6}$$

that supports our data.

Additional constraints on the ε_H for different values of Hubble parameter one can receive from the magnitude of the dipole component, Abbott and Schaefer (1986). The mentioned above constraints on ε_H together with the constraints on ε_H caused by measured dipole component give us constraints on the density, Strukov et al.(1994), Strukov (1994).

You can see from Fig.2 that we can receive constraints on the density

$$0.16 < \Omega < 0.5.$$

If we don't tolerate a nonzero cosmological constant $\Omega = 1$ and the dipole gives more stringent constraints on the magnitude of ε_H. This magnitude is less than low limit of the mentioned above experiments.

In this case we can conclude that the part of anisotropy is caused by anisotropy of the galaxy background. It is interesting to make rough comparison of the RELICT-1 and COBE maps and comparison of COBE and low-frequency maps, Haslam et al. (1982). At comparison we used an information about transform and weighting function RELICT-1 mission.

We propose that cross-correlation between the maps is

$$\rho_{31}{}^\frown{}_{37} \approx 1; \quad \rho_{31}{}^\frown{}_{37} \approx 0.7; \quad \rho_{31}{}^\frown{}_{37} \approx 0.3.$$

Than we can compare proposed (for measured signal/noise ratio) and measured cross-correlation between the different channels of COBE data and RELICT-1 data for the different smoothing parameters. Results of comparison (at smoothing function $\exp(-\varphi/2\varphi_0)$ and $b > 20°$) are tabulated, see Table 2.

In the Table 3 cross-correlation between COBE, 408 MHz and 19 GHz maps for $b > 20°$ and for different smoothing parameters is represented.

You can see a rather strong correlation between COBE and mentioned above maps, especially for $\varphi_0 = 24°$. This shows that galaxy emission is strong enough and one need a precise methodic for the separation of the galaxy and cosmic background anisotropies.

There are two ways to separate galactic and cosmic emission. COBE team method based on the recount synchrotron emission from 408 MHz to 31.5, 53 and 90 GHz by using of measured electron energy distribution and determination of variation of magnetic field strength received from

50

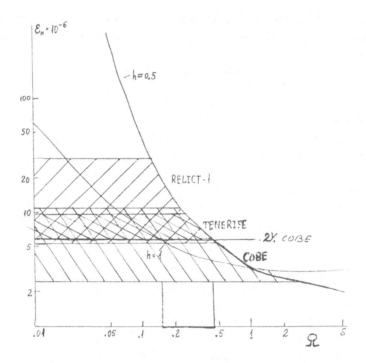

Figure 2.   Constrains on the density

$\varphi_0 = 12°$	$\nu = 19$					
		31A	31B	31Σ	53Σ	90Σ
RELICT-1	prop.	0.27 ± 0.2	0.27 ± 0.2	0.33 ± 0.2	0.3 ± 0.19	0.12 ± 0.23
	meas.	0.29	-0.06	0.11	0.24	-0.29
$\varphi_0 = 24°$	$\nu = 9$					
RELICT-1	prop.	0.30 ± 0.3	0.35 ± 0.3	0.42 ± 0.27	0.4 ± 0.28	0.18 ± 0.3
	meas.	0.30	-0.01	0.12	0.34	-0.35
$\varphi_0 = 12°$	$\nu = 50$					
COBE 31B	prop.	0.36 ± 0.12	$^{1.00}_{1.00}$		0.37 ± 0.12	0.17 ± 0.12
	meas.	0.28			0.54	0.42
$\varphi_0 = 24°$	$\nu = 12$					
COBE 31B	prop.	0.2 ± 0.27	$^{1.00}_{1.00}$		0.34 ± 0.22	0.13 ± 0.28
	meas.	0.22			0.47	0.43

TABLE 2.   *Comparison results*

spatial variation in β_{synch} (408, 1420 MHz), Bennett et al. (1992) or taking $\beta_{synch} = 2.75$. Accuracy of such methodic may be not very good, because of

	$\varphi_0 = 3°$		$\varphi_0 = 12°$		$\varphi_0 = 24°$	
	408 MHz	19 GHz	408 MHz	19 GHz	408 MHz	19 GHz
31Σ	0.45	0.65	0.56	0.68	0.92	0.90
53Σ	0.18	0.25	0.38	0.35	0.65	0.65
90Σ	0.20	0.15	0.24	0.21	0.48	0.48

TABLE 3. *Cross-correlation between COBE, 408 MHz and 19 GHz maps*

very high frequency difference and instrumental and systematic noise of 408 and 1420 MHz maps and difference of angular resolution at low frequency and COBE maps.

Separation of the galactic and cosmological background is complicated because of similarity of their spatial frequencies spectra. Our analysis showed that spectral index $n = 1.27$ for total 408 MHz map and $n = 1.33$ for 20° cutting.

Our method is based on the possibility to do very precise measurements of the large scale anisotropy in background emission at different frequencies . But we must improve the equipment sensitivity at least by factor 10. So one of the main goals in time of preparation of the RELICT-2 mission was an essential improvement of the radiometers sensitivity.

In case of cosmic nature of the anisotropy of the background radiation that has been discovered by RELICT-1 and COBE teams and if its spectrum is the Harrison-Zeldovich spectrum a minimal magnitude of the rms per 10° field of view would be about 30 μK, Strukov et al.(1992, 1992), Smoot et al.(1992).

The quadrupole components of the dust emission are rather small (at 53 GHz rms$\approx$ 2 μK), Bennett et al.(1992). So we can not take into account the dust emission in the frequency band below 60 GHz. That gives the possibility to use very precise measurements of the background radiation anisotropy at 22, 34.5 and 60 GHz for separation of the cosmic and galactic emission. But that calls for the use of much more sensitive radiometers than were employed in RELICT-1 and COBE experiments.

Only after such separation we can determine magnitude of spectral index n with accuracy about 10% . We can receive 6% accuracy by improving angular resolution to 3°, Sazhin et al.(1994?).

In spite of the high apogee of the RELICT-1 experiment orbit and rather low level of its antennas sidelobes we observed an existence of a large contribution of the Moon's and Earth's thermal emission to antenna temperature. Thus the RELICT-2 experiment can not have an orbit such as the orbit of the RELICT-1.

3. Status of the RELICT-2 Mission

3.1. WAYS OF A SENSITIVITY IMPROVING

The main goals of the preparation of the RELICT-2 mission till now were:

1. Determination of essential improvement of the radiometers sensitivity to reach a signal/noise ratio $\sigma_s/\sigma_n > 1$.
2. Investigation of the possibility to decrease the size of the halo-orbit about Sun-Earth L2 libration center.
3. Adjustment and test of the radiometers of the RELICT-2 engineering module.

In order to design and build extremely sensitive radiometers working at 80 K we have made the investigation in the following directions: i) finding methods for improving of quality of solid- state devices; ii) investigation of the possibility to decrease the losses in input devices; design of the low-losses switches; iii) choice of the optimal scheme of the radiometer; iiii) finding methods to provide a normal operation of the radiometers at ambient (300 K) and cryogenic (80 K) temperature. This is important especially for radiometers with a degenerated parametric amplifier.

First of all it was necessary to choose the type of receiver for each frequency band. In order to simplify the engineering testing before the launch the cooled radiometers must work at the room and cryogenic temperature without tuning. This makes additional difficulties in design and adjustment of the radiometers. The degenerated paramp testing during the PROGNOZ-9 flight and our comprehensive investigation of its behaviour have shown its excellent noise and operating performances.

We have proposed a such paramp-doubler configuration that provides an extremely high stability. It provides 10 times less gain sensitivity to the changing of power and frequency of the paramp's pump in comparison with RELICT-1. In order to realize this configuration we are needed to improve varactors' quality by a factor 2. Besides that we have designed and built and tested room and cryogenic temperature frequency doublers for 22-150 GHz frequency band with efficiency more than 50%.

Degenerated paramps are the most sensitive mm wavelengths devices today. Fig.3 shows as a function of the input frequency, noise measure of the degenerated paramps at different ambient temperatures.

Fig.4 shows that recent degenerated paramps have significantly better noise measure that futuristic MODFET.

3.2. SPACECRAFT AND ORBIT

The RELICT-2 configuration is shown in Fig.5. The LIBRIS spacecraft is a cylindrical structure (1). House keeping systems and electronic units

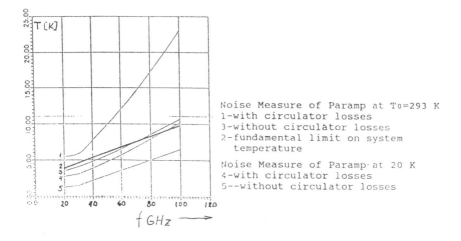

Noise Measure of Paramp at To=293 K
1-with circulator losses
3-without circulator losses
2-fundamental limit on system
  temperature

Noise Measure of Paramp at 20 K
4-with circulator losses
5--without circulator losses

Figure 3.   Noise measure

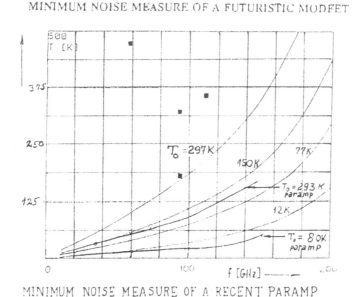

Figure 4.   Comparison of the recent degenerated paramp and futuristic MODEFET

of the scientific payload are installed inside the spacecraft in a hermetic
volume on two frames. Most of the scientific instruments, sunsensors of the

54

orientation system (3), spherical gas containers of the orientation system (4), the propulsion system (5), and the antenna for communication (6) are located on a cylindrical structure with four solar array panels (2) extended like petals of a flower to provide power.

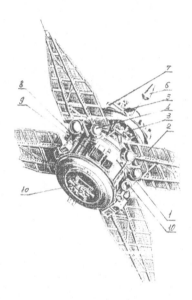

Figure 5.   LIBRIS spacecraft with the installed radiometers

The passive cooling system with the radiometers is located on the shadow side of the spacecraft (10). The 22 GHz radiometer is installed on the cylindrical side of the spacecraft (11). The spacecraft is spin stabilized with the spin-axis pointed toward the Sun. The LIBRIS spacecraft will be built by Babakin Space Center in Moscow in collaboration with other industrial enterprises.

Two orbits: a large-amplitude Lissajous orbit and small-amplitude orbit with a lunar-gravity assist maneuver has been investigated, Strukov et al. (1993).

The results of analysis that had been made by a joint US/Russia flight dynamic team show a possibility to put RELICT-2 into a small-amplitude orbit about the Sun-Earth L2 libration point utilizing a lunar swingby, Eismont et al. (1991), Durbeck et al. (1991), Farquhar (1991), Relict-2 ... (1992), Fig.6. Multiple revolutions should be made in the transfer orbit from the parking orbit to the lunar swingby to reduce the size of maneuvers needed to correct TTI errors, as explained in Dunham et al. (1990). The multiple lunar-swingby on the trajectory would provide an extremely small signal from the Moon, Earth and Sun (less than 1 mK). The possibility of the realization of such an orbit, taking spacecraft maneuver execution errors into account, is proved now.

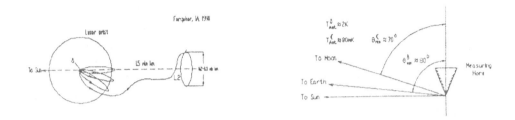

Figure 6.   Scheme of the RELICT-2 transfer trajectory with lunar swingby maneuver

3.3. INVESTIGATION OF THE PARAMETERS OF THE ENGINEERING COMPLEX

The entire engineering instrument consists of four radiometers, an electronic subsystem and star detector OREOL-3 for the definition of three-axial orientation of spacecraft with accuracy 4 s. Three radiometers are passively cooled ($T_0 = 80$ K) for 34.5, 60 and 83 GHz, Strukov et al.(1993). 22 GHz radiometer is an ambient temperature radiometer ($T_0 = 300$ K). Each of the radiometers has an antenna system with the corrugated horns. Their nominal beamwidth is 7 deg full width at half maximum power.

We have chosen the schematic of the 22 GHz and 34 GHz radiometers with the degenerated paramp at the input, Strukov et al. (1993). The output of the Faraday-rotation-switch is alternate between the inputs from the corrugated horn antenna and the ambient temperature load. The nominal output signal is compensated by means of the IF-amplifier gain modulation. The Table 4 represents the results of testing of the engineering complex.

f (GHz)	22	34.5	60	72	83
σ_{real} mK·s$^{-1/2}$	5.5	3	8	-	8
σ_{progn} mK·s$^{-1/2}$	2.4	0.8	< 1	1	-

TABLE 4. *Bench testing and future performances*

3.4. THE RELICT-2 FUNDAMENTAL INSTRUMENT

The entire fundamental instrument for investigation of the large scale anisotropy of cosmic background radiation will consist of four radiometers, electronic subsystem and the star detector, Strukov et al. (1993). We have chosen an ambient temperature (300 K) radiometer with a degenerated paramp for 22 GHz system and three passively cooled radiometers (80 K): one of them for 34.5 GHz and other two for 60 GHz. All configurations of the radiometers have a waveguide switch. In the 22 GHz radiometer output of the waveguide switch will be alternate between the inputs from corrugated horn antenna and the temperature stable load. We hope to receive the radiometers sensitivity 2.4 μK·s$^{1/2}$.

In 34.5 and 60 GHz radiometers with degenerated paramp at the input an output of the waveguide switch will be alternate between two inputs from the corrugated horns antennas. The receiving signal has an orthogonal polarization for these radiometers and the same polarization for their reference horns. That provides the additional possibility for investigation of the cosmic background radiation.

The recent launch date of the RELICT-2 is the end 1996. We hope that the RELICT-2 mission put the last point into an experimental investigation of the large scale anisotropy of microwave background radiation.

4. Future Plans

Now we look for a possibility of collaboration in the field of cosmological and solar-terrestrial investigation. One of possible examples of such collaboration is shown in Fig.7. For such investigation we propose to use two additional LIBRIS-type spacecrafts. One of them will be put in the neighbourhood L2 point and two others - in the vicinity L1 point. The instrumental complex for the anisotropy CMB investigation will be installed on the shadow side of spacecraft and solar-terrestrial equipment - on the Sun-side of the spacecraft. On the first spacecraft will be installed two radiometers at 60 and 72 GHz with angular resolution 1.5 deg and sensitivity 1 μK·s$^{1/2}$. On the second spacecraft will be installed three radiometers at 34, 60 and 72 GHz with angular resolution 3 deg and with the same sensitivity.

If we can receive foreign support then these investigations would be incorporated into RELICT-2 program.

In a future plan of Russian Academy of Sciences was included RELICT-3 mission (Phase A). Main goal of the RELICT-3 mission is investigation of the intermediate scale anisotropy of the CMB. We plan to perform investigations at frequencies from 34 to 90 GHz with an angular resolution near 1 deg and sensitivity 0.5 μK·s$^{1/2}$.

Figure 7. Scheme of cosmological and solar-terrestial investigations

Acknowledgements: We are grateful to all collective of Department 61 of the Space Research Institute of the Russian Academy of Sciences. We acknowledge the support of the RFFI (Grants 93-02-2930 and 93-02-2931), Cosmion and Soros Foundation (Grant M06-000).

References

Abbott, L.F., Schaefer, R,K.: 1986, Ap.J. **308**, p.546.

Bennett, C.L., Smoot, G.F., Hinshaw, G., Wright, E.L., Kogut, A., De Amisi, G.: 1992, Ap.J.Lett. **396**, L7.

Cheng, E.S., Saulson, P.R., Wilkinson, D.T., Corey, B.E.: 1979, Ap.J.Lett. **232**, L139.

Durbeck, M., Hung, J., Regardie, M.: 1991, GMAS User's Guide, Revision 3, Goddard Space Flight Center document 552-FDD-91/020/CSC, Feb. 1991.

Dunham, D.W. et al.: 1990, IAF Paper 90-309, Dresden, Germany, Oct. 1990.

Eismont, N. et al.: 1991, Paper at the 3rd Intern. Symp. on Spacecraft Flight Dynamics, Darmstadt, Germany, Oct. 1991.

Fabbri, R., Guidi, I., Melchiorri, F., Natale, V.: 1980, Phys.Rev.Lett. **44**, p.156.

Farquhar, R.W.: 1991, Acta Astronautica **24**, p.227.

Gorski, K.M., Hinshaw, G., Banday, A.J., Bennett, C.L., Wright, E.L., Kogut, A., Smoot, G.F., Lubin, P.: 1994, COBE Preprint No.94-08.

Gush, H.B., Halpern, M., Wisshnow, E.H.: 1990, Phys.Rev.Lett. **65**, p.537.

Hancock, S. et al.: 1994, Nature **367**, p.333.

Haslam, C.G.T., Salter, C.J., Stoffer, C.J., Wilson, W.E., Thomasson, P.T.: 1982, Astro. and Astrophys. Suppl. Ser. **47**, p.1.

Mather, J.C. et al.: 1990, Ap.J.Lett. **354**, L37.

Reich, P., Reich, W.: 1988, Astron. and Astrophys. **196**, p.211.

Relict-2 Mission Trajectory Design, Goddard Space Flight Center document 554-FDD-92/027/CSC: 1992.

Reynolds R.J.: 1984, Ap.J. **282**, p.191.

Sazhin, M.V. et al.: 1994?, Sov.Astron.Lett. (in prepare)

Smoot, G.F. et al.: 1990, Ap.J. **360**, p.685.

Smoot, G.F. et al.: 1992, Ap.J. Lett. **396**, L1.

Smoot, G.F., Tenorio, L., Banday, A.J., Kogut, A., Wright, E.L., Hinshaw, G., Bennett, C.L.: 1994, COBE Preprint No.94-03.

Strukov, I.A., Skulachev, D.P.: 1984, Pis'ma Astron. Zh. **10**, 3, p.3

Strukov, I.A. et al.: 1984, Abstracts of 25 COSPAR, Graz, Austria, 1984.

Strukov, I.A., Skulachev, D.P.: 1987, SSR/Astrophys. and Space Phys., **vol.6**, ser.E, p.145.

Strukov, I.A., Skulachev, D.P.: 1987, Sov.Astron.Lett. **13(3)**, p.191.

58

Strukov, I.A., Skulachev, D.P., Klypin A.A.: 1988, Acta Astronautica **17**, No.8, p.903.

Strukov, I.: 1991, Thesis on Phys. and Mathem. Dr. Degree.

Strukov, I.A., Brjukhanov A.A., Skulachev, D.P., Sazhin, M.V.: 1992, Pis'ma Astron. Zh. **18**, 5, p.387.

Strukov, I.A., Brjukhanov, A.A., Skulachev, D.P., Sazhin, M.V.: 1992, M.N.R.A.S. **258**, p.37.

Strukov, I.A. et al.: 1993, Adv. Space Res. **13**, No.12, p.(12)425.

Strukov, I., Brjukhanov. A., Skulachev, D., Sazhin, M.: 1993, Phys.Lett.B **315**, p.198.

Strukov, I.: 1994, Proc.Conf. "Astrophysics and Cosmology after Gamov", Odessa, 1994 (in prepare).

FLUCTUATIONS IN THE MICROWAVE SKY

R. B. PARTRIDGE

Haverford College

Abstract. This paper reviews the great progress recently made in searches for and the characterization of anisotropies in the cosmic microwave background. We now have secure detections on some angular scales and improved upper limits on others. As the sensitivity of such searches increases, understanding sources of foreground noise (e.g., Galactic and extragalactic radio emission) becomes more important. Also reviewed are the contributions aperture synthesis (interferometric) observations can make in characterizing cosmic background fluctuations and foreground sources of noise. Some recent results from the Very Large Array at $\lambda = 3.6$ cm are given; these set a limit $\Delta T/T \lesssim 1.4 \times 10^{-5}$ on fluctuations at $\theta \sim 80''$. Recent work on the Sunyaev-Zel'dovich effect is summarized.

The cosmic microwave background radiation (CBR), the remnant of the hot Big Bang phase of the Universe, is the best studied of all cosmic backgrounds. In the 5 years since the last IAU symposium dealing with background radiation fields (IAU Symposium 139, held in Heidelberg), there has been startling progress in studies of the CBR. Other papers in this volume will present the results of much of this recent work. In my contribution, I wish to reflect a bit on the change of mood in the field of CBR studies produced by these recent observational successes, then reemphasize the importance of a clear understanding of foreground sources of noise, and finally to highlight the importance of interferometry as a technique in the study of the angular distribution of the CBR. In the course of the last of these three, I will present some new observational work in which I was involved.

When I reviewed the CBR at Symposium 139, I stressed just how featureless that background is. The spectrum is thermal and, in 1989, no anisotropy had been reliably detected with the exception of the dipole moment due to the motion of the earth. The first of these statements is still true; there is no evidence for spectral distortions of any sort in the spectrum of the CBR, and upper limits on various classes of spectral perturbations have been sharply improved, as summarized by John Mather here. The second situation has changed radically. Anisotropy on a range of angular scales has now been detected or reported. The first such report was by the COBE group (Smoot *et al.*, 1992), but other papers have followed quickly on the heels of that breakthrough letter. The fact that we do now have detections rather than increasingly stringent upper limits has produced a mental phase change in the field. Dare I refer to the work of Thomas Kuhn and call it a "paradigm shift"? Formerly, those of us who spent decades working on the CBR aimed to set tighter and tighter upper limits on the amplitude $\Delta T/T$ of fluctuations in the CBR on various angular scales. The aim of reducing limits on $\Delta T/T$ may have led some of us on occasion to discard or downplay some real signals. The situation is different now—we are now trying to measure an effect known

59

M. Kafatos and Y. Kondo (eds.), Examining the Big Bang and Diffuse Background Radiations, 59–70.
© 1996 *IAU.*

to exist. I would suggest that some of the confusion in the field, particularly the apparently discordant results of those groups working on scales θ ~ 1°, may be due to the understandable turmoil induced by this paradigm shift.

There is another consequence of this change in mood. Formerly, if we encountered a foreground source of noise we did not fully understand, we were safe in lumping it in with the signal when deriving upper limits on ΔT/T. Now that we have detections, we may no longer do so. We must understand the backgrounds in order to correct actual measurements. For just that reason, I will discuss here foreground emission, both Galactic and extragalactic.

1.) Summary of Recent Results

Let me begin with a brief survey of the observational status of searches for anisotropies in the CBR. On scales of ~10° and above, we have an excellent all-sky map provided by the DMR instruments aboard COBE (see Smoot's paper in this volume). Those important satellite results have been confirmed by a balloon-borne experiment (Ganga et al., 1993) and more recently by a ground-based experiment at comparable angular scales but lower frequencies (Hancock et al., 1994; Davies et al. in this volume). The cross-correlation of the results of Hancock et al. with the COBE map allows us to identify regions of real high and low temperature, as well as to calculate rms values of ΔT/T. We thus do now have real "pictures" of the microwave sky on angular scales of ~10°.

The single most active area in observational CBR studies is the search for anisotropies on scales of approximately 1°. Several groups are involved in observations from the ground, generally at the South Pole, or from balloon-borne instruments. The work of some of those groups, at least, will be reviewed by Lubin in this volume. While I believe it is safe to say that fluctuations on degree scales have indeed been detected, the experimental situation is not completely clear, and some of the results appear to be inconsistent as of the summer of 1994. Some groups report only upper limits (e.g. the work of Tucker et al., 1993, on a scale of 0.15°, and the work of Schuster et al., 1993, on a scale approximately ten times larger); other groups report clear detections (e.g. Wollack et al., 1993; Piccirillo and Callisse, 1993; Cheng et al., 1994; deBernardis et al., 1994; Dragovan et al., 1994, all on intermediate scales). As a rough rule of thumb, however, I believe it is fair to claim that ΔT/T is likely to fall in the range 1.5–5 × 10⁻⁵.

On scales ~0.1°, that is a few arcminutes, the best results were from the Owens Valley Radio Observatory (OVRO) until very recently. This group performed two different experiments at a wavelength of 1.5 cm, with a beam size of 1.5'. Both sets of observations were made near the north celestial pole. The first (Readhead et al., 1989) traced out a small circumpolar arc with seven independent samples. The beam switch angle employed was 7.15', and the resulting upper limit on ΔT/T was 1.7 × 10⁻⁵ (see Fig. 1). The second program (Myers et al., 1993) fully sampled a larger circumpolar ring, with each point receiving less integrating time. More points were sampled, but the limit on ΔT/T is lower: ≤10⁻⁴.

Very recently, results from the Ryle telescope at Cambridge have begun to come in (Jones et al., 1993, 1994; Saunders, 1994). Using interferometric techniques, they have mapped the Sunyaev-Zel'dovich signal in several clusters of galaxies with ~100 μK accuracy. The resolution used for these studies is ~80", and the instrument has the potential to work at both higher resolution and somewhat higher sensitivity, i.e. to be able to detect $\Delta T/T \sim 1$–3×10^{-5} fluctuations. On scales below a few arcminutes, as we shall see, the technique of choice is also interferometry, and I will report on that more fully in §3 below.

Before proceeding, I want to draw a few conclusions from the work I've summarized above. The first is that COBE and some of the other experiments have detected fluctuations, and the amplitudes of these fluctuations are, to within an all important factor of 2 or so, consistent with CDM models for structure formation and with a spectral index of the initial density perturbations n = 1. The Tenerife results do hint at a value for this index n slightly greater than unity. There is no evidence (yet) for secondary ionization which would damp the fluctuations at 0.1°–1° (Ostriker and Vishniac, 1986; Bond et al., 1991; but see also Tegmark, Silk and Blanchard, 1994) or for cosmic strings (Bouchet et al., 1988). The Sunyaev-Zel'dovich effect in nearby clusters has been detected (e.g. Birkinshaw et al., 1994; and Jones, 1994), and in some cases we are able to calculate a value for Hubble's constant using the technique first noted by Gunn (1978): $H_0 \sim 40$ km/sec per Mpc. It is worth noting that in order to obtain a larger value for H_0, one would need to show that values of $\Delta T/T$ are substantially *smaller* than currently reported.

2.) Foreground Sources of Noise

As I have noted, now that we have actual detections we need to understand and correct for foreground sources of noise that can mimic or mask $\Delta T/T$ fluctuations in the CBR. There are two kinds of error, statistical (errors that increase the size of the individual error bars in a plot like Fig. 1) and systematic (those that increase the scatter of the points in a diagram like Fig. 1). Needless to say, it is systematic errors that cause the most concern.

Now let me list briefly sources of foreground error. First, there is pickup of emission from the ground or from other nearby "room temperature" sources. This can create both statistical and systematic error; systematic if the amplitude of ground pickup depends on the position of the telescope (see for instance Perrenod and Lada, 1979, or Lake and Partridge, 1980). A second source, primarily of statistical noise, is emission from the earth's atmosphere. To reduce atmospheric noise, many groups now work at the South Pole, at an effective altitude of ~3000 meters, or use balloons to carry their equipment above most of the atmosphere. I shall focus in this paper on patchy emission from our Galaxy, which has the potential for creating substantial systematic errors, and on emission from extragalactic radio sources which of course contributes to the variance of the microwave sky, and hence to systematic errors as defined above.

Galactic Emission

Microwave and millimeter wave emission from our Galaxy is now seen as the most troublesome source of foreground noise and the primary limit to improved measurements of CBR fluctuations. At wavelengths $\gtrsim$ 3–5 mm, bremsstrahlung and

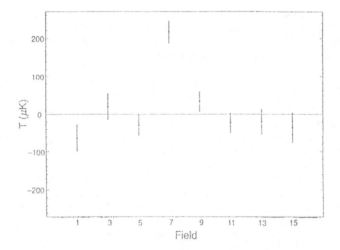

Fig. 1. Results of a typical search for fluctuations in the CBR (in this case, the work of Readhead *et al.*, 1989, with permission). *Statistical* errors, as defined here, increase the size of the error bars (and may also, therefore, increase the point-to-point scatter). Systematic errors instead increase the point-to-point scatter only.

synchrotron radiation dominate, with spectra $T(v) \propto v^{-2.1}$ and $v^{-2.8}$, respectively. At wavelengths $\lesssim$ 3–5 mm, re-emission from Galactic dust with T ~ 24 K dominates; here $T(v) \propto v^{1.6}$ in the Rayleigh Jeans region (Bennett, *et al.*, 1992; Toffolatti, *et al.*, 1994).

Our interest is in the spatial *fluctuations* in foreground Galactic noise as a function of frequency and angular scale. Such calculations are done *either* for specific regions of the sky (e.g. Banday and Wolfendale, 1991a, b) using other astronomical data, such as low frequency radio maps (Haslam *et al.*, 1982; Lawson, *et al.*, 1987), *or* "generically" for typical regions of the sky (see, e.g. Banday and Wolfendale, 1990; Brandt *et al.*, 1994; Partridge, 1994). Most of the work has concentrated on degree scales. The VLA observations to be described below permit us to derive "generic" constraints on θ ~ 0.01° scales as well. Typical "generic" limits for θ ~ 1° are shown in Fig. 2 below, adopted from the useful recent review by Danese *et al.* (1994) and from my 1994 book.

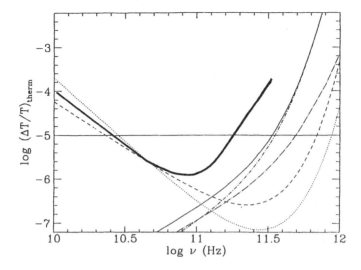

Fig. 2. Models for Galactic foreground noise on ~1° scales. The thin lines are models of Danese *et al.* (1994), and represent synchrotron fluctuations (dotted) and free-free emission (dashed) at low ν, and dust re-emission at high ν (solid). The two dot–dash lines show the contributions from hot and cool dust. The heavy lines are comparable models from my 1994 book. My (very conservative) model for dust emission assumes up to 100% amplitude fluctuations in the dust emission.

Extragalactic Foregrounds

This problem has been extensively investigated by the Padua group; see, for instance, Franceschini *et al.* (1989) and Danese *et al.* (1994). At wavelengths greater than a centimeter or so, substantial point-to-point fluctuations are introduced by radio galaxies and QSO's. At shorter wavelengths, typically a few millimeters or below, it is the emission from dusty galaxies (e.g. IRAS galaxies) that will dominate. As the work of Franceschini *et al.* shows, that leaves a nice wavelength "window" for such observations: see Fig. 3, adopted from their work.

In order to estimate the fluctuation level at various angular scales and various frequencies, both the counts (number per square degree) and the spectra of different classes of radio sources are needed. At wavelengths greater than, say, 3 cm, both are well known (see Wall here; Gregory and Condon, 1991; and Windhorst *et al.*, 1993). Thus the left half of Fig. 3 is fairly well understood. The right half of the diagram is considerably less certain. Source counts are more difficult at higher frequencies; spectra are less well known, and large k-corrections are needed for sources at substantial redshifts because of the positive spectral index of dust emission.

ν (GHz)

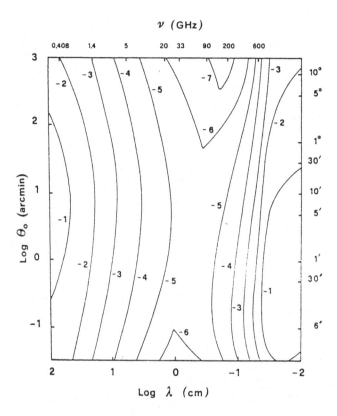

Fig. 3. Estimates of the rms fluctuation level introduced by foreground extragalactic radio sources (from Franceschini *et al.*, 1989). In this generic model, contours of $\Delta T/T$ are shown as a function of observing scale and wavelength.

For measurements on degree scales and above, I would claim, our knowledge of the distribution and spectra of radio sources is probably good enough to allow us to eliminate foreground noise due to radio galaxies and QSO's; for instance, use of the formula $S = 2\, kT\Omega/\lambda^2$, with $\Omega = 1.13\, \theta_{1/2}^2$ for a gaussian beam, allows us to derive the following limit on the flux density of a source which will produce a $\leq 10\mu$ K signal in observations made with a beam of θ degrees full width at half-maximum:

$$S_c = 0.1 \left(\frac{\lambda}{1cm}\right)^{-2}\left(\frac{\theta}{1°}\right)^2 .$$

Note that the limiting value of S is proportional to the square of the beam half-power width. Sources with flux density above 0.1 Jy can easily be detected by survey instruments so that their contributions can be removed.

The spectrum of most ordinary synchrotron radio sources decreases rapidly with frequency; that is the reason for the sharp falloff in frequency in the left half of Fig. 3. There are, however, occasional sources with inverted spectra; that is, sources whose flux density is $\propto \nu^\alpha$ with $\alpha > 0$. Among these is a class of sources called the gigahertz peaked spectrum sources (see O'Dea, Baum and Stanghellini, 1991). They are, fortunately, rare; and as radio surveys have been pushed to higher frequency, no new population of sources with strongly inverted spectra has so far turned up (see recent surveys such as Aizu et al., 1987; Donnelly et al., 1987; Gregory and Condon , 1991; and Windhorst et al., 1993).

Now for a nightmare scenario, suppose one of these gigahertz peaked spectrum sources does lie in an area studied for CBR fluctuations. Furthermore, suppose it is variable in time so that follow-up observations made elsewhere can not necessarily reveal its flux at the time the CBR measurements were made. That would be truly unpleasant! A year or so ago, purely serendipitously, my colleagues and I discovered a gigahertz peaked spectrum source whose low frequency spectrum was indistinguishable from that of a blackbody. At 3.6 and 6 cm, its flux density was of order 0.2 Jy, i.e. bright enough to cause problems in degree scale experiments, if those experiments had been made at centimeter wavelengths. Fortunately, the spectrum was later confirmed to fall off rapidly at higher frequencies. Thanks to observations kindly made at the Ryle Telescope by Mike Jones and his colleagues, we know that the 2 cm flux is only 0.07 Jy, and the Cambridge observations also suggest that the source has not been substantially variable in the past year. It nevertheless provides a warning. Radio surveys are and have been carried out at wavelengths an order of magnitude or more larger than most searches for CBR fluctuations. The properties of extragalactic radio sources—counts, spectra and variability—at mm wavelengths are less well known than we would like. Nor will better data be easy to obtain. Fortunately, since $\Delta T \propto \lambda^2$ for a fixed value of S, we need be concerned only with quite bright mm wave sources, which can be detected and monitored with available radio telescopes.

3.) Interferometric Observations

It has been recognized for some time that interferometric or aperture synthesis observations of the CBR offer a number of advantages in addition to high angular resolution (see Thompson et al., 1986; or for specific reference to the CBR, Partridge, 1994, and Saunders, 1994). Systematic errors contributed by emission from the ground and from the atmosphere are largely canceled out in interferometric observations, for instance (see Knoke et al., 1984; Timbie and Wilkinson, 1990).

Until fairly recently, interferometric observations of the CBR were made primarily on sub-arcminute scales (Fomalont et al., 1988; Martin and Partridge, 1988; Hogan and Partridge, 1989; Fomalont et al., 1993) to probe the CBR angular spectrum on scales likely to reveal the effects of reionization (Vishniac, 1987), explosive galaxy formation (Ostriker and Cowie, 1981), or cosmic strings (Bouchet et al., 1988). The advantages of such observations are now being realized in larger-scale observational programs as well. The pioneering effort was by Timbie and Wilkinson (1990) using a two-element array, with a very short spacing that provided angular resolution of ~1°. More sophisticated, multi-element arrays have been constructed by the Cambridge group, and I will report some of their results below. Let me first summarize our own work at 10"-80" scales.

3a.) Observations at the Very Large Array (VLA)

The as-yet unpublished observational results reported in this section were obtained in collaboration with Ed Fomalont and Ken Kellermann of NRAO, Eric Richards of Haverford College and Rogier Windhorst of Arizona State University.

We employed the VLA (Very Large Array) at 3.6 cm wavelength to make a series of deep maps of the microwave sky, each spanning an area of ~5' in radius. In the first round of observations (Fomalont *et al.*, 1993; Windhorst *et al.*, 1993), we amassed ~80 hours of observations on two fields with the VLA in its D configuration, for which the resolution was 10". These results are presented in the papers listed above.

The second round of observations was a further 100 hours of VLA time in the D configuration on a single field included in the Hubble Space Telescope Medium Deep Survey (Griffiths *et al.*, 1994). The observed rms noise of our final map was 1.97 μJy, making it the most sensitive radio image ever made. We identified 28 sources exceeding 9 μJy or 4.5σ. The brightest has S = 278 μJy. Many are now optically identified; see below.

In the months following IAU Symposium 168, my colleagues and I have made additional observations of the same region using the VLA in the C configuration, with ~3" resolution. The higher resolution will improve our sensitivity for the detection of discrete sources and will substantially improve the accuracy of the radio source positions; the latter in turn will make our optical identifications more certain. Richards, Spillar and I are also making follow-up 2μ observations of some of these optical sources at the Wyoming Infrared Observatory.

We have not yet analyzed our most recent data. *Preliminary* results from the first 3.6 cm runs reveal no evidence for CBR fluctuations, with a 95% confidence level upper limit of ~1.4×10^{-5} on $\Delta T/T$ at 1' scale, i.e. ~30% lower than our published upper limit (Fomalont *et al.*, 1993). It is our hope that additional observations now underway will allow us to lower this preliminary upper limit. The absence of detectable CBR fluctuations can set interesting constraints on the evolution of clusters of galaxies (e.g., Partridge, 1995a) and on cosmic strings. Our upper limits correspond to a limit $\mu \lesssim$ few $\times 10^{22}$ gm/cm on strings; that value in turn sets an approximate limit of ~10^{16} GeV on the energy scale of spontaneous symmetry breaking (Partridge, 1995b). The absence of detectable small-scale fluctuations is also consistent with the absence of re-ionization.

Our most recent 3.6 cm map is of a region at $\alpha = 13^h12^m$, $\delta = +42°38'$, lying within the area of the WFPC Medium Deep Survey (Griffiths, *et al.*, 1994) made by the Hubble Space Telescope. We thus have high resolution optical images for many of the 28 radio sources identified in the map. (The radio detection threshold was set at 4.5σ or 9 μJy.) Of 16 radio sources lying within the WFPC frames, 14 are robustly identified optically. Two are quasars and the other 12 faint, often blue, galaxies, several of which are in small groups or show morphological evidence of mergers and other peculiarities. To optical limits of $23^{m}2$ in V and $21^{m}9$ in I, most of the optically identified galaxies are radio sources. The C configuration observations now underway should allow us to improve our identification statistics as well as to explore the radio morphology of identified sources. All of these results will be discussed in more detail in a paper submitted to *Nature* (Windhorst *et al.*, 1995), and in a future, longer paper.

3b Recent work at the Ryle Telescope

Mike Jones of the Mullard Radio Astronomy Observatory, Cambridge, kindly provided me with some results in advance of publication (see Jones *et al.*, 1993; also Jones, 1994; Saunders, 1994). These were obtained with the Ryle Telescope, an interferometer consisting of eight elements operating at $\lambda = 2$ cm. For these observations, the resolution was 80"; scales from ~2" to ~2' could be probed with reasonable sensitivity.

The instrument was used first to search for and characterize the Sunyaev-Zel'dovich (SZ) signal from several nearby clusters of galaxies. The Sunyaev-Zel'dovich (1972) effect arises from inverse Compton scattering of CBR photons from the electrons in the intergalactic plasma in clusters. In the Rayleigh-Jeans region of the CBR spectrum, the observed magnitude of the temperature decrement is proportional to $n_e T_e \ell$, where n_e and T_e are the electron density and temperature, and ℓ is the path length through the cluster.

Fig. 4 shows one of the Ryle results, for cluster Abell 2218. The detection of the SZ signal is robust, and the results agree with earlier work at "conventional" filled-aperture telescopes (see Birkinshaw, 1991, for a review).

As Gunn (1978) among others suggested, a measurement of the SZ effect combined with a measurement of the X-ray luminosity of the same cluster ($\propto n_e^2 T_e^{1/2}$) provides in principle an independent means of finding the distance to the cluster, and hence determining H_0. It is interesting that Jones (1994), on the basis of the Ryle Telescope observations, finds a value of H_0 in better agreement with $H_0 \sim 50$ km/sec per Mpc than with 100 in the same units. Similar, low, values of H_0 have been found by others as well (e.g. Birkinshaw *et al.*, 1991; Herbig *et al.*, 1994) using this method.

I emphasize that these results are preliminary. The power of interferometric observations of the CBR has not yet been fully realized. By the time this volume appears, substantially better results may have been obtained at the Ryle Telescope and elsewhere. And an even more intriguing instrument, the Very Small Array, may finally be under construction. This instrument, long planned by the Cambridge group, is an array of horn antennas specifically designed to detect CBR fluctuations on ~3'-30' scales, which encode most of the cosmologically interesting information in the angular spectrum.

Acknowledgments

My research has been supported in part by two grants from the National Science Foundation, AST 89-14988 and AST 93-20049.

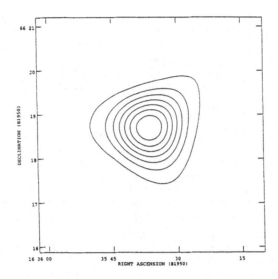

Fig. 4. Map of the SZ effect in Abell 2218, made by the Ryle Telescope (courtesy M. Jones). The contour levels range from -2 μJy to -26 μJy. The "bulge" to the east is present in ROSAT X-ray maps as well.

References

Aizu, K., Inoue, M., Tabara, H., and Kato, T. 1987 in IAU Symposium 124, Observational Cosmology, ed. A. Hewitt, G. Burbidge, and L. Z. Fang, Reidel Publ. Co., Dordrecht, the Netherlands.

Banday, A. J., and Wolfendale, A. W. 1990, *Monthly Not. Roy. Astr. Soc.*, **245**, 182.

Banday, A. J., and Wolfendale, A. W. 1991a, *Monthly Not. Roy. Astr. Soc.*, **252**, 462.

Banday, A. J., Giler, M., Szabelska, B., Szabelski, J., and Wolfendale, A. W. 1991b, *Ap. J.*, **375**, 432, and references therein.

Bennett, C. L. *et al.* 1992, *Ap. J. (Letters)*, **396**, L7.

Birkinshaw, M. 1991, in *Physical Cosmology*, eds. A. Blanchard *et al.*, Editions Frontiers, Gif-sur-Yvette, France.

Birkinshaw, M., Hughes, J. P., and Arnaud, K. A. 1991, *Ap. J.* **379**, 466.

Birkinshaw, M., and Hughes, J. P. 1994, *Ap. J.* **420**, 33.

Bond, J. R., Carr, B. J., and Hogan, C. J. 1991, *Ap. J.* **367**, 420.

Bouchet, F. R., Bennett, D. P., and Stebbins, A. 1988, *Nature*, **335**, 410.

Brandt, W. N., *et al.* 1994, *Ap. J.* **424**, 1.

Cheng, E. S., *et al.* 1994, *Ap. J. (Letters)* **422**, L37.

Danese, L. *et al.* 1994, *Astrophys. Letters and Communications*, in press.

de Bernardis, P., *et al.* 1994, *Ap. J. (Letters)* **422**, L33.

Donnelly, R. H., Partridge, R. B., and Windhorst, R. A. 1987, *Ap. J.*, **321**, 94.

Dragovan, M. *et al.* 1994, *Ap. J.* **427**, L67.

Fomalont, E. B., Kellermann, K. I., Anderson, M. C., Weistrop, D., Wall, J. V., Windhorst, R. A., and Kristian, J. A. 1988, *A. J.*, **96**, 1187.

Fomalont, E. B., Partridge, R. B., Lowenthal, J. D., and Windhorst, R. A. 1993, *Ap. J.*, **404**, 8.

Franceschini, A., Toffolatti, L., Danese, L., and De Zotti, G. 1989, *Ap. J.*, **344**, 35.

Ganga, K., Cheng, E., Meyer, S., and Page, L. 1993, *Ap. J. (Letters)* **410**, L57.

Gregory, P. C., and Condon, J. J. 1991, *Ap. J. Suppl.* **75**, 1011.

Griffiths, R. E. *et al.* 1994, *Ap. J.* **437**, in the press.

Gunn, J. E. 1978, in *Observational Cosmology*, ed. A. Maeder, L. Martinet, and G. Tammann (Geneva: Geneva Observatory), p. 3.

Hancock, S., Davies, R. D., Lasenby, A. N., Guttierez de la Cruz, C. M., Watson, R. A., Rebolo, R., and Beckman, J. E. 1994, *Nature* **367**, 333.

Haslam, C. G. T., Salter, C. J., Stoffel, H., and Wilson, W. E. 1982, *Astron. and Astrophys. Suppl.*, **47**, 1.

Herbig, T., Lawrence, C. R., Readhead, A. C. S., and Gulkis, S. 1994, *Ap. J. (Letters)*, submitted.

Hogan, C., and Partridge, R. B. 1989, *Ap. J. (Letters)*, **341**, L29.

Jones, M. 1994, *Astrophys. Letters and Commun.*, in press.

Jones, M. *et al.* 1993, *Nature* **365**, 320.

Knoke, J. E., Partridge, R. B., Ratner, M. I., and Shapiro, I. I. 1984, *Ap. J.*, **284**, 479.

Lake, G., and Partridge, R. B. 1980, *Ap. J.*, **237**, 378.

Lawson, K. D., Mayer, C. J., Osborne, J. L., and Parkinson, M. L. 1987, *Monthly Not. Roy. Astr. Soc.* **225**, 307.

Martin, H. M. and Partridge, R. B. 1988, *Ap. J.*, **324**, 794.

Myers, S. T., Readhead, A. C. S., and Lawrence, C. R. 1993, *Ap. J.* **405**, 8.

O'Dea, C. P., Baum, S. A., and Stanghellini, C. 1991, *Ap. J.* **380**, 66.

Ostriker, J. P., and Cowie, L. L. 1981, *Ap. J. (Letters)*, **243**, L127.

Ostriker, J. P. and Vishniac, E. T. 1986, *Ap. J. (Letters)*, **306**, L51.

Partridge, R. B. 1994, *3K: The Cosmic Microwave Background Radiation*, Cambridge Univ. Press, Cambridge.

Partridge, R. B. 1995a, in *Extragalactic Background Radiation*, ed. M. Livio *et al.*

Partridge, R. B. 1995b, in *Non-Accelerator Particle Physics*, eds. R. Cowsik and K. R. Sivaraman, World Scientific Publ. Co., Singapore.

Perrenod, S. C., and Lada, C. J. 1979, *Ap. J. (Letters)*, **234**, L173.

Piccirillo, L., and Calisse, P. 1993, *Ap. J.* **411**, 529.

Readhead, A. C. S., Lawrence, C. R., Myers, S. T., Sargent, W. L. W., Hardebeck, H. E., and Moffet, A. T. 1989, *Ap. J.*, **346**, 566.

Saunders, R. 1994, *Astrophys. Letters and Commun.*, in press.

Schuster, J. *et al.* 1993, *Ap. J. (Letters)* **412**, L47.

Smoot, G. F., *et al.* 1992, *Ap. J. (Letters)*, **396**, L1.

Tegmark, M., Silk, J., and Blanchard, A. 1994, *Ap. J.* 420, 484, and erratum in *Ap. J.* **434**, 395.

Thompson, A. R., Moran, J. M., and Swenson, G. W. 1986, *Interferometry and Synthesis in Radio Astronomy*, J. Wiley and Sons, New York.

Timbie, P. T., and Wilkinson, D. T. 1990, *Ap. J.*, **353**, 140.

Toffolatti, L., *et al.* 1994, *Astrophys. Letters and Communications*, in press.

Tucker, G. S., Griffin, G. S., Nguyen, H. T., and Peterson, J. B. 1993, *Ap. J.* **419**, L45.

Vishniac, E. T. 1987, *Ap. J.*, **322**, 597.

Windhorst, R. A., Fomalont, E. B., Partridge, R. B., and Lowenthal, J. D. 1993, *Ap. J.* **405**, 498.

Windhorst, R. A., Franklin, B. E., Pascarelle, S. M., Fomalont, E. B., Kellermann, K. I., Griffiths, R. E., Partridge, R. B., and Richards, E. 1995, *Nature*, submitted.

Wollack, E. J., *et al.* 1993, *Ap. J. (Letters)* **419**, L49.

RADIO GALAXIES AND QUASARS AS COSMOLOGICAL PROBES

ASHOK K. SINGAL
Netherlands Foundation for Research in Astronomy
Postbus 2, 7990 AA Dwingeloo, The Netherlands
e-mail: singal@nfra.nl

Abstract. Various techniques and methods that have been developed for using extragalactic radio sources as cosmological probes of the universe are listed. The discussion is mainly confined to the cosmological tests employing extended radio sizes of powerful radio galaxies and quasars as standard metric rods (in a statistical sense) to figure out the geometry of the universe. Some comments are made on the recent use of the milliarcsec scale sizes of compact radio sources for angular size–redshift tests. It is further pointed out that the estimates of clustering for quasars selected at centimetre wavelengths could be seriously affected by the relativistic beaming.

1. Introduction

Right in the early years of radio astronomy it became apparent that the population of extragalactic radio sources consists of extremely powerful objects, visible up to large cosmological distances. It gave rise to the early hope that by using these sources as deep probes into the universe we might be able to discriminate between various cosmological world-models. These hopes were somewhat belied when the subsequent studies showed that in addition to the presence of a large statistical spread in the intrinsic luminosities and physical sizes of extragalactic radio sources, even the average properties of their population seem to evolve heavily with the cosmic epoch and that the evolution almost completely masks out any distinguishable features of geometry between various world-models. However, a study of the cosmological evolution of the various properties of these radio sources is in itself of interest as it may yield information on the conditions that prevailed

M. Kafatos and Y. Kondo (eds.), Examining the Big Bang and Diffuse Background Radiations, 71–78.
© 1996 IAU.

in the cosmos at different epochs. In addition, some other techniques have now emerged that make use of the radio sources for cosmological investigations without relying critically upon their above evolving properties.

The following techniques and methods exploit radio galaxies and quasars as direct or indirect probes for the study of various cosmological questions.

Classical Methods

1. Angular distribution across the sky : looking for departures from an isotropic distribution on large angular scales – a consistency check for the Cosmological Principle.
2. Hubble diagram : The possibility to identify radio galaxies at high redshifts ($z \geq 1$) and the relatively small scatter seen in their $K-z$ plots could revive a further interest in one of this most classical cosmological test.
3. Radio Source Counts : Yield information on the space distribution of sources and the cosmological evolution of their number densities and/or of their radio luminosities.
4. Radio size distribution across the luminosity–redshift plane : looking for any signature of the curvature of the universe from a variation of the median value of their angular sizes with redshift. Interpretation complicated due to a cosmological evolution of their sizes and/or the presence of a luminosity–size correlation.

Other Methods

5. High redshift radio galaxies and quasars : These provide lower limits on the epoch of galaxy formation and further, yield information on the properties of the intervening matter from the absorption spectra of high redshift quasar.
6. Gravitational lenses: H_0 from the arrival times in gravitationally–lensed multiple images of a source.
7. Superluminal expansions: H_0 (and also q_0!) from the observed proper motions of the VLBI jet–components.
8. Pressure and energy–density in the lobes of giant radio galaxies and quasars: Possibility of putting some constraints on the physical conditions in the intergalactic medium (IGM).

A brief review of most of these topics with references is available in Shaver (1992), which I will not repeat here. Also, some of these topics are being covered in detail by others in the present proceedings. My discussion will mainly be confined to the observed size distributions of radio galaxies and quasars across the luminosity–redshift plane. But first I would like to make a few comments about the amount of clustering inferred for quasars selected at centimetre wavelengths.

Quasars may appear to be the ideal objects for studying very large scale structures in the universe. There have been many such studies based on optically selected quasar samples (see e.g. Komberg et al. 1994 and the references therein). Some of the investigators have used very heterogeneous samples, which contained a very mixed sample of optically as well as radio selected quasars (see e.g. Shaver 1988 and the references therein). The spatial clustering of quasars seems to have been confirmed at high significant levels (see Shaver 1992). There are even reports that the radio-selected quasars appear to have larger correlation amplitude than the other type of quasars (Chu and Zhu 1988). Here I will like to draw attention to an effect which may cause the two–point correlation function inferred for radio quasars, selected generally at centimetre wavelengths, to be an underestimate. If we do believe that the flat spectrum quasars get selected mostly by their "beamed" cores (as first suggested by Orr and Browne 1982), then we do not expect to see many physically associated close pairs (provided they do exist) to show up in samples selected at high frequencies which are dominated by flat spectrum quasars. It is very unlikely that both members of a pair will be beamed towards the same observer and thus even if large number of quasar pairs do exist the probability of finding them this way appears to be very low. It appears more logical that one should look for such pairs in metre–wavelength selected samples where quasars are selected mostly by their extended lobe emission. The only available such complete sample at metre wavelength is the 3CRR (Laing et al. 1983), unfortunately it contains only a small number of quasars. The absence of sufficient number of known steep–spectrum quasars in metre–wavelength selected samples in the first place is the very reason why people went for flat spectrum quasars. Recently Kapahi et al. (1994) have formed a somewhat larger sample of metre–wavelength selected quasars, called the Molonglo Quasar Sample. Unfortunately this sample is confined to a narrow strip in the sky, but one could still try to look for clustering in such samples.

2. Extended radio sizes as cosmological tools

The usage of the extended sizes of radio galaxies and quasars for cosmological investigations first began with Miley (1968, 1971) and Legg (1970) who showed that there appears to be a deficiency of large sized radio quasars at higher redshifts. In fact the upper envelope of the angular size–redshift $(\theta-z)$ distribution in Miley's (1971) plot seems to fit well with a function $\theta \propto z^{-1}$, which in the literature has often been "loosely" described as what would be expected in a static–Euclidean universe. I may point out here that in a static–Euclidean universe, by definition, there is no cosmological expansion. Consequently, the observed redshift of a source has no simple cosmological interpretation and it may bear no direct relation to the "distance" of the source (except perhaps in a "tired–light effect" type model). Therefore, a relation of the type $\theta \propto z^{-1}$ is not something to be normally expected in a static–Euclidean universe.

In an alternative approach, Swarup (1975) and Kapahi (1975) investigated the variation in the median value of the angular size distribution with a decreasing flux density level of the source sample. One of the main advantages of this approach at that time was that the sample was not necessarily limited to a small number of known–redshift sources as the redshift information for the whole sample is not a prerequisite in this technique. However, for a proper interpretation one does need to assume a luminosity function for the radio source population, adding to the uncertainties.

Both kinds of above studies showed the inadequacies of constant physical size models for the population of extragalactic radio sources and the inference generally drawn was that the sources had smaller physical sizes at earlier epochs. However, a suitable luminosity–linear size (P–l) correlation among the radio source population could also explain these observations, without invoking the size evolution. And there is a possibility that both the size evolution and the luminosity–size correlation could be present. The only reliable way to separate these two effects is to investigate the size distribution in the luminosity–redshift plane, where one could not only examine the l–z relation for a given luminosity class, but could also check for a P–l correlation in a given redshift bin. In addition, one also avoids the need to know an appropriate luminosity function.

The main requirement for undertaking such an investigation is to have a large enough sample so that relevant portions of the P–z plane could be populated with statistically significant number of sources in various bins. One needs information on the flux–density, spectral index, angular size and redshift for each source in the sample. Moreover the sample has to be "fare" in the sense that no selection bias based on a prior information on the angular sizes of the souces has crept in. We may point out here that contrary to the view expressed sometimes in the literature it is neither needed nor desired to have a 'radio complete' sample for the above purpose. In fact, paradoxical as it may sound, the condition of the strict radio–completeness of the chosen sample, a must for number counts and other such studies, may somewhat be of a disadvantage for the present purpose. Such a flux–limited radio complete sample covers only a narrow band in the P–z plane. In order to fill the luminosity–redshift plane more evenly, we are forced to employ data from many sub-samples selected at different flux–density limits, which will make our final sample to be far from a radio–complete one.

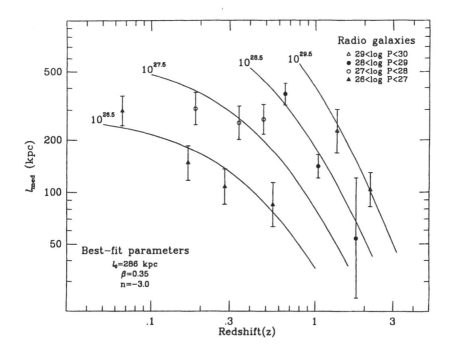

Figure 1. The change in l_{med} with redshift for radio galaxies. The family of curves is drawn according to the relation $l = l_0(P/10^{26.5})^{\beta}(1+z)^n$, for the best-fit parameter values $l_0 = 286$ kpc, $\beta = 0.35, n = -3$ (Singal 1993).

Following such an approach, many independent workers (although their data used were not entirely independent) showed (Oort et al. 1987; Singal 1988; Kapahi 1989) that the physical sizes of the radio galaxies appear to evolve rapidly with redshift ($l \propto (1+z)^{-3}$). In addition, Singal (1988) also pointed out that the size evolution of quasars, if any, appears to be much weaker (see also Barthel and Miley 1988). This unexpected difference between radio galaxies and quasars was later confirmed by Singal (1993) from a larger sample of sources that included many more galaxies at high redshifts. Using a multiple regression analysis with the functional form $l \propto P^{\beta}(1+z)^n$ on a large sample of 789 powerful ($P_{408} \geq 10^{26}$W Hz^{-1} for $H_0 = 50$km s^{-1} Mpc^{-1}, $q_0 = 0$) radio sources, Singal (1993) arrived at the following conclusions (see Figures 1 and 2 here):

1. There is a direct correlation between luminosity and size for radio galaxies ($\beta \simeq 0.35$) for a given redshift.

2. The radio size of galaxies in a given luminosity range falls sharply ($n \simeq -3.0$) with the cosmic epoch.

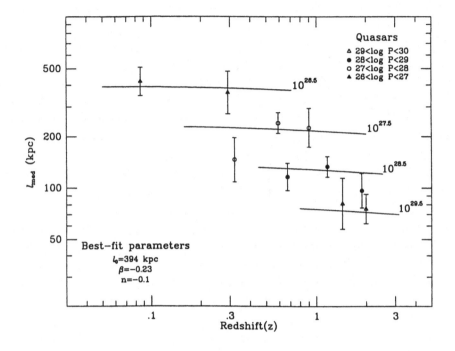

Figure 2. The size evolution for quasars. The family of curves is drawn according to the relation $l = l_0(P/10^{26.5})^\beta(1+z)^n$, for the best-fit parameter values $l_0 = 394$ kpc, $\beta = -0.23, n = -0.1$ (Singal 1993).

3. There is an inverse correlation between luminosity and size ($\beta \sim -0.2$) for quasars, in contrast with the case of galaxies.

4. For quasars, the cosmic size evolution, if any, is only marginal ($n \sim -0.2$).

The large differences seen between radio galaxies and quasars in the redshift and luminosity dependences of their sizes, provide strong evidence against the simple orientation based unified–scheme models (Barthel 1989). Further, the task of explaining the size evolution and/or the luminosity–size correlation by a physical model (see e.g. Gopal-Krishna and Wiita 1991) becomes even more difficult. We may here add that there appeared to be some confusion in the above picture when Nilsson et al. (1993) claimed that in their studies they find no significant difference between the radio sizes of radio galaxies and quasars. They also reported the presence of a *negative* correlation between radio size and power among high luminosity radio galaxies, a result in contradiction with almost all other previous studies. A comparison of Figure 3 of Nilsson et al. (1993) with Figures 1 and 2 here (as reproduced from Singal 1993) does not show much discordant data between the two samples in the regions of overlap but it does show clearly that the conclusions of Nilsson et al. suffer from a lack of data on high redshift galaxies. Further, it is not clear how much bias could have got introduced in the size distribution due to their insistence on retaining only "well-defined" doubles in their final source sample, as such a selection effect is likely to discriminate against small-sized but high redshift sources. Recently Kapahi et al. (1994) have investigated the size distribution in their Molonglo Quasar Sample and their results indicate a clear difference in the luminosity–size correlation between radio galaxies and quasars, in agreement with the conclusions of Singal (1993).

3. Compact radio sources

The failure of the extended structures of extragalactic sources to provide good yard–sticks for measuring the geometry of the universe has led to a recent growth of interest in the use of milli-arcsec sizes of the compact radio sources for determining the deceleration parameter q_0 (Kellermann 1993; Gurvits 1994). The main argument here is that since these compact structures lie well within the host galaxy environment, their sizes should not be directly affected by a change in the density of the IGM with the cosmological expansion and as a result are less likely to be evolving with redshift. Using a sample of 82 compact radio sources, taken from the literature, Kellermann (1993) found that his angular size–redshift plot supports the Einstein-de Sitter universe with $q_0 = 0.5$. On the other hand Gurvits (1994) finds the data on milli-arcsec sizes to be somewhat more consistent with an open universe.

Unfortunately, the usage of milli-arcsec structures of compact sources brings its own bigger bag of problems. The first and the foremost difficulty is that unlike in the case of extended structures, where hot–spots could be seen, the milli-arcsec scale VLBI jets do not possess well recognizable extremities and it becomes very difficult to give an unambiguous size estimate which is independent of the observing technique (frequency of observations, the achieved sensitivity etc.). It is not even clear whether the measured size reflects the same intrinsic property that is stable from source to source (Pearson et al. 1994). In this context it is not clear whether a sizable fraction of sources which may remain unresolved should be included as upper limits on sizes. The altogether absence of a VLBI scale jet is not the same thing as the jet being smaller in size than the observing beam. We can not even be very sure that evolution plays no significant role in VLBI scale jets (even if these may not be directly influenced by the IGM). Then there is a strong likelihood of the relativistic beaming introducing large selection effects. Any small changes in the orientation with respect to the observer may introduce large changes in the estimates of the projected sizes. In the angular size–redshift plot of Pearson et al. (1994) there is no evidence of any dependence of angular size with redshift. In fact their data disagree with those of Kellermann (1993) in the lowest-redshift bin which unfortunately happens to provide the most crucial point in Kellermann's estimate of q_0. To be fare, it is not clear how much this disagreement may be arising from the difference in the luminosity levels of the two samples. It is to be hoped that in the next few years the large number of VLBI maps expected to come out from the VLBA and EVN (European VLBI Network) may help resolve some of these important issues.

References

Barthel, P. D., 1989, ApJ, 336, 606
Barthel, P. D., Miley, G. K., 1988, Nature, 333, 319
Chu, Y., Zhu X, 1988, A&A, 205, 1
Gopal-Krishna, Wiita, P. J., 1991, ApJ, 373, 325
Gurvits, L. I., 1994, ApJ, 425, 442
Kapahi V. K., 1975, MNRAS, 172, 513
Kapahi, V. K., 1989, AJ, 97,1
Kapahi V. K., Athreya R. M., Subrahmanya C. R., Hunstead R. W., Baker J. C., McCarthy P. J., van Breugel W., 1994, JA&A, in press
Kellermann, K. I., 1993, Nature, 361, 134
Komberg, B. V., Kravtsov, A. V., Lukash, V. N., 1994, A&A, 286, L19
Laing, R. A., Riley, J. M., Longair, M. S., 1983, MNRAS, 204, 151
Legg, T. H., 1970, Nature, 226, 65
Miley, G. K., 1968, Nature, 218, 933
Miley, G. K., 1971, MNRAS, 152, 477
Nilsson, K., Valtonen M. J., Kotilainen J., Jaakkola T., 1993, ApJ, 413, 453
Oort, M. J. A., Katgert, P., Windhorst, R. A., 1987, Nature, 328, 500
Orr, M. J. L., Browne, I. W. A., 1982, MNRAS, 200, 1067
Pearson, T. J., Xu, W., Thakkar, D. D., Readhead, A. C. S., Polatidis, A. G., Wilkinson, P. N., 1994, in Compact Extragalactic radio sources, Proc. Socorro Workshop, eds. Zensus, J. A., Kellermann, K. I., NRAO, p. 1
Shaver, P. A., 1988, in Large Scale Structure of the Universe, ed. Audouze J., Pelletan, M.-C., Szalay, A., Kluwer, Dordrecht, p. 359
Shaver, P. A., 1992, Physica Scripta, T43, 51
Singal, A. K., 1988, MNRAS, 233, 87
Singal, A. K., 1993, MNRAS, 263, 139
Swarup, G., 1975, MNRAS, 172, 501

High Redshift Radio Galaxies

James S. Dunlop
Astrophysics, Liverpool John Moores University, Byrom Street, Liverpool L3 3AF, UK

Abstract. The potentially important role of jet-cloud interactions in determining the appearance of high-redshift radio galaxies is discussed and investigated via new 3-dimensional simulations of off-axis jet-cloud collisions. The results indicate that the most powerful radio sources are likely to be observed during or shortly after a jet-cloud interaction, and that such interactions can explain both the radio structures and the spatial association between optical and radio light found in powerful radio galaxies at high redshift. It is argued that, due to the radio-power dependence of such complicating effects, the optical-infrared colours and morphologies of very radio-luminous high-redshift galaxies can tell us essentially nothing about their evolutionary state. Either one must study much less radio-luminous sources in which the AGN-induced contamination is minimised, or one must attempt to determine what fraction of the baryonic mass of the radio galaxy has been converted into stars at the epoch of observation. Recent observations aimed at performing the latter experiment on two well-known high-redshift radio galaxies (4C 41.17 & B2 0902+34) are described. It is concluded that at present there exists no clear evidence that either of these famous galaxies is 'primæval'; on the contrary, the continued low-dispersion of the infrared Hubble diagram at $z > 2$ points toward a much higher redshift of formation for elliptical galaxies.

Key words: cosmology: observations, galaxies: active, galaxies: formation

1   Introduction

The evolved stellar populations and low gas masses of present-day elliptical galaxies imply that they formed most of their stars in a rapid burst at high redshift, but an unambiguous example of a 'primæval' elliptical has yet to be discovered. Since radio galaxies are identified exclusively with ellipticals at low redshift, their high-redshift ($z > 2$) counterparts are obvious candidate primæval ellipticals. However, although recent studies have suggested that at least some high-redshift radio galaxies may be very young (Eales *et al.* 1993; Eales & Rawlings 1993), uncertainities remain because of the mounting evidence that high radio luminosity couples to distorted ultraviolet and optical properties through processes associated with nuclear activity (Dunlop & Peacock 1993). In this paper I consider one possible cause of this radio-luminosity dependent distortion, namely the interaction between the radio-source jet and dense gas clouds in the intra/inter-galaxy medium. I then discuss whether, given such complications, there exists any clear evidence that high-redshift radio galaxies really are primæval, focusing on new observations of two famous high-redshift galaxies and concluding with a re-appraisal of the radio-galaxy infrared Hubble diagram at $z > 2$.

M. Kafatos and Y. Kondo (eds.), Examining the Big Bang and Diffuse Background Radiations, 79–87.
© 1996 IAU.

2 The interaction between the radio source and its environment

2.1 OBSERVATIONAL EVIDENCE

Following the discovery that the optical and radio structures of many high-redshift radio galaxies are closely aligned (McCarthy *et al.* 1987; Chambers, Miley & van Breugel 1987) it has become increasingly clear that the optical and radio properties of such objects are much more closely linked than had been previously supposed. In fact a considerable body of evidence already existed to suggest that interaction with inhomogeneities in the surrounding medium was important in determining the appearance of many high-redshift radio sources (*e.g.* Lonsdale & Barthel 1986; Barthel 1987), but the alignment effect prompted consideration of the potential reciprocal ability of the radio jet to influence the surrounding medium (*e.g.* Rees 1989). Since then the discovery of a number of important correlations has led to the suggestion that the environment may be the dominant* factor in determining the appearance of powerful radio galaxies (Dunlop & Peacock 1993). Of particular note are i) the discovery by McCarthy, van Breugel & Kapahi (1991) that the radio and optical line-emission asymmetries in high-redshift radio galaxies are correlated in the sense that the brighter line emission occurs on the side of the nearer radio-lobe; ii) the discovery by Liu & Pooley (1991) that the radio lobe with the steeper spectral index is virtually always the nearer and more depolarized lobe, and iii) the discovery that both the level of blue light and the strength of the alignment effect are correlated with a mix of radio power and spectral index (Dunlop & Peacock 1993). The implication is that alignments arise because of a selection effect, with higher radio luminosities and steeper spectra arising when a radio jet encounters dense material (Eales 1992; Dunlop & Peacock 1993).

2.2 NUMERICAL SIMULATION OF JET-CLOUD INTERACTIONS

From theoretical considerations, a radio source of given beam power is certainly expected to produce a higher luminosity in a high-density environment (Williams 1985), and observationally the most luminous radio sources are known to reside in regions of high galaxy density (Yates, Miller & Peacock 1989; Hill & Lilly 1990).

The effect of off-axis jet-cloud collisions has already been investigated via 3-dimensional numerical simulations by De Young (1991). However, these simulations used the 'beam scheme' (Sanders & Prendergast 1974) which is only 1st-order accurate, and has proved unsuitable for some astrophysical problems (van Albada *et al.* 1982). In addition they did not involve proper monitoring of the separate progress of jet and ambient material; tracer particles were used to provide some indication of the behaviour of the jet but calculation of synchrotron emission from jet material was not possible.

At present, therefore, the detailed effects of jet-cloud interactions in extra-galactic radio sources have not been simulated with sufficient sophistication to enable meaningful comparison with the growing body of observational evidence described above. In an attempt to fill this important gap, I, in collaboration with

Tim O'Brien and Steve Higgins at Liverpool, have initiated a new programme of 3-dimensional simulations the first results of which are presented here.

The simulations are carried out with a numerical hydrodynamics code based on the second-order Godunov method of Falle (1991) which solves the inviscid Euler equations with an adiabatic equation of state in a 3-dimensional cartesian co-ordinate system. In addition to the usual dynamical variables, we have also computed a parameter representing the advection of jet material through the grid. This allows us to easily distinguish jet material from ambient material. We then assume that the magnetic field is frozen into the jet plasma, so that the field energy density U_{mag} is related directly to the density of the jet material. If, in addition, we assume that the energy density of particles accelerated to relativistic energies in shocks can be simply related to the thermal energy density in the gas, we can then calculate the intensity of the expected synchrotron emission in a straightforward manner. These simplifying assumptions enable us to produce synthetic radio maps directly from the hydrodynamic results (Higgins, O'Brien & Dunlop 1995).

Example results from our initial simulations are shown in Figure 1. In this case the jet has a velocity of 0.1c, a radius of 0.05 kpc, is in pressure balance with the ambient medium and is at the same density. The cloud is also in pressure balance, and has a density of 100 times the ambient density, and a radius of 0.7kpc. Figure 1 gives synthetic radio map contour plots (after convolution with a gaussian representing the observing beam) and corresponding greyscale plots of density at four different epochs during the interaction of the jet with the gas cloud ($t = 0.16$, 0.32, 0.48 and 0.64 million years).

The preliminary results from these simulations display several interesting features. First, this simple off-axis collision with a single cloud can produce a rather complex radio structure which clearly evolves through the simulation. There are three main features – a hotspot at the deflection point, a hotspot at the head of the jet and a secondary hotspot – but the relative brightness of these features changes rapidly with time. Second, as a consequence of the jet-cloud interaction, the radio source becomes much more luminous. Third, after collision the gas cloud becomes elongated and closely aligned to the radio structure.

2.3 IMPLICATIONS FOR HIGH-REDSHIFT RADIO GALAXIES

As expected, the radio source brightens dramatically from its pre-impact luminosity when the jet impacts upon the gas cloud. However, the simulations also indicate that increased radio luminosity persists for a considerable time after the initial collision, albeit at a gradually decreasing level. Given the relative time-scales involved it would thus seem reasonable to suggest that samples of ultra-luminous radio sources are likely to be dominated by sources which have 'recently' been involved in a jet-cloud interaction. Such a selection bias favouring sources which are 'just past their best' could be accentuated by the technique of ultra steep spectrum selection which has been so successful in locating luminous sources at high redshift (Rottgering 1992).

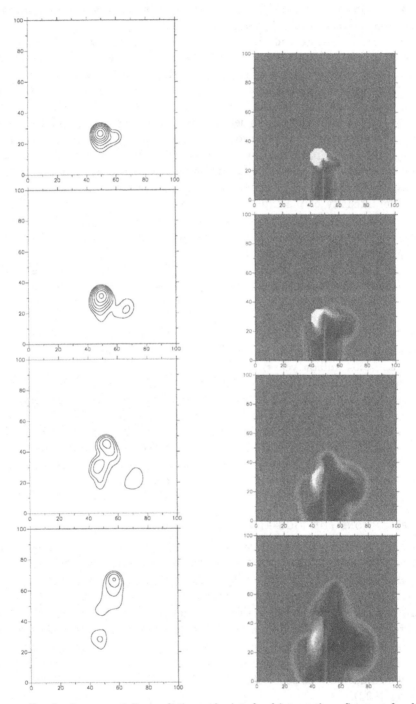

Fig. 1.   Results from new 3-D simulations of a jet-cloud interaction. See text for details.

These simulations also predict that the jets in powerful/steep-spectrum radio sources should be 'bordered' by the elongated remnants of the disrupted gas cloud. If we suppose that such gas cloud fragments are the sites of line emission and scattering of continuum from a hidden quasar, then it is to be expected that samples of such sources will display a strong alignment effect between their optical and radio structures. In addition, the simulations show that the apparent accuracy of the spatial correspondence between the optical and radio structures is expected to vary considerably depending upon the precise epoch of observation. This simple fact can explain some of the apparently different manifestations of the alignment effect observed in different sources. For example, a very close spatial correspondence between the radio and optical structures, such as is observed in the $z = 3.8$ radio galaxy 4C41.17 (Miley et $al.$ 1992) can be produced shortly after the initial collision; in Fig 1c the interaction is producing multiple hotspots such as are observed in 4C41.17 (Carilli, Owen & Harris 1993), and these radio hotspots lie within 1 kpc of the aligned optical light. At later times ($e.g.$ Fig. 1d) the predicted spatial correspondence between the radio and optical structures is much less exact due to fact that the head of the radio jet has moved beyond the cloud, but nevertheless the optical and radio structures remain clearly aligned. Such a structure is observed in, for example, the $z = 1.2$ radio galaxy 3C324 (Rigler et $al.$ 1992; Dickinson 1995). By this stage in the collision the radio and optical axes appear to be misaligned by $\simeq 20°$ due to the combined effects of jet diversion and cloud disruption. Such a apparent optical-radio misalignment is exactly what is observed in 3C324 and indeed is typical of that found in most 'aligned' high-redshift radio galaxies.

Several other observed aspects of the alignment effect can be explained by this type of jet-cloud interaction. First it leads naturally to the observed correlation between radio power/spectral index and the statistical prevalence of the alignment effect (Dunlop & Peacock 1993). Second, unlike some other proposed explanations of the alignment effect ($e.g.$ Daly 1992), these simulations predict that, even in well-aligned radio galaxies, the optical emission will in general not appear precisely co-spatial with the radio emission if viewed with sufficient resolution. Recent HST images of 3C368 have shown that this appears to be the case, with much of the optical emission being resolved into curved filaments which border the radio structure rather than being precisely coincident with it (van Breugel priv. comm.). Third, if the interacting gas cloud is in fact another galaxy, then the (undiverted) stars in this galaxy may contribute to the aligned light, providing one possible explanation of the radio-infrared alignments seen in many high-redshift radio galaxies (Dunlop & Peacock 1993).

3   The ages of high-redshift radio galaxies

3.1   AGN-INDUCED CONTAMINATION

Regardless of whether jet-cloud interactions of the type described in the previous section are the primary cause of optical-radio alignments, it is now clear that the optical/infrared colours and morphologies of powerful high-redshift radio galaxies are of little help in determining the true age of the dominant stellar population in the host galaxy. At optical/ultraviolet wavelengths emission from material around the jet (emission-lines, scattered quasar continuum and starlight, either from jet-induced starformation or from pre-existing stars in an interacting galaxy) will in general make the galaxy appear artificially blue as well as giving it a misleadingly complex morphology. Conversely, while it might be thought that near-infrared observations would yield a more undistorted view of the starlight, the presence of a dust-obscured quasar can make the galaxy appear excessively nucleated, red and luminous at near-infrared wavelengths (*e.g.* McCarthy, Persson & West 1992; Lacy *et al.* 1995).

Given the above it is reasonable to ask what, if anything, can be done to determine the true age of high-redshift radio galaxies. One solution is to study the optical-infrared properties of much less radio-luminous sources in which the effect of the AGN should be insignificant (Dunlop & Peacock 1995). A second, and complementary, approach is to attempt to constrain what fraction of the present-day baryonic mass of a giant elliptical galaxy has been processed into stars in high-redshift radio galaxies at the epoch of observation. In the remainder of this section I describe two recent (and very different) attempts to perform the latter experiment on two famous high-redshift galaxies, both of which have been hailed as primæval galaxies – B2 0902+34, for several years the most distant known galaxy ($z = 3.4$; Lilly 1988), and 4C 41.17, the current holder of this title ($z = 3.8$; Chambers, Miley & van Breugel 1990).

3.2   SUB-MILLIMETRE PHOTOMETRY OF 4C41.17

On the basis of its rather blue ultraviolet-optical spectral energy distribution and complex multi-component morphology, it has been claimed that 4C41.17 may be a genuine example of a primæval elliptical galaxy (Chambers, Miley & van Breugel 1990; Miley *et al.* 1992). However, as stressed above, the contaminating effects of the AGN at ultraviolet-optical wavelengths are expected to be severe for such a luminous steep-spectrum radio source. In addition the discovery that the radio core of this galaxy (Carilli, Owen & Harris 1993) is located at the position of a gap in the ultra-violet emission provides strong circumstantial evidence that its short-wavelength morphology is further complicated by significant dust obscuration. Consequently it seems that the best, and perhaps only way of determining whether this galaxy really is primæval is to determine the mass of gas which has yet to be converted into stars at the epoch of observation. We have recently attempted to do this via a determination of dust mass based on sub-millimetre photometry

(Dunlop *et al.* 1994), and have succeeded in detecting 4C41.17 at $800\mu m$ using the JCMT (Figure 2). As discussed by Dunlop *et al.* (1994) this detection combined with an upper limit at $450\mu m$ leads to an estimate of the gas mass of $10^{11} M_\odot$, or $\sim 10\%$ of the stellar mass of the most luminous elliptical galaxies. Taken at face value this result suggests that while 4C41.17 may well be involved in a spectacular interaction/starburst, it need not be genuinely primæval. In fact none of the currently available data on 4C41.17 can exclude the possibility that the bulk of its stellar population formed at much higher redshifts.

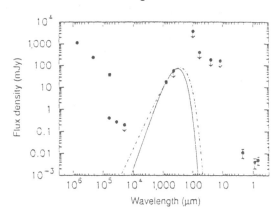

Fig. 2. The radio-optical spectrum of 4C41.17 showing the far-infrared excess due to dust emission as detected by sub-millimetre photometry.

3.3 DEEP K-BAND IMAGING OF B2 0902+34

At the time of its discovery, the apparently red optical-infrared colour of the radio galaxy B2 0902+34 generated great interest because it implied that the stellar population in this $z = 3.4$ radio galaxy was already $\sim 1 - 2$ Gyr old (Lilly 1988). Since then this result has been turned on its head by the discovery that the true 4-arcsec aperture magnitude of 0902+34 is a magnitude fainter than measured by Lilly (1988), and that even this fainter K-band luminosity is in fact dominated by emission lines (Eales & Rawlings 1993). 0902+34 has thus re-entered the limelight, but this time cited as perhaps the best candidate for a primæval galaxy on account of its line-corrected flat-spectrum spectral energy distribution, and its apparently low optical luminosity (Eales *et al.* 1993).

Nevertheless some confusion has remained because, independent of his 1988 infrared image, Lilly (1989) reported three single-element large-aperture (7.4 – 12.4 arcsec) photometry measurements of 0902+34 between 1983 and 1987, all of which yielded $K < 19$. This led Eisenhardt & Dickinson (1992) to attempt to investigate the K-band morphology of 0902+34 but they concluded that this could not be reliably attempted until the advent of larger format arrays.

Figure 3 shows a new deep (162-minute) K-band image of 0902+34 obtained using the new large-format (256×256) IRCAM3 camera at UKIRT in May 1994. This image is, of course, contaminated by emission lines, but interestingly it yields an 8-arcsec software aperture magnitude of $K = 18.7 \pm 0.05$, consistent with the original value determined by Lilly, and a 4-arcsec magnitude of $K = 19.7 \pm 0.05$ consistent with the value measured by Eales *et al.* (1993). This shows that the apparent disagreement of a magnitude between the original aperture magnitude and the more recent 4-arcsec imager-determined magnitude is due to the fact that 0902+34 is very extended at K. In fact, as illustrated in Fig. 3, the de Vaucouleurs scale-length r_e appears to be $\simeq 100$ kpc (for $\Omega_0 = 1$, $H_0 = 50 \, \mathrm{km \, s^{-1}}, \mathrm{Mpc^{-1}}$). If as concluded by Eisenhardt & Dickinson (1992) the [OIII] emission only dominates within the central 3–4 arcsec, this would suggest that 0902+34 may not in fact be under-luminous, merely very extended.

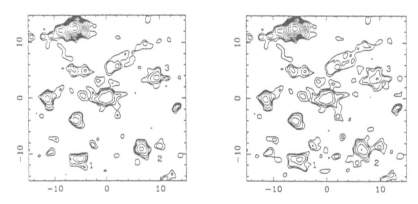

Fig. 3.   A new deep K-band image of B2 0902+34. The galaxies numbered 1, 2 and 3 are synthesized galaxies of scale-length 30 kpc in the LH plot and 100 kpc in the RH plot.

4   Conclusion – the K–z diagram for radio galaxies

If there exists no clear evidence that high-redshift radio galaxies are primæval, is there any counter-evidence that they are old? One such piece of evidence is provided by the continued low dispersion of the radio-galaxy $K - z$ diagram at high redshift. Although Eales & Rawlings (1993) have demonstrated that the scatter in this relation appears to rise dramatically at $z > 2$, if one adopts larger aperture (8-arcsec) measurements at high-redshift this effect seems to disappear (Figure 4). The difference is undoubtedly in part due to the increased sensitivity of the smaller apertures (4 arcsec) used by Eales & Rawlings to different quasar contributions in different sources. It may, however, also be telling us that significant dynamical evolution is occurring at $z \simeq 3$. It will be of interest to see whether comparably deep infrared imaging of other high-redshift radio galaxies shows them to be as unusually extended as 0902+34.

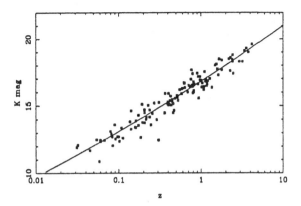

Fig. 4. The K–z diagram for powerful radio galaxies using large aperture data at high z.

References

Barthel, P.D. & Miley, G.K., 1987. *Nature*, **333**, 319.
Carilli, C.L., Owen, F. & Harris, D.E., 1993. *AJ*, **107**, 480.
Chambers, K.C., Miley, G.K. & van Breugel, W.J.M., 1987. *Nature*, **329**, 624.
Chambers, K.C., Miley, G.K. & van Breugel, W.J.M., 1990. *ApJ*, **363**, 21.
Daly, R.A., 1992. *ApJ*, **386**, L9.
De Young, D.S., 1991. *ApJ*, **371**, 69.
Dickinson, M., 1995. In: *'Galaxies in the Young Universe'*, Ringberg Workshop Sep 1994, in press.
Dunlop, J.S. & Peacock, J.A., 1993. *MNRAS*, **263**, 936.
Dunlop, J.S. & Peacock, J.A., 1995. In: *'Galaxies in the Young Universe'*, Ringberg Workshop Sep 1994, in press.
Dunlop, J.S. et al., 1994. *Nature*, **370**, 347.
Eales, S.A., 1992. *ApJ*, **397**, 49.
Eales, S.A. & Rawlings, S., 1993. *ApJ*, **411**, 67.
Eales, S.A. et al. 1993. *Nature*, **363**, 140.
Eisenhardt, P. & Dickinson, M., 1992. *ApJ*, **399**, L47.
Falle, S.A.E.G., 1991. *MNRAS*, **250**, 581.
Higgins, S.W., O'Brien, T.J. & Dunlop, J.S., 1995. *MNRAS*, in preparation.
Hill, G.J. & Lilly, S.J., 1990. *ApJ*, **367**, 1.
Lacy, M. et al., 1995. *MNRAS*, in press.
Lilly, S.J., 1988. *ApJ*, **333**, 161.
Lilly, S.J., 1989. *ApJ*, **340**, 77.
Liu, R. & Pooley, G., 1991. *MNRAS*, **249**, 343.
Lonsdale, C.J. & Barthel, P.D., 1986. *ApJ*, **303**, 617.
McCarthy, P.J. et al. 1987. *ApJ*, **321**, L29.
McCarthy, P.J., van Breugel, W. & Kapahi, V.K., 1991. *ApJ*, **371**, 478.
McCarthy, P.J., Persson, S.E. & West, S.C., 1992. *ApJ*, **386**, 52.
Miley, G.K. et al. 1992. *ApJ*, **401**, L69.
Rees, M.J., 1989. *MNRAS*, **239**, 1P.
Rigler, M.A. et al. 1992. *ApJ*, **385**, 61.
Rottgering, H., 1993. *PhD Thesis*, University of Leiden.
Sanders, R.H. & Prendergast, K.H., 1974. *ApJ*, **188**, 489.
van Albada, G.D., van Leer, B. & Roberts, W.W., 1982. *A&A*, **108**, 76.
Williams, A.G., 1985. *PhD Thesis*, University of Cambridge.
Yates, M.G., Miller, L. & Peacock, J.A., 1989. *MNRAS*, **240**, 129.

COLD MATERIAL IN DISTANT RADIO GALAXIES

HUUB RÖTTGERING

MRAO/IoA, Cambridge, UK

Abstract. Dust, H I and molecular gas must be important components of distant radio galaxies. Recent observations to detect these constituents are discussed.

1. Introduction

Radio galaxies are important probes of the early universe. During the last few years extensive surveys for distant radio galaxies have been carried out using different selection techniques (eg. McCarthy 1993; Miley 1994). At the moment more than 60 radio galaxies are known at redshifts larger than two. The main advantage of studying these galaxies, as opposed to studying other objects known at $z > 2$, such as absorption systems and quasars, is that radio galaxies contain three extended components that are easily resolved with conventional techniques. These components are the radio synchrotron plasma, the ionised gas and the IR/optical/UV continua. Studying these components and their relation to one another has revealed a great deal about the nature of these systems. One of the most interesting parameters is the maturity of the stellar population. Unfortunately, it is difficult to use the observations of the three different components just mentioned to probe the stellar content directly. The main reason for this is that it is difficult to determine the relative contribution from stars and non-thermal emission from the IR/optical/UV continuum data. In nearby galaxies an important probe of star formation is molecular gas. The sensitivity of present day instruments is good enough that at least attempting to detect cold material in distant radio galaxies seems a worthwhile exercise.

M. Kafatos and Y. Kondo (eds.), Examining the Big Bang and Diffuse Background Radiations, 89–93.
© 1996 *IAU.*

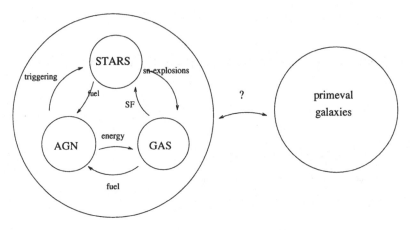

Figure 1. This diagram shows schematically the relation between the various constituents together with their interactions within a distant radio galaxy. One of the main questions is what such systems tell us about primeval galaxies.

2. The importance of cold gas in distant radio galaxies

Figure 1 is an illustration of how complicated the systems we call distant radio galaxies are. The three main constituents of these objects are the active nucleus, stars and gas in various phases. All three constituents are related to one another. It is clear that the stars in these systems have to be formed from gas and that stellar mass loss, mostly in the form of supernovae explosions, will transfer material from stars back to the ISM. This symbiotic relation between stars and gas is complicated by the presence of the AGN. The AGN is being fuelled by surrounding material (eg. gas/stars) and as a result it inputs energy in various forms to the gas. In some models of distant radio sources it is assumed that the radio jet emerging from the AGN triggers star formation in the ISM.

It is clear that we have to understand these processes in a certain amount of detail before we can decide what the relation is between distant radio galaxies and young and forming galaxies. Questions that should be answered include how profound the impact of the AGN is on these objects and what age these stellar systems are.

Cold material such as dust and cold gas ($T \ll 10^4$ K) must play an important role in these systems. Since in some models the radio galaxies are producing stars at a prodigious rate, this is the phase of the ISM that provides the building material for stars. In nearby AGN such as Seyferts the observation of molecular material shows that a large fraction of the CO gas, for example, is concentrated within one kpc of the active nucleus. This suggests that at least some of the molecular gas will ultimately fuel the AGN.

In the rest of this contribution I will concentrate on the observational evidence for this gas phase in distant radio galaxies.

3. Dust

There are several ways of probing the dust content in high redshift radio galaxies. One of the first attempts was to measure the Lyα/Hα ratio. For two radio galaxies McCarthy et al. (1992) found values within a factor of 2 of case B recombination, implying E(B-V)=0.1. Cimatti et al. (1993) found dust masses of order $10^8 M_\odot$ from modelling the optical polarisation of radio galaxies. The most spectacular measurement of dust is probably the detection of submillimetre emission from the radio galaxy 4C41.17 (z=3.8) implying dust masses of order a few times $10^8 M_\odot$ (Dunlop et al. 1994). Van Ojik et al. (1994) discovered that the radio galaxy 0211−122 ($z = 2.34$) has a peculiar emission line spectrum, with Lyα very much fainter with respect to the high ionisation lines than in typical high redshift radio galaxies. They suggest that this galaxy is undergoing a vigorous starburst producing enough dust to attenuate the Lyα emission.

4. Neutral Hydrogen

There are two important methods of detecting H I associated with radio galaxies. The first method is to measure the redshifted 21 cm absorption line against the radio continuum. The only distant radio galaxy for which this has been measured is 0902+34 (z=3.4). Uson et al. (1991) found an absorber in this system with a column density of 4.4×10^{22} atoms cm^{-2}, assuming a spin temperature of 10^4 K. The existence of this absorber was confirmed by Briggs et al. (1993).

The second method is to measure absorption against the Lyα emission of high-z radio galaxies. The main reason that this is possible is that the Lyα emission from distant radio galaxies can be very bright ($> 10^{-15}$ erg s^{-1}). We therefore started a programme to detect such absorption in distant radio galaxies. The high resolution AAT spectra (1.5 Å) of the Lyα region of the $z = 2.9$ radio galaxy 0943−242 reveal a complex emission line profile which is dominated by a black absorption trough centered 250 km s^{-1} blueward of the emission peak (Figure 2, see also Röttgering et al. 1994). The main H I absorber has a column density of 1×10^{19} cm^{-2}.

One of the fundamental limitations in the study of quasar absorption lines is that quasars are point sources so that in general no information can be obtained about the spatial scale of the absorber. In the case of radio galaxies however, the Lyα emission is extended, in some cases even up to 10 arcsec. The main absorption in the Lyα emission from 0943−242 is black and it covers the entire Lyα emission region which has an angular scale of

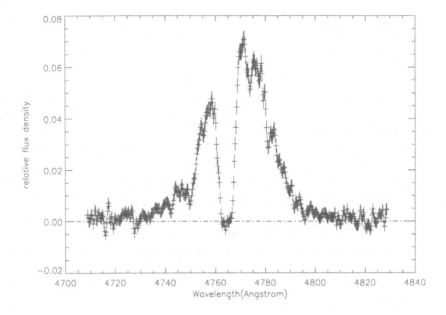

Figure 2. Part of the high resolution AAT spectrum spectra (1.5 Å) of the Lyα region of the $z = 2.9$ radio galaxy 0943−242. (see also Röttgering et al. 1994).

1.7 arcsec. Its linear size is then at least 13 kpc, making this the first direct measurement of the spatial scale of an absorber with a column density of $\sim 10^{19} \, \text{cm}^{-2}$.

5. Carbon Monoxide

Since the CO molecule is a good tracer of star formation in galaxies, studying this molecule at cosmological distances is important for understanding the evolution of young galaxies. To date the two objects at high redshift that have a confirmed detection of a CO line are the ultra luminous IRAS galaxy F10214+4724 at z=2.3 (Solomon et al. 1992) and the cloverleaf quasar (Barvainis et al. 1994). The good news is that these two detections demonstrate convincingly that studying CO in the early universe is possible. The bad news is that the IRAS galaxy F10214+4724 is the only IRAS galaxy known at these distances. Studying a sample of such objects is therefore not (yet) possible. Since the cloverleaf quasar is a gravitationally lensed quasar for which the amplification factor is not well known, the estimate for the intrinsic CO luminosity of this object is uncertain.

The significant samples of quasars and radio galaxies that exist are a good starting point for a further search for CO in the early universe. The expected width of the CO emission lines is a few hundred km s^{-1} (FWHM) and the bandwidth of the receivers and backend combination as used in major sub-millimetre observatories is limited to about 1000 km s^{-1}. It is therefore important that objects to be studied have well determined redshifts, preferably better than a few hundred km s^{-1}. Since the FWHM of the emission lines of radio galaxies is of order 1000 km s^{-1}, a factor of 5 − 10 smaller than in quasars, there is some advantage using radio galaxies rather than quasars for a survey of CO in the distant universe. We have started a programme to detect CO emission from distant radio galaxies using the JCMT and IRAM telescopes. From our spectra we have clear indications that we are indeed detecting CO lines from some of these distant objects. Currently we are working to confirm these detections using new observations. If these lines are real, they indicate that the density of CO gas is in the range $10^4 - 10^5$ cm^{-3} and that the temperature is of order 50 − 100 K. Using galactic conversion factors the total H_2 masses are then $10^{11} M_\odot$, comparable to the mass obtained for F10214+4724. If correct, these detections would support the idea that distant radio galaxies are undergoing vigorous starbursts. They might, therefore, be producing the bulk of their stellar content, as expected for a forming galaxy.

Acknowledgements. I would like to thank my collaborators, Dick Hunstead, George Miley, Rob van Ojik, Paul van der Werf and Mark Wieringa for continuous discussion.

References

Barvainis, R., Tacconi, L., Antonucci, T., Alloin, D., and Coleman, P.: 1994, Nature 371, 586

Briggs, F. H., Sorar, E., and Taramopoulos, A.: 1993, ApJ 415, L99

Cimatti, A., di Serego Alighieri, S., Fosbury, R., lvati, M. S., and Taylor, D.: 1993, MNRAS 264, 264

Dunlop, J. S., Hughes, D. H., Rawlings, S., Eales, S. A., and Ward, M. J.: 1994, Nature 370, 347

McCarthy, P. J.: 1993, ARA&A 31, 639

McCarthy, P. J., Elston, R., , and Eisenhardt, P.: 1992, ApJ 387, L29

Miley, G. K.: 1994, in G. V. Bicknell, M. A. Dopita, and P. J. Quinn (eds.), The First Stromlo Conference: The Physics of Active Galaxies, Vol. 54, p. 385

Röttgering, H., Hunstead, R., Miley, G. K., van Ojik, R., and Wieringa, M. H.: 1994, Spatially Resolved Lyα Absorption in the z = 2.9 Radio Galaxy 0943−242, MN: submitted

Solomon, P. M., Downes, D., and Radford, S. J. E.: 1992, ApJ 398, L29

Uson, J., Bagri, D. S., and Cornwell, D. S.: 1991, Phys. Rev. Letter 67, 3328

van Ojik, R., Röttgering, H., Miley, G., Bremer, M., Macchetto, F., and Chambers, K.: 1994, A&A 289, 54

A SEARCH FOR GRAVITATIONAL MILLI–LENSES

P.N. WILKINSON, D.R. HENSTOCK AND I.W.A BROWNE
University of Manchester
NRAL, Jodrell Bank, Macclesfield,
Cheshire, SK11 9DL, UK

AND

A.C.S. READHEAD, G.B. TAYLOR, R.C. VERMEULEN,
T.J. PEARSON AND W. XU
California Institute of Technology
OVRO, Robinson 105–24, Pasadena, CA 91125 USA

Abstract. We have searched for gravitational milli–lens systems by examining VLBI maps of ~ 300 flat–spectrum radio sources. So far we have followed up 7 candidates, with separations in the range 2–20 mas. None have been confirmed as lenses but several of them can not yet be definitively ruled out. If there are no milli–lenses in this sample then uniformly–distributed black holes of 10^6 to 10^8 $M_\odot$ cannot contribute more than $\sim 1\%$ of the closure density.

1. Introduction

A few lens systems have been found with sub–arcsecond image separations but lensing by mass concentrations below $\sim 10^9 M_\odot$ has not yet been investigated because the image separation has been below the resolution limit of the VLA and the HST. VLBI imaging is the only direct way to push the image separation limit below ~ 100 mas.

Searches for lensed systems with image separations in the range 1–100 mas (so–called "milli–lenses"), are of particular cosmological interest. Such separations correspond to lensing masses in the range $\sim 10^6$–$10^8 M_\odot$, comparable with the expected masses of pre–galactic compact objects (PCOs). PCOs arise in a wide range of cosmogonic scenarios with a natural mass scale, set by the Jeans mass, of order $10^6 M_\odot$ (e.g. Carr 1990). Uniformly

M. Kafatos and Y. Kondo (eds.), Examining the Big Bang and Diffuse Background Radiations, 95–98.
© 1996 IAU.

distributed PCOs could provide a significant fraction of the closure density of the universe and can only be detected by their gravitational lensing effects. Press & Gunn (1973) first calculated the optical depth of the Universe to lensing by point masses and their calculations have been extended by Ostriker & Vietri (1986), Nemiroff and Bistolas (1990) and Kassiola, Kovner and Blandford (1991). Gnedin & Ostriker (1992) have recently suggested that radiation from an early generation of massive stars ($10^{6.5} M_\odot$), forming somewhat after decoupling, may have altered the light element abundances. Their hypothesis allows a larger amount of baryonic dark matter and does away with the need for non–baryonic forms. The massive stars collapse to black holes and with the required density, $\Omega \sim 0.15$, $\sim 5\%$ of high-redshift quasars should be milli–lensed.

2. The VLBI Surveys

A series of VLBI surveys have now been made with resolutions of ~ 1 milliarcsec (mas). The most recent was the second Caltech-Jodrell Bank (CJ2) VLBI survey—a global MkII snapshot VLBI survey of 193 flat–spectrum radio sources at $\lambda 6$ cm (Taylor et al. 1994; Henstock et al. 1995). The aims of the CJ2 survey were to extend the morphological classification of the PR (Pearson & Readhead 1988) and CJ1 (Polatidis et al. 1994; Thakkar et al. 1994; Xu et al. 1994) VLBI surveys and to address several new cosmological questions. Amongst these was whether or not there is evidence for gravitational lensing on scales 2–200 mas.

On the mas scale the great majority of flat–spectrum radio sources are asymmetric core–jets. Simulations of the lensing effect of a PCO show that the typical image configuration is likely to consist of the core-jet primary with the demagnified, hence fainter, secondary off to one side. When the images are convolved with a 1 mas restoring beam the secondary may appear unresolved even if the primary is well resolved. Any source with a compact companion component no larger than the strongest component was therefore treated as a lens candidate.

About 300 flat–spectrum radio sources have now been mapped in the PR, CJ1 and CJ2 VLBI surveys; the typical resolution is 1 mas and the dynamic range is $> 100 : 1$. We are confident of detecting compact lensed components 30 times weaker than the brightest component in the map out to ± 200 mas. However, for image separations < 1.5 mas, i.e. comparable with the beam size, we are only sensitive to compact components with flux ratios $< 10 : 1$.

3. The Redshift Distribution

The majority ($\sim 80\%$) of the ~ 300 flat–spectrum sources in the PR, CJ1 and CJ2 surveys now have measured redshifts. The redshift distribution is shown in Fig. 1. The fraction of high redshift sources is a strong function of the limiting flux density of the sample. In the combined PR sample ($S_{6cm} > 1.3$ Jy) and CJ1 sample (1.3 Jy $\leq S_{6cm} \leq 0.7$ Jy) only 3 objects ($\sim 3\%$) have $z > 2.5$ whereas in the CJ2 sample (0.7 Jy $\leq S_{6cm} \leq 0.31$ Jy) 18 objects ($\sim 9\%$) have $z > 2.5$. As yet we have redshifts for only $\sim 70\%$ of the CJ2 sample but expect to increase this to $\sim 90\%$ with further optical observations of the faint objects.

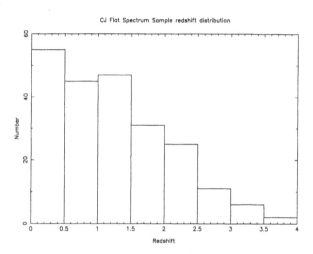

Figure 1. The redshift distribution of flat–spectrum sources in PR+CJ1+CJ2

4. Follow–up Observations

For the seven lens candidates we are made initial follow–up observations with 6 telescopes of the VLBA during February and March 1994. We made total intensity maps at λ 18cm, λ 3.6cm and λ 2cm to complement the existing λ 6cm maps. Lensed sources *must* contain two or more mas components which have simply–related structures and identical radio spectra. This multi–wavelength attack enabled us to show that about half of the candidates are conventional core–jet sources. For example Fig. 2 shows the CJ2 survey map of 0740+768; our higher resolution VLBA map at λ 2cm reveals the compact double source to be a core with a jet in p.a. $-100°$. A few candidates cannot yet be definitively ruled out. For these we must await the results of higher sensitivity VLBA observations at $\lambda 2$ cm scheduled for the Winter of 1994/5.

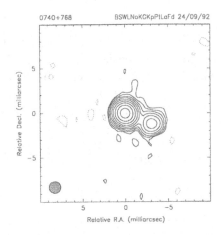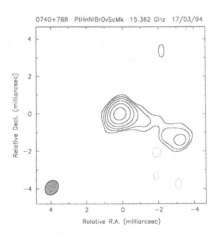

Figure 2. The lens candidate 0740+768: left) CJ2 map at λ6cm (resolution ~ 1 mas) made with a "global" array; right) VLBA map at λ2 cm (resolution ~ 0.5 mas)

It we fail to detect milli–lensing in ~ 300 flat–spectrum radio sources then uniformly–distributed PCOs can only contribute < 1% to the closure density. In order to place tighter limits on the cosmological density of PCOs, and specifically to confront Gnedin and Ostriker's (1992) prediction that ~ 5% of high redshift quasars might be lensed, at least 1000 radio–selected flat–spectrum sources must be searched. The VLBA is well–suited to this new surveying task.

References

Carr, B. J. 1990, *Comments on Astrophysics*, **14**, 257.

Gnedin, N. Yu., Ostriker, J. P. 1992, *Astrophys. J.*, *400*, 1.

Henstock, D. R., Wilkinson, P. N., Taylor, G. B., Readhead, A. C. S. and Pearson, T. J., *Astrophys. J. Suppl.*, submitted.

Kassiola, A., Kovner, I., Blandford, R. D. 1991, *Astrophys. J.*, **381**, 6.

Nemiroff, R. J. and Bistolas, V. G. 1990, *Astrophys. J.* **358**, 5.

Ostriker, J. P. and Vietri, M. 1986, *Astrophys. J.*, **300**, 68.

Pearson, T. J., Readhead, A. C. S. 1988, *Astrophys. J.*, **328**, 114.

Polatidis, A. G., Wilkinson, P. N., Xu, W., Readhead, A. C. S. and Pearson, T. J., 1994, *Astrophys. J. Suppl.*, in press.

Press, W. H., Gunn, J. E. 1973, *Astrophys. J.*, **185**, 397.

Taylor, G. B., Readhead, A. C. S., Pearson, T. J., Henstock, D. R., and Wilkinson, P. N., *Astrophys. J. Suppl.*, in press.

Thakkar, D.D., Xu, W., Readhead, A. C. S., Pearson, T. J., Polatidis, A. G. and Wilkinson, P. N. 1994, *Astrophys. J. Suppl.*, in press.

Xu. W., Readhead, A. C. S., Pearson, T.J., Polatidis, A. G., Wilkinson, P. N. 1994, *Astrophys. J. Suppl.*, in preparation.

THE COBE DIRBE SEARCH FOR THE COSMIC INFRARED BACKGROUND

MICHAEL G. HAUSER

Code 680, Laboratory for Astronomy and Solar Physics
NASA Goddard Space Flight Center, Greenbelt, MD 20771
USA

Abstract. The Diffuse Infrared Background Experiment (DIRBE) on the Cosmic Background Explorer (*COBE*) satellite is designed to conduct a sensitive search for isotropic cosmic infrared background radiation over the spectral range from 1.25 to 240 μm. The cumulative emissions of pregalactic, protogalactic, and evolving galactic systems are expected to be recorded in this background. The DIRBE instrument has mapped the full sky with high redundancy at solar elongation angles ranging from 64°to 124°to facilitate separation of interplanetary, Galactic, and extragalactic sources of emission. Conservative limits on the isotropic infrared background are given by the minimum observed sky brightnesses in each DIRBE spectral band during the 10 months of cryogenic operation. Extensive modeling of the foregrounds is under way to isolate or strongly limit the extragalactic infrared component. The current approach to these modeling efforts is described and representative present residuals are reported.

1. Introduction

The search for cosmic infrared background (CIB) radiation is a relatively new field of observational cosmology. Measurement of this distinct radiative background, expected to arise from the cumulative emissions of pregalactic, protogalactic, and galactic systems, would provide new insight into the cosmic 'dark ages' following the decoupling of matter from the cosmic microwave background radiation (Partridge & Peebles 1967; Low & Tucker 1968; Peebles 1969; Harwit 1970; Kaufman 1976; Bond, Carr, & Hogan 1986, 1991; Franceschini *et al.* 1991; Franceschini *et al.* 1994). Observation-

M. Kafatos and Y. Kondo (eds.), Examining the Big Bang and Diffuse Background Radiations, 99–108.
© 1996 *IAU*.

ally, there have been no corroborated direct detections of the CIB, though possible evidence for an isotropic infrared background, or at least limits on emission in excess of galactic foregrounds, has been reported in data from rocket experiments (Matsumoto *et al.* 1988; Matsumoto 1990; Noda *et al.* 1992; Kawada *et al.* 1994). Stringent limits have been set on any extragalactic background from 500 to 5000 μm wavelength in excess of that in the cosmic microwave background (CMB) radiation (Mather *et al.* 1994; Wright *et al.* 1994). The status of direct attempts to detect the CIB was recently reviewed by Hauser (1995). Indirect limits on the extragalactic infrared background have been inferred from possible attenuation of TeV γ-rays due to pair-production by photon-photon interactions (De Jager, Stecker, & Salamon 1994; Dwek & Slavin 1994; Biller *et al.* 1995).

The Diffuse Infrared Background Experiment (DIRBE) on the *COBE* spacecraft is the first satellite instrument designed specifically to carry out a systematic search for the CIB. The *COBE* FIRAS instrument, designed primarily to make a precise measurement of the spectrum of the cosmic microwave background radiation, is also a powerful instrument for the CIB search in the submillimeter range (Mather, this Symposium). The search for the CIB is the most exploratory of the *COBE* objectives. Even from a spaceborne instrument, this cosmic fossil is far more difficult to discern than the CMB radiation. Whereas the CMB is the dominant celestial radiation over most of the sky at millimeter wavelengths, the local infrared foregrounds from interplanetary dust and the Galaxy are far brighter than the CIB.

Conservative upper limits on the CIB based on preliminary reduction of DIRBE measurements of the south ecliptic pole brightness, and a comparison of DIRBE data with those from instruments on sounding rockets and the *Infrared Astronomical Satellite (IRAS)*, were presented by Hauser *et al.* (1991). The data collection phase of the DIRBE investigation is now completed, and the data from the period of cryogenic operation of the instrument (18 Nov 89 to 21 Sept 90) are fully reduced to calibrated sky maps. In this paper we report updated CIB limits in the DIRBE spectral range based upon the faintest sky brightnesses observed at each wavelength and a more recent calibration of the DIRBE data than that used by Hauser *et al.* (1991). We also describe the initial modeling of the solar system and Galaxy foreground contributions to the infrared sky brightness by the DIRBE team, and indicate the nature of the residuals.

2. The Observational Challenge

The search for the CIB is impeded by two fundamental challenges: there is no unique spectral signature of such a background, and there are many local contributors to the infrared sky brightness at all wavelengths, often quite bright. The lack of a distinct spectral signature arises in part because so many different sources of primordial luminosity are possible (e.g., Bond, Carr, & Hogan 1986), and in part because the primary emissions are then shifted into the infrared by the cosmic red-shift and dust absorption and re-emission. Hence, the present spectrum depends in a complex way on the characteristics of the luminosity sources, on their cosmic history, and on the dust formation history of the Universe.

Setting aside the difficult possibility of recognizing the CIB by its angular fluctuation spectrum (Bond, Carr, & Hogan 1991), the only identifying CIB characteristic for which one can search is an isotropic glow. One must solve the formidable observational problem of making absolute brightness measurements in the infrared. One must then discriminate and remove the strong signals from foregrounds arising from one's instrument or observing environment, the terrestrial atmosphere, the solar system, and the Galaxy. Particular attention must be given, of course, to possible isotropic contributions from any of these foreground sources. Finally, if an extragalactic isotropic residual remains, one must evaluate the contribution from galaxies over their luminous lifetime to distinguish their light from that of pregalactic or protogalactic sources. The DIRBE investigation is presently at the stage of modeling the known solar system and Galactic foregrounds and assessing the uncertainties in the models and measurements so as to determine whether evidence for an isotropic residual remains.

3. The COBE Diffuse Infrared Background Experiment

The primary aim of the Diffuse Infrared Background Experiment (DIRBE) is to conduct a definitive search for an isotropic CIB, within the constraints imposed by the local astrophysical foregrounds. The experimental approach was to obtain absolute brightness maps of the full sky in 10 photometric bands at 1.25, 2.2, 3.5, 4.9, 12, 25, 60, 100, 140, and 240 μm. In order to facilitate discrimination of the bright foreground contribution from interplanetary dust, linear polarization was also measured at 1.25, 2.2, and 3.5 μm. Because of the Earth's motion within the interplanetary dust cloud, the diffuse infrared brightness of the entire sky varies over the course of a year. To monitor this variation, the DIRBE field of view was offset from the *COBE* spacecraft spin axis by 30°. This produced a helical scan of the sky during each *COBE* orbit. Over the course of six months this yielded observations of each celestial direction hundreds of times at all accessible solar elongation angles (depending upon ecliptic latitude) in the range 64° to 124°. All spectral bands viewed the same instantaneous field-of-view, and the helical scan allowed the DIRBE to map 50% of the celestial sphere each day. The instrument was designed to achieve a sensitivity at each wavelength for each $0.7° \times 0.7°$ field of view of approximately $\nu I_\nu = 1$ nW m^{-2} sr^{-1} (1 σ, 1 year). This is well below the actual sky brightness, and below many of the predictions for the CIB, at all wavelengths.

The DIRBE instrument is an absolute radiometer, utilizing an off-axis folded Gregorian telescope with a 19-cm diameter primary mirror. The optical configuration (Magner 1987) was carefully designed for strong rejection of stray light from the Sun, Earth limb, Moon or other off-axis celestial radiation, or parts of the *COBE* payload (Evans 1983). The instrument, which was maintained at a temperature below 2 K within the *COBE* superfluid helium dewar, measured absolute brightness by chopping between the sky signal and a zero-flux internal reference at 32 Hz. Instrumental offsets were measured about five times per orbit by closing a cold shutter located at the prime focus. A radiative offset signal in the long wavelength detectors arising from JFETs (operating at about 70 K) used to amplify the detector signals was identified and measured in this fashion. Internal radiative reference sources were used to stimulate all detectors when the shutter was closed to monitor the stability and linearity of the instrument response. The highly redundant sky sampling and frequent response checks provided precise photometric closure over the sky and reproducible photometry to $\sim$ 1% or better for the duration of the mission. Calibration of the photometric scale was obtained from observations of isolated bright celestial sources. Comparison of the DIRBE and FIRAS absolute calibrations at 100, 140, and 240 μm shows excellent agreement between these two independent instruments on *COBE*. A more detailed description of the *COBE* mission has been given by Boggess *et al.* (1992), and the DIRBE instrument has been described by Silverberg *et al.* (1993).

4. The Character of the Infrared Sky

The DIRBE sky brightness maps show the dominant anticipated features of Galactic starlight and zodiacal light at short wavelengths, emission from the interplanetary dust dominating at 12 and 25 μm, and emission from the interstellar medium dominating at longer wavelengths (Hauser 1993; DIRBE Sky Maps at Elongation 90°, available from the National Space Science Data Center). The brightness of the sky in any fixed direction varies roughly sinusoidally over the year, though the detailed time dependence and amplitude of variation depends upon direction in the sky (Kelsall *et al.* 1993), indicating rather complex features of the interplanetary dust cloud. The DIRBE polarization maps reveal the expected strong linear polarization arising from the zodiacal light (Berriman *et al.* 1994). These data dramatically illustrate both the challenge of distinguishing the CIB from signals arising in our local cosmic environment, and the value of the DIRBE data for studies of the interplanetary medium and the Galaxy.

The most credible direct observational limits on the CIB are the minimum observed sky brightnesses. Since publication of the spectrum of a nominally dark direction, the south ecliptic pole, which was based on a quick-look reduction and initial calibration of the DIRBE data (Hauser et al. 1991), the complete data set acquired while the DIRBE was cryogenically cooled has been processed with an improved calibration. In each DIRBE weekly sky map, the faintest direction was determined for each wavelength. The smallest of these values at each wavelength over the duration of the mission provides a refined 'dark sky' limit to the CIB. Table 1 lists these dark sky values. At wavelengths where interplanetary dust scattering or emission is strong, the sky is darkest near the ecliptic poles. At wavelengths where the interplanetary cloud signal is rather weak (i.e., longward of 100 μm), the sky is darkest near the galactic poles or in minima of HI column density. The error shown for each value is the present absolute calibration uncertainty.

TABLE 1. UPPER LIMITS ON THE COSMIC INFRARED BACKGROUND

λ μm	νI_ν (nW m^{-2} sr^{-1}	Reference
1.25	393±13	DIRBE dark sky
2.2	150±5	"
3.5	63±3	"
4.9	192±7	"
12	2660±310	"
25	2160±330	"
60	261±22	"
100	74±10	"
140	57±6	"
240	22±2	"
500-5000	680/$\lambda(\mu$m)	Mather et al. (1994)

As described by Mather et al. (1994), the CMB spectrum in the wavelength range 0.5 - 5 mm deviates from a 2.726 K blackbody shape by less than 0.03% of the peak intensity. Taking this as an upper limit to an additional cosmic infrared background implies $\nu I_\nu < 340/\lambda(\mu$m) nW m^{-2} sr^{-1}. To make some allowance for systematic errors in separating the Galactic signal from the FIRAS CMB signal, the above limit has been doubled in Table 1.

5. Discriminating the Infrared Foregrounds

In order to determine whether an isotropic signal is present, thus measuring or setting more stringent limits upon the CIB, one must address the problem of discriminating the various contributions to the measured sky brightness. The DIRBE team is presently approaching this in a sequential process: (1) find a parameterized model of the interplanetary dust (IPD) cloud; (2) integrate this IPD model along the line of sight to calculate the sky brightness for each celestial direction for each week of the DIRBE mission; (3) subtract the calculated IPD map from each weekly DIRBE map and average the residual weekly maps; (4) blank bright sources discernible in the residual averaged maps, both stellar and more extended sources; (5) use a detailed model of discrete Galactic sources to calculate a map of the integrated sky brightness due to sources fainter than those blanked in the previous step, and subtract this map from the DIRBE residual map; (6) determine an interstellar medium emission map at each wavelength using the DIRBE 100 μm map as a template, and subtract this map from the residual map; (7) test the final residual maps for isotropy.

The initial interplanetary dust model is very similar to that used by Good in creating the *IRAS Sky Survey Atlas* (Wheelock *et al.* 1994). The cloud is assumed to have an intrinsically static, Sun-centered density distribution, with a symmetry plane inclined to the ecliptic plane. The combined effects of symmetry plane inclination and eccentricity of the Earth's orbit produce the annual variations of sky brightness in the model. The model is extended to the short-wavelength DIRBE bands by introducing particle albedos and phase functions which are sums of Henyey-Greenstein functions. Our first test model includes only a main 'smooth' IPD cloud. The IPD dust bands discovered in the IRAS data (Low *et al.* 1984) and confirmed in the DIRBE data (Spiesman *et al.* 1995) and the dust ring containing particles resonantly trapped just outside 1 A. U. (Dermott *et al.* 1994; Reach *et al.* 1995) will be incorporated in future versions. The polarization data, and corresponding parameters to characterize particle properties, are also ignored in the initial model. The 14 free parameters of the model are determined by simultaneously fitting the time variation of the sky brightness at several hundred lines of sight over the 10 months of cryogenic operation. Though this simple model represents the IPD signal fairly well, typically to several percent of the brightness in the ecliptic plane, there are clearly systematic artifacts in the residuals which can be reduced by more refined models. The model shows that the IPD signal is a large fraction of the sky brightness at wavelengths short of 100 μm, even at high galactic latitude. The uncertainty in the residual sky maps at these wavelengths will be dominated by uncertainties in the IPD signal removal until more refined models are developed.

The sky brightness contribution from faint Galactic sources is obtained by integrating the source counts in the model of Wainscoat et al. (1992), as elaborated by Cohen (1993; 1994). The model includes 87 types of Galactic sources, distributed in spatial components including the disk, bulge, halo, spiral arms, and the molecular ring. Galactic extinction is included in the model. Sky brightnesses in each DIRBE band from 1.25 to 25 μm were calculated for 238 sky zones (M. Cohen, private communication to the DIRBE team). At high galactic latitudes these sources contribute 5 - 20% of the observed sky brightness from 1.25 to 4.9 μm; at 12 and 25 μm their contribution is less than 0.1%.

The initial model of interstellar medium (ISM) emission assumes uniform color, at least at high galactic latitude. The 100 μm DIRBE residual map is used as a spatial template for the ISM because of its high signal-to-noise ratio. The color for each band is determined from the slope of the correlation between the brightness in that band and the 100 μm brightness. The 100 μm brightness not associated with any apparent ISM material is found by correlating 100 μm brightness at high latitude with 21-cm HI emission (Stark et al. 1992), and extrapolating this correlation to zero 21-cm emission. This assumes either that HI gas is a good spatial tracer of all interstellar gas, or that infrared emission from other gas phase components is negligible. Tests made using available high latitude CO and HII maps show that this is a reasonable first approximation. This method for estimating the ISM contribution has been used for wavelengths of 12 μm and longer. A modified procedure which distinguishes extinction and emission effects is under study for determining the shorter wavelength ISM contribution. Rough estimates suggest that the ISM signal at the short wavelengths is small compared to the present uncertainties in the IPD model. At wavelengths of 140 and 240 μm the ISM signal is the largest foreground contribution to the sky brightness at high latitudes.

6. Status of the CIB Search

The process of refining the DIRBE foreground models, assessing the uncertainties in the models and the measurements, and testing the residuals for isotropy is not yet complete. Hence, no conclusions regarding DIRBE detection of, or limits upon, the CIB more restrictive than the dark sky limits in Table 1 can yet be stated.

Some insight into the present status is provided by looking at the residual brightnesses in several high latitude regions. For this purpose, we have examined the residuals in $10° \times 10°$ fields at the Galactic poles and ecliptic poles, and a $5° \times 5°$ field in the 'Lockman Hole' (Lockman, Jahoda, & McCammon 1986; Jahoda, Lockman, & McCammon 1990), the region of minimum HI column density at $(\ell, b) \sim (148°, +53°)$. Table 2 lists the minimum and maximum mean residual brightnesses for these 5 patches after all of the foreground removal steps described above. While the range of values at each wavelength is still substantial, typically a factor of 2 or more, comparison with Table 1 shows that we are now dealing with residuals which are small fractions, approaching 10% at wavelengths shortward of 100 μm, of the 'dark sky' limits. However, the fact that the residuals are brightest in the region of peak IPD thermal emission, 12 to 25 μm, strongly suggests that significant foreground emission still remains, at least in the middle of the DIRBE spectral range.

TABLE 2. RANGE OF SKY BRIGHTNESS RESIDUALS AT HIGH LATITUDE

λ μm	νI_ν (nW m^{-2} sr^{-1})
1.25	59 - 164
2.2	15 - 71
3.5	9 - 25
4.9	15 - 42
12	104 - 330
25	121 - 287
60	27 - 45
100	23 - 41
140	24 - 50
240	12 - 25

As a complement to the direct searches for the CIB by diffuse infrared background measurements, galaxy counts in the infrared allow estimation of lower limits to the total extragalactic infrared background. For example, Cowie et al. (1990) estimated the integrated contribution of galaxies at 2.2 μm to be $\nu I_\nu = 5$ nW m^{-2} sr^{-1} on the basis of deep galaxy counts. Hacking & Soifer (1991) used galaxy luminosity functions derived from *IRAS* data to predict minimum backgrounds (integrated to z=3) at 25, 60, and 100 μm of 1, 2, and 4 nW m^{-2} sr^{-1} respectively. Beichman & Helou (1991) used synthesized galaxy spectra, also based largely on *IRAS* data, to estimate the diffuse infrared background due to galaxies. At 300 μm, their minimum estimated brightness (integrated to $z=3$) is 2 nW m^{-2} sr^{-1}. The integrated galaxy far-infrared background contribution may exceed these estimates substantially if there has been evolution in galaxy luminosity or space density: deeper counts from future space infrared observatories such as *ISO* and *SIRTF* will improve these estimates. Even these minimum extragalactic background contributions should be detectable if the foreground contributions to the *COBE* measurements can be modeled to about the 1% level, a difficult but perhaps achievable goal.

7. Conclusion

Measurement of the extragalactic infrared background radiation will advance our understanding of cosmic evolution since decoupling. The high quality and extensive new measurements of absolute infrared sky brightness obtained with the DIRBE and FIRAS instruments on the *COBE* mission have already set significant limits upon this elusive background.

When foreground models are completed and uncertainties carefully assessed, the *COBE* data will certainly provide cosmologically valuable constraints upon, or perhaps even direct detection of, the extragalactic infrared background to the limits imposed by our bright astrophysical environment.

8. Acknowledgments

The author gratefully acknowledges the contributions to this report by the many participants in the *COBE* Project, to his colleagues on the *COBE* Science Working Group, and especially the many scientists, analysts, and programmers engaged in the DIRBE investigation. Those particularly engaged in the work reported here include E. Dwek, T. Kelsall, H. Moseley, R. Silverberg, E. Wright, R. Arendt, G. Berriman, B. Franz, H. Freudenreich, C. Lisse, M. Mitra, N. Odegard, W. Reach, J. A. Skard, T. Sodroski, S. Stemwedel, G. Toller, and J. Weiland. The National Aeronautics and Space Administration/Goddard Space Flight Center (NASA/GSFC) is responsible for the design, development, and operation of the *COBE*. Scientific guidance is provided by the *COBE* Science Working Group. GSFC is also responsible for the development of the analysis software and for the production of the mission data sets.

References

Beichman, C. A. & Helou, G., 1991, *ApJ*, **370**, L1
Berriman, G. B., *et al.* , 1994, *ApJ*, **431**, L63

Biller, S. D., et al. , 1995, ApJ, to be published

Boggess, N. et al. , 1992, ApJ, **397**, 420

Bond, J. R., Carr, B. J., & Hogan, C. J., 1986, ApJ, **306**, 428

Bond, J. R., Carr, B. J., & Hogan, C. J., 1991, ApJ, **367**, 420

Cohen, M., 1993, AJ, **105**, 1860

Cohen, M., 1994, AJ, **107**, 582

Cowie, L. L., et al. , 1990, ApJ, **360**, L1

De Jager, O. C., Stecker, F. W. & Salamon, M. H., 1994, Nature, **369**, 294

Dermott, S. F. et al. , 1994, Nature, **369**, 719

Dwek, E. & Slavin, J., 1994, ApJ, **436**, 696

Evans, D. C., 1983, SPIE Proc., **384**, 82

Franceschini, A. et al. , 1991, A&A SupplSer, **89**, 285

Franceschini, A. et al. , 1994, ApJ, **427**, 140

Hacking, P. B. & Soifer, B. T., 1991, ApJ, **367**, L49

Harwit, M., 1970, Rivista del Nuovo Cimento, **II**, 253

Hauser, M. G., 1993, in Back to the Galaxy, AIP Conf. Proc. **278**, ed. S. S. Holt & F. Verter (New York: AIP), 201

Hauser, M. G., 1995, in Extragalactic Background Radiation, Space Telescope Sci. Inst. Symp. Ser. **7**, ed. D. Calzetti, M. Livio, & P. Madau, (Cambridge: Cambridge Univ. Press), 135

Hauser, M. G., et al. , 1991, in After the First Three Minutes, AIP Conf. Proc. **222**, ed. S. Holt, C. L. Bennett, & V. Trimble (New York: AIP), 161

Jahoda, K., Lockman, F. J., & McCammon, D., 1990, ApJ, **354**, 184

Kaufman, M., 1976, ApSpSci, **40**, 369

Kawada, M., et al. , 1994, ApJ, **425**, L89

Kelsall, T., et al. , 1993, Proc. SPIE Conf. 2019, Infrared Spaceborne Remote Sensing, ed. M. S. Scholl (Bellingham: SPIE), 190

Lockman, F. J., Jahoda, K., & McCammon, D., 1986, ApJ, **302**, 432

Low, F. J. & Tucker, W. H., 1968, PRL, **22**, 1538

Low, F. J. et al. , 1984, ApJ, **278**, L19

Magner, T. J., 1987, OptEng, **26**, 264

Mather, J.C., et al. , 1994, ApJ, **420**, 439

Matsumoto, T., 1990, in The Galactic and Extragalactic Background Radiation, IAU Symposium **139**, ed. S. Bowyer and C. Leinert
        (Dordrecht: Kluwer), 317

Matsumoto, T., Akiba, M., & Murakami, H., 1988, ApJ, **332**, 575

Noda, M. et al. , 1992, ApJ, **391**, 456

Partridge, R. B. & Peebles, P. J. E., 1967, ApJ, **148**, 377

Peebles, P. J. E., 1969, Phil. Trans. Royal Soc. London, A, **264**, 279

Reach, W. T. et al. , 1995, Nature, to be published

Silverberg, R. F. et al. , 1993, Proc. SPIE Conf. 2019, Infrared Spaceborne Remote Sensing, ed. M. S. Scholl (Bellingham: SPIE), 180

Spiesman, W. J. et al. , 1995, ApJ, **442**, to be published

Stark, A. A. et al. , 1992, ApJS, **79**, 77

Wainscoat, R. J. et al. , 1992, ApJS, **83**, 111

Wheelock, S. L. et al. , 1994, IRAS Sky Survey Atlas Explanatory Supplement, JPL Pub. 94-11, (Pasadena: JPL), Appendix G

Wright, E.L. et al. , 1994, ApJ, **420**, 450

GALAXY FORMATION: ISOCAM COUNTS

C.J. CESARSKY AND D. ELBAZ
Service d'Astrophysique
CEA, DSM, DAPNIA, CE Saclay, F91191 Gif-Sur-Yvette cedex,
France

1. Introduction

Detailed observations in the optical do not allow to discriminate between the following scenarii of galaxy formation:

- synchronous formation at large redshift (z > 7).

- hierarchical merging, down to z ≤ 1, of proto-galactic lumps formed at very high redshift (≈ 10).

- galaxy formation at low z (≈ 2-5), but partially obscured by dust.

More than 1500 galaxies in a field of 2.6'×4.6' were detected by Tyson (1988) with CCD images in the B band. These deep counts of galaxies revealed the presence of an unexpectedly large population of blue galaxies, corresponding to a number excess of 3 to 5 times more galaxies at B>23 than one can expect from the local luminosity function. These objects lie apparently at moderate redshift (z=0.3 for B= 23-24) as shown by spectroscopy of B-selected galaxies (Colless *et al.* 1990).

Multiplexing spectroscopy with the MOS (multi-object spectrograph) at CFHT confirmed this result, but also showed that up to 50% of the faint blue galaxies (B=24) are associated to narrow emission line AGNs (Tresse *et al.* 1994) rather than to starburst galaxies as previously stated.

In their discussion about the origin of this population of blue galaxies at moderately low redshift, Colless *et al.* (1993) reviewed different models of galaxy formation and evolution, and concluded that the current observations do not allow to discriminate between these models, although no-evolution models seemed to be best suited.

A separate issue is that of the intrinsically bright and distant blue galaxies, which were found to be absent in Colless *et al.* (1993) redshift surveys made using the Low Dispersion Survey Spectrograph (LDSS) on the Anglo-

M. Kafatos and Y. Kondo (eds.), Examining the Big Bang and Diffuse Background Radiations, 109–116.
© 1996 *IAU.*

Australian Telescope (AAT), with CCD photometry in B, R and I bands. Tresse *et al.* (1994), Hammer *et al.* (1994) found that these galaxies were brighter by several tens of magnitude when z>0.5, hence showing evidence of evolution.

As pointed out several years ago by Ostriker and Heisler (1984), distant objects may be significantly obscured by dust at optical wavelengths, either by surrounding dust, dust in intervening galaxies, or some combinaison of these. Thus (see also Colless *et al.* 1993), infrared-selected redshift surveys are mandatory to go one step further towards the understanding of galaxy formation and evolution. Two teams intend to carry out deep surveys within the Central programme of the ISO satellite: Cesarsky *et al.* and Taniguchi *et al.* (see references).

2. Infrared Surveys

Even if they go much less deep than Tyson's (1988) survey, surveys in the thermal infrared are decisive in the determination of the process of galaxy fomation. Indeed, as pointed out by De Zotti *et al.* (1994), three effects concur in easing the detection of high-z galaxies at mid-IR wavelengths:

- the decreasing effect of dust absorption as compared to shorter wavelengths.

- the positive K-correction implied by galaxy spectra raising up to ≈ 1 μm.

- the positive luminosity evolution.

In this context, a mid-IR survey will allow to test:

- the presence of "slightly dusty" proto-galaxies, i.e. galaxies where most of the MIR emission comes from the redshifted stellar emission rather than from dust radiation.

- the luminosity evolution of spiral galaxies (either starbursts or not).

- the possibility that ellipticals are formed in a strong burst of star formation at z=2 to 5 (in a phase similar to that observed for IRAS F10214+4724, at z=2.3, Rowan-Robinson *et al.* 1991).

- the density of low-luminosity AGNs (≈ 15 μm).

- the origin of quasars. Are quasars born in strong galaxy interactions, in conjonction with an active starburst phase (Sanders 1988, Lipari *et al.* 1994) ?

3. The Infrared Space Observatory, ISOCAM and ISOPHOT

In 1983, just as the first results of IRAS were presented to the european astronomical community, the decision was taken at the European Space Agency to fly a second generation infrared cryogenic satellite, ISO (the Infrared Space Observatory). While IRAS was scanning the whole sky in

Figure 1. ISOCAM IR camera

four colours, ISO is destined to perform detailed studies of selected regions, with better angular resolution, wider wavelength coverage, enhanced
imaging and spectroscopic capabilities, and a higher sensitivity. ISO will be
launched in September 1995.

Four instruments have been placed in the focal plane of the ISO telescope: a camera (ISOCAM), a photometer (ISOPHOT), and two spectrometers (SWS and LWS), which, together, cover the range 2.5 to 200 μm. Two
of them will be used in the deep surveys of the ISO Central Programme:
ISOCAM and ISOPHOT.

ISOCAM (Cesarsky *et al.* 1994) is a two way camera, featuring two
(32×32) array detectors, one for short wavelengths (SW, 2.5 to 5.5 μm), the
other for long wavelengths (LW, 4 to 17 μm). On each channel, there are two
wheels, one carrying four lenses, allowing four different pixels fiels of view
(1.5", 3", 6", 12"), and the other carrying 10 to 12 fixed filters and one or
two Circular Variable Filters (CVF), allowing to reach a spectral resolution
of about 45. A wheel at the entrance has four positions: 3 polarizers and
a hole. A selection wheel carries Fabry mirrors which can direct the light
beam of the ISO telescope towards one or the other of the detectors, or
illuminate them uniformly with an internal calibration source, for flat field
purposes. The picture of ISOCAM is shown on figure 1.

With exposures of a few minutes, the long wavelength detector will easily detect sources at the sub-mJy level, and the short wavelength channel sources at the mJy range. With longer exposures, the sensitivity of the long wavelength channel will be limited in most filters by the flat field accuracy achievable, given the presence of the zodiacal background; then, it will be necessary, as in ground observations, to use beam switching or microscanning techniques.

ISOPHOT (Lemke *et al.* 1994) is a complex instrument, with three subsystems. Only one of them will be used in the Deep Survey: the camera, which features two arrays made of individual detectors: C100 (3×3 pixels) and C200 (2×2 pixels), and allows to take images of the sky in the wavelength range 50 to 240 μm, with various filters and an angular resolution of 83.9" (d_{Airy}) at 100 μm and 168" at 200 μm.

3.1. THE PHYSICS

The typical continuum emission of galaxies in the IR shows two peaks: a first peak in the NIR, around 1 μm, due to the emission of the red stars, and a second peak in the FIR, around 100 μm and with a flux density one order of magnitude larger, due to the cool dust (<100 K) emission, typical of star forming regions. Between these two peaks, we find MIR emission from hot dust, around 10 μm, and NIR-MIR emission, of PAHs.

Given the relatively coarse resolution that can be obtained, even at diffraction limit in the infrared with a 60 cm telescope, and the great sensitivity obtained with a cryogenic system, the confusion limit may be a concern even with ISOCAM. With a 6" pixel field of view, the 5σ confusion limit for the galaxies in the two most sensitive ISOCAM filters is respectively 20 and 30 μJy for the 6.7 and 12 μm bands, using the count prediction by Franceschini *et al.* (1991).

3.2. THE OBSERVATION PROGAMMES

Two galaxy count programmes have been scheduled in the ISO Central Programme using ISOCAM and ISOPHOT (see table 1):

- Cesarsky *et al.* (see references) Deep Survey programme (CCESARSK. IDSPCO) consists in long and repeated integrations in broad and sensitive CAM (LW2: 5-8 μm, LW3: 12-18 μm) and PHOT (C100: 90 μm) filters. The survey will be performed at two different depths with CAM, with a pixel field of view of 6": a "deep" survey (DS) with both filters, with a 4 σ sensitivity of 34 μJy for LW2 and 13 μJy for LW3, and a "shallow" survey (SS) with only LW3, for a lower sensitivity of 38 μJy. Two fields will be mapped, the "Marano" field (MF, southern field) and the "Lockman" hole (LH, northern field), with a large area for the SS (about 1 square degree) and a smaller area for the DS (0.17 and 0.11 sq. deg. for MF and

TABLE 1.

Programme	CCESARSK. IDSPCO			YTANIGUC. DEEPPGPQ
	SS-LW3	DS-LW3	DS-LW2	LW2
4σ sensitivity (point source) μJy	380	130	34	10
Total time (h)	MF: 20 LH: 14	MF: 20 LH: 13	MF: 30 LH: 20	LH: 30
Number of CAM beams (3'×3')	MF: 350 LH: 250	MF: 60 LH: 40	MF: 60 LH: 40	LH: 4
Area covered (sq. degree)	MF: 1.0 LH: 0.7	MF: 0.17 LH: 0.11	MF: 0.17 LH: 0.11	LH: 0.01

LH respectively). The same pixel of the sky is seen by different parts of the camera during the survey in order to detect any systematic effects.

A long wavelength map of the Southern SS (about 1 square degree in the Marano field, for a total observation time of 20 hours) with ISOPHOT C100 at 90 microns will be achieved in the same observing mode than the one described afterwards for the Taniguchi *et al.* proposal.

- Taniguchi *et al.* (see references) Search for Primeval Galaxies and Quasars programme (YTANIGUC.DEEPPGPQ). The survey will be performed at one single depth, with a pixel field of view of 3" and the LW2 filter, hoping to reach a 4 σ sensistivity of 10 μJy. The area covered will be small: 0.01 square degree. A 90 and 160 μm map will be performed with the oversampling mode (PHT32) of PHOT in about 2.4 square degree in the Lockman Hole. The sensitivity at a 5 sigma level is expected to be 6.7 mJy for the 90 μm filter and 13mJy for the 160 μm filter (for a total observation time of 28 hours). It is slightly below the confusion limit due to galaxies (7 mJy and 10 mJy for respectively 90 and 160 μm) and well above the confusion due to cirrus clouds (because the region is selected in a dark IRAS area). This survey will cover most or all of the area surveyed by Cesarsky *et al.* at shorter wavelengths.

3.3. PREDICTIONS FOR ISOCAM COUNTS

In order to estimate the observational feasibility of this survey, Taniguchi *et al.* estimated their galaxy fluxes using the Model B of Larson (1974) where a protogalactic gas cloud is assumed to collapse in a free-fall time, resulting in an intense star formation phase in the nuclear region; the infrared emission results mainly from the shifted emission of the stars. They expect to detect about 3 primeval galaxies, in the case of 1 mag extinction ($E_{B-V}=1$). About 30 primeval quasars are expected to be found with the PHOT survey.

On the other hand, Cesarsky et al. used the different model of Franceschini et al. (1991, De Zotti et al. 1994) to make predictions on the galaxy counts. This model takes into account the reprocessing of the stellar emission by dust that re-radiates it in the IR. With this approach, they expect to find about 400 galaxies in the whole survey with the LW2 filter (DS only), where the contribution of elliptical galaxies is not dominant. Two possibilities have been tested in this model, for the survey in the LW3 filter, depending on the assumption that ellipticals show evolution or not: the total galaxy counts in the DS (and SS) is equal to 315 (respectively 486), if evolution is assumed, and 81 (respectively 189), if no evolution is assumed. Maffei et al. (1994) have considered the influence of a more detailed modelisation of the PAHs emission, including emission lines, and found a slightly different estimation of the expected galaxy counts (with no galaxy evolution). Due to the K correction, the PAHs emission is shifted towards higher wavelength, increasing the galaxy counts in the LW2 and LW3 filters. When comparing the predictions of Franceschini et al. and Maffei et al.:

- for LW2, the two logN-logS curves cross each other around the sensitivity of S=0.1 μJy, but Maffei's model predicts one order of magnitude more counts for a sensitivity of S=0.1 mJy.

- for LW3, Maffei's model predicts about 3 times more galaxies than Franceschini (1991), whatever the sensitivity.

3.4. WILL THESE SURVEYS REVEAL MORE FAR-AWAY, ULTRALUMINOUS SOURCES ?

With the IRAS survey and its follow up at various wavelengths, the importance of the role of galaxy interactions in the formation of ultraluminous galaxies galaxies was established (Sanders et al. 1988). Moreover, if these objects were more common in the past (Sanders et al. 1988; Rowan-Robinson et al. 1991), we would expect to detect at least several of them at high z with the more sensitive ISO surveys.

Mrk 231 is the nearest object (z=0.0269) of this kind showing extreme IR emission and where the bolometric luminosity is dominated by the IR continuum emission (Lipari et al. 1994). Such a galaxy could be detected as far as z=3 with the DS (CAM LW2 and LW3), z=2 with the SS (for CAM LW3), and z=1, for PHOT at 90 μm.

Similarly, figure 3 shows the depth at which a typical elliptical galaxy with $M_B \approx 20$ could be detected with ISO: the LW2 DS goes deeper than z=1, while for LW2 the DS gets close to z=0.5 and the SS to z=0.2.

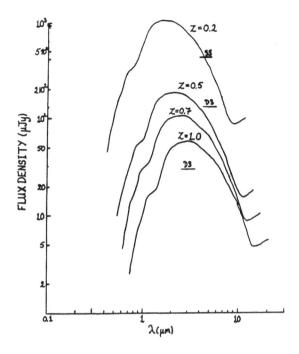

Figure 2. Spectrum of a typical E/SO galaxy with $M_B \approx 20$ transposed to various redshifts with no significant evolution (from S. Chase)

A completely different approach seems to favor the hypothesis that these surveys can also detect very young spheroidal galaxies. Indeed it is now well-known that the intergalactic gas in clusters of galaxies (intracluster medium, ICM) has been enriched by the galaxies with an iron mass at least equal to that contained in the galaxies (Renzini *et al.* 1993). This Fe mass being directly proportional to the total luminosity of ellipticals (E/SO) only, it was suggested to originate from these galaxies (Arnaud *et al.* 1992). To process such a large amount of iron and eject it into the ICM, E/SO must have experienced a strong starburst at the scale of the whole galaxy, with an IMF shifted towards massive stars (Elbaz *et al.* 1994). The SNII origin of the intergalactic iron has recently been confirmed by the detection in X-ray of an overabundance of O over Fe, typical of SNII (Mushotzky 1994, to appear) and a low abundance (solar or less) of the hot corona of three bright ellipticals by ROSAT (Forman *et al.* 1993) and ASCA (Awaki *et al.* 1994), much too low for an SNIa origin of the iron. Such a model has been applied to the hyperluminous galaxy 10214+4724 (Elbaz *et al.* 1992) and predicts an intense and very brief phase (a few hundred million years only) of high-mass stars formation, that should be detectable in the IR at relatively high redshift.

References

Arnaud, M., Rothenflug, R., Boulade, O., Vigroux, L., Vangioni-Flam, E., 1992, A&A 254, 49

Awaki, H., et al., 1994, PASJ 46, L65

Cesarsky, C.J., 1993, Proceedings of the XXVIIIth Rencontre de Moriond: "The Cold Universe", eds. Montmerle, Th., Lada, Ch.J., Mirabel, I.F., Trân Thanh Vân, Ed. Frontieres, p.417

Cesarsky, C.J., Chase, S., Danese, L., Desert, F.X., Franceschini, A., Harwitt, M.O., Hauser, M., Koo, D., Mandolesi, N., Puget, J.-L., ISO Central Progamme: CCESARSK.IDSPCO

Cesarsky, C.J., et al., 1994, Opt. Eng. 33, 751

Colless, M., Ellis, R., Taylor, K., Hook, R., 1990, MNRAS 235, 827

Colless, M., Ellis, R.S., Broadhurst, T.J., Taylor, K., Peterson, B.A., 1993, MNRAS 261, 19

De Zotti, G., Franceschini, A., Mazzei, P., Toffolatti, L., Danese, L., 1994, to be published.

Elbaz, D., Arnaud, M., Casse, M., Mirabel, I.F., Prantzos, N., Vangioni-Flam, E., 1992, A&A 265, L29

Elbaz, D., Arnaud, M., Vangioni-Flam, E., 1994, to appear in A&A

Forman, W., Jones, C., David, L., Franx, M., Makishima, K., ohashi, T., 1993, ApJ 418, L55

Franceschini, A., Toffolatti, L., Mazzei, P., Danese, L., De Zotti, G., 1991, A&AS 89, 285

Hammer, F., Lilly, S., Le Fevre, O., Crampton, D., Tresse, L., 1994, in Proceedings of the ESO workshop: Science with the VLT, Munchen 28th June, Eds. Walsh & Danziger, Springer-Verlag, in press.

Larson, R.B.,1974, MNRAS 166, 585

Lemke, D., et al., 1994, Opt. Eng. 33, 20

Lipari, S., Colina, L, Machetto, F., 1994, to appear in ApJ

Maffei, B., Puget, J.-L., Desert, F.X., 1994, in preparation.

Ostriker, J.P., Heisler, J., 1984, ApJ 278,1

Renzini, A., Ciotti, L., D'Ercole, A., Pellegrini, S., 1993, ApJ 419, 52

Rowan-Robinson, M., et al., 1991, Nature 351, 719

Sanders, D.B., Soifer, B.T., Elias, J.H., Matthews, K., Madore, B.F., 1988, ApJ 325, 74

Taniguchi, Y., Okuda, H., Matsumoto, T., Wakamatsu, K., Kawara, K., Cowie, L., Joseph, R., Sanders, D., Chambers, K., Wynn-Williams G., Sofue, Y., Matsuhara, H., Sato, Y., Desert, X., ISO Central Progamme: YTANIGUC.DEEPPGPQ

Tresse, L., Rola, C., Hammer, F., Stasinska, G., 1994, in preparation.

Tyson, J., 1988, AJ 96, 1

OBSERVATIONS OF DIFFUSE IR BACKGROUND RADIATION BY IRTS AND IRIS

H. OKUDA

ISAS
3-1-1 Yoshinodai, Sagamihara, 229 Japan

1. Introduction

Diffuse background radiation is integrated light which is consisted of various components of intrerplanetary, stellar, interstellar, galactic and intergalactic origins as well as cosmic background radiation, the remnant of the pre-galactic phenomena in the early history of the universe.

Observations of these radiations have been made by variety of techniques in various ranges of wavelengths. Since the discovery of the microwave background radiation by Penzias & Wilson (1965), particular attention has been paid to the observations in submillimeter and millimeter ranges. These observations was culminated by the launch of COBE satellite, in which the FIRAS provided a complete and precise spectrum of the 3K emission (Mather 1994), and the DMR gave the first indication of the presence of anisotropy in the radiation (Smoot 1994), while the DIRBE has provided sky maps of various wavelength bands from near infrared to far infrared (Hauser 1994).

The observed intensties are, however, mixture of radiations of different origins and their separation is not straigtfoward but needs complicated and tedious procedures.

For better segregation of the different components and more comprehensive understanding of the diffuse background radiations, we are planning two space missions to observe the diffuse background radiations in infrared regions. Here we address briefly on the two missions, a currently ongoing mission IRTS and a future plan IRIS.

M. Kafatos and Y. Kondo (eds.), Examining the Big Bang and Diffuse Background Radiations, 117–124.
© 1996 IAU.

2. IRTS (Infrared Telescope in Space)

2.1. TELESCOPE AND OBSERVATIONAL INSTRUMENTS

IRTS is a small space borne telescope cryogenically cooled by liquid helium which will be dedicated for observations of diffuse emissions of galactic and extragalactic origins in infrared region. It will be launched onboard the SFU (Space Flyer Unit), a multi-purpose space platform, by the newly developed H-2 launcher in early 1995 and retrieved by Space Shuttle after the mission completes. IRTS has a telescope with an aperture of 15cm and F.O.V.s of sub-degrees. It has four types of focal plane instruments. Typical characteristics of each instrument are summarized in Table 1. NIRS (Near

TABLE 1. Major characteristics of the focal plane instruments on IRTS

	NIRS	MIRS	FILM	FIRP
Optical system	grating	grating	grating	filters
Wavelength coverage (μm)	1.4 – 4.0	4.5 – 11.7	63([O I]), 158([C II]) 155, 160 (cont.)	140, 230, 400, 700
Resolution ($\Delta\lambda/\lambda$)	15 – 30	20 – 30	400	3
Beam size	$8' \times 8'$	$8' \times 8'$	$8' \times 13'$	$30'$
Detector	InSb $\times 24$	Si:Bi $\times 32$	Ge:Ga $\times 1$, stressed $\times 3$ Ge:Ga (stressed) $\times 3$	0.3K bol. $\times 4$

IR Spectrometer) and MIRS (Middle IR Spectrometer) are grating spectrometers with medium resolution ($\lambda/\Delta\lambda$) of 15 ~ 30: both have a higher spatial resolution ($0.1° ~ 0.2°$) than COBE's DIRBE. FILM (Far IR Line Mapper) is a grating spectrometer specifically designed for measurements of [O I] (63μm) and [C II] (158μm) lines. FIRP (Far IR Photometer) is a multi-band photometer incorporated with bolometers highly sensitized by cooling down to 0.3K with a ^{3}He refrigerator. The spatial resolution is substantially higher than FIRAS and a little better than DIRBE in COBE. The technical details of the instruments are given in the series of papers published in ApJ vol.428, No 1, (Murakami *et al.* 1994).

The sensitivity limit of each instrument estimated from the laboratory calibrations are displayed in Fig 1.

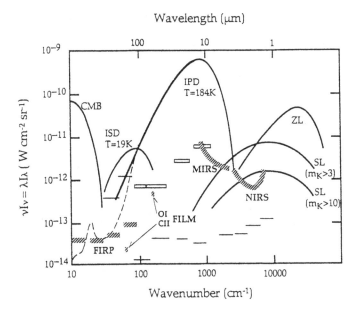

Figure 1. Observational limits of the focal plane instruments of IRTS. Those of COBE DIRBE (thin line) and FIRAS (dashed line) as sell as of IRAS (rectangular box) are also shown for comparison.

2.2. OBSERVATIONS BY IRTS

The SFU will be sent to a circular orbit with an altitude about 450km and an inclination of 28°. The telescope axis is pointed almost perpendicular to the sun's direction and rotated once a revolution of the SFU synchronized with its orbital motion so as to keep off the earth's radiation. Accordingly, it will make a continuous survey of the diffuse radiation along a great circle of the sky. The four focal plane instruments work together and provide spectroscopic information of the diffuse radiation in a wide range of wavelengths simultaneously.

The NIRS will be used for observations of spectroscopic features of diffuse emission between 1.5μm and 4.0μm. By the rocket observations, Nagoya group has detected an excessive emission in this region which is unexplanable simply by the summation of zodiacal light and integrated light of stellar component in the Galaxy (Noda *et al.* 1992). The same level of the emission has been also detected in the observation by DIRBE of COBE (Mather *et al.* 1990). The observed results by the rocket as well as by the COBE's DIRBE are shown in Fig. 2 together with the observational limits of the NIRS. The high sensitivity of the NIRS would reveal detailed

120

characteristics of the emission features and its spatial variation.

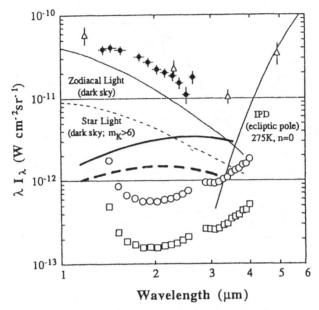

Figure 2. The observed results of near infrared diffuse radiation by the rocket (solid dots, Noda *et al.* 1992) and COBE (open triangles, Hauser 1994). The intensity levels of the theoretical prediction for integrated light of primeval galaxies (Yoshii & Takahara 1988) are also shown by thick solid line (evolutional) and dashed line (non-evolutional). The detection limits of NIRS are indicated by open circles (1F.O.V., 1σ) and by open squares (10sec, 1σ)

Whether the extragalactic component is really present or not should be subject to assessments of the foreground emissions. The better spectroscopic information would be useful for more reliable assessment of the foreground emission. The narrower beam size is certainly advantageous to remove the contribution of stellar light. Stars brighter than 10 mag can be easily removed in the observation. A great number of spectroscopic samples of the galactic stars shall be collected in the course of the survey. They will give a valuable data to estimate the average spectrum of integrated stellar light, which should help better evaluation of the contribution of stellar light.

The MIRS covers the wavelength region between $7 \sim 14\mu$m. This is the best region to study the infrared emission of dust particles, particulaly hot dust (PAH) component in interstellar space which has many characteristic peaks in mid IR region. It also covers the typical spectral band structures of silicates and silicon carbide, which should be present in the interplanetary dust emission. The MIRS will therefore provide very valuable information to assess the contribution of interstellar and interplanetary dust emissions.

The FILM will be used for observations of [O I] (63μm) and [C II] (158μm) lines, mostly of interstellar origin. It is sensitive to very low level emission as cirrus component and thus the observation would be useful for evaluation of the contribution of interstellar dust emission. This is important to check genuine features of the observed anisotropy in extragalactic background emission which is one of the major objectives of the FIRP.

The FIRP is the most relevant channel for observations of cosmic background radiation as well as low temperature component of inerstellar dust emission which has been found in the COBE FIRAS observation (Wright et al. 1992). The detection limit of the FIRP are considerably better than those of FIRAS and DIRBE for far IR channels. The FIRP has a much higher angular resolution than FIRAS and thus it would be able to make deeper search for anisotropies with higher harmonics. It is very interesting that the beam size is just matched for detection of the anisotropy with the similar angular scales where the recent balloon observations claim the presence of anisotropy (Lubin 1994).

The beam size of sub-degrees is a little too large for detection of Sunyaev Zeldovich effect, but could afford a possibility of its detection for relatively large clusters of galaxies or superclusters. The spectral modulation is amplified in the submillimeter region of the Wien side of 3K blackbody spectrum and hence its detection could be easier than the radio wavelength regions so far made by the balloon observations.

The wavelengh bands spreading between 100μm and 800μm are well fit for study of the low temperature dust emission. The higher spatial resoluion and the better detection limit would provide more detailed information on the low temperature dust component than the observations by COBE.

It is a little regret that due to the constraints of sharing of the mission with many other experiments on SFU, the observational period is limited in about three weekes and only about 10% of the sky shall be covered.

3. IRIS (Infrared Imaging Surveyor)

3.1. MISSION CONCEPTS

As an advanced mission following the IRTS, a full scale mission called IRIS is under planning. IRIS will be launched by the new launcher called M-V under development in ISAS. The launching capability of M-V is relatively small; about 1 ton satellite can be sent into a polar orbit at an altitude of about 700km. Under this constraint, we are planning to make survey type observations with a simple instrument and by a simple operation. Although limited in variety of functions, the simple mission could more easily achieve high efficiency and better coordination of observational programs. Moreover, we are planning to maximize its capabilities by introducing new and state of the art techniques in the mission design; mechanical coolers of two-stage Stirling type will be used for better shielding of the liquid helium tank from external heat. As a result, we could accomodate a telescope as large as 70cm in a small and light-mass cryostat. The recent progress of array detectors particularly in near and mid infrared regions would allow to use large format array detectors, which could make it easy to achieve quick and efficient survey in wide areas of the sky.

As the focal plane instrument, two types of observational instruments are under consideration. One is Near/Mid infrared camera with large F.O. V.s, but with fine spatial resolution (2" ~ 5") using large format array detectors. The other is far infrared scanners with multiple bands (3 ~ 4) using one dimensional composite arrays of many single detectors. The general characteristics of the focal plane instruments are summarized in Table 2. Combined with the 70cm telescope, the detection limits of the observations are estimated and illustrated in Fig. 3. They are compared with the intensity levels of some typical targets to be observed.

TABLE 2. Observational instruents on IRIS

	Near/Mid IR Camera		Far IR Scanner	
F.O.V.	16' × 16'		8' × 8'	
Image	2" – 5"		15" – 45"	
Detectors	InSb (3–5μm)	1024 × 1024	Ge:Be (40–60μm)	32 × 2
			Ge:Ga (70–120μm)	24 × 2
	SiAs (6–25μm)	(256 × 256) × 3	Ge:Ga (120–200μm)	12 × 2
			(stressed)	

3.2. OBSERVATIONS BY IRIS

The IRIS is basically designed for survey works with variety of different wavelengths. The surveys are classified into two modes; one is pointing mode, fixing the telescope at a selected direction for relatively long time (maximum 15 min is available in the 900km orbit) and the other is scanning mode, drifting the telescope continuously with the orbital motion. The former is mostly applied to the Near/Mid IR observations and the latter to the Far IR observations.

The followings are typical observatons of each mode;

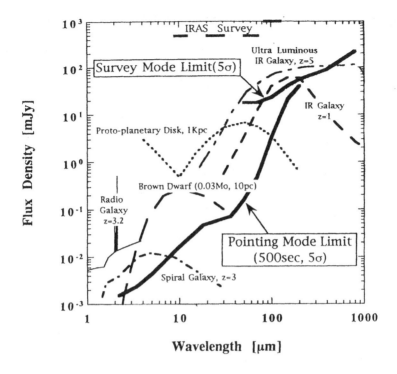

Figure 3. The detection limits of the pointing mode survey and the scanning mode survey. The intensity levels of typical target sources are given for comparison.

1) Observations by Near/Mid IR camera – pointing mode survey
One of the major objectives of the Near/Mid IR survey is a search of primeval galaxies or formation process of normal galaxies in the early history of the universe. Such surveys have been tried from the ground in optical and near infrared regions (Tyson 1988, Cowie *et al.* 1990), In spite of their extensive efforts, however, no evidence has been found yet. This may be explained as due to extinction by dust which should be strong in optical and next infrared regions. If it is the case, observaitons in longer wavelengths should be effective. It can be shown that a primeval galaxy with a moderate luminosity ($10^{12.5}L_{\odot}$) located at cosmological distrance ($z \sim 3$) could be easily found by our survey. The wide range of the wavelength coverage would be useful to identify the observed sources. Among the detected sources, there may be contained brown dwarfs on extra-solar planetary systems. This is another objectives of our survey works with great interests. If we could make such a survey even in relatively small area of several square degrees, substantially large number of the sources should be sampled and used for statistical studies.

2) Observations by Far IR Scanner – scanning mode survey

This observation will be made in scanning mode, that is a continuous sweeping of the sky with one-dimensional arrays. Thus the integration time should be relatively short (0.1sec), but the detecton limit would be improved by an order of magnitude compared with IRAS. The wavelength coverage will be expanded to $200\mu m$. The observation would extend the IRAS survey to much deeper space and fainter sources in the number count of IR galaxies as made by Hacking *et al.* (1987). This would evidently distinguish different models of galaxy evolution in the early history universe.

There have been found extremely luminous object in the IR galaxies such as F10214+4724 or F15307+3252. If such objects prevail to farther distances or earlier phase of the history of the universe, many other sources could be found even at longer distances, $z = 1 \sim 5$. The total number of detectable sources would be of the order of 10^4 for $z > 1$ and 10^2 for $z > 3 \sim 4$ in the full sky survey. A great number of active galaxies as Q.S.O. and Seyfert should be sampled. They would provide valuable data to trace their evolutional trend as well as large scale structure in their distribution.

Both observations by the Near/Mid IR camera and the Far IR scanner should provide cruicial information for studies of formation and evolution of normal galaxies as well as active galaxies and their genetic relation. Large and systematic samplings of the sources would provide the data to assess the contribution of discrete sources to the diffuse background radiation and check whether it is explained solely by the sum of discrete sources or genuinly diffuse component should be left.

The IRIS has been proposed for a scientific mission following the X-ray mission ASTRO-E which is scheduled to be launched in 1999. If it is successfully approved, the IRIS should be launched in early 2000's. A collaboration and coordination with SIRTF mission is under discussion with US groups.

References

Cowie, L. L., Gardner, J. P., Lilly, S. J., & Mclean, I. 1990, ApJ, 360, L1

Hacking, P., Condon, J. J., & Houck, J. R. 1987, ApJ, 316, L16

Hauser, M. G. 1994, this volume

Lubin, P. M. 1994, this volume

Mather, J. C., *et al.* 1990, in IAU colloquium 123, Observatories in Earth Orbit and Beyond, ed. Y. Kondo (Dordrecht: Kluwer), p.9

Mather, J. C. 1994, this volume

Murakami, H., *et al.* 1994, ApJ, 428, 354, and the relevant papers in the same series

Noda, M., *et al.* 1992, ApJ, 391, 456

Penzias, A. A. & Wilson, R. W. 1965, ApJ, 147, 73

Smoot, G. F. 1994, this volume

Tyson, J. A. 1988, ApJ, 96, 1

Wright, E. L., *et al.* 1992, ApJ, 396, L13

Yoshii, Y. & Takahara, F. 1988, ApJ, 326, 1

DEGREE SCALE ANISOTROPY:
CURRENT STATUS

P.M. LUBIN

University of California Santa Barbara
Physics Department
Santa Barbara, CA 93106, USA

AND

Center for Particle Astrophysics
University of California
Berkeley, CA 94720, USA

1. ABSTRACT

The Cosmic Background Radiation gives us one of the few probes into the density perturbations in the early universe that should later lead to the formation of structure we now observe. Recent advances in degree scale anisotropy measurements have allowed us to begin critically testing cosmological models. Combined with the larger scale measurements from COBE we are now able to directly compare data and theory. These measurements promise future progress in understanding structure formation. Because of the extreme sensitivities needed (1-10 ppm) and the difficulties of foreground sources, these measurements require not only technological advances in detector and measurement techniques, but multi spectral measurements and careful attention to low level systematic errors. This field is advancing rapidly and is in a true discovery mode. Our own group has been involved in a series of eleven experiments over the last six years using the ACME (Advanced Cosmic Microwave Explorer) payload which has made measurements at angular scales from 0.3 to 3 degrees and over a wavelength range from 1 to 10 mm. The recent data from these and other measurements will be reviewed as well as some of the challenges and potential involved in these and future measurements.

M. Kafatos and Y. Kondo (eds.), Examining the Big Bang and Diffuse Background Radiations, 125–141.
© 1996 IAU.

126

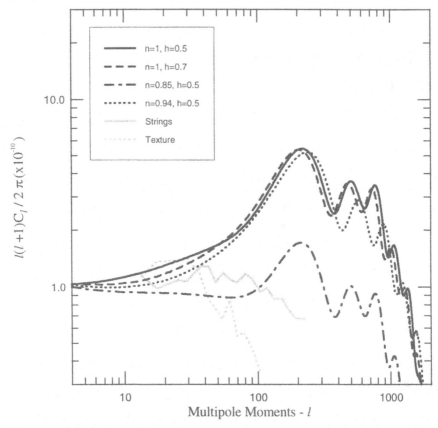

Figure 1. Theoretical CBR Power Spectrum (model results from P. Steinhardt and R. Bond, private communication for a $Q_{rms-ps} = 18\mu K$)

2. Introduction

The Cosmic Background Radiation (CBR) provides a unique opportunity to test cosmological theories. It is one of the few fossil remnants of the early universe to which we have access at the present. Spatial anisotropy measurements of the CBR in particular can provide a probe of density fluctuations in the early universe. If the density fluctuation spectrum can be mapped at high redshift, the results can be combined with other measurements of large scale structure in the universe to provide a coherent cosmological model.

Recent measurements of CBR anisotropy have provided some exciting results. The large scale anisotropy detected by the COBE satellite allows us to normalize the cosmological power spectrum at long wavelengths. The COBE detection at a level of $\Delta T/T = 10^{-5}$ at $10°$ gives us crucial information at scales above 10 degrees about the primordial fluctuations. The

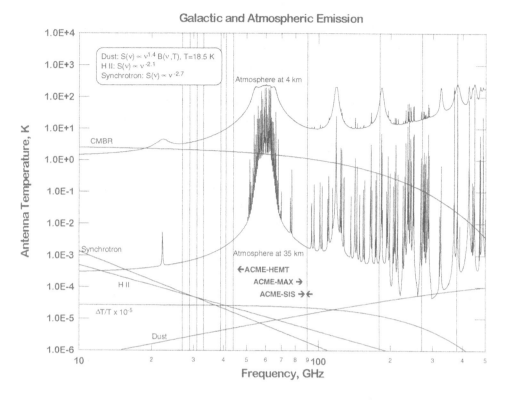

Figure 2. Relevant backgrounds for terrestrial measurements at the South Pole and at 35 Km where ACME observes. Representative galactic backgrounds are shown for synchrotron, Bremsstrahlung and interstellar dust emission as well as the various ACME (center) wavelength bands.

largest scales do not however define the subsequent evolution of the CBR structure in the collapse phase after decoupling. In addition due to the limited number of sky patches available at large scales along with the fact that we are only able to sample our local horizon and not the entire universe (cosmic variance) limits the information available from larger scale measurements. Additional measurements must be made at smaller angular scales. As an example Figure 1 shows the power spectrum expected in a number of models.

3. CBR Anisotropy Measurements

The spectrum of the cosmic background radiation peaks in the millimeter-wave region. Figure 2 shows a plot of antenna temperature vs. frequency, demonstrating the useful range of CBR observation frequencies and the various backgrounds involved. The obvious regime for CBR measurements is in the microwave and millimeter-wave regions.

In the microwave region, the primary extra-terrestrial foreground contaminants are galactic synchrotron and thermal bremsstrahlung emission. Below 50 GHz, both of these contaminants have significantly different spectra than CBR fluctuations. Because of this, multi-frequency measurements can distinguish between foreground and CBR fluctuations (provided there is large enough signal to noise). For example, Figure 2 shows the ACME bands.

Above 50 GHz, the primary contaminant is interstellar dust emission. At frequencies above 100 GHz, dust emission can be distinguished from CBR fluctuations spectrally, also using multi-frequency instruments.

At all observation frequencies, extra-galactic radio sources are a concern. For an experiment with a collecting area of $1 m^2$ (approximately a $0.5°$ beam at 30 GHz for sufficiently under-illuminated optics), a 10 mJy source will have an antenna temperature of about 10 μK, which will produce a significant signal in a measurement with a sensitivity of $\Delta T/T \approx 1 \times 10^{-6}$. Extra-galactic radio sources have the disadvantage that there is no well known spectrum which describes the whole class. For this reason, measurements over a very large range of frequencies and angular scales are required for CBR anisotropy measurements in order to achieve a sensitivity of $\Delta T/T \approx 1 \times 10^{-6}$.

4. Instrumental Considerations

Sub-orbital measurements differ from orbital experiments in at least one important area, namely our terrestrial atmosphere is a potential contaminant. A good ground-based site like the South Pole has an atmospheric antenna temperature of 5 K at 40 GHz, for example. For a measurement to reach an error of $\Delta T/T \approx 1 \times 10^{-6}$, the atmosphere must remain stable over 6 orders of magnitude. In addition to this, the atmosphere will contribute thermal shot noise. At balloon altitudes, atmospheric emission is 3-4 orders of magnitude lower and much less of a concern. In addition, the water vapor fraction is extremely low at balloon altitude. Satellite measurements avoid this problem altogether. Another consideration for CBR anisotropy measurements is the sidelobe antenna response of the instrument. Astronomical and terrestrial sources away from foresight can contribute significant signals if the antenna response is not well behaved. Under-illuminated optical elements and off-axis low blockage designs are typically employed for the task. The sidelobe pattern can be predicted and well controlled with single-mode receivers, but appears to be viable for multi-mode optics as well. Even with precautions, sidelobe response will remain an area of concern for all experiments.

Most of the measurements to be discussed are limited by receiver noise when atmospheric seeing was not a problem. It is possible to build receivers today with sensitivities of 200-400 $\mu K \sqrt{s}$ using HEMTs or bolometers. A balloon flight obtaining 10 hours of data on 10 patches of sky, for example, could achieve a 1 σ sensitivity of 6.7 μK or $\Delta T/T = 2.5 \times 10^{-6}$ *per pixel* using one such detector.

To map CBR anisotropy with a sensitivity of $\Delta T/T = 1 \times 10^{-6}$ requires more integration time, lower noise receivers or multiple receivers. A 14-day, long duration balloon flight launched from Antarctica could result in a per pixel sensitivity of $\Delta T/T = 5 \times 10^{-7}$, if 10 patches could be observed with a single detector element or $\Delta T/T = 5 \times 10^{-6}$ on 1000 patches as another example. Multiple detectors obviously help here.

Measurements from the South Pole are also very promising. The large atmospheric emission (compared to the desired sensitivity level - few million times larger!) is of great concern and based upon our experience, even in the best weather, there is significant atmospheric noise. Estimated single difference atmospheric noise with a 1.5 degree beam is about 1 $mK\sqrt{s}$ at 30 GHz during the best weather. This added noise, as well as the overall systematic atmospheric fluctuations, make ground-based observations challenging but so far possible, and, in fact, yielding the most sensitive results.

Another approach to the problem is to use very low noise receivers and obtain the necessary integration time by flying long duration balloons. These receivers can be tested from ground-based observing sites like the South Pole. Should the long duration balloon effort prove inadequate, the only means toward the goal of mapping CBR anisotropy at this level will be a dedicated satellite. Again, the receivers on such a satellite would have to be low noise. The minimal cryogenic requirements for HEMT (High Electron Mobility Transistor) amplifiers make them an obvious choice for satellite receivers, but bolometric receivers using ADR coolers or dilution refrigerators offer significant advantages at submillimeter wavelengths.

5. History of the ACME Experiments

In 1983, with the destruction of the 3 mm mapping experiment (Lubin et al. 1985), we decided to concentrate on the relatively unexplored degree scale region. Motivated by the possibility of discovering anisotropy in the horizon scale region where gravitation collapse would be possible and with experience using very low noise coherent detectors at balloon altitudes, we started the ACME program. A novel optical approach, pioneered at Bell Laboratories for communications, was chosen to obtain the extreme sidelobe rejection needed. In collaboration with Robert Wilson's group at Bell Labs, a 1 meter off-axis primary was machined. A lightweight, fully-automated, stabilized, balloon platform capable of directing the 1 meter off-axis telescope was constructed. As the initial detector we chose a 3 mm SIS receiver. Starting with lead alloy SIS junctions and GaAs FET pre-amplifiers we progressed to Niobium junctions and a first generation of HEMTs to achieve chopped sensitivities of about 3 mK $\sqrt{s}$ in 1986 with a beam size of 0.5 degrees FWHM at 3 mm.

The first flight was in August 1988 from Palestine, Texas. Immediately afterwards, ACME was shipped to the South Pole for ground-based observations. The results were the most sensitive measurements to date (at that time) with 60 μK errors per point at 3 mm. The primary advantage of the narrow band coherent approach is illustrated in Figure 2 where we plot atmospheric emission versus frequency for sea level, South Pole (or 4 km mountain top) and 35 km balloon altitudes. With a proper choice of wavelength and bandpass, extremely low residual atmospheric emission is possible. (Total < 10 mK. The differential emission, over the beam throw, is much smaller.) Another factor of 10 reduction is possible in the "troughs" in going to 40 km altitude. The net effect is that atmospheric emission does not appear to be a problem in achieving μK level measurements, if done appropriately.

Subsequently, ACME has been outfitted with a variety of detector including direct amplification detectors using HEMT technology. These remarkable devices developed largely for communications purposes are superb at cryogenic temperatures as millimeter wavelength detectors. Combining relatively broad bandwidth (typically 10-40%) with low noise characteristics and moderate cooling requirements (including operation at room temperature) they are a good complement to shorter wavelength bolometers allowing for sensitive coverage from 10 GHz to 200 GHz when both technologies are utilized. The excellent cryogenic performance is due in large part to the efforts of the NRAO efforts in amplifier design (Pospieszalski 1990). We have used both 8-12 mm and 6-8 mm HEMT detectors on ACME, these observations being carried out from the South Pole in the 1990 and 1993 seasons. The beam sizes are 1.5 degrees and 1 degree FWHM for the 8-12 and 6-8 mm HEMTs respectively. Detectors using both GaAs and InP technology have been used. The lowest noise we have achieved to date is 10 K at 40 GHz, this being only 3.5 times the quantum limit at this frequency. These devices offer truly remarkable possibilities. Figure 3 shows the basic experiment configuration.

There have been a total of eleven ACME observations/flights from 1988 to 1994. Over twenty articles and proceedings have resulted from these measurements as well as seven Ph.D. theses. A summary of the various

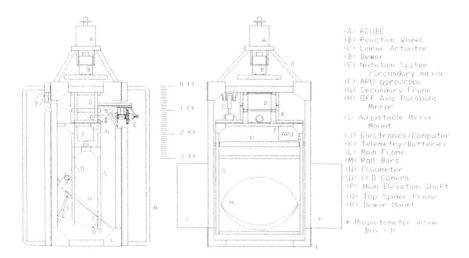

Figure 3. Basic ACME configuration

observations is given in Table I.

TABLE 1. CBR Measurements made with ACME

Date	Site	Detector System	Beam FWHM (deg)	Sensitivity
1988 Sep	Balloon^P	90 GHz SIS receiver	0.5	4 mK s$^{1/2}$
1988 Nov-1989 Jan	South Pole	90 GHz SIS receiver	0.5	3.2
1989 Nov	Balloon^{FS}	MAX photometer (3, 6, 9, 12 cm^{-1}) ^{3}He	0.5	12, 2, 5.7, 7.1
1990 Jul	Balloon^P	MAX photometer (6, 9, 12 cm^{-1}) ^{3}He	0.5	0.7, 0.7, 5.4
1990 Nov-1990 Dec	South Pole	90 GHz SIS receiver	0.5	3.2
1990 Dec-1991 Jan	South Pole	4 Channel HEMT amp (25-35 GHz)	1.5	0.8
1991 Jun	Balloon^P	MAX photometer (6, 9, 12 cm^{-1}) ^{3}He	0.5	0.6, 0.6, 4.6
1993 Jun	Balloon^P	MAX photometer (3, 6, 9, 12 cm^{-1}) ADR	0.55-0.75	0.6, 0.5, 0.8, 3.0
1993 Nov-1994 Jan	South Pole	HEMT 25-35 GHz	1.5	0.8
1993 Nov-1994 Jan	South Pole	HEMT 38-45 GHz	1.0	0.5
1994 Jun	Balloon^P	MAX photometer (3, 6, 9, 14 cm^{-1}) ADR	0.55-0.75	0.4, 0.4, 0.8, 3.0

Sensitivity does not include atmosphere which, for ground-based experiments, can be substantial.

P - Palestine, TX
FS - Fort Sumner, NM

6. The MAX Experiment

During the construction of ACME, a collaboration was formed between our group and the Berkeley group (Richards/Lange) to fly bolometric detectors on ACME. This fusion is called the MAX experiment and subsequently blossomed into the extremely successful Center for Particle Astrophysics' CBR effort. Utilizing the same basic experimental configuration as other ACME experiments, MAX uses very sensitive bolometers from about 1-3 mm wavelength in 3 or 4 bands. Flown from an altitude of 35 km, MAX has had five very successful flights. The first MAX flight (second ACME flight) occurred in June 1989 using ^{3}He cooled (0.3 K) bolometers, and the most recent flight occurred in June 1994 using ADR (Adiabatic Demagnetization Refrigeration) cooled bolometers. All the MAX flights have had a beam size of near 0.5 degrees.

7. Evidence for Structure Prior to COBE

Prior to the COBE launch, ACME had made two flights and one South Pole expedition. Prior to the April 92 COBE announcement, ACME had flown four times and made two South Pole trips for a total of seven measurements. Our 1988 South Pole trip with ACME outfitted with a sensitive SIS (Superconductor-Insulator-Superconductor) receiver resulted in an upper limit of $\Delta T/T \lesssim 3.5 \times 10^{-5}$ at 0.5° for a Gaussian sky. This was tantalizingly close to the "minimal predictions" of anisotropy at the time and as we were to subsequently measure, just barely above the level of detectability. In the fall of 1989, we had our first ACME-MAX flight with a subsequent flight the next summer (so called MAX-II flight). Remarkably, when we analyzed the data from this second flight, we found evidence for structure in the data consistent with a cosmological spectrum. This was data taken in a low dust region and showed no evidence for galactic contamination. The data in the Gamma Ursa Minoris region ("GUM data") was first published in Alsop et al. (1992) **prior** to the announcement of the COBE detections. At the time our most serious concern was of atmospheric stability so we decided to revisit this region in the next ACME flight in June 1991. In the meantime, ACME was shipped to the South Pole in October 1990 for another observing run, this time with both an SIS detector and a new and extremely sensitive HEMT receiver. At scales near 1 degree, close to the horizon size, results from the South Pole using the ACME (Advanced Cosmic Microwave Experiment) with a High Electron Mobility Transistor (HEMT) based detector placed an upper limit to CBR fluctuations of $\Delta T/T \leq 1.4 \times 10^{-5}$ at 1.2° (Gaier et al. 1992). This data set has significant structure in excess of noise with a spectrum that was about 1.4σ from flat (Gaier 1993). This upper limit for a Gaussian autocorrelation function sky was computed from the highest frequency channel. Since the data is taken in a step scan and not as a continuous scan it is not possible to eliminate the possibility that the structure seen is cosmological since the beam size varies from channel to channel. Under the assumption that the structure seen is cosmological, a four channel average of the bands yields a detection at the level of $\Delta T/T \simeq 1 \times 10^{-5}$ (Bond 1993). Interestingly, this is about the same level seen in another SP 91 scan (see next) as well as that observed in the nearly same region of sky observed in the SP 94 data (Gundersen et al. 1995).

Additional analysis of the 1991 South Pole data using another region of the sky and with somewhat higher sensitivity shows a significant detection at a level of $\Delta T/T = 1 \times 10^{-5}$ (Schuster et al. 1993). The structure observed in the data has a relatively flat spectrum which is consistent with CBR but could also be Bremsstrahlung or synchrotron in origin. This data also sets an upper limit comparable to the Gaier et al. upper limit, but can also be used to place a lower limit to CBR fluctuations of $\Delta T/T \geq 8 \times 10^{-6}$, if all of the structure is attributed to the CBR. The 1σ error measured per point in this scan is 14 μK or $\Delta T/T = 5 \times 10^{-6}$. Per pixel, this is the most sensitive CBR measurement to date at any angular scale. Combining these two scans in a multichannel analysis results in a detection level slightly above 1×10^{-5} (Bond 1993). The relevant measurements just prior to the COBE announcement are summarized in Figure 4. With apparent detection and good upper limits at degree scales, what was needed was large scale normalization. This was provided by the COBE data in 1992 and, as shown in Figure 5, the degree scale measurements were consistent with COBE given the errors involved. Without the large scale normalization of the COBE data, it was hard to reconcile the apparently discordant data. However, with the refinement in theoretical understanding and additional data, the pre-COBE ACME data now are seen to be remarkably consistent with the post-COBE data.

8. Results

There have been a total of eleven ACME observations/flights from 1988 to 1994. ACME articles by Meinhold & Lubin (1991), Meinhold et al. (1992), ACME-HEMT articles by Gaier et al. (1992), Schuster et al. (1993), Gundersen et al. (1995) and ACME-MAX articles by Fischer et al. (1992, 1995), Alsop et al. (1992), Meinhold et al. (1993), Gundersen et al. (1993), Devlin et al. 1994 and Clapp et al. 1994 summarize the results to date.

Significant detection by ACME at 1.5 degrees is reported by Schuster et al. (1993) at the 1×10^{-5} level and by Gundersen et al. (1993) at 0.5 degrees at the 4×10^{-5} level in adjacent issues of *ApJ Letters*. The lowest

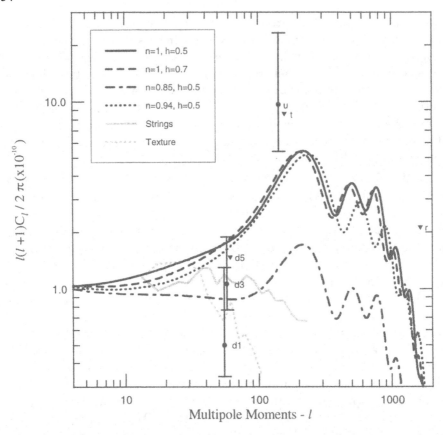

Figure 4. ACME CBR Power Spectrum data prior to the COBE detection. Theoretical curves are from Figure 1. See KEY in Figure 5 caption.

error bar per point of any data set to date is in the Schuster et al. 1.5° data with 14 μK while the largest signal to noise signal is in Gundersen et al.(1993) with about a 6 σ detection (at the peak). Recently Wollack et al. (1993) reported a detection at an angular scale of 1.2 degrees of about 1.4×10^{-5} consistent with Schuster et al., Gaier et al. and the combined 9 + 13 pt. analysis using a detector and beam size nearly identical to ours. Remarkably this result is taken in a completely different region of the sky and at lower galactic latitude and yields similar results. A conspiracy to yield comparable results in very different parts of the sky from point sources or sidelobe spill over is always possible. When one takes into account the similar level of the COBE detection at larger scales it would seem to require multiple conspiracies however. Additional data taken at the South Pole by ACME in 1993/94 ("SP94" data) in a region of the sky close to the SP91 area but with additional HEMT detectors from 25 - 45 GHz in seven bands

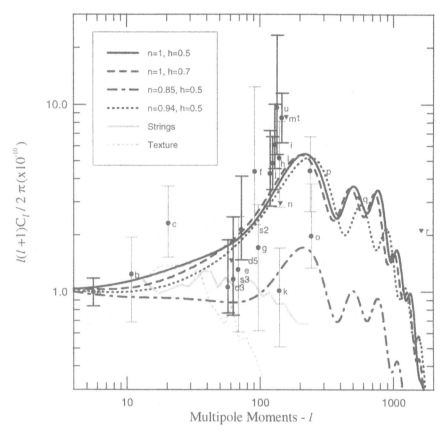

Figure 5. Figure 5: Recent ACME Results (in BOLD) along with results from other groups. Key: a-COBE, b-FIRS, c-Tenerife, d1-SP91 9 pt. 4 channel analysis-Bond 93, d3-SP91 9+13pt 4 channel analysis-Bond 93, d5-SP91 9 pt. Gaier et al. 92, e-Big Plate, f-PYTHON, g-ARGO, h-MAX4-Iota Dra, i-MAX4-GUM, j-MAX4-Sig Herc, k-MSAM2, l-MSAM2, m-MAX3-GUM, n-MAX3-mu Peg, o-MSAM3, p-MSAM3, q-Wh. Dish, r-OVRO7, s2-SP94-Q, s3-SP94-Ka, t-SP89, u-MAX2-GUM, many from Steinhardt and Bond, private communication.

and with beam sizes from 1.0 - 1.7 degrees FWHM, yield results consistent with a CDM model (and others) normalized to COBE. The SP94 data are consistent with the SP91 results (Gundersen et al. 1995) as shown in Figure 5. At 0.5 degrees, the MSAM group reports detection of a "CBR component" at a level of about 2×10^{-5} but with "point like sources" that are being reanalyzed and which may contribute additional power. Our results from the June 1993 ACME-MAX flight give significant detections at the $3 - 4 \times 10^{-5}$ level at angular scales near 0.5 degrees.

The most recent ACME-MAX data have been in low dust regions so that no subtraction of dust was needed. In one scan, the μ-Pegasi region, there was enough dust to provide a good calibration of high galactic latitude interstellar dust emission (Meinhold et al. 1993). Interestingly, the residual "CBR component" was anomolously low compared to the other regions surveyed. Whether this is indicative of other issues, such as non-gaussian fluctuations, or is just due to limited sampling statistics is unclear at this time.

The most recent ACME-MAX flight in June 1994 included two more low dust regions and a revisit of the μ-Pegasi region. The data is currently being analyzed (Lim et al. 1995, Tanaka et al. 1995).

It is remarkable that over a broad range of wavelengths, very different experiments using a variety of technologies and observing in different parts of the sky report degree scale detection at the one to a few $\times 10^{-5}$ level. ACME, in particular, has now been used to measure structure from 25 - 250 GHz and from 0.4 - 2 degrees that is in reasonably good agreement with a CDM power spectrum model. The agreement of the ACME-HEMT data with other experiments (notably Big Plate at Sasskatoon) up to ℓ of about 75 is very good, as can be seen from Figure 5. At 0.5 degree scales (ℓ about 150) the agreement of ACME-MAX and MSAM data is marginal and will hopefully be clarified soon. It is important to keep in mind that the statistical and sampling errors need to be taken into account in any comparison between data sets and between data and theory.

In any case, 1992 and 1993 were clearly historical years in cosmology and CBR studies in particular. The ACME results along with the results of other groups are shown in Figure 5. As can be seen by comparison to Figure 4 which was the data prior to COBE, the pre- and post-COBE data are reasonably consistent given the errors. The deluge of theoretical results and scrutiny that followed COBE was a boon for degree scale results giving us a theoretical insight we lacked just a few years ago. The current ACME degree scale results are summarized in Table II.

9. Detector Limitations - Present and Fundamental

Detectors can be broadly characterized as either coherent or incoherent being those that preserve phase or not, respectively. Masers, SIS and HEMTs are coherent. Bolometers are incoherent. SIS junctions can also be run in an incoherent video detector mode. Phase preserving detectors inherently must obey an uncertainty relationship that translate into a minimum detector noise that depends on the observation frequency, the so called quantum limit. Incoherent detectors do not have this relationship but are ultimately limited by the CBR background itself. At about 40 GHz, these fundamental limits are comparable. Current detectors are not at these fundamental limits, though they are within an order of magnitude for both HEMTs and

TABLE 2. Recent ACME Degree Scale Results

Publication	Configuration	Beam FWHM (deg)	$\Delta T/T \times 10^{-6}$ (GACF)**	ℓ	$C_\ell \ell(\ell+1)/2\pi$* ($\times 10^{-10}$)
Meinhold & Lubin 91	ACME-SIS SP89	0.5	< 35	145	< 8.6
Alsop et al. 92	ACME-MAX-II (GUM)	0.5	45^{+57}_{-26}	143	$9.6^{+13.7}_{-4.2}$
Gaier et al. 92	ACME-HEMT SP91	1.5	< 14	58	< 1.5
Meinhold et al. 93	ACME-MAX-III (μ Peg - upper limit)	0.5	< 25	143	< 2.96
Meinhold et al. 93	ACME-MAX-III (μ Peg - detection)	0.5	15^{+11}_{-7}	143	
Schuster et al. 93	ACME-HEMT SP91	1.5	9^{+7}_{-4}	58	$0.76^{+0.80}_{-0.21}$
Bond 93	SP91 4 channel 9+13 pt. Analysis	1.5		58	$1.06^{+0.83}_{-0.29}$
Bond 93	SP91 4 channel 9 pt. Analysis	1.5		58	$0.5^{+0.80}_{-0.16}$
Gundersen et al. 93	ACME-MAX-III (GUM)	0.5	42^{+17}_{-11}	143	$8.5^{+3.0}_{-2.2}$
Devlin et al. 94	ACME-MAX-IV (GUM)	0.55-0.75	37^{+19}_{-11}	129	$6.1^{+3.9}_{-1.5}$
Clapp et al. 94	ACME-MAX-IV (Iota Draconis)	0.55-0.75	33^{+11}_{-11}	129	$4.9^{+1.9}_{-1.4}$
Clapp et al. 94	ACME-MAX-IV (Sigma Hercules)	0.55-0.75	31^{+17}_{-13}	129	$4.3^{+3.0}_{-1.4}$
Gundersen et al. 95	ACME-HEMT SP94	1		73	$2.14^{+2.00}_{-0.66}$
Gundersen et al 95	ACME-HEMT SP94	1.5		58	$1.17^{+1.33}_{-0.42}$
Lim et al. 94	ACME-MAX-V	0.5	in progress		
Tanaka et al. 94	ACME-MAX-V	0.5	in progress		

* from P. Steinhardt & R. Bond, priv. communication. 1σ errors, upper limits are 95%
** GACF=Gaussian Autocorrelation Function - Upper limits and error bands are 95%

bolometers when used over moderate bandwidths. Here we include all effects including coupling efficiencies. Currently both InP HEMTs and ADR and ^{3}He cooled bolometers exhibit sensitivities of under 500 μK s$^{1/2}$. This assumes no additional atmospheric noise, true at balloon altitudes. For ground-based experiments, even at the South Pole, atmospheric noise is significant however.

Significant advances have been made in recent years in detector technology with effective noise dropping by over an order of magnitude over the past decade. With moderate bandwidths the fundamental limits for detectors are about a factor of 5 below the current values, so fundamental technology development is to be highly encouraged for both coherent and incoherent detectors.

With current detectors, achieving 1 μK sensitivity requires roughly one day per pixel for a single detector. This is appropriate for detector limited, not atmospheric limited, detection. This would be appropriate for balloon altitudes.

Small arrays of detectors are currently planned for several experiments. This should allow μK per pixel sensitivity over 100 pixels in time scales of a few weeks, suitable for long duration ballooning or polar observations. If the fundamental detector limits could be achieved, the effective time would drop to about a day. Factors of 2-3 reduction in current detector noise are not unreasonable to imagine over the next five years, and if they could be achieved, the above time scale would drop to less than a week. Multiple telescopes are also possible. If we are willing to accept a goal of 3 μK per pixel (1 part per million of the CBR) instead of 1 μK then roughly 10 times as many pixels can be observed for the same integration time allowing significant maps to be made from balloon-borne detectors. A 10 μK error per pixel measurement would allow 100 times as many pixels to be measured in the same time. As we learn more about the structure of the CBR and about the nature of low level foreground emission the choice of sensitivity for a given angular scale will become clearer.

10. Spectrum Measurements

A related area of interest that could yield interesting cosmology in the next few years is the long wavelength spectrum. Although the spectrum of the CBR has been extremely well characterized by the COBE FIRAS experiment in the millimeter wavelength range. However, in the range of about 1-100 GHz, where interesting physical phenomenon may distort the spectrum, much work remains to be done; particularly, at the longest wavelengths. Fortunately, the atmospheric emission is quite low over much of this range from both good ground-based sites and extremely low at balloon altitudes. Galactic emission and sidelobe contamination are of primary concern at the longest wavelengths, but it is expected that a number of ground-based and possibly balloon-borne experiments will be performed and should be encouraged.

A recent balloon-borne experiment, Schuster et al. (1994), is an example of what might be done in the future from balloon spectrum experiments. With all cryogenic optics and no windows, this experiment measured $T = 2.71 \pm 0.02$ K at 90 GHz with negligible atmospheric contamination ($\sim$ a few mK) and no systematic corrections. Errors of order 1 mK should be obtainable. The basic configuration could be extended to longer wavelengths where much remains to be done. In particular coherent measurements at 10 - 50 GHz from a balloon could be done. The BLAST (Balloon Absolute Spectrometer)-ARCADE (Absolute Radiometer for Cosmology,

Astrophysics, and Diffuse Emission) experiment, a joint UCSB-Goddard balloon borne experiment will attempt to exploit the low atmospheric emission available from balloon altitudes using coherent HEMT detectors in the 10-30 GHz range. Accuracies of under 1 milliKelvin may be feasible. This will allow extremely sensitive measurements of long wavelength distortions in the CBR should they exist. Since the spectral deviation rises rapidly at long wavelengths as does the galactic emission from synchrotron radiation, measurements in the 5-20 GHz range will be particularly useful.

11. Polarization

Very little effort has been directed towards the measurement of the polarization of the CBR compared to the effort in direct anisotropy detection. In part, this is due to the low level of linear polarization expected. Typically, the polarization is only 1-30% of the anisotropy and depends strongly on the model parameters (Steinhardt 1994). This is an area which in theory can give information about the reionization history, scalar and tensor gravity wave modes and large scale geometry effects. In the future, this may be a very fruitful area of inquiry particularly at degree angular scales.

12. To Space

The question of whether or not a satellite is needed to get the degree scale "answer" is complex. There is no question that the measurements can be done from space and given sufficient funding this is definitely the preferable way. It is unclear at this time what the limitations from sub-orbital systems will be and vigorous work is planned for sub-orbital platforms over the next decade. The galactic and extragalactic background problem remains the same for orbital and sub-orbital experiments. The atmosphere can be dealt with, particularly from balloon-borne experiments, with careful attention to band passes. Per pixel sensitivities in the μK region are achievable with current and new technologies, HEMTs, and bolometers over hundreds to thousands of pixels. The major issue will be control of sidelobes and getting a uniform dataset. Ideally full sky coverage would be best and this is one area where a long term space based measurement would be ideal. In the control of sidelobe response a multi AU orbital satellite would be a major advance. This advantage is lost for near Earth orbit missions, however. European efforts such as SAMBA and COBRAS and US efforts such as PSI, MAP and FIRE are examples of possible future space based efforts. A low cost precursor mission such as the university led COFI satellite is an example of an economical approach to proving HEMT technology in space for a possible future effort. By the end of the millennium, degree scale maps over a reasonable fraction of the sky at the 10^{-6} level should be possible from balloons and the ground. The potential knowledge to be gained is substantial, and I can think of few areas of science where the potential "payoff" to input (financial and otherwise) is so high.

13. Acknowledgements

This work was supported by the National Science Foundation Center for Particle Astrophysics, the National Aeronautics and Space Administration, the NASA Graduate Student Research Program, the National Science Foundation Division of Polar Programs, the California Space Institute, the University of California, and the U.S. Army. Its success is the result of the work of a number of individuals, particularly the graduate students and post doc's involved, in particular Peter Meinhold, Alfredo Chincquanco, Jeffery Schuster, Michael Seiffert, Todd Gaier, Tim Koch, Joshua Gundersen, Mark Lim, John Staren, Thyrso Villela, Alex Wuensche, and Newton Figueiredo. The bolometric portions of the ACME program (MAX) were in collaborations with Paul Richards' and Andrew Lange's groups at UCB and in particular with Mark Fischer, David Alsop, Mark Devlin, Andre Clapp and Stacy Tanaka. The entire ACME effort would not have been possible without the initial support and vision of Nancy Boggess and Don Morris. Dick Bond and Paul Steinhardt supplied much appreciated theoretical input to the data analysis and interpretation. The exceptional HEMT amplifier was provided by NRAO and in particular by Marion Pospieszaski and Michael Balister. Robert Wilson, Anthony Stark, and Corrado Dragone, all of AT&T Bell Laboratories, provided critical support and discussion regarding the early design of the telescope and receiver system as well as providing the primary mirror. I would like to thank Bill Coughran and all of the South Pole support staff for highly successful 88-89, 90-91 and 93-94 polar summers. In addition, I want to acknowledge the crucial contributions of the entire team of the National Scientific Balloon Facility in Palestine, Texas for their continued excellent support. Finally, I would also like to thank my wife Georganne for providing the loving support to make this program a reality.

References

Alsop, D.C., et al. 1992, ApJ, 317, 146
Bond, R. 1993, CMB Workshop, Capri, Italy
Cheng, E.S., et al. 1993, ApJ Lett, submitted
Clapp, A., et al. 1994, ApJ Lett, submitted
Devlin, M., et al. 1994, ApJ Lett, submitted
Fischer, M., et al. 1992, ApJ, 388, 242
Fischer, M., et al. 1995, ApJ, submitted
Gaier, T., et al. 1992, ApJ, 398, L1

Gaier, T. 1993, Ph.D. Thesis, UCSB

Gundersen, J.O., et al. 1993, ApJ, 413, L1

Gundersen, J.O., et al. 1995, ApJ Letters, submitted

Lim, M., et al. 1995, ApJ Letters, in preparation

Lubin, P., et al. 1985, ApJ, 298, L1

Meinhold, P.R., & Lubin, P.M. 1991, ApJ, 370, L11

Meinhold, P., et al. 1992, ApJ, 406, 12

Meinhold, P., et al. 1993, ApJ, 409, L1

Pospieszalski, M.W., et al. 1990, IEEE MTT-S Digest, 1253

Schuster, J., et al. 1993, ApJ, 412, L47

Schuster, J., et al. 1994, in preparation

Smoot, G.F., et al. 1992, ApJ, 396, L1

Steinhardt, P. 1994, Proc. of 1994 Snowmass Workshop

Tanaka, S., et al. 1995, ApJ Letters, in preparation

Wollack, E., et al. 1994, ApJ, 419, L49

GALAXY MOTIONS IN THE NEARBY UNIVERSE

JOHN P. HUCHRA

Harvard-Smithsonian Center for Astrophysics
60 Garden Street, Cambridge, MA 02138 USA

Abstract. In this paper we review the history of the search for and study of the motions of nearby galaxies with respect to the Hubble Flow. The current status of the field is that (1) convincing infall has been detected into dense clusters, especially the Virgo cluster, (2) the microwave background direction is moderately well aligned with the measured flow nearby but not apparently on larger scales, and (3) there is good but not perfect consistency between the nearby density fields and velocity fields. Particular problems exist in the different Ω's required to fit the density field derived from optically selected and IRAS (60μ) selected galaxy samples.

1. HISTORY

The study of the uniformity of the expansion of the universe has a long an varied history which has only achieved a well defined focus in the last decade. Although there were several early attempts to search for distortions in the local flow field (e.g. Rubin 1951), the mythology that the local Hubble flow was basically smooth continued to hold sway in the astronomical community at-large throughout the early 1980's (e.g. Sandage and Hardy 1972). despite the work of Rubin and collaborators (Rubin *et al.* 1976a, 1976b).

This view changed dramatically in the late 1970's. There was a major paradigm shift brought on by two related events. First, the Cosmic Microwave Background (CMB) dipole was convincingly detected by a number of groups (e.g. Cheng *et al.* 1979; Smoot and Lubin 1979). This detection was so convincingly a dipole that, despite attempts to explain it via emission from galaxies or other cosmological effects, there was little doubt that a motion of the Solar sytem of ∼300 km/sec with respect to the CMB frame had been detected and that a transformation of that motion into the

143

M. Kafatos and Y. Kondo (eds.), Examining the Big Bang and Diffuse Background Radiations, 143–155.

144

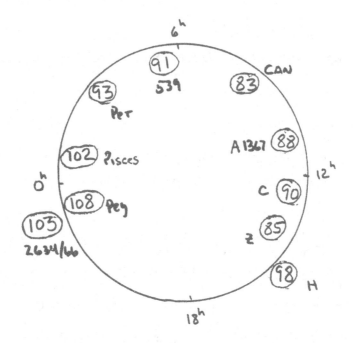

Figure 1. The Hubble constant towards medium distant clusters measured in the Arecibo declination zone. The "dipole" is indicative of a motion of ∼480 km/s towards Virgo — consistent with the CMB vector in this band. (based on Aaronson *et al.* 1980).

reference frame of the Local Group gave a motion w.r.t. the CMB frame of ∼600 km/s.

Second, evidence from high quality measurements of the relative distances of galaxies (made primarily to measure the Hubble Constant) began to indicate that motions in the nearby universe might exist. Aaronson *et al.* (1980) found that Infrared Tully-Fisher distance measurements to medium distant (4-10,000 km/s) clusters of galaxies gave a Hubble constant nearer 95 km/s/Mpc than the ∼65 km/s/Mpc obtained for the nearby Virgo and Ursa Major clusters (Figure 1). This could be explained by an infall of the Local Group into Virgo of a few hundred km/sec relative to a uniform Hubble flow. Furthermore, there was a dipole in the H_0 measurements to the distant clusters that could also be explained by such a motion. This effect was also seen in measurements of relative distances to elliptical galaxies made by Tonry and Davis (1980) as well as other investigators.

The infall into Virgo, which had been predicted by numerous authors since the mid-1950's (de Vaucouleurs 1958; Peebles 1976; Silk 1974) was finally confirmed by Aaronson *et al.* (1982) with a detailed study of the flow field w.r.t. the Virgo Cluster interior to 3,000 km/s (Figure 2). Aaronson

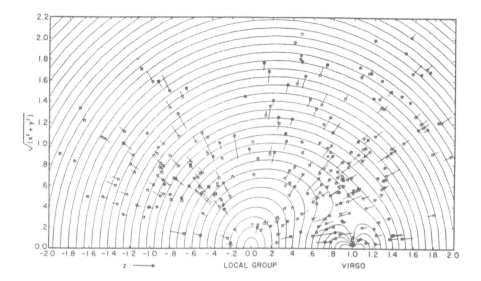

Figure 2. The infall pattern around the Virgo cluster matched to velocity-distance residuals from a uniform Hubble expansion (from Aaronson *et al.* 1982).

et al. detected the LG motion w.r.t. Virgo at the 5+ σ level and were able to split it into two components, a "pattern" infall of $\sim$250 km/s and a random component of $\sim$80 km/s.

All of these results were echoed by the detection of significant anisotropy in the galaxy distribution by the redshift surveys of the time (Sandage and Tammann 1981; Davis *et al.* 1982; Kirshner *et al.* 1981; c.f. Geller and Huchra 1989 for a more detailed review). Combined with the realization that the Universe not empty, i.e. that $\Omega \neq O$ and more likely between 0.2 and 1.0, this was strong proof that flows like the infall to Virgo *must* exist in the nearby Universe.

However, there was still a problem. Simply stated:

$$\overrightarrow{Virgo} \neq \overrightarrow{CMB} \text{ by } \sim 45°,$$

$$\text{and} \quad V_{Virgo} \neq V_{CMB} \text{ by } 350km/s.$$

The Virgo infall velocity and Virgo direction cannot be completely responsible for the CMB motion. This problem caused a number of people (Shaya 1984; Tully and Shaya 1984; Lilje, Yahil and Jones 1986; Tammann and Sandage 1985) to suggest that there was an additional density concentration causing the remaining component of the motion roughly in the direction of the Hydra-Centuarus Supercluster. Most of these authors suggested that Hydra-Cen might even be the gravitational cause of the additional component of our motion.

The solution to this problem seemed to be in sight when, in 1986, the "Seven Sammurai" (Burstein *et al.* 1986; Lynden-Bell *et al.* 1988) claimed a detection of a "Great Attractor" – an gravitating mass causing a <u>bulk</u> flow roughly towards Hydra-Centaurus and the CMB direction, from D_n-σ measurements of $\sim$400 elliptical galaxies. However, the details of the motion were way off expectations — Hydra and Centaurus were themselves moving towards a point not coincident with their center-of-mass and large deviations, in excess of 1000 km/s, were seen in the flow field. Aarosnon *et al.* (1986) continued their work with the IRTF relation to map the flow field using clusters of galaxies rather than individual objects (thus beating down the relative distance errors by $\sqrt{n}$) and also found evidence for both an infall to Virgo and motion towards Hydra-Cen.

These works have set the stage for a large number of studies of local motions aimed at answering a range of fundamental questions related to flows, large-scale structure, Ω and cosmology in general.

2. FUNDAMENTAL QUESTIONS

The fundamental questions we are asking are relatively simple to state:

1. What is the local Flow Field? (a) Does it converge, and if so on what scale? (b) Can we explain the CMB dipole locally? (c) How does it effect the determination of the Hubble Constant, H_0?

2. What is the local Density Field? (a) Can it explain the flow field? (b) If so, what Ω is required to do so?

Despite their simplicity, these are not easy questions to answer. The answer to the first set depends on the existence of accurate relative distance indicators. The best available are the Surface Brightness Fluctuation technique pioneered by Tonry (Tonry and Schneider 1988). This technique is gradually taking the place of the D_n-σ technique for nearby early type galaxies, but is time consuming (telescope time!) and hard to use at large distances (although HST can be used, again, if sufficient time can be made available). Other reasonably accurate methods include the IR and red (usually R band, 6500Å, or I band, 8500Å) Tully-Fisher method, useful for measuring distances of edge-on spirals, the D_n-σ technique and Supernovae (both type Ia's and type II's via the standard-candle approach now modified by light curve fitting (Hamuy *et al.* 1995; Reiss *et al.* 1995) or the Expanding Photospheres Method (Schmidt *et al.* 1994). Application of any or all of these techniques to all-sky flow field maps with relatively dense coverage is a long way away.

Since the mid-1980's there have been a number of studies of the flow field. Most, unfortunately, have concentrated on small, well defined regions of the sky. Mathewson, Ford and Buchhorn (1992) and Mathewson and Ford (1994), have concentrated their studies in the region of the original Great Attractor and find a complicated and non-uniform flow over the region, probably not converging out to 8,000 km/s. Dressler and Faber (1991) dispute the early claims of Mathewson *et al.* and claim to find backside infall into the GA. Willick (1991) investigated the Perseus-Pisces region and found an infall of the Pisces-Perseus region towards us! Courteau (1992) rexamined this region and found evidence for both a bulk flow and a relatively quiet region in the Hubble flow above PP. Courteau *et al.* 1993 systhesize all the available peculiar velocity results to show that there may be a "large-scale parallel streaming" or bulk flow of all galaxies inside a 6000 km/s sphere as well as the traditional GA. This conclusion has been strongly supported by the ground breaking work of Lauer and Postman (1994), who in a manner analogous to Sandage's work on the Hubble Diagram, used first ranked cluster galaxies to probe the velocity field out to 15,000 km/s. They too find evidence for a bulk flow on very large scales but one that is not consistent with the microwave background vector! Lauer and Postman find that the Abell Cluster Inertial Frame that they investigate appears to be moving with a velocity of 689 ± 178 km/s w.r.t. the CMB frame and that the motions (at least of clusters!) inside that ragion is relatively quiescent. This is also seen in the work of Mould *et al.* (1991; 1993) who find motion towards the GA (Figure 3), but a relatively quiet field inside 8,000 km/s except for a "hiccup" at the position of the GA.

The answer to the second question requires unbiased (in the classical sense) samples of galaxies. Attempts have been made (e.g. POTENT Bertschinger and Dekel 1989; Dekel, Bertschinger and Faber 1990) to derive the density field from the velocity field. However, what one really wants to do is to determine the density field and velocity field independently so as to be able to perform the comparison between the two. In that case, one needs to construct an all-sky map of galaxies and measure approximate distances, usually via redshifts, to the galaxies. Unfortunately, the Milky Way gets in the way!

Most work on the density field since the early 1980's has been, almost by definition, on three samples of objects — (1) optically selected galaxies from the Zwicky, UGC and ESO catalogs, (2) infrared (60μ) selected galaxies from the IRAS all sky survey, the point source catalog, and (3) galaxy clusters from the Abell catalog. The optical work, both on galaxies and clusters, is hampered by the effects of absorption by the dust in the Galaxy, but can produce exquisitely dense surveys (Figure 5), while the IRAS samples are much less well sampled and are biased against high den-

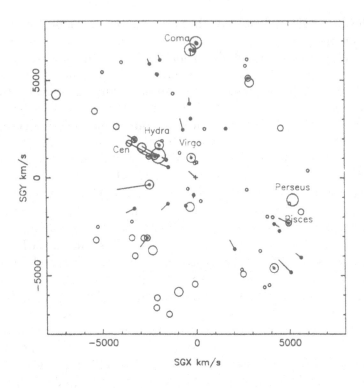

Figure 3. Peculiar velocities in the CMB frame for clusters and groups of galaxies with accurate Tully-Fisher distances. Major superclusters are labeled. The circle size indicates the richness of the cluster. The local group is at the origin, and the lines indicate the amplitude and direction of the peculiar motion. (From Mould *et al.* 1993)

sity regions because they are morphologically biased to large, dusty late type galaxies, but sample through the galactic plane moderately well (e.g. Figure 6).

The major IRAS surveys include the QDOT survey of Saunders *et al.* (1992) and Rowan-Robinson *et al.* (1990), the IRAS 1.936 Jy sample of Strauss *et al.* (1990; 1992a; 1992b), the and the IRAS 1.2 Jy sample of Fisher *et al.* (1995), which is soon to be published (e.g. Figure 6). The major optical galaxy surveys include the compilations of Lynden-Bell and Lahav (1989), Scharf and Lahav (1993) and Hudson (1993; 1994) primarily based on the optical radial velocity catalog of Huchra *et al.* (1992), and the new Optical Redshift Survey of Santiago *et al.* (1995), which defined a complete diameter limited sample *in advance* of obtaining redshifts for the galaxies in it. Both the optical galaxy surveys and IRAS surveys find essentially the same structures (viz Figs 5 and 6), although the density enhancements associated with any structure or cluster depends on the sample

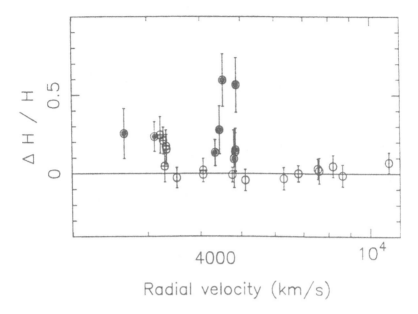

Figure 4. Deviations from a uniform Hubble flow, $\Delta H/H$, in the CMB frame plotted against cluster redshift. The asymptotic H_0 appears to be reached beyond 5,000 km/s. Note the "hiccup" in the flow near the velocity of Hydra-Centaurus. (From Mould *et al.* 1991).

used to define it. All major structures seen inside 10,000 km/s, probably even including the Great Wall (c.f. Mathewson and Ford 1994), appear to have some flows associated with them, albeit weak (Pisces-Perseus, Willick 1991). On very small scales (inside 3,000 km/s), Tully and Fisher's (1987) atlas has been an invaluable tool.

Major cluster surveys have been done by Postman, Huchra and Geller (1992) and Olowin *et al.* (1993), both to map the large-scale distribution of galaxy clusters and measure the amplitude of cluster clustering. Perhaps the most important aspect to the study of the cluster distribution has been the controversy started by Scaramella *et al.* (1989) and Tully and collaborators who have suggested that the bulk flow beyond the Great Attractor, and thus the missing few hundred km/s of motion w.r.t. the CMB frame, is caused by the presence of the Shapley Supercluster on the line-of-sight to the GA but at a redshift of ∼14,000 km/s. By definition, this implies that the flow field doesn't converge till at least 15,000 km/s. Lauer and Postman's (1994) result is consistent with this lack of convergence nearby, but they do *not* find a significant motion w.r.t. the Shapley Supercluster, but it is near the edge of their sample. As usual, more work is needed!

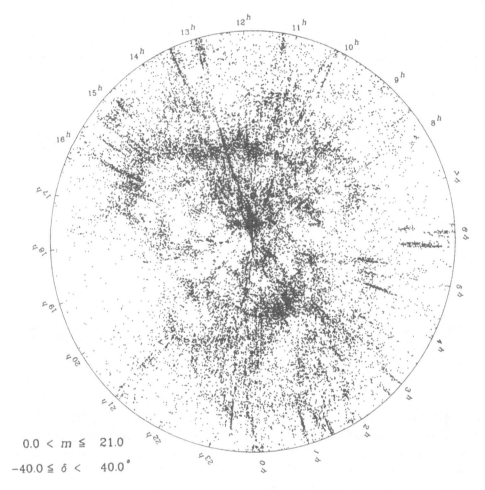

Figure 5. The latest galaxy map from the CfA Redshift Catalogue (Huchra *et al.* 1992) showing all galaxies with measured redshifts inside a wedge between declinations -40° and +40° and inside a radius of 15,000 km/s. The major well known structures such as the Great Wall between 8^h and 17^h at 8,000 km/s, Perseus-Pisces between 0^h and 5^h and at 4,000 km/sec, and Hydra-Cen, between 10^h and 14^h at ~3,800 km/s, are easily seen. Since all galaxies with measured redshifts are plotted, well studied clusters of galaxies, which appear as "Fingers of God," are over represented.

Significantly deeper surveys are now underway. The QDOT2 survey of Saunders and collaborators will contain well over 10,000 galaxies over the whole sky while optical surveys based onthe extension of the Southern Sky Redshift Survey (the SSRS2) and the CfA redshift survey (daCosta *et al.* 1994; Geller and Huchra 1989)

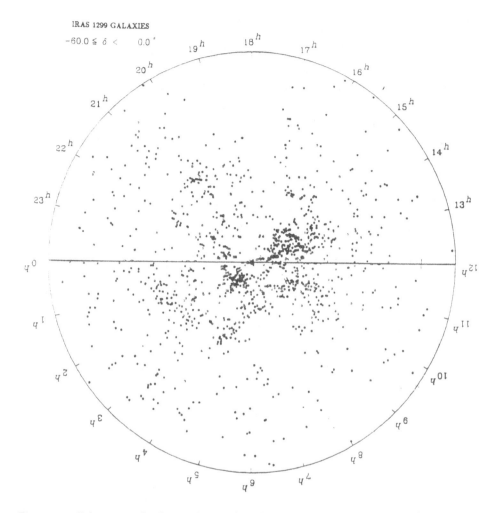

Figure 6. Galaxy map in the southern celestial hemisphere, declinations -60° to 0° and within 12,000 km/s from the IRAS 1.2 Jy survey of Strauss *et al.* 1995). Hydra-Centaurus is easily seen as in Figure 5; the cluster on the other side of the sky (2.5^h and 1,000 km/s) is Fornax.

3. CURRENT STATUS – What Exactly Do We Know?

We can summarize the current status of our knowledge of nearby galaxy motions relatively quickly — although I think you'll be brought more up to date in the next few talks!

On the question of convergence, we don't yet know the answer. There are some indications that convergence has occurred by about 10,000 km/s (to wit the most recent IRAS and POTENT matches), but the results of

Lauer and Postman (1994) as well as the earlier conjectures of Scaramella *et al.* (1989) strongly indicate that we have not yet seen convergence on scales of nearly 15,000 km/s. This question also remains bound up with the determination of the maximum scales (both dimension and mass) on which large scale clustering occurs (c.f. Park *et al. et al.* 1994).

Can we explain *most of* the CMB motion locally? Maybe! Mould *et al.* (1993) find several possible models that fit the data: a Great Attractor model, a Bulk Flow Model, a Bi-infall Model, and a Moving Attractor model all give acceptable fits to the available data with velocities and directions "consistent" with the CMB motion.

What are such galaxy motions effects on the determination of H_0? They are still too large since almost all of the calibrating galaxies for SN, IRTF, etc. are in the flow field (e.g. in the Virgo Cluster core), but this situation is rapidly being improved by HST Cepheid distances (e.g. Freedman *et al.* 1994). There are still significant discrepancies between H_0 derived from different distance indicators, but as better and better cross-calibrations are performed, these differences are likely to go away. Indeed, the most recent such cross-calibration based on the Cephied distance to Virgo (Mould *et al.* 1995) indicates that the best local value of H_0 is near 80 km/s/Mpc independent of the distance measuring technique used to step beyond Virgo.

The density field is somewhat better known, at least out to moderate distances and above the plane. The surveys mentioned above, ORS, CfA2, SSRS2, and QDOT2, are providing reasonable maps to 8,000 km/sec. The IRAS-POTENT match (Dekel *et al.* 1993) is pretty good but very limited because the IRAS catalogs pnly sparsely sample the density field so that flows on linear scales smaller than 1000-1200 km/sec are lost in the smoothing of the data — Virgo flow is only marginally detectable.

Perhaps the biggest problem comes from the discrepancy between the IRAS and optical survey matches of the velocity and density field. From the 1.936Jy survey data, Dekel *et al.* (1993) found $\Omega b(\text{IRAS}) = 1.3^{+0.7}_{-0.6}$ (where Ω is, of course, the ratio of the mass density to the cosmological closure mass density, and b is the biasing factor of Bardeen *et al.* 1986). Preliminary results from the 1.2Jy sample indicate that this number might be somewhat smaller — $\Omega b(\text{IRAS}) \sim 0.8 \pm 0.4$. However the optically selected galaxy density field gives $\Omega b(\text{Optical}) \sim 0.3 \pm 0.1$ (Hudson 1993; 1994), similar to the values derived earlier for Virgo infall (Huchra 1988). This is a problem that absolutely must be understood in order to settle the Ω debate as well as to understand if "biased" galaxy formation is required to fit the observations.

Lastly, it is interesting to point out that new, and possibly massive, nearby objects are being found every year (Kraan-Korteweg *et al.* 1994). Our census of the nearby universe, both in terms of the density field and the velocity field still needs a lot of work. A number of projects are underway to improve this situation, for example the 2 Micron All Sky Survey and DENIS, which will produce significantly better galaxy catalogues for density field studies as well as calibrated IR magnitudes for use in some of the distance estimating programs, and the HST Cephied program which should provide a good cross-calibration of many distance measuring techniques thus producing better maps of the flow field. The "Warpfire" project of Lauer and Postman is proceeding with theuse of clusters to map the flow field, albeit sparsely, to greater and greater distances. We need to complete these programs and get on with the systematic measurement of galaxy distances to $z \sim 0.05$ and beyond.

As has been concluded in the recent review of Dekel (1994), there are large-scale motions in the Universe. We are just beginning to map them in nearby space, but are far from having a complete understanding even nearby. Wish us luck!

This work has been supported in part by NASA grant GO-2227 from StScI and by the Smithsonian Institution. The author would like to thank his collaborators in the Hubble Constant/Large-Scale Flow project, especially J. Mould, W. Freedman, G. Bothun, R. Schommer, M. Postman, T. Lauer, J. Tonry, R. Marzke, M. Geller M. Strauss, A. Yahil, O. Lahav, and M. Davis, as well as fellow travellers A. Dressler, D. Burstein, G. Wegner, D. Lynden-Bell, G. de Vaucouleurs, S. Van den Bergh, D. Mathewson, S. Courteau, J. Willick, A. Sandage and G. Tammann, without whose interactions this work would never have been done. Special thanks go to Marc Aaronson, a friend whose drive and vision, cut short all too early, started this all.

References

Aaronson, M., Bothun, G. Mould, J., Huchra, J. Schommer, R. and Cornell, M. 1986, ApJ302,536.

Aaronson, M., Huchra, J., Mould, J., Schechter, P. and Tully, R. B. 1982,ApJ,258,64.

Aaronson, M., Mould, J., Huchra, J. Sullivan, W., Schommer, R. and Bothun, G. 1980,ApJ,239,12.

Bardeen, J., Bond, J. R., Kaiser, N. and Szalay, A, 1986, ApJ304,15

Bertschinger, E. and Dekel, A. 1989, ApJL,336,L5.

Burstein, D. Davies, R., Dressler, A., Faber, S., Lynden-Bell, D., Terlevich, R. and Wegner, G. 1986, in *Galaxy Distances and Deviations from Universal Expansion*, Madore and Tully, eds. (Dordrecht: Reidel).

Cheng, E., Saulson, P., Wilkinson, D. and Corey, B. 1979,ApJ,232,L139.

Courteau, S. 1992, PhD Thesis, UC Santa Cruz.

Courteau, S, Faber, S., Dressler, A. and Willick, J. 1993, ApJL,412,L51.

daCosta, L., Vogeley, M., Geller, M., Huchra, J. and Park, C. 1994, ApJL, in press.

Davis, M., Huchra, J., Latham, D. and Tonry, J. 1982,ApJ,253,423.

Dekel, A. 1994, Ann. Rev. Astron. and Ap. 32,371.

Dekel, A., Bertschinger, E. and Faber, S. 1990, ApJ364,349.

154

de Vaucouleurs, G. 1958,AJ,63,253.

Dressler, A. and Faber, S. 1991, ApJ,368,54.

Fisher, K. *et al.* 1995, ApJ, in prep.

Freedman, W., Madore, B., Mould, J., Hill, R., Ferrarese, L., Kennicutt, R., Saha, A., Stetson, P., Graham, J., Ford, H., Hoessel, J., Huchra, J., Hughes, S and Illingworth, G. 1994, Nature,371,757.

Geller, M. and Huchra, J. 1989,Science,246,897.

Hamuy, M., Phillips, M., Maza, J., Suntzeff, N., Schommer, R. and Avilés, R. 1995, AJ, in press.

Huchra, J. 1988, in *The Extragalactic Distance Scale*, van den Bergh and Pritchet, eds., p257.

Huchra, J., Geller, M., Clemens, C., Tokarz, S., and Michel, A. 1992, Bull.CDS,41,31.

Hudson, M. 1993, MNRAS,265,72.

Hudson, M. 1994, MNRAS,266,475.

Kirshner, R., Oemler, A., Schechter, P. and Shectman, S. 1981,ApJL,248,L57.

Kraan-Korteweg, R., Loan, A., Burton, W., Lahav, O., Ferguson, H., Henning, P. and Lynden-Bell, D. 1994, Nature,372,77.

Lauer, T. and Postman, M. 1994, ApJ,425,418.

Lynden-Bell, D., Faber, S., Burstein, D. Davies, R., Dressler, A., Terlevich, R. and Wegner, G. 1988,ApJ,326,19.

Lynden-Bell, D. and Lahav, O. 1989, in *Large Scale Motions in the Universe*, Rubin and Coyne, eds, (Princeton: Princeton), p199.

Mathewson, D., Ford, V. and Buchhorn, M. 1992, ApJL,389,L5.

Mathewson, D. and Ford, V. 1994, ApJL,434,L39.

Mould, J., Akeson, R., Bothun, G., Han, M., Huchra, J., Roth, J. and Schommer, R. 1993, ApJ409,14.

Mould, J., Stavely-Smith, L., Schommer, R., Bothun, G., Hall, P., Han, M., Huchra, J., Roth, J., Walsh, W. and Wright, A. 1991, ApJ,383,467.

Mould, J. *et al.*, 1995, ApJ, in prep.

Olowin, R., Huchra, J. and Corwin, H. 1993, in *Observational Cosmology*, Chincarini *et al.*, eds, p104.

Park, C., Vogeley, M., Geller, M. and Huchra, J. 1994,ApJ,431,596.

Peebles, P. J. E. 1976,ApJ,205,318.

Postman, M., Huchra, J. and geller, M. 1992, ApJ,384,404.

Reiss, A., Press, W. and Kirshner, R. 1995,ApJ, in press.

Rowan-Robinson, M., Lawrence, A., Saunders, W., Crawford, J., Ellis, R. Frenk, C., Parry, I., Xiaoyang, X., Allington-Smith, J., Efstathiou, G., and Kaiser, N. 1990, MNRAS,247,1.

Rubin, V. 1951, MSc thesis, Cornell University.

Rubin, V., Ford, W., Thonnard, N., Roberts, M. and Graham, J. 1976a, AJ,81,687.

Rubin, V., Thonnard, N., Ford, W. and Roberts, M. 1976b, AJ,81,719.

Sandage, A. and Hardy, E. 1972,ApJ,172,253.

Sandage, A. and Tammann, G. 1981, *A Revised Shapley Ames Catalog of Bright Galaxies* (Washington: Carnegie Institution).

Santiago, B, Strauss, M., Lahav, O., Davis, M., Dressler, A. and Huchra, J. 1995, ApJ, in press.

Saunders, W., Rowan-Robinson, M., and Lawrence, A. 1992, MNRAS,258,134.

Scaramella, R., Baiesi-Pillastrini, G., Chincarini, G. Vettolani, G. and Zamorani, G. 1989, Nature,338,562.

Scharf, C. and Lahav, O. 1983, MNRAS,264,439.

Schmidt, B., Kirshnerm R., Eastman, R., Phillips, M., Suntzeff, N., Hamuy, M., Maza, J. and Avilés, R. 1994, ApJ,432,42.

Shaya, E. 1984,ApJ,280,470.

Silk, J. 1974,ApJ,193,525.

Smoot, G. and Lubin, P. 1979,ApJ234,L83.

Strauss, M., Davis, M., Yahil, A. and Huchra, J. 1990, ApJ,361,49.

Strauss, M., Davis, M., Yahil, A. and Huchra, J. 1992a, ApJ,385,421.

Strauss, M., Huchra, J., Davis, M., Yahil, A., Fisher, K. and Tonry, J. 1992b, ApJS,83,29.

Tammann, G. and Sandage, A. 1985,ApJ,294,81.

Tonry, J. and Davis, M. 1980,ApJ,246,680.

Tonry, J. and Schneider, D. 1988, AJ,96,807.

Tully, R. B. and Fisher, R. 1987, *Nearby Galaxy Atlas*, Cambridge: Cambridge).

Tully, R. B and Shaya, E. 1984,ApJ281,31.

Willick, J. 1991, PhD Thesis, UC Berkeley.

THE HUBBLE PARAMETER - A STATUS REPORT AT EPOCH 1994.5

SIDNEY VAN DEN BERGH
Dominion Astrophysical Observatory
5071 West Saanich Road
Victoria, B.C., V8X 4M6
Canada

ABSTRACT. New observations obtained during the last two years have slightly shifted the balance of evidence towards larger values of H_0. A compilation of recent determinations shows reasonable agreement and suggests that $H_0 \geq 75$ km s^{-1} Mpc^{-1}.

1. Introduction

The Hubble parameter H_0 is usually expressed in units of km s^{-1} Mpc^{-1}, i.e. it has a dimension (time)$^{-1}$. If $H_0 \geq 1000$ km s^{-1} Mpc^{-1} then $t_H \leq 1$ Gyr. This would not allow enough time to form galaxies, stars, planets and the carbon that is required for all known forms of life. On the other hand for $H_0 \leq 10$ km s^{-1} Mpc^{-1}, corresponding to $t_H \geq 100$ Gyr, almost all gas in galaxies will have been used up long ago to form stars - most of which would by now have exhausted their fuel supply and become white dwarfs. As a result few sites that are congenial to life as we know it would remain in the Universe. Such *a priori* arguments suggest that $10 \leq H_0$ (km s^{-1} Mpc^{-1}) ≤ 1000. A compilation of H_0 values by Okamura & Fukugita (1991) shows that the overwhelming majority of recent determinations of H_0 fall in the narrower range $50 \leq H_0$ (km s^{-1} Mpc^{-1}) ≤ 100.

Turner, Cen & Ostriker (1992) have suggested that H_0(local) might differ significantly from H_0(global). Using first-ranked galaxies in rich clusters van den Bergh (1992) concluded that H_0(global) = (0.94 ± 0.07) H_0(local), in which H_0(local) refers to the region with V < 10000 km s^{-1}. By applying the same technique to a full-sky sample of Abell clusters Lauer & Postman (1992) find that H_0 is globally constant to ± 7% for the region of space with V < 15000 km s^{-1}. Due to local streaming velocities of a few hundred km s^{-1} it is, however, necessary to derive "cosmological" redshifts of local clusters such as Virgo and Fornax, from the Coma redshift and the distance ratio D(Coma)/D(cluster).

2. Meter Sticks and Standard Candles

2.1 GLOBULAR CLUSTERS

Shapley (1953) first pointed out that the mean luminosity of globular clusters might turn out to be a good "standard candle". This conclusion appears to be supported by modern observations of globulars in the Galaxy, M31 and the Large Cloud. It has, however, not yet been proven beyond reasonable doubt that the mean luminosity of globular clusters around distant giant E galaxies is identical to that of the globular cluster systems surrounding spiral galaxies such as M31 and the Milky Way System.

M. Kafatos and Y. Kondo (eds.), Examining the Big Bang and Diffuse Background Radiations, 157–161.

Recently Sandage & Tammann (1994) have compared Virgo globulars with local ones to obtain $H_0 = 55 \pm 5$ km s^{-1} Mpc^{-1}. *Using the same technique* van den Bergh (1994b) finds $H_0 \geq$ 73 km s^{-1} Mpc^{-1}. Reasons for this difference are the following: (1) Sandage & Tammann used the rather bright luminosity calibration of RR Lyrae stars adopted by Sandage (1993), which results in a high luminosity for Galactic globular clusters. On the other had van den Bergh adopts a more conservative value $M_V(RR) = +0.6$. Even fainter luminosities for Galactic RR Lyrae stars have been derived by Storm, Carney & Jones (1994), who use the Baade-Wesselink technique to obtain $M_V(RR) \approx +0.8$. Such faint RR Lyrae magnitudes appear to be supported by Hubble Space Telescope color-magnitude diagrams of two globulars in M31 (Ajhar et al. 1994). However, the issue will remain clouded until it is understood (Walker 1992) why $M_V(RR) = +0.44$ (at [Fe/H] = -1.9) in the Large Cloud. Furthermore (2) Sandage & Tammann (1994) adopt a small cosmic velocity [V(cosmic) = 1179 ± 17 (sic) km s^{-1} Mpc^{-1}] for the Virgo cluster, whereas van den Bergh (1994b) derives a larger value V(cosmic) = 1311 ± 132 km s^{-1} Mpc^{-1} from the Coma velocity and the Coma/Virgo distance ratio. Finally (3) new observations by Fleming (1994) suggest that the peak of the M87 globular cluster luminosity function is 0.2 mag brighter than the value previously obtained by Harris *et al.* (1991). It is not yet known (Harris 1994) whether a similar correction also applies to the globular cluster systems surrounding other Virgo giant E galaxies. All three of the effects discussed above will tend to *increase* the value of H_0 derived from globular clusters.

2.2 GALAXY DIAMETERS

Sandage (1993a) has derived a value $H_0 = 43 \pm 11$ km s^{-1} Mpc^{-1} from a comparison of the diameter of M101 with the diameters of other, more distant, galaxies of DDO type Sc I. However, intercomparison of the Sc I galaxies NGC 309 and M100 (van den Bergh 1992) shows that Sc I galaxies have a range of 2 - 3 in their diameters, i.e. they are not good standard "meter sticks". Similarly Sandage (1993b) compares M31 to other Sc I galaxies at larger redshifts to obtain $H_0 = 45 \pm 12$ km s^{-1} Mpc^{-1}. However, van den Bergh (1994a) finds $H_0 \sim 76$ km s^{-1} Mpc^{-1} by comparing M31 with similar objects in the Virgo cluster. A comparison of M33 with Sc II-III galaxies in Virgo yields $H_0 \sim 124$ km s^{-1} Mpc^{-1}. From the examples cited above it is concluded that the dispersion of diameters of spirals of any given luminosity class is too large for galaxy diameters to be useful tools for precision distance determinations.

2.3 SUPERNOVAE OF TYPE Ia

In a recent review Branch & Tammann (1992) conclude that the intrinsic scatter in $M_B(max)$ for supernovae of Type Ia (SNe Ia) is less than 0.25 mag. This would make such objects excellent standard candles. Just after their paper was submitted SN 1991bg, which was subluminous by 2.5 mag in B, exploded in the Virgo elliptical M84. Furthermore the Type Ia supernova 1991T was found to probably be of above-average luminosity. More recently Branch, Fisher & Nugent (1993) and Branch et al. (1994) have found that SNe Ia with strongly deviant luminosities are spectroscopically abnormal and can therefore be weeded out. However, Maza et al. (1993) find that the supernovae 1992bc and 1992bo, which were both spectroscopically normal, differed in luminosity by 0.8 ± 0.2 mag. The use of SNe Ia as standard candles is also suspect because recent high quality spectra obtained at Lick (Lynch et al. 1992) and at Cerro Tololo (Phillips 1993a) reveal a range of lightcurve and spectroscopic characteristics that had not previously been appreciated. An additional problem (Branch & van den Bergh 1993) is that the expansion velocities of SNe Ia in galaxies that have experienced recent star formation appear, on average, to be higher than those of SNe Ia in galaxies with an old stellar population.

Phillips (1993b) has suggested that M_V(max) and the rate of decline of supernovae are correlated. However, observations of SN 1992bc (Hamuy et al. 1994) and of S Andromedae (van den Bergh 1994a) indicate that the dispersion around Phillips' luminosity versus rate of decline relation may be large. Finally Phillips (1993c) also finds evidence which suggests that the Si II $\lambda5979/\lambda6355$ intensity ratio near maximum may be closely correlated with luminosity. If this suspicion is supported by future observations, then SNe Ia might yet fulfil their promise as standard candles that could be used to calibrate the extragalactic distance scale.

2.4 SUPERNOVAE OF TYPE II

Recently Schmidt (1993) has used the expanding photosphere (Baade-Wesselink) method, in conjunction with detailed model atmosphere calculations, to determine distances to individual supernovae of Type II (SNe II). For the two most distant objects in their sample (SN 1990ae, cz=7800 km s^{-1} and SN 1992am, cz = 14500 km s^{-1}), for which the effects of deviations from a smooth Hubble flow should be small, Schmidt (1994) and Schmidt et al. (1994) obtain D = 115 Mpc and D = 180 Mpc, respectively. The corresponding values of the Hubble parameter are H_0 = 68 km s^{-1} Mpc^{-1} and H_0 = 81 km s^{-1} Mpc^{-1}. The good agreement between distances derived to individual SNe II, and the Tully-Fisher distances to their parent galaxies, indicates that expanding photosphere distance determinations have probably not been significantly affected by deviations from spherical symmetry in expanding supernova envelopes.

2.5 THE TULLY-FISHER RELATION

Recent applications of the Tully-Fisher relation (e.g. Pierce & Tully 1988) to spiral galaxies generally yield relatively large values of H_0. The fact that the dispersion in the luminosity-line width relation is small for linewidths $\geq$ 200 km s^{-1} suggests that the effects of Malmquist bias ($\Delta M = 1.38\ \sigma_M^2$) on determinations of H_0 from giant and supergiant spirals are probably small. From studies of galaxies in two volume elements, one in the direction of (but beyond) the Virgo cluster, and another in the opposite direction Lu, Salpeter & Hoffman (1994) find H_0 = 84 ± 8 km s^{-1} Mpc^{-1}. This value includes a small correction for Malmquist bias and is, to first order, independent of the adopted infall velocity (retardation) of the Local Group into the Virgo cluster.

2.6 SURFACE BRIGHTNESS FLUCTUATIONS

Tonry & Schneider (1988) have shown that surface brightness fluctuations in galaxies containing old stellar populations can be used to determine the distances to such objects. Using this technique Jacoby et al. (1992) derive a distance modulus $(m-M)_0$ = 30.88 ± 0.2 for the elliptical-rich core of the Virgo cluster, from which H_0 = 87 ± 12 km s^{-1} Mpc^{-1}. Since different early-type galaxies may have experienced differing evolutionary histories the brightest stars in different old galaxies might not all have exactly the same luminosities. This may account for the fact that Lorentz et al. (1993) find a small range in Virgo distance moduli for individual galaxies. These distance moduli appear to correlate with the Mg_2 indices observed for individual objects.

2.7 PLANETARY NEBULAE

The luminosity function of planetary nebulae appears to have the same shape in all galaxies in which it has so far been studied. Planetary nebulae therefore provide a powerful tool (see Jacoby et al. 1992 for a review and references) for the determination of extragalactic distances. Possible problems with the use of planetaries as distance indicators have recently been reviewed by Tammann (1993). It is, however, encouraging that McMillan, Ciardullo & Jacoby (1993) find no evidence for a correlation between galaxy luminosity (and hence metallicity) and the *differences* between Fisher-Tully distances to objects in four clusters of galaxies and the distances to these same galaxies derived from the luminosity functions of planetary nebulae. Jacoby, Ciardullo & Ford (1990) find $82 \leq H_0$ (km s^{-1} Mpc^{-1}) ≤ 94 from observations of planetary nebulae in Virgo galaxies. A somewhat smaller value $H_0 = 75 \pm 8$ km s^{-1} Mpc^{-1} has been obtained by McMillan, Ciardullo & Jacoby (1993) from observations of planetary nebulae in the Fornax cluster.

Table 1. Recent Determination of the Hubble Parameter.

Method	H_0 (km s^{-1} Mpc^{-1})
SNe Ia	75 ($\pm$ 12?)
SNe II	68 - 81
Tully-Fisher	84 $\pm$ 8
Surface brightness fluctuations	87 $\pm$ 12[a]
Planetary nebulae (Virgo)	86 $\pm$ 18[a]
Planetary nebulae (Fornax)	75 $\pm$ 8
Galaxy diameters	76 - 124
Globular clusters (Virgo)	$\geq$ 73[b]
Surface brightness profiles (Fornax)	99 $\pm$ 16
Gravitational lens	< 87
Sunyaev-Zel'dovich effect	~ 55 $\pm$ 17
Compact radio sources	~ 100

[a] A 10% uncertainty in the cosmological distance of the Virgo Cluster has been added in quadrature.

[b] Reduction of the mean luminosity of Virgo globulars (Fleming 1994) would increase the value of H_0.

3. Summary and conclusions

A compilation of recent determinations of the Hubble parameter is given in Table 1. In view of the difficulty of extragalactic distance determinations the agreement between the individual values of H_0 listed in this table is encouraging. Taken at face value these data appear to indicate that H_0

≥ 75 km s^{-1} Mpc^{-1}. Successful observations of Cepheid variables in Virgo spirals with the Hubble Space Telescope would greatly add to our confidence in the reliability of the extragalactic distance scale. However, such observations will probably not prove to be a panacea. This is so because (1) the cosmological redshift of the Virgo cluster remains uncertain at the ~10% level, and (2) the zero point (and possibly the metallicity dependence) of the Cepheid period-luminosity relation remains uncertain. Observations with Hipparcos should, however, reduce such uncertainties in the not too distant future.

References

Branch, D., Fisher, A., & Nugent, P. 1993, AJ, 106, 2383

Branch, D., & Tammann, G.A. 1992, ARAA, 30, 359

Branch, D., & van den Bergh, S. 1993, AJ, 105, 2231

Branch, D., Vaughan, T., Perlmutter, S., & Miller, D. 1994, in preparation

Fleming, D. 1994, MSc Thesis, McMaster University

Hamuy, M., Phillips, M.M., Suntzeff, N.B., Avilés, R. & Maza, J. 1993, BAAS, 25, 1340

Harris, W.E. 1994, private communication

Harris, W.E., Allwright, J.W.B., Pritchet, C.J., & van den Bergh, S. 1991, ApJS, 76, 115

Jacoby, G.H., Ciardallo, R., & Ford, H.C. 1990, ApJ, 356, 332

Jacoby, G.H., Branch, D., Ciardullo, R., Davies, R., Harris, W., Pierce, M.J., Pritchet, C.J., Tonry, J.L., & Welch, D.L. 1992, PASP, 104, 599

Lauer, T.R., & Postman, M. 1992, ApJ, 400, L47

Lorenz, H., Böhm, P., Capaccioli, M., Richter, G.M., & Longo, G. 1993, A&A, 277, L15

Lu, N.Y., Salpeter, E.E., & Hoffman, G.L. 1994, ApJ, 426, 473

Lynch, D.K., Erwin, P., Rudy, R.J., Rossano, G.S., & Puetter, R.C. 1992, AJ, 104, 1156

Maza, J., Hamuy, M., Phillips, M.M., Suntzeff, N.B., & Avilés, R. 1994, ApJ, 424, L107

McMillan, R., Ciardullo, R., & Jacoby, G.H. 1993, ApJ, 416, 62

Okamura, S., & Fukugita, M. 1991 in Primordial Nucleosynthesis and Evolution of the Early Universe, eds. K. Sato and J. Audouze (Dordrecht: Kluwer), p. 45

Phillips, M.M. 1993a, BAAS, 25, 834

Phillips, M.M. 1993b, ApJ, 413, L105

Phillips, M.M. 1993c, NOAO Newsletter, Dec. 1993 p.5

Pierce, M.J., & Tully, R.B. 1988, ApJ, 330, 579

Sandage, A. 1993a, ApJ, 402, 3

Sandage, A. 1993b, ApJ, 404, 419

Schmidt, B.P. 1993 unpublished Harvard Ph.D. thesis

Schmidt, B.P. 1994 private communication

Schmidt, B.P. et al. 1994, AJ, 107, 1444

Shapley, H. 1953, Proc. Nat. Acad. Sci., 39, 349

Tammann, G.A. 1993, in Planetary Nebulae = IAU Symposium No. 155, eds. R. Weinberger and A. Acker (Kluwer: Dordrecht), p.515

Tonry, J., & Schneider, D.P. 1988, AJ, 96, 807

Turner, E.L., Cen, R., and Ostriker, J.P. 1992, AJ 103, 1427

van den Bergh, S. 1992, PASP, 104, 861

van den Bergh, S. 1994a, ApJ, 424, 345

van den Bergh, S. 1994b, PASP, in preparation

THE LOCAL VELOCITY FIELD AND THE HUBBLE CONSTANT

G.A. Tammann
Astronomisches Institut der Universität Basel
Venusstrasse 7, CH-4102, Binningen, Switzerland
and
Allan Sandage
The Observatories of the Carnegie Institution of Washington
813 Santa Barbara Street, Pasadena CA 91101

> *"When you reach a fork in the road, take it"*
> (American folklore, sometimes attributed to Yogie Berra).

ABSTRACT
The methods are reviewed that give a distance modulus to the core of the Virgo cluster of m - M = 31.64 ± 0.08 (D = 21.3 ± 0.8 Mpc). It is shown that the cosmic velocity of the cluster core is 1179 ± 17 km s^{-1}, which, when combined with the distance gives H_0 = 55 ± 3 km s^{-1}Mpc^{-1} from the Virgo cluster data alone. Nine independent methods are reviewed that confirm that H_0 = 50 ± 2. Discussion is made why all methods that are said to give the short distance scale ($H_0 \sim 85$) are incorrect.

1. Introduction

The debate continues on the value of the Hubble constant despite the multiple evidence from many experiments that its value is near H_0 = 50 km s^{-1}Mpc^{-1}. The purpose of this lecture is to list the evidence for the long distance scale. For those at the fork in the road that wish to still take it, we show why the short distance scale ($H_0 \sim 85$ to 100) is not supported by any unbiased data and analysis.

The plan of the report is to (1) show that the local value of H_0 is close to the global value because there is no step in the Hubble diagram that separates the nearby expansion field from the Machian (global) field relative to the microwave background (CMB), (2) show that the global expansion velocity of the Virgo cluster is v(cosmic) = 1179 ± 17 km s^{-1} tied to the kinematic frame of the CMB, (3) show that six distance indicators give the distance modulus of the Virgo cluster as m - M = 31.64 ± 0.08 (D = 21.3 ± 0.8 Mpc), (4) show thereby that H_0 = 55 ±

M. Kafatos and Y. Kondo (eds.), Examining the Big Bang and Diffuse Background Radiations, 163–173.
© 1996 IAU.

2 km s^{-1}Mpc^{-1} from the Virgo cluster alone, (5) show that the result is confirmed by nine independent methods, the most powerful of which is through distant supernovae of type Ia calibrated in absolute magnitude via Cepheids, (6) justify that the local distance scale from Cepheids is confirmed to within 5% from the independent calibrations via the old population II objects (RR Lyrae variables, globular clusters, red giant stars), and (7) show how it can be understood that all methods that are said to support the short distance scale ($H_0 \sim 85$) are incorrect.

2. The Local Velocity Field Tied to the CMB

The two best established motions of the Local Group relative to the Machian frame of the cosmic CMB are (1) the perturbation of the free (cosmic) expansion of the Local group from the Virgo region due to the mass of the Virgo complex, and (2) the larger-scale, nearly bulk motion of the "local bubble" of size $\leq$ 6000 km s^{-1} relative to the CMB, carrying the Local Group and the Virgo complex with it. The model for the salient features is still that set out elsewhere (Tammann & Sandage 1985), given as Figure 1 here. Refinements on this picture, tying the "local

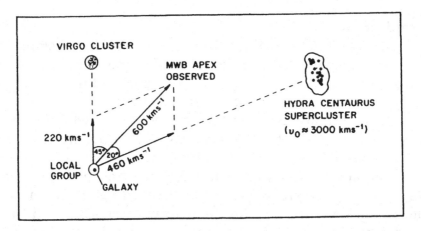

Fig. 1. Vector diagram showing the "infall" (actually retarded expansion) of the Local Group toward the Virgo cluster center plus the supposed motion in the direction of Hydra, which must be caused by the clumpy mass distribution within $\leq$ 6000 km s^{-1}. The two vectors together explain the observed dipole motion toward the warm pole of the CMB.

bubble" into the Machian cosmic frame can be made using the extensive data of Mathewson, Ford, & Buchhorn (1992). Analyses by Federspiel et al. (1994, FST) show that the nearly bulk peculiar motion of the local bubble containing the Local Group, the Virgo complex, and the local region gradually peters out beyond ~ 6000 km s^{-1}, merging gradually into the unperturbed Hubble flow. Figures 15 to 19 of FST are decisive on this point.

That the effect of the peculiar dipole motion toward the CMB within the Local Supercluster and beyond is so small and can be neglected in the determination of H_0 is shown by the lack of a *step* in the Hubble diagrams at the "edge" of the Local Supercluster. The lack of an effect can be made quantitative. Figure 2 shows the Hubble diagram (*m* vs. *log cz*) for nearby clusters and groups to 10,000 km s^{-1} from both hemispheres.

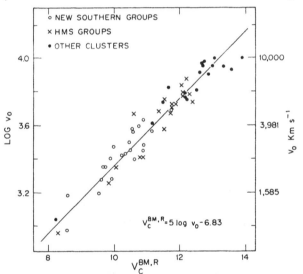

Fig. 2. Hubble diagram for nearby groups and clusters using first-ranked E galaxies corrected for richness and contrast effects. Diagram from Sandage (1975). No steps or large-scale streaming motions are visible.

The data show no step nor any other large-scale streaming motions at the level of more than ~ 500 km s^{-1}. With this limit applied at 5000 km s^{-1}, the typical effect of perturbations on the Hubble flow is <10%. The error on the Hubble constant due to streaming motions will also be equal to or less than this.

3. The Cosmic Velocity of the Virgo Cluster Core Freed From All Streaming Motions

The most direct method to sample the true expansion field devoid of all effects of local streaming motions is to tie the Virgo cluster core to remote clusters that themselves have known redshifts relative to the kinematic frame of the CMB (Sandage & Tammann 1990; Jerjen & Tammann 1993). The result, shown in Figure 3, is that the cosmic (global, Machian) redshift of the

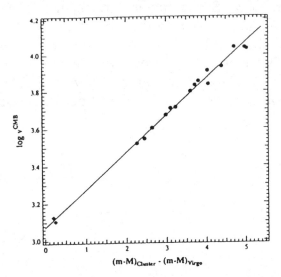

Fig. 3. The Hubble diagram using redshifts in the kinematic frame of the CMB vs. the differences in distance moduli between each of the 17 clusters and the E galaxies in the Virgo cluster core. The cosmic redshift at the distance of the Virgo cluster core, read from the Hubble line of slope 5 at zero modulus difference, is 1179 ±17 km s⁻¹.

Virgo cluster core is v(cosmic) = 1179 ±17 km s⁻¹, devoid of all streaming motions. The result can be used to determine the Hubble constant once the distance to the Virgo cluster is known.

Part of the disagreement in the Hubble constants derived by others is their use of too high a value for v(cosmic) for Virgo. The decisive result from Figure 3 with its error of only 1.4% removes this one source of the differences between our results and others.

4. Six Methods for the Distance to the Virgo Cluster Core

Before giving the synopses of six methods to find the distance of the Virgo cluster core, we give the summary of the results in Table 1. Results of the two methods (planetary nebulae and surface brightness fluctuations) that are said (Jacoby et al. 1992) to give $(m - M)_{Virgo} = 30.9$, are outside the range of the Table, and are not shown. Methods and data that apparently support the short distance scale are mentioned in section 8.

TABLE 1			
Method	(m - M)	Galaxy Type	Calibrators
Globular clusters	31.64 ± 0.25	E	RR Lyraes
Novae	31.57 ± 0.43	S	M31, galactic novae
Supernovae	31.63 ± 0.25	E,S	Cepheids, model
Dn - σ	31.85 ± 0.19	E	Galaxy, M31, M81
21 cm line-widths	31.60 ± 0.15	S	13 nearby calib.
Scale length of ScIs	31.50 ± 0.20	S	MW and M31 sizes
mean	31.64 0.08		(D = 21.3 ± 0.08 Mpc)

(a) *Globular clusters*: The maximum of the globular cluster luminosity function, calibrated via RR Lyrae distances for MW globular clusters using a new calibration (Sandage 1993c) and via Cepheids for M31 clusters, is applied to the observed luminosity turnover in the cluster samples for five E galaxies associated with the Virgo cluster core (Harris et al. 1991). The modulus first obtained by Harris (1988) of m - M = 31.7 is confirmed (Sandage & Tammann 1994). The new precepts introduced by Secker & Harris (1993) that led them to a short distance scale are criticized there (cf. also McLaughlin et al. 1994).

(b) *Normal novae*: Pritchet & van den Bergh (1987) discovered nine normal novae in NGC 4472, the brightest E galxay in Virgo subcluster B, obtaining a modulus difference between M31 and NGC 4472 of 6.8 ± 0.4 mag. Using a corrected M31 modulus as argued by Sandage & Tammann (1988) gives $(m - M)_{NGC\ 4472} = 31.57 \pm 0.43$.

(c) *Supernovae*: Following Tammann (1988, 1992) and Branch & Tammann (1992), the Type Ia supernovae observed in Virgo cluster galaxies give $(m - M) = 31.54 \pm 0.22$ using a calibration of

M_B = -19.6 based on Cepheids (Sandage et al. 1994; Saha et al. 1994b). Schmidt et al. (1992) obtained "expanding envelope" distances for two SNe II in the Virgo cluster of $(m - M)_{Virgo}$ = 31.71± 0.26 mag before they incorrectly (Branch 1994) changed (Schmidt et al. 1994a,b) their value of the dilution (Wagoner) factor. The average of the determinations using SNe Ia and SNe II is shown in the Table.

(d) $D_n - \sigma$: Combining the known variation of surface brightness (SB) of E galaxies with absolute magnitude (Oemler 1974, 1976; Kormendy 1977; Sandage & Perelmuter 1991) with the Minkowski (1962) relation (later called the Faber-Jackson relation) between absolute magnitude and central velocity dispersion gave a relation between SB, M, and velocity dispersion (Dressler et al. 1987; Dressler 1987), known now as the $D_n - \sigma$ relation. A calibration of the Dressler relation using the bulges of M31, M81 and the MW (Tammann 1988) gives a Virgo modulus of (m - M) = 31.85 ± 0.19.

(e) *21 cm line widths*: Kraan-Korteweg, Cameron, & Tammann (1988), Fouqué et al. (1990), Teerikorpi (1987), Bottinelli et al. (1987), Sandage (1988a,b, 1994a,b); Federspiel et al. (1994) and undoubtedly others demonstrate that observational selection causes all distances in flux-limited samples to be distorted toward too small values, giving Hubble constants that are too large when using the Tully-Fisher method. Theory and application of the correction for this type of bias (cf. Tammann 1988; KKCT 1988; Fouqué et al. 1990) gives $(m - M)_{Virgo}$ = 31.60 ± 0.15 for the Virgo cluster, as listed in the Table.

(f) *Size of the MW and M31 relative to Virgo spirals*: van der Kruit (1986) compares scale lengths of Virgo Sb and Sc galaxies with the known absolute scale lengths of the MW and M31 disks to derive a *lower limit* to the Virgo cluster distance of 20 Mpc, listed as a real value in the Table.

5. The Hubble Constant From the Virgo Cluster Distance Itself

Dividing the cosmic velocity of the Virgo cluster core of v = 1179 ± 17 km s[-1] by the adopted distance from Table 1 gives

$$H_0 \text{ (cosmic)} = 55 \pm 2 \text{ km s}^{-1}\text{Mpc}^{-1},$$

using the Virgo cluster data alone. Six methods that are

independent of the Virgo cluster method, set out in the next section, also support this long distance scale.

6. Nine Independent Astronomical and Astrophysical Ways to H_o

The results of nine independent methods to H_o, in addition to the Virgo method, are listed in Table 2. Only the salient literature references are given here. Details of the methods are justified in these references. The principal references are:

TABLE 2: Summary of the Various Methods to H_o	
Method	H_o
Virgo Distance	55 ± 2
ScI Hubble diagram	49 ± 15
M101 look-alike diameters	43 ± 11
M31 look-alike diameters	45 ± 12
Tully-Fisher field galaxies	48 ± 5
Tully-Fisher cluster data	55 ± 8
Supernovae Ia (B)	52 ± 8
Supernovae Ia (V)	55 ± 8
Supernovae Ia expansion parall. & ^{56}Ni	50-60
Sunyaev-Zeldovich effect	38 ± 17
Magn.variations of lensed double QSO	< 70
mean	~ 50

Virgo distance: last section; *ScI galaxy Hubble diagram*: The Hubble diagram itself is set out in Sandage & Tammann (1975). The calibration using the absolute magnitude of the only ScI galaxy with a Cepheid distance (as of 1993), i.e. M101 at m - M = 29.3, is in Sandage & Tammann (1974). The bias properties of the sample and the way to correct for them is in Sandage (1988a); *M101 look-alike diameters*, calibrated with M101 and corrected for bias is in Sandage (1993a), following the method of van der Kruit (1986); *M31 look-alike diameters* calibrated by M31 and corrected for bias is discussed in Sandage (1993b); *Tully-Fisher field galaxies* in the distance-limited sample of Kraan-Korteweg & Tammann (1979), calibrated with local galaxies by Richter & Huchtmeier (1984), give the bias-free result in Table 2 (Sandage 1994b); *Tully-Fisher cluster* data corrected for the "cluster-population-incompletness bias" of Teerikorpi (1987, 1990), is analyzed by Sandage, Federspiel, & Tammann (1995); *The supernovae data* for the first two HST calibrations via Cepheids are discussed in Sandage et al. (1992, 1994) and Saha et al. (1994a, 1994b). - Purely physical distance determinations are: *Type Ia supernovae* from ^{56}Ni-powered light curves and expanding-photosphere models (Branch & Khokhlov 1994); from the *Sunyaev-Zeldovich effect* (Birkinshaw 1993; Jones 1994); and from *gravitational double quasars* (Dahle et al. 1994).

7. The Local Distance Scale is Stable

The value of H_0 is no better than the reliability of the local distance scale upon which the secondary calibrators rest (involving M31 and all other members of the Local group, used for example by Richter & Huchtmeier (1984) for the calibration of the TF relation, as well as M81, M101, and others in nearby groups). The reliability of the data for the Local Group is discussed by Tammann (1987, 1992), by Madore & Freedman (1991), and by Sandage (1995). The agreement on these various distance scales is within a few hundreths of a magnitude in the mean.

The agreement is particularly significant by noting that the distances to four of the local calibrators determined from RR Lyrae stars (LMC, IC 1613, M31, and M33) using the new RR Lyrae calibration given elsewhere (Sandage 1993c) agree with the distances from Cepheids to within less than 0.1 mag (Lee et al. 1993; Tammann & Sandage 1995).

8. Criticisms of the Short Distance Scale

A series of papers have appeared that show how the bias properties of flux-limited samples compared with bias-free distance-limited samples always lead to an incorrect short distance scale and to too large a Hubble constant. Entrance to the literature can be had via Teerikorpi (1987, 1990), Kraan-Korteweg et al. (1988), Sandage (1988a,b; 1994a,b), Bottinelli et al. (1988), Fouqué et al. (1990), Federspiel et al. (1994). All show that proper correction for selection bias reduces the uncorrected Hubble constants near 85 to the range centered near 50. We do not repeat the arguments here.

The two remaining methods that are said to give the short distance scale (H_o = 85) are the planetary nebulae (PNe) and the surface brightness fluctuation method (SBFM). It is useful to mention our reservations about them here.

The PN method relies on a cutoff of their luminosity function in the λ 5007Å light. PNe do not have an infinitely sharp luminosity function at the bright end. Therefore the method is susceptable to population bias. Large samples will have brighter first-ranked PNe than small samples. The nine brightest PNe in Jacoby's et al. (1990) sample are brighter than the vertical "asymptote" of the luminosity function finally adopted in Jacoby et al. (1992). The consequences of the non-sharp bright end to the LF are discussed by Bottinnelli et al. (1991), by Tammann (1993), and also by Méndez et al. (1993).

Comparison of relative distances determined by the SBFM and the D_n - σ method show that the former are smaller by an average of 25% (Tammann & Sandage 1995). More seriously, the individual distances for 13 E and S0 galaxies in the Virgo cluster (Tonry, Ajhar, & Luppino 1990) show a large scatter in their individual distances (12 to 23 Mpc), but also that these distances are strongly correlated with metallicity (Tammann 1992), indicating uncorrected systematics in the method itself. Consequently Tonry (1991) introduced the V - I color as an additional free parameter, but some metallicity dependence yet remains (Lorenz et al. 1993), with the faint, metal poor ones being (artificially) nearer. In addition, in the one case where a direct comparison with a Cepheid distance is possible (NGC 5253), the SBFM gives 2.5 Mpc whereas the Cepheid distance is 3.9 Mpc, i.e. a distance ratio of a factor of 1.6.

Finally, it is necessary to record our disbelief concerning the announcement of the solution to H_o by Pierce et al. (1994), which also is the communication by van den Bergh in this volume. Our concerns center on (1) the technical aspects of their data, and their displayed P-L relation, and (2) their precept that the distance, even if correct, of the one spiral that is the most easily resolved in the total spiral sample of Virgo "associates" has a connection to the distance of the E galaxy Virgo core.

Support of the Swiss National Sci. Foundation is gratefully acknowledged.

REFERENCES

Birkinshaw, M. 1993, Harvard-Smithsonian Center of Ap., preprint no. 3725.

Bottinelli, L., Fouqué, P., Gouguenheim, L., Paturel, G., & Teerikorpi, P. 1987, A&A, 181, 1

Bottinelli, L., Gouguenheim, L., Paturel, G., & Teerikorpi, P. 1988, ApJ, 328, 4

Bottinnelli, L., Gouguenheim, L., Paturel, G., & Teerikorpi, P. 1991, A&A, 252, 550

Branch, D. 1995, *Proc. Seventh Marcel Grossman Conf.*, eds. M. Kaiser, & R. Jantzen (World Scientific), in press

Branch, D., & Khokhlov, A.M. 1994, Physics Reports, in press

Branch, D. & Tammann, G.A. 1992 Ann. Rev. A&A, 30, 359

Dahle, H., Maddox, S.J., & Lilje, P.B. 1994, ApJ Letters, in press

Dressler, A. 1987, ApJ, 317, 1

Dressler, A. et al. 1987, ApJ, 313, 42

Feast, M. 1994, MNRAS, 266, 255

Federspiel, M., Sandage, A., & Tammann, G.A. 1994, ApJ, 430, 29

Fouqué, P., Bottinelli, L., Gougueheim, L., & Paturel, G. 1990, ApJ, 349, 1

Harris, W.E. 1988, in *The Extragalactic Distance Scale*, ed. S. van den Bergh & C.J. Pritchet, ASP Conf. Ser. Vol. 4, p 231

Harris, W.E., Allwright, J.W.B., Pritchet, C.J., & van den Bergh, S. 1991, ApJS, 76, 115

Jacoby, G., Ciardullo, R., & Ford, H.C. 1990, ApJ, 356, 332

Jacoby, G. et al. 1992, PASP, 104, 599

Jerjen, H., & Tammann, G.A. 1993, A&A, 276, 1

Jones, M. 1994, Ap.Letters & Comm., in press

Kormendy, J. 1977, ApJ, 218, 333

Kraan-Korteweg, R.C., Cameron, L.M., & Tammann, G.A. 1988, ApJ, 331, 620

Kraan-Korteweg, R.C.& Tammann, G.A. 1979, Astron. Nach., 300, 181

Lee, M.G., Freedman, W.L., & Madore, B.F. 1993, ApJ, 417,

Lorenz, H., Böhm, P., Capaccioli, M., Richter, G.M., & Longo, G. 1993, A&A, 277, L15

Madore, B.F. & Freedman, W.L. 1991, PASP, 103, 933

Mathewson, D.S., Ford, V.L., & Buchhorn, M. 1992, ApJS, 81, 413

McLaughlin, D.E., Harris, W.E., & Hanes, D.A. 1994, ApJ 422, 486

Méndez, R.H., Kudritzki, R.P., Ciardullo, R., & Jacoby, G.H. 1993, A&A 275, 534

Minkowski, R. 1962, in *Problems in Extragalactic Research*, IAU Symp. 15, ed. G.C. McVittie, (Macmillan Co.), p 112

Oemler, A. 1974, ApJ, 194, 1

Oemler, A. 1976, ApJ, 209, 693

Pierce, M. et al. 1994, Nature, 371, 385

Pritchet, C.J., & van den Bergh, S. 1987, ApJ, 318, 507

Richter, O.-G., & Huchtmeier, W.K. 1984, A&A, 132, 253

Saha, A. et al. 1994a, ApJ, 425, 14

------------ 1994b, ApJ, in press

Sandage, A. 1975, ApJ, 202, 563

--------- 1988a, ApJ, 331, 583

---------- 1988b, ApJ, 331, 605

---------- 1993a, ApJ, 402, 3

--------- 1993b, ApJ, 404, 419

--------- 1993c, AJ, 106, 703

---------- 1994a, ApJ, 430, 1

---------- 1994b, ApJ, 430, 13

--------- 1995, the 23rd Sass Fee Course of the Swiss Astron. Soc., eds. B. Binggeli & R. Buser (Springer Verlag), in press

Sandage, A., Federspiel, M., & Tammann, G.A. 1995, ApJ submitted

Sandage, A. & Perelmuter, J.-M. 1991, ApJ, 370, 455

Sandage A., & Tammann, G.A. 1974, ApJ, 190, 525

-------------------------- 1975, ApJ, 197, 265

-------------------------- 1988, ApJ, 328, 1

-------------------------- 1990, ApJ, 365, 1

-------------------------- 1994, ApJ, in press

Sandage, A., Saha A., Tammann, G.A., Panagia, N., & Macchetto, D. 1992, ApJ, 401, L7

Sandage, A., Tammann, G.A., Labhardt, L., Schwengeler, H., Panagia, N., & Macchetto, F.D. 1994, ApJ, 423, L13

Schmidt, B.P., Kirshner, R.P., & Eastman, R.G. 1992, ApJ, 395,

Schmidt, B.P. et al. 1994, AJ, 107, 1444

Secker, J., & Harris, W.E. 1993, AJ, 105, 1358

Tammann, G.A. 1987, in Relativistic Astrophysics, ed. M. Ulmer, 13th Texas Symp. World Scientific (Singapore), p. 8

------------ 1988, in The Extragalactic Distance Scale, ed. S. van den Bergh, & C.J. Pritchet, ASP. Conf. Ser. No. 4, p 282,

------------ 1992, Physica Scripta, T43, 31

------------ 1993, in Planetary Nebulae, IAU Symp. 155, eds. R. Weinberger, & A. Acker (Dordrecht: Kluwer), p 515

Tammann, G.A. & Sandage, A. 1985, ApJ, 294, 81

-------------------------- 1995, Lectures at the 3rd International School "D. Chalonge", Erice, Nov. 1994, in press

Teerikorpi, P. 1987, A&A, 173, 39

------------ 1990, A&A, 234, 1

Tonry, J.L. 1991, ApJ, 373, L1

Tonry, J.L., Ajhar, A.E., & Luppino, G.A. 1990, AJ, 100, 1416

van der Kruit, P. 1986, A&A, 157, 230

LARGE-SCALE FLOWS IN THE LOCAL UNIVERSE

D.S. MATHEWSON and V.L. FORD

Mount Stromlo and Siding Spring Observatories,
The Australian National University, ACT 2611, Australia

Abstract. Peculiar velocity measurements of 2500 southern spiral galaxies show large-scale flows in the direction of the Hydra-Centaurus clusters which fully participate in the flow themselves. The flow is not uniform over this region and seems to be associated with the denser regions which participate in the flow of amplitude about 400km/s. In the less dense regions the flow is small or non-existent. This makes the flow quite asymmetric and inconsistent with that expected from large-scale, parallel streaming flow that includes all galaxies out to 6000km/s as previously thought. The flow cannot be modelled by a Great Attractor at 4300km/s or the Centaurus clusters at 3500km/s. Indeed, from the density maps derived from the redshift surveys of "optical" and IRAS galaxies, it is difficult to see how the mass concentrations can be responsible particularly as they themselves participate in the flow. These results bring into question the generally accepted reason for the peculiar velocities of galaxies that they arise solely as a consequence of infall into the dense regions of the universe. To the N. of the Great Attractor region, the flow increases and shows no sign of diminishing out to the redshift limit of 8000km/s in this direction. We may have detected flow in the nearest section of the Great Wall.

1. Introduction

The results of a typical N-body simulation of the formation of structure in our Universe and the predicted peculiar velocity field using the standard CDM theory are shown in Figure 2 of Nusser et al. (1991). As expected the flows are towards the mass-centers produced by their simulation. Their working hypotheses concern the nature of the primordial density fluctuations, the process of gravitational instability, the value of Ω/b where b is the biassing factor and that the galaxy and mass-density fluctuations are proportional at each point.

These hypotheses can be tested by measurements of, 1) the cosmic microwave background (CMB), its irregularities and dipole, 2) the distribution of galaxies in redshift-space from which the galaxy-density field may be obtained to make peculiar velocity (Vpec) predictions and 3) the Vpec of galaxies and the construction using POTENT (Bertschinger & Dekel 1989) of the mass-density field. A comparison of the density fields obtained from redshifts to those obtained from Vpec is the crucial test of the current theories of the evolution of the universe.

The 1990s have seen the disappearance of the concept of the Great Attractor (GA) introduced by Lynden-Bell et al. (1988) as the mystery object responsible for the observed flows in the local universe and the reintroduction of large-scale bulk flows of amplitude about 400km/s. What drives these bulk flows is the subject of much debate.

M. Kafatos and Y. Kondo (eds.), Examining the Big Bang and Diffuse Background Radiations, 175–182.
© 1996 IAU.

Willick (1990) detected a flow of galaxies from the direction of Perseus-Pisces toward the Local Group. Mathewson, Ford and Buchhorn (1992a, b) from the Vpec measurements of 1355 southern spiral galaxies, found that there was evidence of a bulk flow of about 600km/s towards the direction of the GA region on a scale of at least 6000km/s. This, when added to the bulk flow of 450km/s found by Willick in the opposite part of the sky, suggested that there is bulk flow in the supergalactic plane over very large scales greater than 13000km/s. More recently Courteau et al. (1993), using a new set of Tully-Fisher (TF) distances for northern spirals and the data from Mathewson et al. (1992b), support the previous findings and conclude that there is a large-scale, parallel streaming flow of 360km/s towards $l = 294°$, $b = 0°$ that includes all galaxies out ιo at least 6000km/s radius around the Local Group.

In an attempt to probe deeper into the local universe to see if the scale of this bulk flow can be determined, we extended the survey of 1355 galaxies by Mathewson et al. (1992b) and made Vpec measurements of an additional 1200 galaxies. The combined data set of about 2500 galaxies has been used to examine these large-scale flows and the preliminary results are presented here (also see Mathewson & Ford 1994).

2. The Data

The extended sample of spiral galaxies was selected from the ESO-Uppsala Catalog of types Sb - Sc, diameters between 1'.0 and 1'.6, velocities in general between 4000km/s and 14000km/s inclinations greater than 40° and |b|>11°. Altogether over 1200 galaxies were observed. These included 90 galaxies with redshifts less than 8000km/s N. of the GA region (l = 250° to 360°, b = 45° to 70°) from a list kindly supplied by Dr John Huchra.

The Hα rotation curves of the spiral galaxies were measured with the 2.3m telescope at Siding Spring Observatory using the same procedures described in Mathewson et al (1992b). Photometry was obtained in the Kron-Cousins I passband with a GEC CCD on the 1m telescope. The I magnitudes were found using methods outlined in the first survey. The "zero-velocity" HI was observed in the direction of each galaxy with the Parkes radiotelescope and used to measure the extinction in our Galaxy following Burstein and Heiles (1978).

The complete data set when combined with the 1355 galaxies measured by Mathewson et al (1992b) consists of about 2500 galaxies with 1200 galaxies lying in the direction of the GA region defined as 260°<l<360°, -40°<b<45°. The aim is to investigate the flow already detected in the direction of the GA with this combined data set.

3. The TF Relation and Selection Bias

The method is to assume that the galaxies outside the GA region (apart from galaxies to the N. of b = 45° and in the Perseus-Pisces region) are at sufficiently large angles to the flow that their velocities only reflect pure Hubble flow i.e., their redshifts are their distances. These galaxies in this control zone are then used to calibrate the TF relation which is used to estimate the Vpec of the galaxies in the direction of the flow.

The luminosity distribution of the sample changes with increasing redshift, showing that observational selection bias exists. The peak magnitude progressively moves to brighter values as the redshift increases. This, coupled with the dispersion ($\sigma = 0.44$) of the TF relation combines to make the absolute magnitude of galaxies a function of redshift as well as rotational velocity, Vrot. (also see Federspiel et al. 1994). The dispersion of the TF relation increases markedly for Vrot<63km/s and these galaxies have been omitted from the analysis.

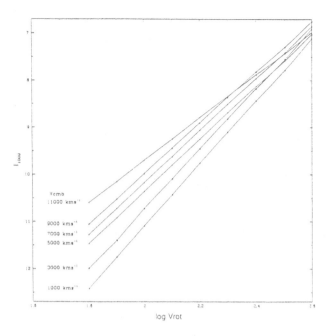

Fig. 1. The Tully-Fisher relation as a function of redshift (Vcmb) for the galaxies in the control region. I 1000, the magnitude of the galaxies at a distance of 1000km/s, is plotted against log Vrot, the velocity of rotation for galaxies in redshift bins 2000km/s wide centred on the values given at the bottom of each plot.

Using the galaxies in the control zone, Figure 1 was constructed, which gives I 1000 as a function of Vrot and redshift (distance). I 1000 is the magnitude of the galaxies at a distance of 1000km/s. In this paper all redshifts are relative to the cosmic microwave background, Vcmb. The values of I 1000 for the galaxies in the GA region are found using Figure 1. As the luminosity distribution of galaxies in the GA region is very similar to those in the control zone, this process should remove most of the selection bias and allow measurement of the Vpec of galaxies in the GA region relative to the control zone. The accuracy of an individual measurement of distance is about 22%.

To improve the accuracy of the detection of large-scale flows, the galaxies were divided into redshift bins of 1000km/s and averaged. With the present sample, this procedure would allow flows of about 300km/s or more to be detected at distances of 11000km/s. At greater distances, the fluctuations increase rapidly due to errors which increase with distance and small number statistics and the results have small significance.

As a check, Figure 2 plots Vpec vs Vcmb for the 1100 galaxies in the control zone. The galaxies were divided into redshift bins of 1000km/s and averaged. The averaged points lie close to the Vpec = 0 line with mean value -57 ± 49km/s consistent with no net flow.

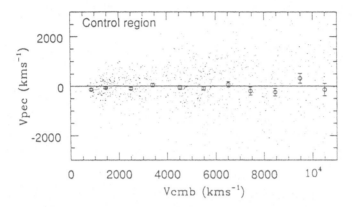

Fig. 2. Vpec vs Vcmb for the control zone which is the region outside the Great Attractor box in Fig. 3 (apart from galaxies to the N. of that box and in the Perseus-Pisces region). This is assumed to be a region of quiet Hubble flow. The open circles are the averages of the galaxies (dots) in intervals of 1000km/s Vcmb. A total of 1100 galaxies are plotted.

4. Results

Figure 3 shows the distribution in galactic longitude and latitude of the galaxies used in this analysis. The GA region has been divided into four areas and Figure 4 shows Vpec vs Vcmb plots for the galaxies within them. The galaxies have been placed in 1000km/s bins of Vcmb and averages taken. The three areas in Centaurus (Fig. 3, area a), Hydra (Fig. 3, area b) and Pavo (Fig. 3, area c) show well defined flows of approximately 300 - 400km/s out to distances of 6000 - 7000km/s when the flow vanishes into the increasing noise at greater distances. The amplitudes of the flows in each area are marked against each box in Figure 3. Figure 5 shows the distribution in redshift space of the galaxies in the GA region. The flow appears to be associated with the denser regions (Figs. 3 and 5 and Fig. 1 of Hudson 1994) which participate in the flow. In the less dense areas in the GA region, i.e., beyond 6500km/s and in the direction of area d in Fig. 3, the flow is not so well defined.

Figure 6 shows the combined plot of Vpec vs Vcmb for the three areas in the GA region showing·flow. The galaxies have been divided into bins of 1000km/s Vcmb and averages taken in each bin denoted by the open circles. The flow of 370±16km/s is clearly defined up to 6000km/s but from Vcmb = 7000km/s to 11000km/s, the mean value of Vpec is only 3±128km/s which is consistent with no flow. As the control region (Fig. 2) has Vpec effectively zero from Vcmb = 0 to 11000km/s and if it can be regarded as a region of quiet Hubble flow, then the above estimated flow values represent the real flow values of the GA region relative to the CMB. Our flow value of 370km/s agrees closely with that found by Courteau et al (1993) over the same distance range who used as a control, a region almost diametrically opposed to ours.

Figure 7 shows the plot of Vpec vs Vcmb for the region $l = 250° - 360°$, $b = 45° - 70°$, to the N. of the GA region (see Fig. 3). There is strong bulk flow of 489±61km/s which shows no sign of diminishing out to 8000km/s, the redshift limit of the sample in this direction.

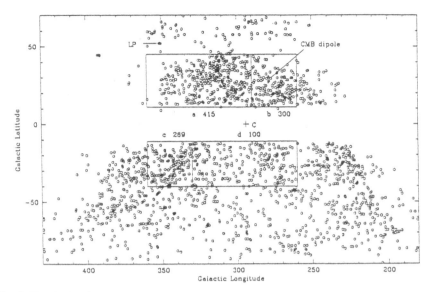

Fig. 3. The distribution in galactic longitude and latitude of the 2500 southern spiral galaxies used in the analysis of peculiar velocities. The GA region is outlined and split into four areas against which is indicated the amplitude of the flow derived using Fig. 4. The directions of the CMB dipole and the bulk flow directions denoted C and LP of Courteau et al (1993) and Lauer and Postman (1994) respectively, are shown.

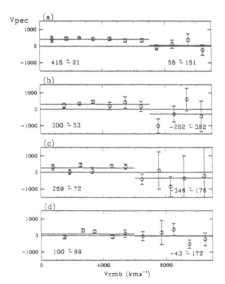

Fig. 4. Vpec vs Vcmb for the four areas in the GA region shown in Fig. 3. Each plot is labelled with a letter to identify which area it refers to in Fig. 3. The open circles give the average Vpec of the galaxies in intervals of 1000km/s Vcmb. The mean of Vpec above and below Vcmb = 7000km/s for (a) and (b) and 6000km/s for (c) and (d) are shown on the RHS and LHS of each plot respectively. The amplitude of the flows is shown against each box in Fig. 3.

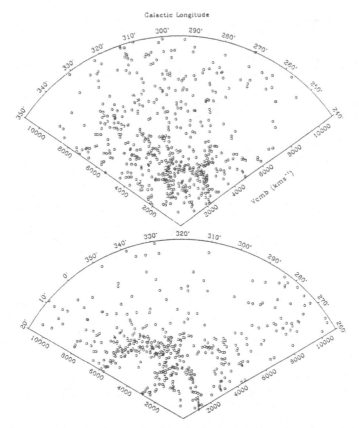

Fig. 5. The distribution in redshift-space for the galaxies in the GA region; a) +ve latitudes, b) -ve latitudes.

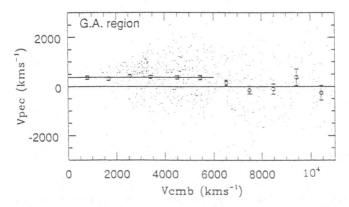

Fig. 6. The combined Vpec vs Vcmb plot for the three areas (a, b, c) which show strong flow in the GA region. The open circles give the average Vpec of galaxies (dots) in intervals of 1000km/s Vcmb. The mean of Vpec below and above Vcmb = 6000km/s is 370±16km/s and 3±128km/s respectively.

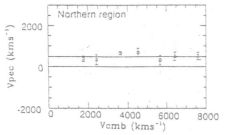

Fig. 7. Vpec vs Vcmb plot for the region to the N. of the GA region. The open circles give the average Vpec of the galaxies (dots) in intervals of 1000km/s Vcmb. The redshift limit of the sample in this direction is 8000km/s. The mean Vpec over this redshift range is 489±61km/s.

5. Discussion

Because the flow in the GA region is associated with its denser regions, it is natural to suspect that the TF relation is dependent on environment which may create the illusion of a flow. However the flow would seem to be genuine because:-

1. northern hemisphere observers, using the TF relation, have found that the Perseus-Pisces supercluster is flowing towards us, i.e., in the opposite direction to the GA region which is moving away. If there was some effect artificially producing flows, it would need to change sign with direction, which is most unlikely,

2. Dressler et al (1987), using the σ-D relation for ellipticals, have shown outflow in the Centaurus-Hydra region of about the same amplitude as we have shown using the TF relation for spirals. However the σ-D and TF relations are quite different physical relations, the former tells us about galaxy equilibrium while the latter tells us about galaxy formation. M/L constancy does enter both relations but the stellar populations in the two galaxy types are quite different so it would be remarkable if M/L variations in both types were the same,

3. the strongest flow in the GA region of 450km/s is centred on l = 310°, b = 30° in Centaurus only 30° from the CMB dipole at l = 276°, b = 30° (Kogut et al 1993) which is itself in a region of strong flow (Fig 3). This provides independent support for our Vpec measurements as there is no doubt now that the dipole is due to a Vpec of the Local Group of about 600km/s (Fixsen et al 1994).

It is very difficult to model the observed asymmetric flow with a bulk flow of amplitude 360km/s toward l = 294°, b = 0° over a distance of 6000km/s (Courteau et al 1993) when the adjacent area d, Fig. 3, shows a flow of only 100±89km/s instead of a predicted flow of at least 300km/s. Also there is no sign of the expected "S-curve" from a GA at 4200km/s in the direction l = 307°, b = 9° (Lynden-Bell et al 1988). Likewise there appears little chance of salvaging the GA by calling it Hydra-Centaurus (Federspiel et al 1994) because these clusters at 3500km/s do not have an overly large effect on the flow in which they themselves fully participate. Courteau et al (1993) suggested that large-scale, low amplitude density fluctuations are responsible for the flow of about 400km/s from the Perseus-Pisces region to the GA region, over a distance of 12000km/s. However this does not explain the fall to zero around 6500km/s in the GA region (Fig. 4) nor the asymmetric nature of the flow about its apex, nor the large amplitude of the flow compared with the apparent small effect on the flow of the mass concentrations, e.g., Centaurus. Shapley 8 at 14000km/s does not seem to exert much power on the flow as at 11000km/s, the flow is still small. It is apparently not the Giant Attractor as proposed by Scaramella et al (1989).

Hudson (1994) derived the density field out to a depth of 8000km/s using redshift surveys of galaxies selected from ESO and UGC catalogs and predicted the peculiar velocities of the galaxies. He finds that the Centaurus-Hydra-Virgo and Pavo supercluster complex is not primarily responsible for the large streaming motions of galaxies. Hudson believes that most of the bulk motion of the 405km/s which is required so as to agree with the predicted motion of the Local Group, is due to sources beyond 8000km/s. However the next large mass concentration encountered in the flow direction is Shapley 8 at 14000km/s and this does not appear to have much effect on the flow.

It is not just a question of whether light traces mass but whether the standard theory of the formation of large-scale structure with the consequent Vpec flows is correct (c.f. Silk 1987). In this regard the main difficulty that the standard theory faces is to explain why the large visible mass centers of the Local Universe do not appear to produce large scale flows but instead fully participate in the flows themselves. This finding does not negate the result of Lauer and Postman (1994) of bulk flows of 689km/s out to distances of 15000km/s as flows may be a property of the denser regions in which are embedded their observed clusters. In this respect, it is intriguing that the bulk flow is strongest in the region N. of the GA region and shows no sign of abating out to the survey limit in this direction of 8000km/s (Fig. 7). This region encompasses Lauer and Postman's flow direction of l = 343°, b = 52°, marked in Fig. 3. It is interesting that the nearer parts of the Great Wall lie around l = 330°, b = 70° at a distance of about 7000km/s (Geller and Huchra, 1989 and also see Fig. 10 of Hudson, 1993). Has the flow of the Great Wall been detected? There is an obvious need for more observations in this area.

The authors thank the staff of Siding Spring Observatory for maintaining the telescopes in excellent condition throughout the observations.

References

Bertschinger, E. and Dekel, A.: 1989, ApJ. 336, L5.

Bursteain, D. and Heiles, C.: 1978, ApJ. 225, 40.

Courteau, S., Faber, S.M., Dressler, A. and Willick, J.A.: 1993, ApJ. 412, L51.

Dressler, A., Faber, S.M., Burstein, D., Davies, R.L., Lynden-Bell, D., Terlevich, R.J. and Wegner, F.: 1987, ApJ. 313, L37.

Federspiel, M., Sandage, A. and Tammann, G.A.: 1994, ApJ. 430, 29.

Fixsen, D.J. et al: 1994, ApJ. 420, 439.

Geller, M.J. and Huchra, J.P.: 1989, Science. 246, 857.

Hudson, M.J.: 1993, MNRAS. 265, 43.

Hudson, M.J.: 1994, MNRAS. 266, 475.

Kogut, A. et al: 1993, ApJ. 419, 1.

Lauer, T.R. and Postman, M.: 1994, ApJ. 425, 418.

Lynden-Bell, D., Faber, S.M., Burstein, D., Davies, R.L., Dressler, A., Terlevich, R.J. and Wegner, G.: 1988, ApJ. 326, 19.

Mathewson, D.S., Ford. V.L. and Buchhorn, M.: 1992a, ApJ. 389, L5.

Mathewson, D.S., Ford. V.L. and Buchhorn, M.: 1992b, ApJS. 81, 413.

Mathewson, D.S. and Ford, V.L.: 1994, ApJ (in press)

Nusser, A., Dekel, A., Bertschinger, E. and Blumenthal, G.R.: 1991, ApJ. 379, 6.

Scaramella, R., Baiesi-Pillastrini, G., Chincarini, G., Vettolani, G. and Zamorani, G.: 1989, Nature. 338, 562.

Silk, J.: 1987, "Observational Cosmology", IAU Symposium No. 124, p.391.

Willick, J.A.: 1990, ApJ. 351, L5.

SPIRAL GALAXIES AND THE PECULIAR VELOCITY FIELD

RICCARDO GIOVANELLI AND MARTHA P. HAYNES
Cornell University, Ithaca, NY, USA

PIERRE CHAMARAUX
Observatoire de Meudon, Meudon, France

LUIZ N. DA COSTA
*Observatorio Nacional, Rio de Janeiro, Brazil
and Institut d'Astrophysique, Paris, France*

WOLFRAM FREUDLING
ESO ST–European Coordinating Facility, Garching, Germany

JOHN J. SALZER
Wesleyan University, Middletown, CT, USA

AND

GARY WEGNER
Dartmouth College, Hanover, NH, USA

Abstract.
We report results of a redshift–independent distance measurement survey that extends to all sky and out to a redshift of approximately 7500 km s^{-1}. Tully–Fisher (TF) distances for a homogeneous sample of 1600 late spiral galaxies are used to analyze the peculiar velocity field. We find large peculiar velocities in the neighborhood of superclusters, such as Perseus–Pisces (PP) and Hydra–Centaurus, but the main clusters embedded in those regions appear to be virtually at rest in the CMB reference frame. We find no compelling evidence for large–scale bulk flows, whereby the Local Group, Hydra–Cen and PP would share a motion of several hundred km s^{-1} with respect to the CMB. Denser sampling in the PP region allows a clear detection of infall and backflow motions, which can be used to map the mass distribution in the supercluster and to obtain an estimate of the cosmological density parameter.

M. Kafatos and Y. Kondo (eds.), Examining the Big Bang and Diffuse Background Radiations, 183–191.

1. Introduction

The discovery of the dipole anisotropy of the Cosmic Microwave Background (CMB) has been widely interpreted as a Doppler effect, resulting from the motion of the Local Group with respect to the comoving reference frame at a velocity of 620 km s^{-1} in the direction $l = 270 \pm 5$, $b = 30 \pm 3$ (Kogut et al. 1993). The independent measurement of redshifts and distances for individual galaxies, which can be assumed to be fair tracers of the peculiar velocity field, provides a test for the Doppler interpretation of the CMB dipole and an estimate of the distribution of masses whose gravitational effects give rise to the peculiar motion field. After the early measurements of the infall of the Local Group in the Local Supercluster (Aaronson et al. 1982), the existence of a larger attractor was independently postulated by Shaya (1984), Tammann and Sandage (1985) and Lilje et al. (1986) to coincide with the Hydra–Centaurus supercluster. The measurements of Lynden–Bell et al. (1987) suggested the existence of an even larger, somewhat more distant mass perturber, at an approximate $cz \simeq 4300$ km s^{-1} . Scaramella et al. (1989) made the radical suggestion that much of the perturbation responsible for the LG motion could arise from the Shapley Supercluster at $cz \simeq 14,000$ km s^{-1} . The TF distance measurements reported by several groups (Willick 1991, Mathewson et al. 1991, Courteau et al. 1993) and the dipole of the distribution of cluster brightest ellipticals (Lauer and Postman 1994) lent credence to a scenario where the large–scale bulk flows extend over distances comparable to that of the Shapley Supercluster. These results were in conflict with the relatively smaller convergence depth predicted from the redshift distribution of IRAS galaxies (Strauss et al. 1992). It is also somewhat disconcerting that, amidst reported bulk flows of several hundred km s^{-1} and the dramatic density contrast in the light distribution, as emphasized by redshift surveys, very small gradients in the peculiar velocity flow would be detected within the 7,000 km s^{-1} or so radius of the sphere sampled by TF studies. In this paper, we shall report results that are discrepant with previous TF studies, favoring a picture of the local universe characterized by a small convergence depth.

2. Samples Used

The results presented in the following sections are drawn from several data sets, namely: (a) a sample of 1500 Sbc–Sc galaxies of Dec. $> -45°$ and $cz < 7500$ km s^{-1} , restricted to objects with blue visual diameters $>$ 1.3′ (hereinafter referred to as the Sc sample); (b) a survey of the PP supercluster extending to $cz \simeq 12,000$ km s^{-1} and angular size $> 0.8′$ which includes an additional 300 objects to those in sample (a) (the PP sample);

(c) a cluster sample, including spiral galaxies in 20 clusters to $cz \simeq 12,000$ km s^{-1}, and the addition of 300 objects to those in samples (a) and (b). For each of these galaxies, we have obtained I band CCD images and high quality velocity widths. For samples (a) and (b), velocity widths derive from 21cm observations made at Arecibo, Nançay, Green Bank and Effelsberg; for sample (c), additional velocity widths were obtained from optical spectra taken at the 5m telescope on Mt. Palomar, in collaboration with T. Herter and N. Vogt. The I band observations were carried out at MDM, Kitt Peak, CTIO and ESO telescopes. More details on the selection criteria of the (a) and (b) samples, including their mean photometric characteristics, can be found in Giovanelli et al. (1994, 1995).

In order to achieve full sky coverage with the Sc sample, our data were combined with those of Mathewson et al. (1992) in the South polar cap, after selection through a homogeneous set of criteria that would allow smooth merger into our sample and reprocessing of their data with the same procedures applied to ours. After filtering the sample for duplicate entries, objects with inadequate parameters for TF use, poor photometry, dwarf and interacting systems, etc, we obtain an all sky Sc sample of 1600 galaxies.

3. Corrections Applied to Observed Parameters

The main parameters entering our estimate of a TF distance are the total I band magnitudes, disk inclinations and velocity widths. Several corrections need to be applied to the observed quantities, before they can be used as diagnostic tools, namely those for galactic, cosmological and internal extinction effects to magnitudes; cosmological, instrumental, turbulent and inclination corrections for the widths; seeing and bulge contamination for disk inclinations.

While typical photometric accuracy inferred from CCD images is on the order of a few percent, corrections as applied by different groups may differ by several tenths of a mag. In Giovanelli et al. (1994) we have reviewed the procedures for the internal extinction corrections, and in Giovanelli et al. (1995) we present a luminosity dependent solution for that problem. In fig. 1, we illustrate an important result of those studies: if the correction to the observed total magnitude m for conversion to a standard $m°$ face–on aspect is $m - m° = \gamma \log(a/b)$, where a/b is the disk axial ratio inferred from the ellipticity of the outer isophotes, then γ is found to be luminosity dependent. Its value varies between 0.5 and 1.2, higher for the more luminous galaxies. These are substantially higher values than adopted in most previous TF applications. It is clear that a luminosity dependence of γ will affect the inferred slope of the TF relation. In Giovanelli et al. (1995) we discuss how

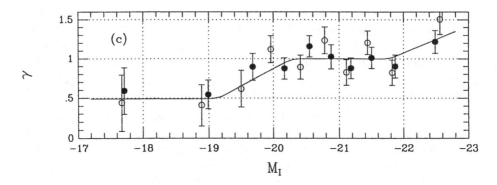

Figure 1. Luminosity dependence of the slope coefficient γ for the internal extinction law: $m - m^\circ = \gamma \log(a/b)$.

an inadequate extinction correction law can produce unwanted effects in the computation of peculiar velocity, such as spurious dependences of the latter on both luminosity and inclination.

Particular attention has been applied to the measurement and the processing of velocity widths, as the associated errors are a very important component of observational scatter in the TF relation, especially for galaxies of smaller velocity widths. Twenty–one cm velocity widths reported in the literature are often measured on heavily smoothed line profiles, as they pertain to low S/N observations optimized for redshift determination. We have reobserved a large fraction of those in order to obtain reliable widths, and have optimized width measurement algorithms that take into consideration line shape, S/N and resolution of the data.

4. Template for the TF Relation

The adoption of a TF template relation is perhaps the most delicate step in a peculiar velocity measurement program. A relation with the wrong offset or slope will produce a spurious peculiar velocity field with a characteristic geocentric signature. In a sample with partial sky coverage, it can easily mimic a bulk flow. The fit to a single cluster of galaxies is subject to the vagaries of small number statistics, and when the resulting TF relation is used as a template, it is likely to produce spurious results. We have thus chosen to use our cluster sample (see list in Table 1) to produce a TF template relation. The first step in such process regards cluster membership. Two different set of membership criteria were applied, producing a "strict

TABLE 1

CLUSTERS USED IN TEMPLATE

Cluster	R.A.	Decl.	V_{cmb} km s^{-1}	V_{pec} this paper	V_{pec} HM [a]	V_{pec} MFB [b]	N_{tot}	N_{in}
N507	012000.0	+330400	4741	+207±165	+76±296		22	10
A 262	014950.0	+355440	4574	0±140	-576±391		27	13
A 400	025500.0	+055000	6842	-224±190	-130±537		27	9
Fornax	033634.0	-353642	1260	-305±45		-58	35	17
Cancer − a	051353.7	+062441	4886	+20±90	-104±377		48	13
Antlia	102745.0	-350411	3134	+318±98	+324±223	+781	20	18
Hydra=A1060	103427.7	-271626	4085	-75±115	+184±296	+311	35	19
A1367	114154.0	+200700	6750	+94±112	-255±431		39	25
UMa	115400.0	+493000	1091	+523±50			31	26
Cen30	124606.0	-410200	3378	+266±110	+509±213	+997	22	22
Coma=A1656	125724.0	+281500	7225		+80±428		53	30
ESO508	130954.0	-230854	3276	+355±170	+654±206	+671	16	8
Kl 27=A3574	134606.0	-300900	4888	+207±160		+368	22	9
A2199	162654.0	+393800	9058	0±255			23	9
Pavo I	201300.0	-710000	4104	-19±160		+661	23	
Pegasus	231742.6	+075557	3958	-290±105	-588±285		34	12
A2634	233554.9	+264419	9483	-206±200	-906±639		41	19

[a] Han & Mould 1992, *Ap. J.* **396**, 453

[b] Mathewson, Ford & Buchhorn 1992, *Ap. J. Suppl. Ser.* **81**, 413

members sample" whereby only galaxies within 2 Abell radii and within a conservative $\Omega = 0.25$ caustic from the cluster center were chosen, and an "extended sample" which includes nearby supercluster members out to 4 Abell radii. The strict sample contains 260 galaxies and the extended one, 525. The corresponding TF relations differ very little from each other. TF diagrams $(M, \log W)$ were built for each individual cluster and combined into a composite by shifting in M, taking into account each cluster's incompleteness correction. The magnitude shift for each cluster necessary to match the composite template yields a relative distance modulus and was χ^2-computed simultaneously with the determination of the optimal TF slope for the composite sample. The magnitude scatter in the composite TF relation is 0.29 mag.

The slope of the TF relation is best determined by a sample that maximizes the dynamic range in $(M + 5 \log h, \log W)$, i.e. one that preferentially includes nearby clusters. On the other hand, the magnitude offset ("zero point", a relative concept as long as the value of h isn't specified) of the relation is best obtained from a sample of distant clusters for which a motion of given amplitude in km s^{-1} translates into a small magnitude shift and for which the physical size of the cluster is small in comparison to its mean distance.

The magnitude zero–point of our TF relation was adopted as that of the mean relation obtained from the subset of most distant clusters in the sample ($cz > 4000$ km s^{-1}) as seen by an observer at rest with respect to the CMB. This is our closest operational definition of a CMB rest reference frame. Even if the clusters partook of very large scale bulk motion, their mean TF relation would define a reference frame at rest with respect to the CMB, provided that their sky distribution is relatively isotropic. The use of the distant clusters rather than the whole cluster sample for this purpose allows for the smallest margin of error in the approximation of a CMB rest frame by our TF template. Individual cluster peculiar velocities with respect to that frame and associated errors (estimated using galaxies in the extended sample) are listed in Table 1. These are preliminary values, and small changes resulting from refinements in the analysis are likely to apply. For those clusters for which other estimates of peculiar velocity by Han and Mould (1992) or Mathewson et al. (1992) are available, a comparison is possible. Note that if the distant clusters partook of large scale bulk flow, their individual velocities would reflect it. Instead, indications are that they are individually at rest with respect to the CMB. Large velocities are measurable for nearby clusters, indicating that they partly share in the LG motion with respect to the CMB.

5. The Peculiar Velocity Field of the Sc Sample

Figure 2 gives two representations of the peculiar velocity field as obtained from the all sky Sc sample. In the top panel, the peculiar velocities are given in the Local Group reference frame, i.e. $V_{pec} = V_{LG} - V_{TF}$, where V_{LG} is the radial velocity measured with respect to the LG, and V_{TF} that predicted by the velocity width and apparent magnitude of the galaxy. In the bottom panel, peculiar velocities are given in the CMB reference frame. The redshift depth of the Sc sample is 7500 km s^{-1} . In the top panel we see the motion of the LG with respect to the CMB reflected in the prevalence of positive peculiar velocities in the southern galactic hemisphere and of negative ones in the northern hemisphere. The apex of the CMB dipole is at roughly ($10.5^h, -25°$). The dipole signature largely disappears in the bottom panel, when the peculiar velocity field is referred to the CMB. This immediately tells us that the majority of galaxies in the sample do not partake of the LG motion with respect to the CMB.

In a quantitative way, the immediately preceding statement is illustrated in Table 2, where the apex location and the amplitude of the dipole of the peculiar velocity field is displayed separately for galaxies within different redshift shells. The first line refers to all galaxies in the all sky Sc sample: a

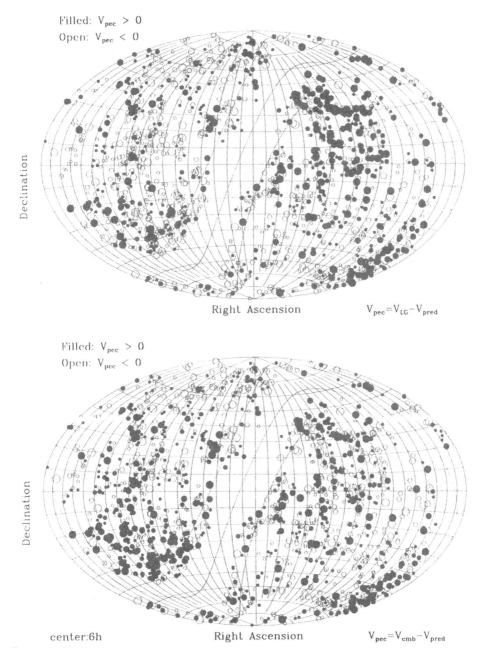

Filled: $V_{pec} > 0$
Open: $V_{pec} < 0$

Declination

Right Ascension $V_{pec} = V_{LG} - V_{pred}$

Filled: $V_{pec} > 0$
Open: $V_{pec} < 0$

Declination

center:6h   Right Ascension $V_{pec} = V_{cmb} - V_{pred}$

Figure 2. Peculiar velocities of galaxies in the all sky Sc sample. Each symbol represents a galaxy, of which 1600 are plotted. Filled symbols are positive (receding) and open symbols negative velocities. The coordinate system is equatorial, with (R.A., Dec.)=(6^h, $0°$) at the center; "meridians" are each 40^m and "parallels" each $10°$. Dotted lines are plotted for galactic $b = 0°$ and $b = \pm20°$. The location of the apex of the CMB dipole is approximately at (R.A., Dec.)=(10.5^h, -25°). In the top panel velocities are referred to the LG reference frame; in the bottom, to the CMB.

TABLE 2

Dipoles of the Peculiar Velocity Field

V_{cmb} Window	$V_{pec}(LG)$	l	b	# Gals.
All	417±28	253±9	37±3	1585
2000–3000 km s^{-1}	333±41	268±16	41±8	235
3000–4000 km s^{-1}	437±57	242±20	24±5	303
4000–5000 km s^{-1}	551±62	236±24	37±5	372
5000–6000 km s^{-1}	566±81	281±15	24±25	270
CMB	627±22	276±3	30±3	*

*KOGUT et al. 1993, *Ap. J.* **419**, 1

net bulk flow of about 200 km s^{-1} is detectable. As the sample is separated in redshift shells, we see that both the apex and the amplitude of the motion converge towards that implied by the CMB temperature anisotropy. By the time the shell has reached a radius of about 5000 km s^{-1}, galaxies contained within it are essentially at rest with respect to the CMB.

In contrast, the peculiar velocity field in the vicinity of large concentrations of galaxies, such as the Hydra–Centaurus and the Perseus–Pisces superclusters appears quite disturbed. Infall and backflow velocities are seen rising to about 1300 km s^{-1}, at distances of about 2000 km s^{-1} of the center of those features. At the same time, the centers of mass of those agglomerates appear to be virtually at rest with respect to the CMB. The amplitude of the peculiar motions around superclusters is consistent with expectations based on redshift surveys and suggest that the light and matter distributions are not strongly biased. Preliminary estimates of the cosmological density suggest values of $\Omega^{0.6}/b \sim 1$, where b is the bias parameter between the galaxian and mass density distributions.

The motion of the LG with respect to the CMB arises from the joint attraction of the large mass concentration represented by Hydra–Cen, neighboring clusters and Virgo, and the virtual repulsion of the large void extending between the LG and the PP supercluster. The Hydra–Cen region does not exhibit a significant motion with respect to the CMB.

In summary, our results indicate that the region enclosed within a 7500 km s^{-1} radius exhibits a mild bulk flow of about 200 km s^{-1}; this arises from the asymmetry of the mass distribution in the local universe and the location of the LG in a region characterized by a large mass density gradient. The central regions of the large "attractors" — the Hydra–Cen and PP superclusters — are at rest with respect to the CMB and *do not* partake of large scale bulk flow. The convergence depth is on the order of 5000 km s^{-1} .

The discrepancy of our results with those of other TF studies arise principally from the merits of (a) access to an all–sky, homogeneous sample, (b) a different TF template relation, which is based on a very extensive study of clusters, and (c) an internal extinction correction that allows for larger flux corrections and is luminosity dependent.

This work was supported by NSF grants AST91-15459, AST92-18038 and AST90-23450. It is based on observations made at the National Astronomy and Ionosphere Center, the National Radio Astronomy Observatory and the National Optical Astronomical Obervatories, which are operated respectively by Cornell University, Associated Universities, Inc. and Associated Universities for Research in Astronomy, under cooperative agreements with the National Science Foundation.

References

1. Aaronson, M., Huchra, J., Mould, J., Schechter, P.L. and Tully, R.B. (1982) *Ap. J.* **258**, 64
2. Courteau, S., Faber, S.M., Dressler, A. and Willick, J.A. (1993), *Ap. J. (Letters)* **412**, L51
3. Giovanelli, R., Haynes, M.P., Salzer, J.J., Wegner, G., da Costa, L.N. and Freudling, W. (1994), *Astron. J.* **107**, 2036
4. id. (1995), *Astron. J.* submitted
5. Han, M. and Mould, J. (1992), *Ap. J.* **396**, 453
6. Kogut, A. et al. (1993), *ApJ* **419**, 1
7. Lauer, T.R. and Postman, M. (1994), *ApJ* **425**, 418
8. Lilje, P.B. Yahil, A. and Jones, B.T. (1986)*Ap. J.* **307**, 91
9. Lynden–Bell, D., Faber, S.M., Burstein, D., Davies, R.L., Dressler, A., Terlevich, R.J. and Wegner, G. (1988), *Ap. J.* **326**, 19
10. Mathewson, D.S., Ford, V.L. and Buchhorn, M. (1992) *Ap. J. Suppl.* **81**, 413
11. Mathewson, D.S., Ford. V.L. and Buchhorn, M. (1991) *Ap. J. (Letters)* **389**, 1L5
12. Scaramella, R., Baiesi–Pillastrini, G., Vettolani, G. and Chincarini. G. L. (1989) *Nature*, **338**, 562
13. Shaya, E. (1984), *Ap. J.* **280**, 470
14. Strauss, M.A., Davis, M., Yahil, A. and Huchra, J.P. (1992), *Ap. J.* **361**, 49
15. Tammann, G. and Sandage, A. (1985) *Ap. J.* **294**, 81
16. Willick, J. (1991), *Ap. J. (Letters)* **351**, L5

FORMATION OF THE SUPERCLUSTER-VOID NETWORK

J. EINASTO
Tartu Astrophysical Observatory, EE-2444, Estonia

1. Evidence

Already first redshift surveys, made 10 – 15 years ago, demonstrated that galaxies and clusters of galaxies are located along patchy sheets and filaments, and the space between is empty. Sheets and filaments form large aggregates – superclusters of galaxies. Between superclusters there are large voids with no clusters at all, in such supervoids we observe only poor galaxy filaments (Lindner *et al.* 1994).

Deep pencil-beam survey in the direction of the North and South galactic poles has found that high-density peaks in the distribution of galaxies are spaced in a surprisingly regular pattern with distance of about 128 h^{-1} Mpc between them (Broadhurst *et al.* 1990). More recently, an ESO Key Project to survey faint galaxies in a two-dimensional sheet suggests the presence of circular high-density regions which surround low-density areas. The mean diameter of these structures is also about 128 h^{-1} Mpc (Vettolani *et al.* 1994). The power spectra of the density distributions in these one- and two-dimensional surveys have maxima at about 128 h^{-1} Mpc (Broadhurst *et al.* 1990, Vettolani *et al.* 1994).

These one- and two-dimensional surveys have been complemented by three-dimensional surveys of clusters of galaxies. These studies have shown that superclusters and supervoids form a fairly regular network of high- and low-density regions, traced by rich clusters of galaxies (Bahcall 1991, Tully *et al.* 1992, Einasto *et al.* 1994b). This is illustrated in Figure 1, which shows the distribution of rich clusters in a 600 h^{-1} Mpc rectangle in supergalactic coordinates (h is the Hubble constant in units of 100 km/s/Mpc). The mean separation between high-density peaks in this distribution is approximately 130 h^{-1} Mpc. High-density peaks are located in a fairly rectangular grid which forms a three-dimensional chessboard (Tully *et al.* 1992).

193

M. Kafatos and Y. Kondo (eds.), Examining the Big Bang and Diffuse Background Radiations, 193–200.
© 1996 IAU.

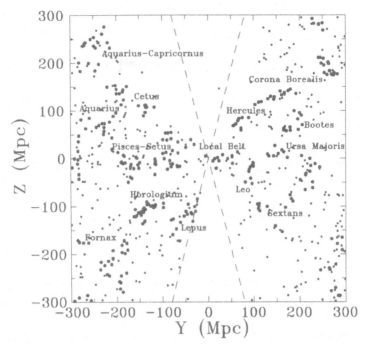

Figure 1. The distribution of clusters of galaxies in supergalactic coordinates, the sheet is taken in between supergalactic $-100 \leq X < 100\ h^{-1}$ Mpc. Clusters belonging to rich superclusters (containing at least 4 clusters) are plotted with filled circles, clusters located in poor superclusters as small dots. Superclusters are identified by common names (from constellations where they are located). Dashed lines mark the zone of avoidance near the galactic plane.

The presence of a 130 h^{-1} Mpc scale in the distribution of rich clusters of galaxies has been demomstrated also by Mo *et al.* (1992a, b) by the study of the correlation function on large separations.

Below we concentrate to the explanation of this semiregular supercluster-void structure. We report on simulations of the formation of the large-scale structure using CDM-type models and models with simple double power spectra. We compare the distribution of simulated galaxies and clusters of galaxies in models with the observations. We compare also the power spectra and the cluster correlation functions of simulations with observations. Our principal conclusion is that CDM-type models, both high- and low-density, are incompatible with available observational data.

2. Models

The observed structure is formed from initially small density perturbations. The amplitude of density perturbations on different scales is characterized by the power spectrum of the density field. The power spectrum can be approximated by a double power law with one power index n_l on large scales, another index n_s on small scales, and with the transition between power indices at a wavelength λ_t,

$$P(k) = \begin{cases} A\kappa^{n_s}, & \text{if } k \geq k_t \\ A\kappa^{n_l}, & \text{if } k < k_t. \end{cases}$$

Here A is the amplitude, $\kappa = k/k_t$, and the wavenumber k is related to the wavelength as follows: $\lambda = 2\pi/k$.

We also used initial spectra from theoretical models, such as the Cold Dark Matter (CDM) scenario of structure formation,

$$P(k) = A\frac{k}{\left(1 + \frac{(ak)^2}{\log(1+bk)}\right)^2}.$$

In this formula the parameters are as follows: $a = 4.0/(\Omega h^2)$, $b = 2.4/(\Omega h^2)$, h is the Hubble constant, and Ω is the density parameter in units of the critical closure density. The values of constants a and b correspond to wavelength λ, expressed in h^{-1} Mpc.

CDM-type spectra in combination with power-law spectra enable us to investigate the impact of the character of the transition between large and small wavelengths. The essential difference between those models is the scale length over which this transition occurs. In the power-law models the transition is sudden, and is much more smooth in the CDM models.

We performed numerical simulations of the formation and evolution of the large-scale structure of the universe by changing initial parameters of the models. A detailed description of all models shall be published elsewhere (Saar et al. 1994, Frisch et al. 1994). Parameters of principal models are given in Table 1. Models were calculated using a PM code using 128^3 particles and cells. In the table we give the size of the computational box L, the density parameter Ω (for the low-density CDM model we assume the presence of a cosmological term Ω_Λ, where $\Omega + \Omega_\Lambda = 1$), and the rms amplitude of density perturbations on 8 h^{-1} Mpc scale σ_8. In CDM-models we have adopted the Hubble constant $h = 0.5$. Models which are not shown in Table 1 include power-law spectra with different values of power indexes n_l and n_s, and models using a single power-law as the initial spectrum.

The evolution of the distribution of particles in the N-body code represents the evolution of the dark matter distribution. Clustered particles, i.e. particles associated with galaxies and their dark haloes, were selected according to the density smoothed in 1 h^{-1} Mpc scale: all particles located at density higher than the mean density were considered as clustered. Clusters of galaxies were identified as high-density peaks in the distribution of dark matter smoothed with a Gaussian window with a dispersion of 1 h^{-1} Mpc.

TABLE 1. Data on models

Model	Ω	h	n_l	n_s	λ_t h^{-1} Mpc	L h^{-1} Mpc	σ_8
N2p	1	0.5	1	−1.5	128	512	0.90
CDM1	1	0.5	1	−0.5	64:	512	0.70
CDM2	0.2	0.5	1	−1.5	500:	512	0.85

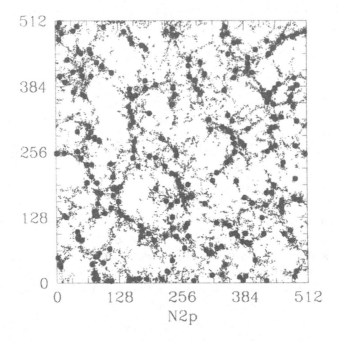

Figure 2. The distribution of simulated galaxies (dots) and clusters of galaxies (filled circles) in the double power law simulation N2p. Note the concentration of galaxies and clusters to ring-like structures separated by about 128 h^{-1} Mpc.

The number density of clusters has been taken equal to the number density of Abell clusters. Superclusters of galaxies were identified as high-density regions smoothed with 8 h^{-1} Mpc smoothing length.

3. Cosmography of galaxies and clusters

An example of the final structure in a model with a sharp transition in the power spectrum is shown in Figure 2. Both, the distribution of simulated galaxies and clusters of our model N2p are plotted.

Figure 2 shows that the distribution of galaxies is well ordered. Rich

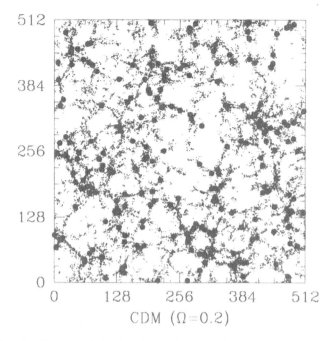

512

384

256

128

0

0   128   256   384   512

CDM ($\Omega = 0.2$)

Figure 3. The distribution of simulated galaxies and clusters in the low-density model CDM2. The location of high-density regions is much less regular than in the model N2p.

clusters of galaxies are located in high-density regions (superclusters) which surround regions of low density (supervoids). High-density regions have circular shapes. The mean distance of high-density regions across supervoids is equal to the scale of the maximum, λ_t. Simulated galaxies form filaments both in superclusters and in supervoids as in the real Universe (Lindner *et al.* 1994). The density of filaments is different: in superclusters they consist of chains of rich clusters, in supervoids of chains of galaxies.

For comparison we show in Figure 3 the distribution of simulated galaxies and clusters of galaxies in a low-density model CDM2. In this model, the power spectrum of density fluctuations has a maximum just at the wavelength equal to the size of the simulation box. Due to the flat maximum high-density regions (superclusters and rich clusters of galaxies) are much less organized than in the previous model. Diameters of voids between superclusters have a large scatter.

Our conclusion from these simulations and a number of others not shown here is that a semiregular pattern of high- and low-density regions forms only in models with spectra which show a well-defined maximum, i.e. a rather rapid transition between positive and negative spectral indexes. By contrast, models with shallow maxima are less organized than the observed distribution of galaxies in space. The wavelength of the maximum defines the scale of the pattern: in models with different position of the maximum, λ_t, the scale of the structure changes and is always equal to the scale of the maximum. The location of high-density regions is irregular if there is no maximum within the computational box, or if the maximum is very flat.

4. Correlation function and power spectra

One possibility to express quantitatively the presence of a semiregular pattern of high- and low-density regions is to use mean diameters of voids defined by rich clusters (Einasto et al. 1994b). This statistics can be combined with the characteristic size of superclusters. We can also use the correlation function of rich clusters of galaxies. In the presence of a regular pattern of high- and low-density regions the correlation function has a minimum on scales $70 - 100\ h^{-1}$ Mpc due to anticorrelation between superclusters and supervoids, and a secondary maximum corresponding to cluster pairs in superclusters across supervoids. Figure 4 gives the correlation function for clusters of galaxies (Einasto et al. 1993), and for simulated clusters in the double power-law model and in CDM models. In CDM models the correlation function of clusters of galaxies has no secondary maximum, because the spacing of superclusters is less regular. Therefore, the correlation function confirms our previous conclusion, that the models which best represent the observations are those with a sharp transition in the initial spectra.

Experimentation with simple toy models has shown, that a secondary maximum is observed in any semiregular model, with clusters located either on grid corners, edges or surfaces. The more regular is the displacement of these structures the stronger is the secondary maximum.

Our calculations show that perturbations on large scale ($\lambda > \lambda_t$) modulate small-scale structures. If there are no large-scale perturbations, then superclusters form a very regular rectangular pattern and all superclusters have approximately equal masses. However, if the amplitude of perturbations continues to grow at scales larger than λ_t with approximately the same rate as on small scales, then the spacing of superclusters becomes irregular and the masses of superclusters differ considerably (Frisch et al. 1994). The best fit to the observed distribution is achieved with a Harrison-Zeldovich index $n_l = 1$ favored also by COBE observations (Smoot et al. 1992).

The amplitude of density fluctuations on large scales can be fixed by COBE observations. The power spectra of models are shown in Figure 4. We see that our model N2p has on large scales an amplitude within the error box permitted by COBE observations. High-density CDM model has on scales $\approx 100\ h^{-1}$ Mpc an amplitude considerably lower than the observed amplitude. Similar discrepancy was found for the correlation function on large scales by Maddox, Efstathiou and Sutherland (1990). The low-density

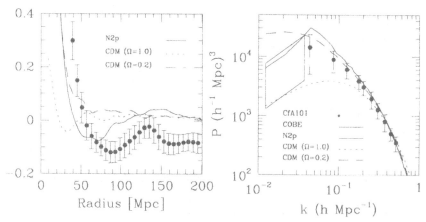

Figure 4. Left panel: The correlation function. Dots are for the observed correlation function for rich clusters of galaxies (Einasto *et al.* 1993) for the 300 h^{-1} Mpc sample in the Northern supergalactic hemisphere; the error corridor is indicated. Solid and dashed lines are for simulated clusters of the models N2p, CDM1 and CDM2. Right panel: The power spectra. Dots show the observed spectrum for galaxies (Park *et al.* 1994), the error box determined by COBE observations includes errors of both $\Omega = 1$ and $\Omega = 0.2$. Solid and dashed lines are for the spectrum of mass fluctuations of models.

CDM-model fits the spectrum of galaxies well, but on large scales it lies considerably higher than suggested by COBE data.

5. Conclusions

Difficulties with the high-density CDM model are well known (Maddox, Efstathiou and Sutherland 1990, Bahcall, Cen 1992). This is a problem of all models with a low effective spectral index, $n_s \approx -0.5$, on small scales. Also the mass distribution function of clusters of galaxies differs considerably from the observed function (Bahcall, Cen 1993, Frisch *et al.* 1994).

This problem does not arise in low-density CDM models since over the whole wavelength interval covered by ordinary galaxies $(1 - 100 \ h^{-1} \ \text{Mpc})$ the spectrum has a mean index $n_s \approx -1.5$ as observed. Most quantitative tests applied to this model show that it agrees well with observations (Bahcall, Cen 1992, 1993, Bahcall, Cen, Gramann 1993). However, in this case the amplitude of the spectrum continues to increase towards long wavelengths, the maximum is located at $\lambda_t \approx 500 \ h^{-1} \ \text{Mpc}$ and is very flat. In this model the location of superclusters is much less regular than in models with a rapid transition between positive and negative spectral indexes (compare Figure 2 and Figure 3). Also this model has problems with the COBE normalization as mentioned above. To overcome the normal-

200

ization problem a high bias, b, (and low value of σ_8) is adopted, but this assumption contradicts data on the fraction of matter in voids, F_v, since $b = 1/F_v \approx 1.15$ (Einasto et al. 1994b).

Thus we must conclude that CDM-type models, both high- and low-density, have two problems: they produce large-scale distribution of superclusters and voids which is less regular than the observed distribution, and it is difficult to reproduce simultaneously the observed power spectrum defined by galaxies and COBE data, taking into account also possible limits of the biasing factor.

ACKNOWLEDGEMENTS. Results reported here have been obtained by a collaboration between Tartu and Göttingen Observatories and European Southern Observatory. I thank my collaborators Heinz Andernach, Maret Einasto, Wolfram Freudling, Klaus Fricke, Patrick Frisch, Mirt Gramann, Ulrich Lindner, Veikko Saar, and Erik Tago for fruitful team work and permission to use our joint results before publication. This work was partly supported by the grant LDC 000 of the International Science Foundation.

References

Bahcall, N. A. 1991. *Astrophys. J.* **376**, 43.
Bahcall, N. A., Cen, R. Y. 1992. *Astrophys. J.* **398**, L81.
Bahcall, N. A., Cen, R. Y. 1993. *Astrophys. J.* **407**, L49.
Bahcall, N.A., Cen, R.Y., Gramann, M. 1993. *Astrophys. J.* **408**, L77.
Broadhurst, T.J., Ellis, R.S., Koo, D.C., Szalay, A.S. 1990. *Nature* **343**, 726.
Einasto, J., Gramann, M. 1993. *Astrophys. J.* **407**, 443.
Einasto, J., Gramann, M., Saar, E., Tago, E. 1993. *Mon. Not. R. astr. Soc.* **260**, 705.
Einasto, J., Saar, E., Einasto, M., Freudling, W., Gramann, M. 1994a. *Astrophys. J.* **429**, 465.
Einasto, M., Einasto, J., Tago, E., Dalton, G.B., Andernach, H. 1994b. *Mon. Not. R. astr. Soc.* **269**, 301.
Frisch, P., Einasto, J., Einasto, M., Freudling, W., Fricke, K.J., Gramann, M., Saar, V., Toomet, O. 1994. *Astron. Astrophys.* (in the press).
Lindner, U., Einasto, J., Einasto, M., Freudling, W., Fricke, K., Tago, E., 1994. *Astron. Astrophys.* (in the press).
Maddox, S.J., Efstathiou, G., Sutherland, W. 1990. *Mon. Not. R. astr. Soc.* **246**, 433.
Mo H.J., Deng Z.G., Xia X.Y., Schiller P., & Börner G. 1992a: *Astron. Astrophys.* **257**, 1.
Mo H.J., Xia X.Y., Deng Z.G., Börner G., & Fang, L.Z. 1992b: *Astron. Astrophys.* **256**, L23.
Park, C., Vogeley, M.S., Geller, M.J., and Huchra, J.P., 1994. *Astrophys. J.* (in the press).
Saar, V., Einasto, J., Einasto, M., Gramann, M., Tago, E., Fricke, K., Frisch, P., Lindner, U., Freudling, W., 1994. *Astrophys. J.* (submitted).
Smoot, G.F. et al. 1992. *Astrophys. J.* **396**, L1.
Tully, R. B., Scaramella, R., Vettolani, G., Zamorani, G. 1992. *Astrophys. J.* **388**, 9.
Vettolani, G. et al. 1994. In *Astronomy from Wide Field Imaging*, ed. MacGillivray, H.T., (in the press).

FAINT FIELD GALAXY COUNTS, COLORS, AND REDSHIFTS

DAVID C. KOO
University of California Observatories, Lick Observatory
Board of Studies in Astronomy and Astrophysics
University of California, Santa Cruz, CA 95064 USA

Abstract. This paper aims to provide the non-specialist a brief overview and major highlights in the rapidly changing research area of faint field galaxy evolution. Special emphasis is given to number counts, colors, and redshifts more recent than 1992 and their interpretation, both conventional and exotic. We close with previews of new directions that exploit sizes and velocity widths to explore the nature of faint blue galaxies, a puzzle that has defied solution for 15 years. The mystery continues.

1. Brief History and Overview

The interest in the evolution of faint galaxies was ignited by the watershed Yale Conference in 1977 when Kron presented the first credible evidence that field galaxies appeared numerous and quite blue. This result was a surprise, since the naive expectation is that fainter galaxies are redshifted and thus redder. Moreover, fainter galaxies should have number counts that declined rapidly relative to Euclidian slopes, since the volume per unit area drops with distance in Friedmann cosmologies. By the early 1980's, several groups had confirmed the large counts. There even appeared to be a consensus that the numerous, faint blue field galaxies could be explained by evolutionary models in which younger galaxies were undergoing more rapid star formation and were thus bluer than their descendents today. Furthermore, the large number counts could be understood because these galaxies were also brighter, more easily visible to higher redshifts ($z \sim 1$-3), and thus counted within larger volumes per unit area of the sky.

By the late 1980's, confusion permeated the field, despite the confirmation to fainter magnitudes with CCD's of the older photographic results

M. Kafatos and Y. Kondo (eds.), Examining the Big Bang and Diffuse Background Radiations, 201–208.
© 1996 *IAU.*

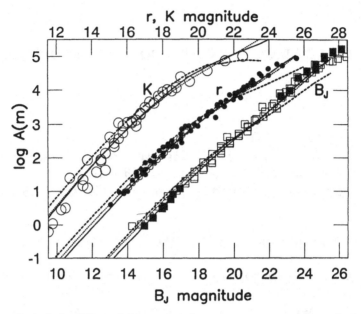

Figure 1. Log of the differential counts A (per mag per square degree) of faint field galaxies versus magnitude in the indicated bands from Gronwall and Koo (1995). B_J and Gunn r band observations (open squares and solid circles) are compilations of data made by Koo & Kron (1992); the infrared K band counts (large open circles) are a compilation from Gardner *et al.* (1993). Solid squares show the bright B_J counts corrected for the systematic errors discovered by Metcalfe *et al.* (1994a) as well as the new faint CCD counts of Metcalfe *et al.* (1994b); small open circles show the new r band counts of Weir (1994). The new mild evolution (with reddening) predictions are shown as solid curves while the equivalent best-fitting no-evolution predictions from Gronwall and Koo (1995) are shown as dotted curves. The short-dash curve shows a more typical no-evolution model, in this case from Rocca-Volmerange and Guiderdoni (1990).

of high counts and very blue colors. The key issue became whether the *redshifts* of faint galaxies are low or high. On the one hand, the Durham (UK) redshift surveys and cold dark matter (CDM) models supported the low side (z < 1). On the other, Cowie and Lilly (1988) claimed to find a z = 3.38 field galaxy and a significant population of faint galaxies showing the Lyman break in a 3400Å filter, both results implying high redshifts z > 2.7. Together with numerous discoveries of high redshift (z > 2) radio galaxies, damped Lyman alpha lines, and QSO's, the high redshift end (z > 1) remained equally plausible . Today, consensus has not yet been reached in understanding the nature of faint blue galaxies, despite an enormous increase in data and theory. The following sections will try to convey the current status of the field and will provide an update on the key issues being addressed by researchers today.

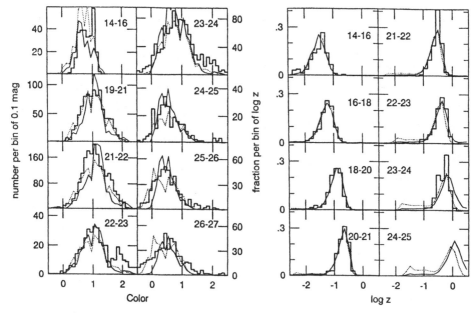

Figure 2. Left hand set of panels: Color distributions versus indicated magnitude intervals are shown as thick lines for the observations and thin solid lines for the mild evolution model predictions and dotted lines for the NE predictions from Gronwall and Koo (1995). The $B = 14 - 16$ interval shows $B - V$ color while the other panels show $B_J - R_F$ color distributions in the given B_J magnitude intervals. Note the trend to bluer colors between galaxies in the $23 < B_J < 24$ panel to the $24 < B_J < 25$ panel. *Right hand set of panels:* Normalized histograms versus redshift ($\log z$) for magnitude (B_J) intervals as indicated (Gronwall and Koo 1995). Bin size is 0.143, since original bin size was 1.0 in 7 $\log z$. The evolutionary model predictions are shown as thin solid lines while the NE model predictions are shown as dotted lines; observations shown as thick lines are compilations from different surveys.

2. Observations and Theories

The status of the observational evidence for evolution of faint field galaxies was reviewed by Koo and Kron (1992), to which the reader is referred for references prior to 1992 and further details. When compared to predictions of no-evolution (NE) models, these data showed number counts in the optical bands to exceed the expected counts by large factors, up to 5 to 15 times by B $\sim$ 25. In contrast, near-infrared counts in the K band (2.2μ) appeared to be so low as to suggest a closed universe with little evolution. Figure 1 shows the current status of these faint counts in the blue (B_J), red (Gunn r), and near-IR (K). CCD photometry also confirmed that these faint (B > 20) galaxies were quite blue (Figure 2, left hand panels), but more importantly, various new multi-object spectroscopic surveys from the Durham and Hawaii groups yielded redshift histograms that were surprisingly well fit by NE models, despite the count excess (Figure 2, right hand panels). In other words, there was no evidence for the high redshift ($z > 1$) tail expected from the "traditional" mild-evolution models of the 1980's. Further evidence for recent enhancement of star formation came from the strength and frequency of emission lines in these spectroscopic samples, while suggestions for a new population of objects at faint magnitudes were suggested by the low amplitudes found in the clustering (correlation function) among faint galaxies.

To explain these data, a number of new ideas have been proposed. These can be roughly divided into two camps. One set of theories is "exotic" in the sense that one or more basic assumptions of the traditional models, such as galaxy number conservation, is modified and predominately affects the observations. The exotic ideas include proposals that: dwarf galaxies today were much brighter in the recent past; galaxies have been undergoing extensive mergers in the recent past; the universe has a non-zero cosmological constant; or a new population of galaxies at moderate redshifts have disappeared by today. The reader is referred to Lilly (1993) for references and further details.

The other "conservative" explanations propose a combination of uncertainties in the observations or local galaxy properties to explain *most* of the faint galaxy data, leaving only minor extraordinary components.The conservative explanations include the possibility that the faint end slope of the *local* luminosity function (LF) of galaxies is steeper than commonly adopted in models (Driver *et al.* 1994, Marzke *et al.* 1994, Koo *et al.* 1993, Treyer &Silk 1994); that low surface brightness galaxies detected in faint samples have been systematically missed in more local surveys (McGaugh 1994, Ferguson and McGaugh 1994); that the local LFs as a function of color are poorly known (Shanks 1990, Metcalfe *et al.* 1991, Koo *et al.* 1993); that the local density remains uncertain due to large scale structure (Picard 1991) or to systematic magnitude errors (Metcalfe *et al.* 1994a, Weir 1994); and that some traditional models, with minor modifications, are good fits (Yoshii and Peterson 1991, Wang 1991, Gronwall and Koo 1995). Figures 1 and 2 show that new traditional models can provide good fits to the counts, colors, and redshifts and thus exotic explanations may not be necessary.

3. Recent Highlights

The observations have continued to expand dramatically since the review of Koo and Kron (1992). Besides the numerous aforementioned references supporting the conservative view, optical counts and colors to very faint

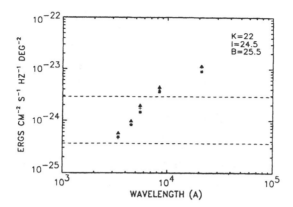

Figure 3. Extragalactic background light versus wavelength as copied from figure by Cowie *et al.* (1994). The triangles show the light from K < 22 galaxies; the crosses from I < 24.5 galaxies; and the boxes from B < 25.5 galaxies. The dashed lines bracket the range of blue background light predicted from star formation needed to produce the heavy elements found in luminous matter today.

limits have been measured by several new, independent groups, but whether a turnover exists in the steep optical number counts is not yet settled (Steidel & Hamilton 1993, Metcalfe *et al.* 1994b). Near-IR (K) counts have been improved at both the bright and faint limits (Gardner *et al.* 1993, Soifer *et al.* 1994, Djorgovski *et al.* 1994, Cowie *et al.* 1994, McLeod *et al.* 1994). More importantly, several new redshift surveys to various depths and selected by optical or near-IR bands have been completed (Lilly 1993, Colless *et al.* 1993, Tresse *et al.* 1993, Glazebrook and Ellis 1994, Songaila *et al.* 1994). Soon to be published will be the Canada-France Redshift Survey with its impressive sample of 1000 objects to a limit of I $\sim$ 22 (B $\sim$ 24.5). Of most relevance to the topic of this symposium is the very recent estimate of the extragalactic background light from deep optical and near-IR counts by Cowie *et al.* (1994) and shown in Figure 3. Their main conclusion from this figure is that the background has a red spectrum, but its blue part can be plausibly explained by the star formation needed to create the heavy elements we see in luminous matter today.

The scientific conclusions from all these new observations remain controversial. Several groups claim that the traditional luminosity evolution models are excluded because no high redshift tail is found in the deepest redshift surveys (Broadhurst *et al.* 1992, Glazebrook and Ellis 1994). Indeed, the trend is so strong that there may actually be a lack of redshift z $\sim$ 1 galaxies compared to even the no-evolution estimates (Colless *et al.* 1993, Koo *et al.* 1993, Songaila *et al.* 1994). This could result from galaxies having been in smaller subunits prior to merging. One area of apparent consensus is that the K band counts no longer appear to be incompatible with the optical blue counts (Gardner *et al.* 1993, Koo *et al.* 1993) when the bluer colors of fainter galaxies are included in the models. Several efforts have recently been directed to the explicit determination of the LF of galaxies as a function of redshift. The trends appear consistent among the work of three groups (Eales 1993, Lonsdale & Chokshi 1993, Colless 1994) — the LF appears to have a steeper faint end slope and/or a higher normalization (Φ^*) at higher redshifts when compared to local estimates. On the other hand, analysis of galaxies associated with foreground Mg II absorption lines seen in background QSOs indicate no evidence for density, luminosity, or color evolution (Steidel, Dickinson, & Persson 1994). Nor do the observations of emission line galaxies show any evolution (Boroson *et al.* 1993). A key question to be settled is whether a large exotic component at the level of 200% or more is needed (Lilly 1993, Babul & Rees 1992, Broadhurst *et al.* 1992) or whether, as shown by the good fits in Figures 1 and 2, a smaller amount at the level of 10's of percent or less will suffice, if the more conservative view is closer to the truth (Gronwall and Koo 1995).

4. New Directions

Several programs will soon make major contributions to our understanding of faint galaxies. Of particular note is the high spatial resolution now available with the refurbished Hubble Space Telescope. Analyses of such images are yielding sizes, surface brightnesses, morphologies, color gradients, signatures of nuclear starbursts or interactions, etc., for faint field galaxies on scales of kpc for redshifts beyond $z \sim 0.2$. Several different groups are pursuing this approach to study faint field galaxies; the reader is referred to the contribution in this volume by R. Griffiths who describes some of the results from the Medium Deep Survey, an HST Key Project.

Another direction is the extraction of velocity widths or rotation curves of distant galaxies. Combined with sizes and inclinations from HST data, masses can be derived. Initial efforts in these directions have already been undertaken by several independent groups (Franx *et al.* 1992, Vogt *et al.* 1993, Colless 1994, Koo *et al.* 1995). Figure 4 gives an example from using the echelle spectrograph on the Keck 10-m telescope; the wide range in velocity widths for a sample of otherwise similarly luminous galaxies supports the need for kinematics as a powerful new tool to study the nature of distant galaxies. With improved signal-to-noise spectra, spectral signatures of age and metallicity from emission and absorption lines, rather than just redshifts, become feasible for analysis. Initial forays using emission lines are beginning to be published (Smetanka 1993, Tresse *et al.* 1994).

We close with what may not be obvious, namely the need for improved data on the properties of nearby galaxies. In a few years, the Sloan Digital

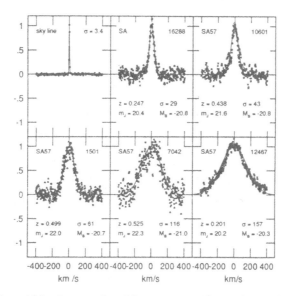

Figure 4. Velocity widths for sample of five compact blue galaxies (Koo *et al.* 1995). The upper left panel shows the instrumental profile from a typical night sky line for the echelle (HIRES) on the Keck 10-m telescope. The solid line in each panel is the best fit Gaussian curve. The name of each object is given, as is the redshift (z), the sigma of the fitted Gaussian in km-s^{-1}, the apparent blue (B_J) magnitude, and the blue luminosity (M_B) assuming a Hubble constant of 50 km-s^{-1}-Mpc^{-1} and $q_0 = 0.1$.

Sky Survey will provide a quantum leap in such vital information (Gunn & Knapp 1992). This survey will use a dedicated 2.5-m telescope to image 10,000 square degrees of sky in 5 optical bands (roughly UBVRI) to a limit of B $\sim$ 24, as well as to secure spectra of about 1,000,000 field galaxies to B $\sim$ 19.5. If the progress since 1992 is any indication, this review will be out of date by the time it is published. Such rapid changes are regarded by some as a sign of confusion within the field; others consider such instability as a sign of a healthy and exciting research area. The mystery of faint blue galaxies has withstood the onslaught of many dedicated observers and theorists for over 15 years; the mystery is not likely to vanish immediately.

5. Acknowledgements

I wish to thank L. Cowie, K. Glazebrook, C. Gronwall, R. Guzmán, S. Lilly, B. McLeod, and A. Phillips for their inputs into this review and the organizers for their invitation. This work was supported by NSF grants AST 88-58203 and AST 91-20005.

208

References

Babul, A., & Rees, M. J. 1992, MNRAS, 255, 346

Boroson, T. A., Salzer, J. J., & Trotter, A. 1993, ApJ, 412, 524

Broadhurst, T. J., Ellis, R. S., & Glazebrook, K. 1992, Nature, 355, 55

Charlot, S., & Bruzual A., G. 1991, ApJ, 367, 126

Colless, M., Ellis, R. S., Broadhurst, T. J., Taylor, K., & Peterson, B. A. 1993, MNRAS, 261, 19

Colless, M. 1994, in *Wide Field Spectroscopy and the Distant Universe*, eds. S. Maddox and A. Aragón-Salamanca (World Scientific Pub.)

Cowie, L. L., Songaila, A., & Hu, E. M. 1991, Nature, 354, 460

Cowie, L. L., Gardner, J. P., Hu, E. M., Hodapp, K. W., & Wainscoat, R. J. 1994, ApJ, in press.

Djorgovski, S., Soifer, B. T., Pahre, M. A., Larkin, J., *et al.* 1994, ApJ, in press

Driver, S. P., Phillips, S., Davies, J. I., Morgan, I., & Disney, M. J. 1994, MNRAS, 266, 155

Eales, S. 1993, ApJ, 404, 51

Ferguson, H. C., & McGaugh, S. S. 1994, ApJ, in press

Franx, M. 1993, ApJ, 407, L5

Gardner, J. P., Cowie, L. L., & Wainscoat, R. J. 1993, ApJ, 415, L9

Glazebrook, K., & Ellis, R. S. 1994, preprint

Gronwall, C., & Koo, D. C. 1995, ApJ, in press.

Gunn, J. E., & Knapp, G. R. 1993, A. S. P. Conf. Ser. No. 43, 267

Koo, D. C., Gronwall, C., & Bruzual A., G. 1993, ApJ, 415, L21

Koo, D. C., Guzmán, R., Faber, S. M., Illingworth, G. D., Bershady, M. A., Kron, R. G., & Takamiya, M. 1995, ApJ, submitted

Koo, D. C., & Kron, R. G. 1992, ARAA, 30, 613

Lilly, S. J. 1993, ApJ, 411, 501

Lilly, S. J., Cowie, L. L., & Gardner, J. P. 1991, ApJ, 369, 79

Lonsdale, C. J., & Chokshi, A. 1993, AJ, 105, 1333

Marzke, R. O., Huchra, J. P., & Geller, M. J. 1994, ApJ, in press

McGaugh, S. S. 1994, Nature, 367, 568

McLeod, B. A., Bernstein, G. M., Rieke, M. J., Tollestrup, E. V., & Fazio, G. G. 1994, ApJS, in press

Metcalfe, N., Fong, R., & Shanks, T. 1994a, MNRAS, submitted

Metcalfe, N., Shanks, T., Fong, R., & Jones, L. R. 1991, MNRAS, 249, 498

Metcalfe, N., Shanks, T., Fong, R., & Roche, N. 1994b, MNRAS, submitted

Picard, A. 1991, ApJ, 368, L11

Rocca-Volmerange, B., & Guiderdoni, B. 1990, MNRAS, 247, 166

Shanks, T. 1990, in IAU Symp. 139, *The Galactic and Extragalactic Background Radiation*, eds. S. Bowyer & C. Leinert, (Kluwer: Dordrecht), p. 269

Smetanka, J. J. 1993, Ph.D. thesis, Univ. of Chicago

Soifer, B. T., *et al.* 1994, ApJ, 420, L1

Songaila, A., Cowie, L. L., Hu, E. M., & Gardner, J. P. 1994, ApJ, in press.

Steidel, C. C., Dickinson, M., & Persson, S. E. 1994, ApJ, in press

Steidel, C. C., & Hamilton, D. 1993, AJ, 105, 2017

Tresse, L., Hammer, F., LeFevre, O., & Proust, D. 1993, Astron. & Astrophys., 277, 53

Tresse, L., Rola, C., Hammer, F., & Stasinska, G. 1994, in *Wide Field Spectroscopy and the Distant Universe*, eds. S. Maddox and A. Aragón-Salamanca (World Sci. Pub.)

Treyer, & Silk, J. 1994, ApJ, in press

Vogt, N. P., Herter, T., Haynes, M. P., and Courteau, S. 1993, ApJ, 415, L95

Wang, B. 1991, ApJ, 383, L37

Weir, N. 1994, Ph.D. thesis, California Institute of Technology

Yoshii, Y., & Peterson, B. A. 1991, ApJ, 372, 8

COSMOLOGICAL APPLICATIONS OF GRAVITATIONAL LENSING

PETER SCHNEIDER
Max-Planck-Institut für Astrophysik
Postfach 1523, D-85740 Garching, Germany

1. Introduction

It was recognized very early that the gravitational lens effect can be used as an efficient cosmological tool. Of the many researchers who foresaw the use of lensing, F. Zwicky and S. Refsdal should be explicitly mentioned. The perhaps most accurate predictions and foresights by these two authors are as follows: Zwicky estimated the probability that a distant object is multiply imaged to be about 1/400, and thus that the observation of this effect is "a certainty"[73] – his value, which was obtained by a very crude reasoning, is in fact very close to current estimates of the lensing probability of high-redshift QSOs. He predicted that the magnification caused by gravitational light deflection will allow a "deeper look" into the universe – in fact, the spectroscopy of very faint galaxies which are imaged into giant luminous arcs have yielded spectral information which would be very difficult to obtain without these 'natural telescopes'. And third, Zwicky saw that gravitational lenses may be used to determine the mass of distant extragalactic objects[72] – in fact, the mass determination of clusters masses from giant luminous arcs is as least as accurate as other methods, but does not rely on special assumptions (like spherical symmetry, virial or thermal equilibrium) inherent in other methods, and the determination of the mass within the inner 0.9 arcseconds of the lensing galaxy in the quadruple QSO 2237+0305 to within 2% [52] is the most accurate extragalactic mass determination known. Refsdal predicted the use of gravitational lenses for determining cosmological parameters and for testing cosmological theories [48][49] – we shall return to these issues below.

The discovery of the first gravitational lens in 1979 [68] marked the begin of 'modern' gravitational lens research. During the past 15 years this field has grown at an ever increasing rate, with about 13 known multiply imaged

M. Kafatos and Y. Kondo (eds.), Examining the Big Bang and Diffuse Background Radiations, 209–217.
© 1996 IAU.

QSOs, 5 ring-like radio sources ('Einstein rings'), and a large number of arcs and arclets known today, and more than 1000 publications on that subject. Here I will describe some recent developments in the application of lensing for cosmology; for a more complete overview of the field, the reader is referred to our monograph [58], three recent review articles [8, 20, 50], and the proceedings of the most recent gravitational lens conference [65].

2. An update on H_0 through lensing

To determine the Hubble constant from a gravitational lens system, the time delay Δt between two (or more) images must be measured,

$$\Delta t = F/H_0, \tag{1}$$

where F is a dimensionless function which depends on the observed configuration of the QSO images, the redshift of source and lens, the cosmological parameters Ω (and λ), and, most crucially, on the mass model for the lens. Note that (1) can be inferred on dimensional grounds – the time delay is the only dimensional observable. Hence, the determination of H_0 requires two steps, the measurement of Δt from the varying flux of the QSO images and the construction of a 'correct' lens model. Both of these steps are much more difficult than anticipated originally.

The first double QSO 0957+561 has been monitored in the optical and radio bands since its discovery (i.e., for nearly 15 years); however, no generally accepted value for Δt has emerged. The analysis of the optical [55, 67] and radio [33] data by different statistical methods [47, 45] yielded no generally accepted results, with $\Delta t \sim 410$ or 540 days being preferred values – but at most one of those can be correct. Hence, the issue is currently not decided, despite the huge observational effort. There are clear signs of microlensing in at least one of the images.

Even if Δt for this system is eventually measured, the determination of H_0 will be highly uncertain, due to the complexity of the lens, which consists of a giant elliptical galaxy, embedded in a cluster at $z = 0.36$, with an additional concentration of galaxies close to the line-of-sight at $z \sim 0.5$. Certainly, the number of unknowns in a realistic lens model is larger than the (already large) number of observational constraints which come mainly from detailed VLBI imaging. If the lens indeed produces an arc [7, 16], earlier detailed lens models [19, 27] become insufficient, and this system will be unsuitable for the determination of H_0.

The system 0218+35.7 [43] is almost certainly the best lens system known to determine H_0; it consists of two compact components separated by 0.3 arcseconds and a radio ring. The small image separation suggests that the lens is a single (spiral) galaxy, and the Einstein ring will allow the construction of a detailed lens model, once it is properly imaged and resolved in width, and the compact components are variable. In fact, from polarized flux variations, a preliminary value of $\Delta t \sim 15$ days is suggested (I. Browne, P. Wilkinson, D. Walsh, private communication). Unfortunately, the source redshift in this system is still unknown.

In conclusion, H_0 is not yet determined from gravitational lensing, but reasonable estimates from the two systems mentioned lie in the range between 30 and 80 km/s/Mpc. Though this looks not like a very impressive achievement, the agreement with the values obtained by other (local) values provides a strong consistency check on our cosmological model, and strongly supports the cosmological interpretation of QSO redshifts. Furthermore, the estimates will improve and yield a single-step determination of H_0 on a truly cosmic scale, i.e., independent of local peculiar velocities and the distance ladder.

3. Lensing statistics, the cosmological constant, and galaxy merging

In recent years, several gravitational lens surveys have been performed in the optical [64, 71, 15, 37] and radio [10, 44] wavebands. In addition to finding new gravitational lens systems, these surveys can be analyzed statistically and compared with theoretical expectations. The results from such analyses [30, 38] can be summarized as follows: the frequency of multiple images in the existing surveys, and their distribution in redshift and apparent magnitude is fully compatible with standard assumptions on the cosmological model parameters, the commonly used parameters in the Schechter luminosity function for galaxies, the Tully-Fischer-Faber-Jackson relation, and the QSO luminosity function. The agreement of the lensing statistics with models does not significantly improve if the standard values for the parameters are allowed to vary [30]. Constant mass-to-light ratio models for early-type galaxies can be ruled out [38], since they would be in conflict with the observed lensing rate and image splitting, but these galaxies must have a dark (isothermal) halo. Moreover, the cosmological constant can be constrained to be ≤ 0.7, otherwise too many multiple images would be produced (a way to avoid this conclusion is obscuration in the lens galaxies with redshift above ~ 0.5 [22]). Also, there is not much room for compact 'dark matter' with $M \geq 10^{11} M_{\odot}$, since these objects would also from multiple images (from the fact that in more than half the lens systems, the lens is seen and thus not 'dark', one concludes that the mass density of compact dark objects in the relevant mass range cannot exceed that of galaxies). Lensing statistics can also be used to probe galaxy evolution and merging scenarios [36, 53], with the result that no-evolution models are statistically preferred; mild evolution cannot be ruled out, but strong merging scenarios are incompatible with the existing data.

One caveat should be kept in mind about these studies: all optical lens surveys have taken their target QSOs from existing catalogs, which do in no sense form unbiased samples. Existing QSO catalogs may be biased against multiply imaged QSO, though this effect is probably small [28]. However, if there exists a class of very red QSOs, and if the reddening is due to dust in *intervening* material, the results from the statistical interpretation of optical lens surveys may yield misleading results. The highly reddened quadruple QSO 0414+0534 [32] may provide a cornerstone for this important question, i.e., whether the reddening occurs in the lensing galaxy. Radio lens surveys are of course largely unaffected by this effect, but are hampered by the success rate of optical identification of source and lens.

The future of lensing statistics looks particularly bright, since the Sloan Digital Sky Survey will most likely find several hundred lens candidate systems, all detected with the same instrument, so that the selection function (which enters the lensing statistics in a significant way) can be well understood.

4. The size of Lyα clouds

The Lyα forest, seen in all high-redshift QSOs, is believed to be caused by intervening photoionized 'clouds'. The size of these clouds can be probed by multiply imaged QSOs or close QSO pairs, by studying the cross-correlation of the spectra of both QSO images: if most absoption lines are found in both spectra, and if these lines have comparable equivalent width, the size of the clouds must be significantly larger than the separation of the two light beams at the redshift of the cloud. A recent summary of this technique is found in [63], from which the following results are quoted: The double QSO UM673 shows a strongly correlated absorption spectrum, and only two (high S/N) lines are found in image A which do not occur in image B. If these two lines are indeed Lyα lines, the size of the clouds is estimated to be $12 \leq R/$ kpc ≤ 160, whereas if they are a MgII doublet, then $R \geq 23$kpc. The absorption spectrum of the recently discovered lens candidate system HE1104−1805 yields a preliminary lower limit on the cloud size of $R \geq$ 50kpc, whereas the QSO pair UM 680/681 yields $R \leq 750$kpc. As reported by Bechtold (this volume), the QSO pair 1343+266 yields an estimate for the size of the clouds of $R \sim 90$kpc.

5. Bounds on dark compact matter

Gravitational lensing is particularly suited to detect, or put upper limits on the density of compact objects in the universe. There are mainly two effects of gravitational light deflection which can be used for this purpose: multiple imaging and magnification (for a recent summary of limits on baryonic dark matter, see [12]). With current VLBI techniques, image separations down to a few tenths of a milliarcsecond can be probed for multiple images, corresponding to about $10^5 M_\odot$ [46]. Hence, imaging surveys with the VLBA, MERLIN, and the VLA can rule out a significant cosmological density of compact dark objects with masses larger than this value. If gamma-ray bursts are at cosmological distances, they can also be multiply imaged; though the image splitting cannot be observed, the time delay between different images will cause a repeating of the burst, separated by that time delay. Given that the shortest rise time of the bursts is about one millisecond, one can discover by this means objects with mass in excess of $10^3 M_\odot$. For smaller mass objects, the magnification must be used. Following the suggestions in [11], upper limits on the density of compact objects with $10^{-3} \leq M/M_\odot \leq 10^2$ have been obtained, by investigating the statistics of line-to-continuum flux ratios in QSOs [17] and upper limits on the variability of QSOs caused by these 'microlenses' [56]. The first of these methods is based on the assumption that the continuum emitting region is sufficiently small to be magnified by these lenses, whereas the broad line region is too large to be affected, whereas the second method relies on the assumption that the relative alignement of source, lens and observer must change in time due to peculiar velocities of all three members.

The limits one gets from these methods are interesting, e.g., $\Omega_c \leq 0.4$ in the mass range $10^7 \leq M/M_\odot \leq 10^9$ [26], and $\Omega_c \leq 0.2$ for $10^{-3} \leq M/M_\odot \leq 10^2$, and further significant tightening of these bounds will become available soon.

6. Arcs in clusters: $\Omega_0 \not\ll 1$ and cluster core radii

A recently completed survey for arcs in X-ray selected clusters of galaxies (Gioia, this volume) has shown that they occur at a fairly high rate. Theoretical expectations of arc statistics from simple (e.g., spherical or slightly elliptical) lens models [70] fall short by a huge factor. However, the predictions of arc statistics from simple models severely underestimate the true rate of occurrence [6], if compared with realistic mass distributions of clusters (obtained from a CDM simulation). The basic reason is that these numerically generated clusters, in accordance with observational results, show much more asymmetry and substructure than 'simple' model can account for, and that the 'central' surface mass density in such asymmetric clusters need not exceed the critical surface mass density [58] to produce arcs, in contrast to symmetric models. Hence, the large number of arcs in a complete sample of clusters provides strong evidence for clusters being unrelaxed and thus young (although the individual galaxies in the cluster generate an 'asymmetric' mass profile, the corresponding small-scale graininess appears to be insufficient to explain the large frequency of arc formation). This in turn implies that structure formation is still going on, i.e., that Ω_0 cannot be very much smaller than unity [51, 2].

In addition to the determination of the cluster mass, arcs can be used to constrain the mass profile in clusters (see also next section). In particular, the core radius of clusters as estimated from detailed lens models (e.g. [39]) is significantly smaller than the core radii as determined from X-ray observations of clusters, thus providing us with an interesting and potentially important discrepancy (e.g., [40, 34, 35]). Detailed observations of the clusters A370 [31] and MS2137−23 [39] yield stringent constraints on the core radius of the clusters, the ellipticity and orientation of the dark mass distribution; the corresponding lens models obtain their credibility from predicting multiple images where there are actually seen.

7. Cluster lens reconstruction

The perhaps 'hottest' topic in gravitational lensing today is the determination of the (surface) mass density of clusters from the image distortions they apply on distant faint galaxies [66]. The fundamental relation for this inversion was obtained in [25]:

$$\kappa(\vec{\theta}) = \frac{1}{\pi} \int_{R^2} d^2\theta' \, D(\vec{\theta} - \vec{\theta}') \, \gamma(\vec{\theta}'), \tag{2}$$

where κ is the (normalized) surface mass density (using the notation of [58]), γ is the shear, which describes the tidal effects of light deflection, and D is a known (and simple) kernel. Hence, if γ can be determined from the observed image distortions, then the surface mass density can be reconstructed. The basic assumption of this technique is that the intrinsic orientations of the faint galaxies are distributed randomly. First attempts to apply (2) to observational data [18, 62] are promising; since the accuracy of this method depends on the observable density of faint background objects, new observations by the Hubble Space Telescope will allow great progress to be made very soon.

It should be noted that the shear γ is not an observable [59], but the only observable from image distortions is a combination of γ and κ. Dispite of this difficulty, the inversion equation (2) can be applied, with γ being determined iteratively, as demonstrated in [60]. There exists a second problem with (2), namely that (2) is exact only if the integral is extended over the whole lens plane, whereas observational data will be confined to a (small) region determined by the size of the CCD. Using a differential relation between κ and γ, derived in [24], I have obtained an inversion equation which is exact for data given only on a finite region in the lens plane [57]. This modification of the inversion technique can yield quite substantial changes in the predicted surface mass density.

8. Cosmological coherent weak distortions and arcmin-scale QSO-galaxy associations: The power spectrum and the bias factor

Mass distributions less compact and more massive than clusters can lead to weak lensing effects, two of which should be mentioned here. Weak distortion of faint galaxy images caused by light deflection of the large scale mass distribution may be detectable [9, 23]. The two-point correlation function of the image ellipticities depends on the redshift distribution of the faint galaxies and on the power spectrum $P(k)$ of the density fluctuations. Hence, if this correlation function could be measured, one would have an integral constraint on $P(k)$, on the scales of about one degree. The rms image polarization predicted from a CDM model are about 3%. A pilot project [41] has recently yielded an upper limit of 5% image polarizations. More important than this number is the fact that such weak effects can now be measured. In a series of papers, it was recently shown that there is a statistically significant correlation between high-redshift radio QSOs and Lick galaxies [21, 3], IRAS galaxies [4] and X-ray photons from the ROSAT All Sky Survey [5]. The interpretation of these correlations as lensing by the large scale structure with which the galaxies are associated remains to be verified; as shown in [1], this interpretation yields for the two-point correlation function of high-redshift QSOs and galaxies

$$\xi(\theta) = (\alpha - 1)b \int dk \, P(k) \int dz \, G(k, z, \theta), \qquad (3)$$

where α is the effective local slope of the QSO source counts, b the bias factor, and G depends on the redshift distribution of QSOs and the galaxies in a flux limited sample. The data used in the above quoted papers are not sufficient for measuring $\xi(\theta)$, but an analysis with a somewhat deeper galaxy catalog will enable us to measure ξ and thus to check the lensing interpretation of these correlations. In addition, with sufficient data, the power spectrum and the bias factor can be tested with weak lensing. Note that there are also indications for an overdensity of high-redshift QSO around Zwicky clusters [54, 61], which are not easily explained quantitatively by a cluster lensing model.

9. Final remarks

Due to the shortness of space, this review has to be selective; I have not discussed several very interesting developments in the field. For example, the search for microlensing in our galaxy has been much more successful than has been anticipated only one year ago, though it may turn out that the results from these searches tell us much more about stellar variability and the structure of the inner part of our galaxy than over cosmic dark matter. The constraints provided by the 'missing odd image' on the core size of lensing galaxies has also not been mentioned.

From the discussion here it is evident that gravitational lensing has developped from an 'exotic' subject to a quantitative tool in observational cosmology within the last few years. It provides the strongest constraints on the cosmological constant [13, 29] and (still) promises a way to determine H_0 (but is nearly blind with respect to q_0). Numerical simulations of the mass distribution in the universe soon will cover sufficiently dynamic range that they can be tested with respect to their predictions about multiple imaging of QSOs and weak image distortions; first results [14, 69] rule out standard CDM models, in agreement with analytical estimates [42]. The determination of the mass distribution in clusters through weak distortions has just begun to be explored; if one recalls that a deep HST image contains well in excess of 10^6 faint galaxy images per square degree, it becomes clear that this method will probably yield more robust results than any other method known. This in fact should not come as a surprise – the gravitational lensing effect depends only on gravity, which we think we understand very well, and not on more complicated and uncertain physical processes. Probably this simple but important thought led Zwicky, Refsdal, and others to their visionary foresights, which have turned out to be remarkable precise.

I would like to thank Hans-Walter Rix for his constructive comments on this manuscript

References

1. Bartelmann, M. (1994) A&A, submitted.
2. Bartelmann, M., Ehlers, J. & Schneider, P. (1993) A&A, 280, 351.
3. Bartelmann, M. & Schneider, P. (1993) A&A, 271, 421.
4. Bartelmann, M. & Schneider, P. (1993) A&A, 284, 1.
5. Bartelmann, M., Schneider, P. & Hasinger, G. (1994) A&A, in press.
6. Bartelmann, M., Steinmetz, M. & Weiss, A. (1994), A&A, submitted.
7. Bernstein, G.M., Tyson, J.A. & Kochanek, C.S. (1993) AJ, 105, 816.
8. Blandford, R.D. & Narayan, R. (1992) ARAA, 30, 311.
9. Blandford, R.D., Saust, A.B., Brainerd, T.G. & Villumsen, J.V. (1991) MNRAS, 251, 600.
10. Burke, B.F. (1990), Lect. Notes in Physics, 360, 127.
11. Canizares, C.R. (1982), ApJ, 263, 508.
12. Carr, B.J. (1994), ARAA, in press.
13. Carroll, S.M., Press, W.H. & Turner, E.L. (1992) ARAA, 30, 499.
14. Cen, R., Gott, J.R., Ostriker, J.P. & Turner, E.L. (1994) ApJ, 423, 1.
15. Crampton, D., McClure, R.D. & Fletcher, J.M. (1992) ApJ, 392, 23.
16. Dahle, H., Maddox, S.J. & Lilji, P.B. (1994), preprint.
17. Dalcanton, J.J. et al. (1994) ApJ, 424, 550.
18. Fahlman, G.G., Kaiser, N., Squires, G. & Woods, D. (1994) preprint.
19. Falco, E.E., Gorenstein, M.V. & Shapiro, I.I. (1991) ApJ, 372, 364.
20. Fort, B. & Mellier, Y. (1994) A&AR, 5, 239.

21. Fugmann, W. (1990) A&A, 240, 11.
22. Fukugita, M. & Peebles, P.J.E. (1994), preprint.
23. Kaiser, N. (1992) ApJ, 388, 272.
24. Kaiser, N. (1994) ApJ, submitted.
25. Kaiser, N. & Squires, G. (1993) ApJ, 404, 441.
26. Kassiola, A., Kovner, I. & Blandford, R.D. (1991) ApJ, 381, 6.
27. Kochanek, C.S. (1991) ApJ, 382, 58.
28. Kochanek, C.S. (1991) ApJ, 379, 517.
29. Kochanek, C.S. (1992) ApJ, 384, 1.
30. Kochanek, C.S. (1993) ApJ, 419, 12.
31. Kneib, J.-P., Mellier, Y., Fort, B. & Mathez, G. (1993) A&A, 273, 367.
32. Lawrence, C.R., Elston, R., Jannuzi, B.T. & Turner, E.L. (1994) preprint.
33. Lehár, J., Hewitt, J.N., Roberts, D.H. & Burke, B.F. (1992) ApJ, 384, 453.
34. Loeb, A. & Mao, S. (1994) preprint.
35. Loewenstein, M. (1994) ApJ, 431, 91.
36. Mao, S. & Kochanek, C.S. (1994) MNRAS, 268, 569.
37. Maoz, D. et al. (1993) ApJ, 409, 28.
38. Maoz, D. & Rix, H.-W. (1993), ApJ, 416, 425.
39. Mellier, Y., Fort, B. & Kneib, J.-P. (1993) ApJ, 407, 33.
40. Miralda-Escudé, J. & Babul, A. (1994) preprint.
41. Mould, J. et al. (1994) preprint.
42. Narayan, R. & White, S.D.M. (1988) MNRAS, 231, 97p.
43. Patnaik, A.R. et al. (1993), MNRAS, 261, 435.
44. Patnaik, A.R. (1994), in [65], p.311.
45. Pelt, J. et al. (1994) A&A 286, 775.
46. Press, W.H. & Gunn, J.E. (1973) ApJ 185, 397.
47. Press, W.H., Rybicki, G.B. & Hewitt, J.N. (1992) ApJ, 385, 404 & 416.
48. Refsdal, S. (1964) MNRAS, 128, 307.
49. Refsdal, S. (1966) MNRAS, 132, 101.
50. Refsdal, S. & Surdej, J. (1994), Rep. Prog. Phys., 56, 117.
51. Richstone, D., Loeb, A. & Turner, E.L. (1993) ApJ, 393, 477.
52. Rix, H.-W., Schneider, D.P. & Bahcall, J.N. (1992) AJ 104, 959.
53. Rix, H.-W., Maoz, D., Turner, E.L. & Fukugita, M. (1994), ApJ, in press.
54. Rodrigues-Williams, L.L. & Hogan, C.J. (1994), AJ, 107, 451.
55. Schild, R. (1990) AJ, 100, 1771.
56. Schneider, P. (1993) A&A, 279, 1.
57. Schneider, P. (1994) A&A, submitted.
58. Schneider, P., Ehlers, J. & Falco, E.E. (1992) *Gravitational Lenses*, Springer, New York.
59. Schneider, P. & Seitz, C. (1994) A&A, in press.
60. Seitz, C. & Schneider, P. (1994) A&A, submitted.
61. Seitz, S. & Schneider, P. (1994) A&A, submitted.
62. Smail, I., Ellis, R.S., Fitchett, M.J. & Edge, A.C. (1994) preprint.
63. Smette, A. (1994), in [65], p.147.
64. Surdej, J. et al. (1993) AJ, 105, 2064.
65. Surdej, J. et al. (1994) *Gravitational Lenses in the Universe*, Liège.
66. Tyson, J.A., Valdes, F. & Wenk, R.A. (1990) ApJ, 349, L1.
67. Vanderriest, C. et al. (1989) A&A, 215, 1.
68. Walsh, D., Carswell, R.F. & Weymann, R.J. (1979) Nature 279, 381.
69. Wambsganss,J., Cen, R., Ostriker, J.P. & Turner, E.L. (1994) Nature, submitted.
70. Wu, X.-P. & Hammer, F. (1993) MNRAS, 262, 187.
71. Yee, H.K.C., Filippenko, A.V. & Tang, D. (1993) AJ, 105, 7.
72. Zwicky, F. (1937a) Phys. Rev. 51, 290.
73. Zwicky, F. (1937b) Phys. Rev. 51, 679.

THE HST MEDIUM DEEP SURVEY:
GALAXY MORPHOLOGY AT HIGH REDSHIFT

R. E. GRIFFITHS, K. U. RATNATUNGA, S. CASERTANO, M. IM AND
L. W. NEUSCHAEFER
JHU, Homewood Campus, Baltimore, MD 21218, USA

R. S. ELLIS, G. F. GILMORE, R. A. W. ELSON, K. GLAZEBROOK AND
B. SANTIAGO
Institute of Astronomy, U. Cambridge

R. A. WINDHORST, S. P. DRIVER, E. J. OSTRANDER AND S. B. MUTZ
ASU, Tempe, AZ

D. C. KOO, G. D. ILLINGWORTH, D. A. FORBES AND A. C. PHILLIPS
Lick Obs., UCSC

R. F. GREEN
NOAO, Tucson, AZ

J. P. HUCHRA
Harvard-Smithsonian CfA, Cambridge, MA

AND

A. J. TYSON
AT&T Bell Labs., NJ

1. ABSTRACT

With HST and WFPC2, galaxies in the Medium Deep Survey can be reliably classified to magnitudes $I_{814} \lesssim 22.0$ in the F814W band, at a mean redshift $\bar{z} \sim 0.5$. The main result is the relatively high proportion ($\sim 40\%$) of objects which are in some way irregular or anomalous, and which are of relevance in understanding the origin of the familiar excess population of faint galaxies. These diverse objects include compact galaxies, apparently interacting pairs, galaxies with superluminous starforming regions and diffuse low surface brightness galaxies of various forms. The 'irregulars' and 'peculiar' galaxies contribute most of the excess counts in the I-band at our limiting magnitude, and may explain the 'faint blue galaxy' problem.

M. Kafatos and Y. Kondo (eds.), Examining the Big Bang and Diffuse Background Radiations, 219–227.
© 1996 IAU.

At least half of the faint galaxies, however, appear to be similar to regular Hubble-sequence examples observed at low redshift. Furthermore, the relative proportion of spheroidal and disk systems of normal appearance is as expected from nearby samples, indicating that the bulk of the local galaxy population was in place at half the Hubble time. Little or no evolution in the properties of these galaxies has been observed.

2. INTRODUCTION

The Medium-Deep Survey (MDS) is an HST Key Project which relies exclusively on parallel observations of random fields taken with the Wide Field and Planetary Cameras. The goals include the statistical studies of the properties of a large sample of faint stars and galaxies. The pre-refurbished results suggested the enormous potential of high resolution imaging as a major tool in tackling one of the outstanding problems in observational cosmology: the nature of the abundant population of faint blue galaxies (for a review, see Koo & Kron 1992, and Koo in these proceedings). Ground-based observations have thus far been unable to discriminate between hypotheses based on a fading population of dwarf galaxies (e.g. Broadhurst, Ellis and Shanks 1988; Cowie, Songaila and Hu 1991; Babul & Rees 1992) and those based on galaxy merging (Rocca-Volmerange and Guiderdoni 1990; Broadhurst, Ellis and Glazebrook 1992). The very first MDS observations (Griffiths et al. 1994a) indicated a deficit of large galaxies (half-light radius $\gtrsim 1\rlap{.}''0$) and an excess of compact galaxies at or near the HST resolution limit.

Morphologies can now be studied with a precision adequate for classification on the normal Hubble scheme to a limiting magnitude of $I_{814} \sim$ 22.0 mag. Cruder information (e.g. scale lengths) can be determined to considerably fainter limits. Unlike the pre-refurbished images, morphological classifications for the WFPC2 objects are not based on parametric fits, nor do they rely on any deconvolution technique.

The MDS team is in the process of correlating the morphological properties of HST-selected galaxies with redshift and star formation rates derived from ground-based follow-up spectroscopy and photometry. Redshift surveys of other fields to this depth have been undertaken by Lilly (1993) and Tresse et al. (1993) and indicate a median redshift $\overline{z} \simeq 0.5$ with a high fraction of objects within the range $0.4 < z < 0.7$. Consequently, we expect our sample to be representative of the field galaxy population at a lookback time of $\simeq 5$–7 Gyr ($H_0 = 50$ km s^{-1} Mpc^{-1} assumed throughout).

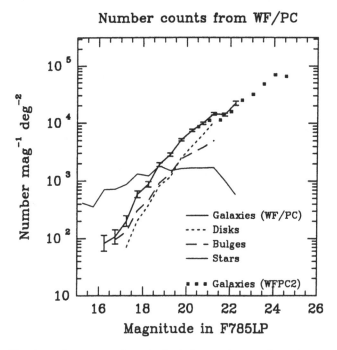

Figure 1. Total number counts for spirals, bulges and stars; from Casertano et al. (1995)

3. OBSERVATIONS

Typical MDS observations range from 600 to 2000 seconds per exposure, and may consist of 1–20 exposures of the same field. Parallel exposures may be registered or not, depending on the needs of the observer using the primary HST instrument. In some cases, up to 12,000 seconds have been accumulated at a single pointing. Following HST refurbishment, the improvement in spatial resolution and read noise of WFPC2 have resulted in 4–5 times better sensitivity for the faintest galaxies. In Cycles 1–3, 146 fields were observed with the WFC in a total exposure time of 240 hours. To September 30 1994, Cycle 4 MDS observations had been made in 113 fields, with a total of 84 hours in the two predominant filters: F606W, somewhat redder and broader than Johnson V, and F814W, close to Kron-Cousins I. Of these fields, 46 were exposed to both filters, and 12 had at least three exposures in each. The filters have been chosen to maximize the sensitivity for typical galaxies at intermediate redshifts ($z \sim 0.5$). Magnitudes in these filters are referred to as V_{606} and I_{814}, respectively. A typical WFPC2 MDS field contains 50–400 detectable sources within the Wide Field Camera (WFC) to a limiting magnitude of $I_{814} = 24$–26 depending on the total exposure. The precision with which morphological properties can be ascer-

Magnitude–size relation

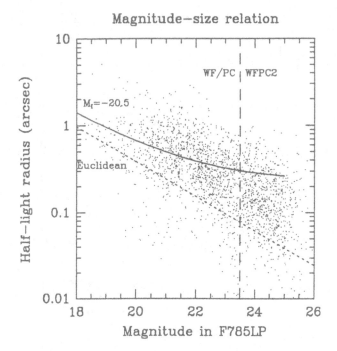

Figure 2. Half-light radius vs. mag. for galaxies in WF/PC and WFPC2 data. The Euclidean line is the extrapolated median for low-z galaxies; the predicted median for $M_I = 20.5$ is also shown – adapted from Casertano et al.(1995)

tained is a function of both angular size and apparent magnitude.

The cycle 1–3 HST MDS object catalog (11,000 objects) is available via the STScI's Electronic Information System, STEIS, in the directory observer/catalogs/mds. The 64x64 pixel images of all objects will be available on the HST archive in early 1995. Super-sky flat fields for WF/PC and WFPC2 are available on STEIS (Ratnatunga et al. 1994a).

4. SUMMARY OF EXTRAGALACTIC RESULTS

(i) Hubble-type morphological classification can be achieved to $I_{814} = 22$ (Griffiths et al. 1994b, Forbes et al. 1995, Driver et al. 1995, Glazebrook et al. 1995). Spectroscopy is in progress for large subsets of these galaxies

(ii) Statistical properties of galaxies are measured to I = 24 with WF/PC and 25 with WFPC2 (figs. 1 – 4). For the pre-refurbishment WF/PC images, the structural parameters of about 13,000 objects are presented by Casertano et al. (1995), using data taken from about 112 fields. Sizes, magnitudes, colors and crude classifications are based on two-dimensional model fitting to undeconvolved images (Ratnatunga et al. 1994b). Number counts

Galaxy magnitude vs. color from WF/PC

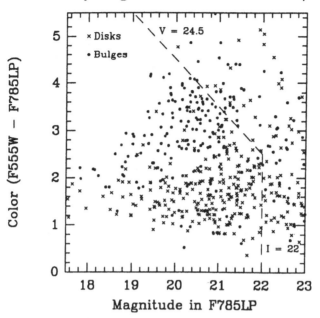

Figure 3. The color-magnitude diagram for galaxies in WF/PC data; completeness lines are drawn at I = 22 and V = 24.5. The 'bulges' appear redder towards fainter magnitudes; from Casertano et al. (1995)

in the range $18 < I < 22$ exceed the ground-based numbers by about 50%, a range over which many ground-based objects may have been misclassified as stars. The ellipticals become redder for $I_{814} > 20$ mag, but the spirals show no trend of color with magnitude, indicating that they are intrinsically bluer at higher redshift when K-corrections are taken into account.

(iii) The local universe is dwarf-rich. The marginal distribution of size vs. magnitude can be compared with the predicted distributions based on various galaxy evolution models such as the no-evolution model, the merger model, and the dwarf-rich model (fig. 5). The mild-evolution or no-evolution models predict luminous galaxies of large angular size at high redshift: such objects are not seen in the data, which are instead consistent with the dwarf-rich models and a dwarf luminosity function with a steep faint-end slope. The MDS results also rule out models in which the faint, excess number counts are caused by large, low-surface brightness galaxies (e.g. Ferguson & McGough 1995) since we do not see the predicted numbers of large objects. The median half-light radius for all galaxies to $I_{814} = 24$ is 0.4 arcsec (Im et al. 1995a).

(iv) The excess number counts are not explained by 'giant' spirals or

224

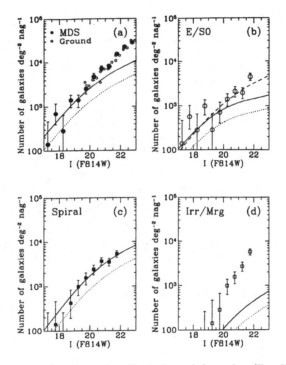

Figure 4. Number counts for spirals, ellipticals and Irregulars/Peculiars in WFPC2 data; from Glazebrook et al. (1995) and Driver et al.(1995)

ellipticals, which are observed to have little or no evolution in terms of number counts (Glazebrook et al. 1995), size vs. redshift (Mutz et al. 1994), or structural parameters (Windhorst et al. 1994; Phillips et al 1995): the bulk of the local (giant) population was therefore in place at half the Hubble time. Furthermore, this population has either undergone relatively little merging (about 10%), or else the mergers have been of the 'minor' kind (with gas-rich dwarfs) and have not caused major disruptions (Driver et al. 1995). For those galaxies which do show evidence of merger activity, photometry shows bluer colors and thus increased star formation (Forbes et al. 1994)

(v) Spiral galaxies are, however, bluer in the past when their K-corrections are taken into account; their apparent (V-I) color does not change between I=18 and I=22 (fig. 3). The microJansky radio population is linked to these spiral galaxies with enhanced star formation, especially those showing evidence of interaction (Windhorst et al. 1995)

(vi) WF/PC data show that elliptical galaxies have a higher angular correlation than spirals, but this difference in correlation amplitude is smaller than that observed in local samples (Neuschaefer et al. 1995). The two-point correlation function (all galaxies, irrespective of morphology) shows

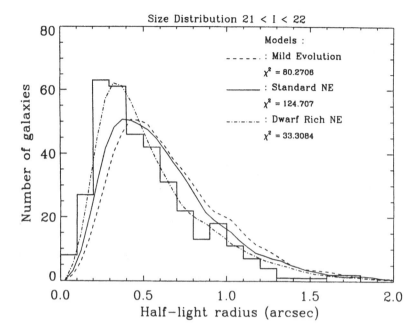

Figure 5. Number counts vs. size, compared with galaxy evolution models; from Im et al.(1995a)

a constant slope down to arcsec scales, with no substantial evidence for the excess galaxy pairs that might result from a high merger rate

(vii) The excess number counts in V and I are explained by the high fraction of compact objects and irregulars/peculiars. The irregulars show a steeply rising number count with magnitude (Glazebrook et al. 1995 Driver et al. – see fig. 4). Multiple cores are evident within 40% of them, and a third of these show evidence of being mergers.

(viii) The irregulars/peculiars also include minority populations which are diverse morphologically, including low surface brightness galaxies (some nucleated), and galaxies with superluminous starburst regions or knots (e.g. the peculiar object at $z = 0.7$ – Glazebrook et al. 1994)

(ix) The small, predominantly 'dwarf' population is dominated by objects with exponential profiles but with the axis ratios of ellipticals (Im et al. 1995b). The axis ratio distribution of faint bulge-like galaxies is like that of local ellipticals, showing no evolution in their properties. The axis ratio distribution of large disk-type galaxies ($r_{hl} > 0\rlap{.}''6$) is similarly consistent with local samples, but the same is not true for the small galaxies with exponential profiles – these have an axis ratio distribution which is like the ellipticals

(x) At I = 22, 15% of all galaxies have 'satellite' galaxies which are fainter than their 'parents' by at least 1 or 2 mags. Such observations may represent evidence in support of the 'minor merger' hypothesis (e.g. Mihos and Hernquist 1994)

5. CONCLUSIONS

Parallel HST operations have steadily improved in efficiency through observing Cycles 1–4, and parallel observations with WF/PC and WFPC2 have been highly successful in providing a database which is extremely useful for achieving the goals of the Medium Deep Survey: the measurement of field galaxy properties to I=22 and fainter, number counts as a function of morphology, studies of the evolution of galaxies to at least z =0.5–0.7, and study of morphology as a function of environment. The detailed morphological studies include the search for compact and multiple nuclei, the frequency of mergers, interactions and groups, and the frequency of irregularities, starburst knots, bars, arms and rings, etc.

A picture is emerging whereby the familiar problem of the excess number counts is being solved in terms of the evolution of low-luminosity systems which have a space density which is higher than expected. This is manifested by a large number of irregular and peculiar galaxies in WFPC2 data, some of which seem to be merger remnants. Ground-based spectroscopy is underway to elucidate the properties of each of the classes of these enigmatic galaxies.

References

Babul, A., & Rees, M. J. (1992), *Mon. Not. R. Astr. Soc.,* **255**, 346

Broadhurst, T., Ellis, R. S., & Shanks, T. (1988), *Mon. Not. R. Astr. Soc.,* **235**, 827

Broadhurst, T., Ellis, R. S., & Glazebrook, K. (1992), *Nature,* **355**, 55

Casertano, S., Ratnatunga, K. U., Griffiths, R. E., Neuschaefer, L. W., & Windhorst, R. A. (1995), "Structural Parameters of Faint Galaxies: Results from a Catalog of Galaxies from the HST Medium Deep Survey with WF/PC", *Astrophys. J., in press*

Cowie, L. L., Songaila, A., & Hu, E. M. (1991), *Nature,* **354**, 460

Driver, S. P., Windhorst, R. A. & Griffiths, R. E. (1995) "The properties of Irregular field galaxies found in HST Medium Deep Survey images", *Astrophys. J., submitted*

Ferguson, H. C. & McGough, S. S. (1995), *Astrophys. J., in press*

Forbes, D. A., Elson, R. A. W., Phillips, A. C., Illingworth, G. D., & Koo, D. C. (1995), "The Nuclear Colors and Morphology of Field Galaxies at Moderate Redshift", *Astrophys. J. Lett., in press*

Glazebrook, K., Lehar, J., Ellis, R., Aragon-Salamanca, A., & Griffiths, R. (1994), "An unusual high redshift object discovered with HST: peculiar starburst galaxy or new gravitational lens", *Mon. Not. R. Astr. Soc., ,* **270**, L63

Glazebrook, K., Ellis, R. S., Santiago, B. and Griffiths, R. E. (1995) "The Distribution of Faint Galaxy Morphologies to I = 22", *Nature, submitted*

Griffiths, R. E., et al. (1994a), "The HST Medium Deep Survey with WF/PC: I. Methodology and Results on the Field Near 3C273", *Astrophys. J.,* **437**, 67

Griffiths, R. E., et al. (1994b) "The Morphology of faint galaxies in Medium Deep Survey Images using WFPC2", *Astrophys. J. Lett.*, **435**, L49

Im, M., Casertano, S., Griffiths, R. E., Ratnatunga, K. U. & Tyson, J. A. (1995a) "A Test of Galaxy Evolutionary Models Via Angular Sizes", *Astrophys. J., in press*

Im, M., Casertano, S., Griffiths, R. E., & Ratnatunga, K. U., (1995b) "The Axis Ratio Distribution of Faint Galaxies and their Morphological Properties", *Astrophys. J., submitted*

Koo, D. C., & Kron R. G. 1992, *Ann. Rev. Astr. Astrophys.*, **30**, 613

Lilly, S. J. 1993, *Astrophys. J.*, **411**, 501

Mihos, J. C., & Hernquist, L. (1994), *Astrophys. J.*, **425**, L13

Mutz, S. B., et al. (1994) "The Θ-z relation for HST bulges and disks out to z≃0.8", *Astrophys. J. Lett.*, **434**, L55

Phillips, A. C., Bershady, M. A., Forbes, D. A., Koo, D. C., Illingworth, G. D., Reitzel, D. B., Griffiths, R. E., & Windhorst, R. A., (1995) "Structure and Photometry of an I < 20.5 Galaxy Sample from the HST Medium Deep Survey", *Astrophys. J., in press*

Neuschaefer, L. W., Casertano, S., Griffiths, R. E. & Ratnatunga, K. U. (1995), "Galaxy Clustering Statistics in HST Medium Deep Survey Images", *Astrophys. J., submitted*

Ratnatunga, K. U., Griffiths, R. E., Casertano, S., Neuschaefer, L.W., & Wyckoff, E. W. (1994b), "Calibration of the HST Wide Field Camera for Quantitative Analysis of Faint Galaxy Image Parameters", *Astron. J., in press*

Ratnatunga, K. U., Griffiths, R. E. and Casertano, S. (1994b), "Maximum likelihood estimation of galaxy Morphology: faint HST WFC images", in *The Restoration of HST Images and Spectra II*, eds Hanisch, R. J., White, R., Proc. Space Telescope Science Institute Workshop, p. 333.

Rocca-Volmerange, B. and Guiderdoni, B. (1990), *Mon. Not. R. Astr. Soc.*, **247**, 166

Tresse, L., Hammer, F., Le Fevre, O., & Proust, D. (1993), *Astron. & Astrophys.*, **277**, 53

Windhorst, R. A., et al. (1994) "The HST Medium Deep Survey: Deconvolution of WFC field galaxy images in the 13H+43deg field", *Astron. J.*, **107**, 930

Windhorst, R. A., et al. (1995) "The nature of microJansky radio sources: Deep HST imaging of an ultradeep VLA 8.4 GHz survey field", *Nature, in press*

THE FAR-ULTRAVIOLET BACKGROUND

PETER JAKOBSEN

Astrophysics Division
Space Science Department of ESA
ESTEC, 2200 AG Noordwijk
The Netherlands

1. Introduction

Extensive reviews of the diffuse background at ultraviolet ($\lambda 1000 - 2500$ Å) wavelengths have been given by Davidsen, Bowyer & Lampton (1974), Paresce & Jakobsen (1980), and most recently by Bowyer (1992) and Henry (1992). Since many astronomical sources emit radiation with photon energies in the range $h\nu \simeq 10 - 20$ eV through various thermal and non-thermal emission processes, the diffuse UV background is of relevance for a range of topics. These include the study of the properties of interplanetary and interstellar dust grains, thermal line emission from the interstellar and intergalactic gas, the integrated light of galaxies, and radiative decay of exotic cosmological particles.

On the observational side, there is today good evidence that the diffuse UV background is largely dominated by dust scattering and other interstellar emission processes occurring within the Milky Way. The intensity of any quasi-isotropic and therefore possibly *extragalactic* component to the background in the $\lambda 1300 - 2000$ Å range is presently only known to within a factor of $\sim 2 - 3$, but is approximately $I_\lambda \approx 100$ photons s^{-1}cm^{-2}sr^{-1}Å^{-1} (Martin & Bowyer 1990, Martin, Hurwitz & Bowyer 1990, Hurwitz, Bowyer & Martin 1991).

Three extragalactic sources of UV radiation that could conceivably produce a diffuse background flux of this intensity: i) diffuse thermal emission from the intergalactic medium; ii) the integrated UV light of galaxies and quasars; and iii) radiative decay of massive neutrinos. This paper briefly summarizes what is currently known about the first two of these emission

229

M. Kafatos and Y. Kondo (eds.), Examining the Big Bang and Diffuse Background Radiations, 229–236.
© 1996 *IAU.*

sources. For a discussion of the latter, more speculative, source, the reader is referred to the paper presented by Dr. Sciama at this symposium.

2. Diffuse UV Emission from the Intergalactic Medium

One of the original motivations behind the first attempts to measure the diffuse background radiation at far-ultraviolet wavelengths was the hope that such observations might reveal the existence of a dense ($\Omega \approx 1$), "lukewarm" (10^3 K$< T < 10^6$ K) intergalactic medium (IGM) through the redshift-smeared HI Lyα and HeII λ304 line emission that such a medium would emit (cf. Kurt & Sunyaev 1967, Davidsen, Bowyer & Lampton 1974, Paresce & Jakobsen 1980).

Since then, a wealth of information about the nature, composition and physical properties of intergalactic gas has been obtained through the study of quasar absorption lines. Thanks to the extensive surveys of the discrete absorption lines that appear in quasar spectra, we today not only have a reasonably accurate measure of how much matter intergalactic space contains, but also a good picture of the ionization processes at work in this gas. Armed with this new information, it is possible to show that diffuse thermal emission from intergalactic gas is almost certainly not a significant contributor to the diffuse far-UV background. In the following, the main arguments leading to this conclusion are briefly outlined. More complete discussions can be found in Martin, Hurwitz & Bowyer (1991) and Jakobsen (1994).

2.1. REDSHIFTED LYα AND HEII λ304 Å LINE EMISSION

From the study of absorption lines in quasar spectra, it is known that intergalactic space contains significant amounts of baryonic matter, in the form of an abundant population of hydrogen "Lyman forest" clouds (Sargent et al. 1980, Murdoch et al. 1986, Sargent 1988, Hunstead 1988), and most likely also in the form of a diffuse "intercloud" IGM (Jakobsen et al. 1994).

Both the so-called Lyman forest clouds and the ambient IGM are believed to be kept highly photoionized by a metagalactic flux of ionizing radiation, presumably arising from the ionizing output from quasars and young galaxies. One key piece of evidence for this is found in the so-called "proximity effect" (Carswell et al. 1982; Murdoch et al. 1986; Bajtlik, Duncan & Ostriker 1988; Lu, Wolfe & Turnshek 1991, Bechtold 1994), i.e. the statistical underabundance displayed by the forest systems as the emission redshift of a given quasar is approached. This effect is interpreted as being caused by the radiation field of the background quasar enhancing the total ionizing flux above the metagalactic level. Since the quasar flux can be estimated from the magnitude and spectrum of the quasar, an estimate of the background metagalactic ionizing background intensity can be derived from the measured contrast of the effect. Through this technique one infers that the intergalactic ionizing background in the redshift range $1.7 \lesssim z \lesssim 3.8$ has an intensity at the Lyman limit of order $I_{\nu_H} \approx 10^{-21} \mathrm{erg\ s^{-1} cm^{-2} sr^{-1} Hz^{-1}}$. At lower redshifts, the proximity effect is less pronounced and the inferred extragalactic ionizing background is approximately two orders of magnitude lower (Kulkarni & Fall 1993); and consistent with limits on the local extragalactic ionizing flux derived from Hα observations of high-latitude and extragalactic HI clouds (Reynolds et al. 1986; Songaila, Bryant & Cowie 1989; Kutyrev & Reynolds 1989)

If ionized, both the Lyman forest clouds and the ambient IGM will emit Lyα λ1216Å and HeII λ304 line radiation through the process of recombination. A firm upper limit on the diffuse background contribution due to this redshifted recombination emission can be placed from the following simple argument. Since a photoionized gas effectively acts as a simple photon down-converter (emitting for example ~ 0.7 Lyα photons per photoionization event), it follows that the intensity of the resulting recombination emission can never exceed that of the ionizing input flux. Therefore the background at observed wavelength λ due to redshift-smeared recombination radiation in a line at wavelength λ_l must obey

$$I_\lambda(\lambda) \ll \left[\frac{I_{\nu_H}(z)}{2\pi\hbar\lambda_H (1+z)^4} \right] \qquad (1)$$

where the expression on the right is merely the ionizing background intensity $I_{\nu_H}(z)$ at the Lyman limit, λ_H, converted to photon units and redshifted to $z = 0$ from the appropriate redshift, $z = \lambda/\lambda_l - 1$.

In the case of the Lyα line, emission from redshifts $0 \lesssim z \lesssim 0.6$ will give rise to diffuse background radiation at wavelengths $\lambda \simeq 1200 - 2000$ Å. The ionizing flux at these redshifts of $I_{\nu_H} \approx 10^{-23}$ erg s^{-1}cm^{-2}sr^{-1}Hz^{-1} yields per eq.(1) an upper limit on the observable background of $I_\lambda \ll 2$ photons s^{-1} cm^{-2} sr^{-1} Å^{-1}. Consequently, redshifted intergalactic recombination emission in Lyα can at most contribute a few percent of the nominal observed extragalactic UV flux.

This simple, but powerful, argument applies equally well to the case of redshifted HeII λ304 Å recombination from higher redshifts $3 \lesssim z \lesssim 5$. Since the spectrum of the ionizing background is likely to decrease with decreasing wavelength, the ionizing background at the $\lambda_{HeII} = 228$ Å photoionization threshold of singly ionized helium is almost certainly not higher than that at $\lambda_H = 912$ Å. The nominal ionizing background at $z \gtrsim 3$ of $I_{\nu_H} \approx 10^{-21}$ergs s^{-1}cm^{-2}sr^{-1}Hz^{-1} corresponds in this case to a redshifted background of $I_\lambda \ll 1$ photons s^{-1}cm^{-2}sr^{-1}Å^{-1}. Hence, it is safe to conclude that redshifted recombination emission in HeII $\lambda 304$ Å also occurs at far too low an intensity to be of any practical importance.

The alternative possibility of *collisionally* excited Lyα emission – as would arise in a shock-heated IGM – can also be ruled out at a high level of confidence. In this case the intensity of collisionally excited Lyα emission constrained to unobservably low levels by the firm upper limit on the total amount of neutral hydrogen present in intergalactic space as determined from quasar absorption line statistics and the Gunn-Peterson test (see Jakobsen 1994 for details).

The equivalent argument does not apply in the case of collisionally excited HeII $\lambda 304$Å emission, since the sole detection of strong HeII Gunn-Peterson absorption in a high redshift quasar obtained so far only provides a *lower* limit on the amount of intergalactic HeII present at $z \sim 3$ (Jakobsen et al. 1994). If the IGM is sufficiently dense ($\Omega \sim 1$) and collisionally ionized, it could in principle emit strongly in HeII $\lambda 304$Å. However, even so, the resulting background flux would still not be observable because the universe is optically thick to HeII $\lambda 304$Å radiation at large redshift.

2.2. THE LYMAN CONTINUUM OPACITY OF THE UNIVERSE

Any emission originating at high redshift at wavelengths below the Lyman limit is subject to photoelectric absorption by neutral hydrogen encountered along its path. Although it has been known for some time that the classical Gunn-Peterson test demonstrates that the Lyman continuum opacity of any *smoothly* distributed IGM is negligible, it has only recently been fully appreciated that the statistics of quasar absorption lines imply the that the accumulated absorption out to moderate and high redshift from the *clumped* component is quite substantial.

Detailed discussions of this topic have been given by Møller & Jakobsen (1990) and Zuo & Phinney (1993), who show that the cumulative absorption is dominated by the densest so-called "Lyman limit" types of quasar absorption systems (Tytler 1982; Bechtold et al. 1984; Lanzetta 1988; Sargent, Steidel & Boksenberg 1989; Bahcall et al. 1993) In the particular case of HeII $\lambda 304$Å emission stemming from redshifts $z \gtrsim 3$, such radiation will be attenuated by approximately two orders of magnitude of accumulated photoelectric absorption out to such high redshift (cf. Jakobsen 1994, Figure 3).

3. The Integrated UV Light of Galaxies and Quasars

The brightness of the extragalactic sky at visible and infrared wavelengths carries important information on the history of star formation and galaxy evolution throughout the age of the Universe. As first emphasized by Tinsley (1973), the diffuse background at the shorter UV wavelengths is also an important element of the modern day version of "Olbers' paradox" – but with several key differences with respect to the situation in the adjacent spectral regions.

For one, the observed integrated far-UV spectra of the different classes of galaxies are not particularly well modelled or understood at present. This is especially true of the UV emission from ellipticals and the bulges of spirals, which is believed reflect complex stages of late stellar evolution (cf. Burnstein et al. 1988; King et al. 1993 and references therein). However, the global galaxy luminosity function at UV wavelengths is almost entirely dominated by the emission from massive O and B stars contained in the star-forming regions of spirals and irregulars. Consequently, in the UV late type galaxies account for close to 90% of the local galaxy luminosity density and therefore dominate the integrated galaxy background (cf. Armand & Milliard 1994 for a recent discussion).

Another important difference with respect to the case at visible and longer wavelengths is that the integrated UV light of galaxies is accumulated over a relatively modest cosmological pathlength. Late type galaxies are generally surrounded by large halos of neutral hydrogen which permit little or no radiation to escape below the Lyman limit at $\lambda \leq 912$ Å. At an observed wavelength of $\lambda 1500$ Å, the Lyman limit is reached at $z = 0.64$, corresponding to a look-back time of $\Delta t \simeq 3h^{-1}$ Gyr. This modest look-back time, combined with the short life times of main sequence OB stars, leads to the integrated UV light of galaxies being primarily a measure of the level of on-going star formation in the relatively local Universe.

Three rather different observational and theoretical approaches have so far been employed in assessing the possible contribution of galaxies to the extragalactic UV background.

3.1. INTEGRATION OF THEORETICAL GALAXY EVOLUTION MODELS

Following the pioneering work of Tinsley (1972), several groups have in recent years developed very elaborate and physically self-consistent models for galaxy evolution based on stellar formation and evolution theory (e.g. Guiderdoni & Rocca-Volmerange 1991, Bruzual & Charlot 1993). Although the primary motivation for these models is to explain faint galaxy counts and colors observed in the visible, as first emphasized by Tinsley (1973), the predicted integrated background spectrum provides an important observational constraint on the models.

A recent discussion with emphasis on the ultraviolet has been given by Martin, Hurwitz & Bowyer (1991), who show that the current evolution models predict a rather wide range of spectral shapes for the background.

Nonetheless, the predicted far-UV intensities generally span the $I_\lambda \simeq 40 - 240$ photons $s^{-1}cm^{-2}sr^{-1}\text{Å}^{-1}$ range – suggesting that the integrated UV light of galaxies should be within reach observationally.

3.2. UV BACKGROUND FLUCTUATION MEASUREMENTS

Martin & Bowyer (1989) have attempted to detect the small-scale fluctuations in the far-UV background expected from the integrated light of galaxies. This technique, which was first pioneered by Schectman (1974) at visible wavelengths, consists of fitting the measured power spectrum of the UV background to that expected from galaxies calculated on the basis of observed visible light correlation functions and assumed models for galaxy spectral evolution. Based on their measurements, Martin & Bowyer concluded that the integrated light of galaxies can at most contribute $\approx 20\%$ of the total (galactic and extragalactic) background observed, corresponding to an intensity of $I_\lambda \simeq 40 \pm 13$photonss$^{-1}cm^{-2}sr^{-1}\text{Å}^{-1}$. Subsequent analysis, however, has questioned the validity of the fluctuation results due to contamination from UV starlight scattered off the *IRAS* cirrus (Sasseen et al. 1993).

3.3. EXTRAPOLATION OF ULTRAVIOLET GALAXY COUNTS

The most convincing and direct demonstration that galaxies must provide a significant contribution to the extragalactic UV background has recently been given by Armand, Milliard & Deharveng (1994). These authors base their analysis on observed UV ($\simeq 2000$ Å) galaxy counts obtained with a balloon borne UV telescope. Their UV counts are complete down to a UV magnitude of $m \simeq 18.5$, which in itself yields an integrated galaxy background of $I_\lambda \simeq 30$ photons $s^{-1}cm^{-2}sr^{-1}\text{Å}^{-1}$. Armand et al. extrapolate this resolved portion of the background to fainter magnitudes by use of theoretical galaxy evolution models. Because of the various observational and theoretical uncertainties, the total background can only be predicted with certainty to lie in the range $I_\lambda \simeq 40 - 130$ photons $s^{-1}cm^{-2}sr^{-1}\text{Å}^{-1}$. Nonetheless, this flux is in good agreement with those obtained through the other less direct approaches above, and clearly demonstrates that galaxies must be a significant source of extragalactic UV background radiation.

3.4. THE INTEGRATED UV LIGHT OF QUASARS AND AGNS

As opposed to the situation at higher X-ray energies, quasars and active galactic nuclei play only a marginal role in the case of the background in the UV (Martin & Bowyer 1989; Martin et al. 1991 and references therein). The reason that quasars are a dominant background source in the X-rays, but not in the adjacent UV, can be traced to the fact that the nominal extragalactic far-UV background intensity of $I_\lambda \approx 100$ photons $s^{-1}cm^{-2}sr^{-1}\AA^{-1}$ is considerably brighter in terms of energy per octave in frequency than the extragalactic background flux in the X-rays.

4. Summary and Conclusions

Of the various sources that have been proposed as contributors to the d-iffuse extragalactic background at UV wavelengths, the integrated light of galaxies is the only component that is likely to be detectable in practice. Direct galaxy counts in the UV down to a limiting magnitude of $m \simeq 18.5$ have already "resolved" the extragalactic UV background to an intensity of $I_\lambda \simeq 30$ photons $s^{-1}cm^{-2}sr^{-1}\AA^{-1}$, which extrapolated to fainter magnitudes by means of galaxy evolution models suggests that the integrated UV light of galaxies is the all-dominant source of extragalactic UV background radiation.

Thermal emission from the intergalactic medium is almost certainly an entirely negligible contributor to the UV background. The intensity of the metagalactic ionizing background at high redshifts inferred from the so-called "proximity effect" displayed by the Lyman forest absorption lines implies that the redshift-smeared backgrounds due to $Ly\alpha$ and HeII $\lambda304\AA$ recombination radiation from photoionized intergalactic gas are to be found at intensities far below current observational limits on the extragalactic background flux. Similarly, the results of the Gunn-Peterson test at low redshifts derived from UV quasar spectra constrain the intensity of redshifted collisionally excited $Ly\alpha$ emission from a shock-heated IGM component to an equally low intensity. Lastly, the frequency of optically thick quasar absorption systems assures that any collisionally excited HeII $\lambda304\AA$ emission stemming from $z \gtrsim 3$ will remain effectively hidden from view because of strong accumulated Lyman continuum absorption.

References

Armand, C, & Milliard, B. 1994, A&A 282,1
Armand, C, Milliard, B., & Deharveng, J. M. 1994, A&A 284,12
Bahcall, J. N. et al. 1993, ApJS 87,1
Bajtlik, S., Duncan, R. C., & Ostriker, J. P. 1988, ApJ 327,570
Bechtold, J. 1994, ApJS 91,1
Bechtold, J., Green, R. F., Weymann, R. J., Schmidt, M., Estabrook, F. B., Sherman, R. D., Wahlquist, H. D., & Heckman, T. M. 1984, ApJ 281,76
Bowyer, S. 1992, ARA&A 29,59
Bruzual A., G. & Charlot, S. 1993, ApJ 405,538
Burnstein, D., Bertola, F., Buson, L. M., Faber, S. M., & Lauer, T. R. 1988, ApJ 328,440

236

Carswell, R. F., Whelan, J. A. J., Smith, M. G., Boksenberg, A., & Tytler, D. 1982, MNRAS, 198,91

Davidsen, A., Bowyer, S., & Lampton, M. 1974, Nat 247,513

Guiderdoni, B., & Rocca-Volmerange, B. 1991, A&A 252,435

Henry, R. C. 1992, ARA&A 29,89

Hunstead, R. W. 1988, in QSO Absorption Lines: Probing the Universe, ed. C. Blades, D. Turnshek, & C. Norman, (Cambridge: Cambridge University Press), p71

Hurwitz, M., Bowyer, S., & Martin, C. 1991, 372,167

Jakobsen, P. 1994, in Extragalactic Background Radiation, eds. M. Livio, M. Fall, D. Calzetta, & P. Madau, Cambridge University Press (in press).

Jakobsen, P., Boksenberg, A., Deharveng, J. M., Greenfield, P., Jedrzejewski, R., & Paresce, F. 1994, Nature 370,35

King, I. R. et al. 1993, ApJ 397,L35

Kulkarni, V. P., & Fall, M. S. 1993, ApJ 413,L63

Kurt, V. G., & Sunyaev, R. A. 1967, Cosmic Research, 5,496

Kutyrev, A. S., & Reynolds, R. J. 1989, ApJ, 344,L9

Lanzetta, K. M. 1988, ApJ 332,96

Lu, L., Wolfe, A. M., & Turnshek, D. A. 1991, ApJ 367,19

Martin, C., & Bowyer, S. 1989, ApJ 338,667

Martin, C., & Bowyer, S. 1990, ApJ 350,242

Martin, C., Hurwitz, M.,& Bowyer, S. 1990, ApJ 345,220

Martin, C., Hurwitz, M.,& Bowyer, S. 1991, ApJ 379,549

Milliard, B., Donas, J., Laget, M., Armaud, C., & Vuillemin, A. 1992, A&A 257,24

Murdoch, H. S., Hunstead, R. W., Pettini, M., & Blades, J. C. 1986, ApJ, 309,19

Møller, P., & Jakobsen, P. 1990, A&A 228,299

Paresce, F., & Jakobsen, P. 1980, Nature 288,119

Reynolds, R. J., Magee, K., Roesler, F. L., Scherb, F., & Harlander, J. 1986, ApJ 309,L9

Sargent, W. L. W., 1988, in QSO Absorption Lines, Probing the Universe, ed. J. C. Blades, D. Turnshek & C. A. Norman (Cambridge: Cambridge University Press), p1

Sargent, W. L. W., Steidel, C. C., & Boksenberg, A. 1989, ApJS 69,703

Sargent, W. L. W., Young, P. J., Boksenberg, A., & Tytler, D. 1980, ApJS 42,41

Sasseen, T. P., Bowyer, S., Wu, X., & Lampton, M. 1993, BAAS 25,822

Songaila, A., Bryant, W.,& Cowie, L. L. 1989, 345,L71

Tinsley B. M. 1972, A&A 20,383

Tinsley B. M. 1973, A&A 24,89

Tytler, D. 1982, Nature 298,427

Zuo, L., & Phinney, E. S. 1993, ApJ 418,28

EVOLUTION OF THE EUV BACKGROUND FROM QUASAR ABSORPTION LINE STUDIES

JILL BECHTOLD
Steward Observatory
University of Arizona
Tucson, AZ 85721

Abstract. The integrated extreme ultraviolet (EUV) radiation from quasars and other high redshift sources provides an ambient ionizing radiation field which may photoionize the gas seen as quasar absorption lines. In particular, the observed evolution of the Lyα forest clouds probably results in part from the evolution of the EUV metagalactic field. Estimates of the EUV field as a function of redshift can be made from measuring the "proximity effect" in quasar spectra; uncertainties in these estimates may be large. Given the uncertainties, the estimated EUV field at z≈3 derived from the proximity effect is in reasonable agreement with the expected contribution from luminous quasars.

1. Introduction

Rees & Setti (1970) first pointed out that quasars are likely to be a significant source of hard UV photons at high redshift, and worked out estimates of how the field should evolve with redshift. Photoionization by this intergalactic field may be important in understanding the intergalactic medium (e.g. Jakobsen, these proceedings), the Lyα forest (e.g. Ikeuchi and Turner 1991; Charlton, Salpeter and Hogan 1993) and optically thin metal-line systems (e.g. Reimers 1995), and the gas in the outermost regions of the Milky Way (e.g. Savage 1995). In addition, evidence for other sources of radiation at high redshift may be implied if independent measures of the EUV field imply values larger than known sources (e.g. Bajtlik, Duncan & Ostriker 1988). The reader is referred to recent reviews of the EUV background by Bechtold (1993,1995) and Madau (1995).

M. Kafatos and Y. Kondo (eds.), Examining the Big Bang and Diffuse Background Radiations, 237–244.
© *1996 IAU.*

2. Predicted Sources of EUV Background

The expected contribution of a population of objects to the mean ambient radiation field can be described by the specific intensity J_ν seen by an observer at redshift z, and frequency ν, and can be computed by integrating over the luminosity function of all sources at higher redshifts. Other inputs include the EUV spectrum of individual objects (since photons at higher energies are redshifted and contribute to the background at lower energies at any redshift), the cosmological parameters (q_o and H_o), and the amount and distribution of dust and gas which attenuates the EUV radiation.

Luminous quasars are likely an important source of photons (Rees & Setti 1970; Sargent *et al.* 1980; Bechtold *et al.* 1987; Miralde-Escude & Ostriker 1990; Madau 1992). Less luminous AGNs and obscured quasars may be important, particularly at low redshift (Ostriker & Heisler 1984; Wright 1986; Terasawa 1992; Fall & Pei 1993). Hot stars in high-redshift galaxies which are expected to have higher star-formation rates than present-day galaxies may contribute a substantial background, if the UV photons can leak out of the star-forming regions (Rees & Setti 1970; Tinsley 1972; Code & Welch 1982; Bechtold *et al.* 1987; Miralde-Escude & Ostriker 1990). In all cases, the effect of attenuation by intervening absorption is likely important, so that the radiation field an observer "sees" at any redshift is generated relatively locally (e.g. Madau 1995 and references therein).

Other sources have been suggested, including the decay of massive particles left over from a hot Big Bang (Cowsik 1977; DeRujula & Glaskow 1980; Kimble, Bowyer & Jakobsen 1981; Melott & Sciama 1981). Decays result in line emission, which is subsequently smeared out with redshift into a background continuum radiation field (e.g. Sciama, these proceedings). This explanation for the EUV background continues to survive observational tests (e.g. Miralde-Escude & Ostriker 1992; Sciama 1995).

3. Empirical Measures of the EUV Background

3.1. Z≈0

Even if the source of EUV photons turns off at redshift z, J_ν will decay as $(1+z)^4$ in the absence of absorption. Thus local limits on J_ν place constraints on the field at higher z.

Hα emission from 21-cm clouds gives a limit on the ionizing flux, assuming Case B recombination theory. Reynolds (1987) and Songaila, Bryant, & Cowie (1989) report detections of Hα from high velocity 21-cm clouds in the Milky Way. Such measurements are difficult since this emission is weak compared to the much brighter background of Hα emission from the warm, ionized ISM (Reynolds 1987). In addition, estimates depend on how far the high velocity clouds are above the plane of the Milky Way, and hence how important other sources of ionizing radiation such as O and B stars are. Similarly, limits can be placed on Hα emission from isolated extragalactic 21-cm clouds, which appear to be far from stars, so one expects the metagalactic field to dominate (Reynolds *et al.* 1986; Stocke *et al.* 1991; Donahue *et al.* 1994). The strongest limits so far are reported by Vogel *et al.* (1994) who observed part of the Haynes-Giovanelli cloud HI 1225+01 (Giovanelli and Haynes 1989), and derive $J_\nu < 1 \times 10^{-22}$ erg cm^{-2} s^{-1} Hz^{-1} sr^{-1}.

The sharp cut-off in the HI disks in spiral galaxies may result from photo-ionization by the EUV background (Silk & Sunyaev 1976; Bochkarev & Sunyaev 1977). Sharp edges have been inferred from 21-cm studies of a number of galaxies (e.g. Corbelli & Schneider 1990, van Gorkom 1993, van Gorkom *et al.* 1994), and have been modeled by Maloney (1993), Charlton, Salpeter & Hogan (1993), and Dove and Shull (1994). Assuming that tidal truncation is not important, the values for J_ν are quite low, $J_\nu \approx 4 \times 10^{-23}$ erg cm^{-2} s^{-1} Hz^{-1} sr^{-1}.

Direct measurement of the extragalactic background can be attempted just longward of the Lyman limit (e.g. Paresce 1990, Bowyer 1991, Henry 1991; Martin *et al.* 1991). These observations are difficult, and the limit on the extragalactic component depends on the large and unfortunately highly uncertain subtraction of Galactic starlight which is scattered by dust. Several speakers at this symposium discuss the complexities in the related issue of understanding the far-IR background. Hopefully progress in modeling the dust emission will help in deriving more secure limits on the EUV background as well.

3.2. RESULTS FROM THE LYMAN ALPHA FOREST

The number of Lyα forest lines near the quasar being used to measure the absorption is smaller than expected from statistics of lines with z(abs) << z(em) (Weymann, Carswell & Smith 1981; Murdoch *et al.* 1986). This so-called "proximity effect" may result from photo-ionization of the Lyα clouds by the EUV photons from the quasar itself. Given the strength of the proximity effect as a function of redshift from the quasar, one can estimate the ambient field far from the quasar (e.g. Bajtlik, Duncan & Ostriker 1988). Carswell *et al.* (1987) first pointed out that the lack of a strong effect in the z=3.78 quasar PKS 2000-33 implied that the EUV background is too large to be from quasars. Since then, estimates of the contribution from quasars have increased substantially since many high redshift quasars have been discovered (see Schmidt, these proceedings) so that the descrepency is not very large, perhaps only a factor of three. Bajtlik, Duncan & Ostriker

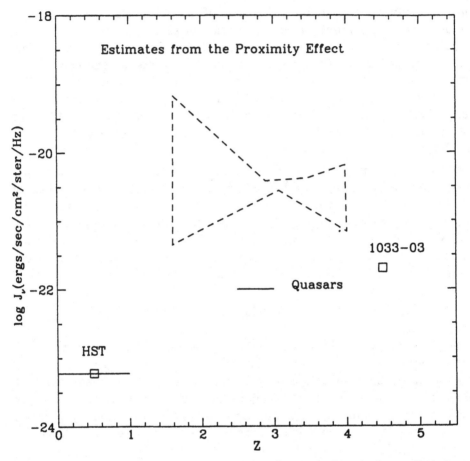

Figure 1. Estimates of the intergalactic radiation background at the Lyman limit from the proximity effect. Dashed lines shows allowed region from Bechtold (1994). HST estimate is from Kulkarni and Fall (1993), and point labeled 1033-03 is from Williger *et al.* (1994). The estimated contribution from known quasars is indicated.

(1988) have emphasized that if the proximity effect estimates are larger than the integrated background expected from quasars then another source of ionizing radiation is implied – quasars obscured by dust, or hot stars in young galaxies for example (see also Fall & Pei 1993).

The most recent estimates of the proximity effect based on optical observations of the Lyα forest (so $z \approx 1.6$-4) are given by Bechtold (1994). Williger *et al.* (1994) derive a value for J_ν at higher redshift from the spectrum of one very high redshift quasar at $z=4.5$. At low redshift, Kulkarni & Fall (1993) used the FOS data from the quasar absorption line key project of Bahcall *et al.* (1993) to derive J_ν at $z \approx 0$-1. All authors use the simply photoionization model described by Bajtlik, Duncan & Ostriker (1988).

The results are shown in Figure 1. The relatively large uncertainty at $z\sim2$ results from the fact that there are very few quasars with published spectra in existing samples (see Bechtold 1995, Bechtold *et al.* 1995). However the general trend that the EUV field is smaller at $z\approx0$ and $z\approx4.5$ than at $z\approx3$ is probably secure.

A number of sources of *systematic* uncertainties in the results of Figure 1 need also be considered. Most effects tend to imply that the background has been *overestimated* by the simple models by plausibly as much as a factor of 10 (see Bechtold 1994, 1995 for complete reviews). One interesting and ultimately tractable effect is that the redshifts of the quasars themselves are probably systematically underestimated for the high redshift objects, since they rely on redshift of Lyα and C IV emission. These lines are probably systematically shifted with respect to the narrow lines and Balmer lines, which for $z=2$-3 quasars are in the near-IR (Espey 1993; Bechtold 1994). The relative shift in low redshift objects can be as high as 1500 km/sec. If the quasars are really at higher redshift, then the observed Lyα forest clouds are farther away from the quasar than assumed, and the importance of the quasar radiation has been overestimated; hence the "true" J_ν is lower than derived above. Substantial advances in IR array technology recently (see Figure 2) allow the rest-frame optical emission lines of large numbers of high redshift quasars to be studied in detail for the first time. The preliminary evidence is that the shifts between Lyα and narrow [O III] may be quite substantial and vary from object to object. Overestimates of factors of 3-9 for J_ν are then implied. Within a year, a large enough sample should be in hand to sort out this effect.

4. Acknowledgements

This research was supported by NSF grant AST-9058510. Travel to the IAU was supported by an IAU travel grant, and a travel grant from the American Astronomical Society.

242

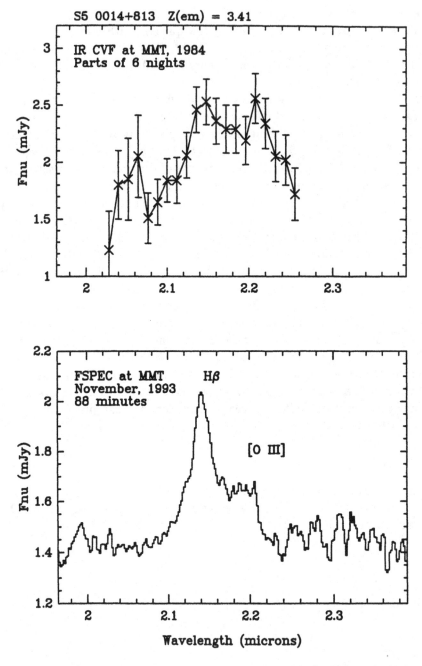

Figure 2. Spectra of the region near rest-frame H-beta and [O III] for the high-redshift quasar S5 0014+813. Top is from Kuhr *et al.* (1984); bottom is from Bechtold, Kuhn and Rieke (1995).

References

Bahcall, J.N. *et al.* (1993), *ApJS*, 87, 1.

Bajtlik, S., Duncan, R. C. and Ostriker, J. P. (1988), *ApJ*, 327, 570.

Bechtold, J. (1993), in the *Third Teton School on the Evolution of Galaxies and Their Environment*, ed. J. M. Shull and H. A. Thronson, (Dordrecht:Kluwer), p. 559.

Bechtold, J. (1994), *ApJS*, 91, 1.

Bechtold, J. (1995), in proceedings of ESO Workshop on QSO Absorption Lines, Garching, ed. G. Meylan, in preparation.

Bechtold, J., Weymann, R. J., Lin, A. and Malkan, M. (1987), *ApJ*, 315, 180.

Bechtold, J., Kuhn, O. and Rieke, M. (1995), in preparation.

Bechtold, J., Dobrzycki, A., Scott, J., Foltz, C. and Lesser, M. (1995), in preparation.

Bochkarev, N. G. and Sunyaev, R. A. (1977), *Astron.Zh.*, 54, 957; translated in *Soviet Astronomy*, 21, 542.

Bowyer, S. (1991), *ARAA*, 29, 59.

Carswell, R. F., Webb, J. K., Baldwin, J. A. and Atwood, B. (1987), *ApJ*, 319, 709.

Charlton, J. C., Salpeter, E. E., and Hogan, C. J. (1993), *ApJ*, 402, 493.

Code, A. D. and Welch, G. A. (1982), 256, 1.

Corbelli, E. and Schneider, S. E. (1990), *ApJ*, 356, 14.

Cowsik, R. (1977), *Phys. Rev. Lett.*, 39, 784.

De Rejula, A. and Glashow, S. (1980), *Phys. Rev. Lett.*, 45, 942.

Donahue, M., Aldering, G., and Stocke, J. (1994), paper 60.05 presented at May 1994 AAS meeting, Minneapolis.

Dove, J. B. and Shull, J. M. (1994), *ApJ*, 423, 196.

Espey, B. R. (1993), *ApJ*, 411, L59.

Fall, S. M. and Pei, Y. C. (1993), *ApJ*, 402, 479.

Giovanelli, R. and Haynes, M. P. (1989), *ApJ*, 346, L5.

Henry, R. C. (1991), *ARAA*, 29, 89.

Ikeuchi, S. and Turner, E. L. (1991), *ApJ*, 381, L1.

Kimble, R., Bowyer, S. and Jakobsen, P. (1981), *Phys. Rev. Lett.*, 46, 80.

Kuhr, H., McAlary, C. W., Rudy, R. J., Strittmatter, P. A. and Rieke, G. H. (1984), *ApJ*, 275 L33.

Kulkarni, V. P. and Fall, S. M. (1993), *ApJ*, 413, L63.

Madau, P. (1992), *ApJ*, 389, L1.

Madau, P. (1995), in proceedings of the ESO Workshop on QSO Absorption Lines, Garching, ed. G. Meylan, in preparation.

Maloney, P. (1992), *ApJ*, 398, L89.

Martin, C., Hurwitz, M. and Bowyer, S. (1991), *ApJ*, 379, 549.

Melott, A. L., and Sciama, D. W. (1981), *Phys. Rev. Lett.*, 46, 1369.

Miralde-Escude, J. and Ostriker, J. P. (1990), *ApJ*, 350, 1.

Miralde-Escude, J. and Ostriker, J. P. (1992), *ApJ*, 392, 15.

Murdoch, H. S., Hunstead, R. W., Pettini, M. and Blades, J. C. (1986), *ApJ*, 309, 19.

Ostriker, J. P. and Heisler, J. (1984), *ApJ*, 278, 1.

Paresce, F. (1990), in *The Galactic and Extragalactic Background Radiation*, ed. S. Bowyer, and C. Leinert, p. 307.

Rees, M. J. and Setti, G. (1979), *A&A*, 8, 410.

Reimers, D. (1995), in proceedings of the ESO Workshop on QSO Absorption Lines, Garching, ed. G. Meylan, in preparation.

Reynolds, R. J. (1987), *ApJ*, 323, 553.

Reynolds, R. J., Magee, K., Roesler, F. L., Scherb, F. and Harlander, J. (1986), *ApJ*, 309, L9.

Sargent, W. L. W., Young, P. J., Boksenberge, A. and Tytler, D. (1980), *ApJS*, 42, 41.

Savage, B. (1995), in proceedings of the ESO Workshop on QSO Absorption Lines, Garching, ed. G. Meylan, in preparation.

Sciama, D. (1995), in proceedings of the ESO Workshop on QSO Absorption Lines,

Garching, ed., G. Meylan, in preparation.

Silk, J. and Sunyaev, R. A. (1976), *Nature*, 260, 508.

Songaila, A., Bryant, W. and Cowie, L. L. (1989), *ApJ*, 345, L71.

Stocke, J. T., Case, J., Donahue, M., Shull, J. M. and Snow, T. P. (1991), *ApJ*, 374, 72.

Terasawa, N. (1992), *ApJ*, 392, L15.

Tinsley, B. M. (1972), *ApJ*, 178, 319.

van Gorkom, J. H. (1993), in *The Third Teton Conference on The Environment and Evolution of Galaxies*, ed. J. M. Shull and H. A. Thronson (Dordrecht:Kluwer), p. 343.

van Gorkom, J. H., Cornwell, T., van Albada, T. S. and Sancisi, R. (1994), in preparation.

Vogel, S. N., Weymann, R., Rauch, M. and Hamilton, T. (1994), preprint.

Weymann, R. J., Carswell, R. F. and Smith, M. G. (1981), *ARAA*, 19, 41.

Williger, G. M., Baldwin, J. A., Carswell, R. F., Cooke, A. J., Hazard, C., Irwin, M. J., McMahon, R. G., Storrie-Lombardi, L. J. (1994), *ApJ*, 428, 574.

Wright, E. L. (1986), *ApJ*, 311, 156.

THE X-RAY BACKGROUND

G. HASINGER
Astrophysikalisches Institut Potsdam
An der Sternwarte 16, 14482 Potsdam, Germany
& Universität Potsdam
Am Neuen Palais 10, 14469 Potsdam, Germany

Abstract.
The recent progress in the measurement and understanding of the X-ray background is reviewd here. Particular emphasis is put on a discussion of the partially discrepant measurement of the X-ray background spectrum in the 0.5-3 keV range. New and important constraints on large scale structure are obtained from measurements of the smoothness of the XRB. Recently the first discovery of a signal in the angular correlation function of the XRB could be announced. Finally, various X-ray surveys and their identification content is summarized. The role of optically inactive galaxies as a major contributor to the faint X-ray source population which might produce a substantial fraction of the XRB is clarified.

1. Introduction

The X-ray background (XRB), discovered as the first cosmic background radiation (Giacconi et al., 1962) well in advance of the Cosmic Microwave Background (CMB), presented one of the long-standing puzzles of modern astrophysics. At higher energies and on scales larger than about 10 degrees its celestial distribution is very isotropic, apart from a weak dipole anisotropy (Shafer & Fabian, 1987), indicating its cosmological origin. Major steps have been taken in the past few years towards an understanding of its nature. Originally the XRB spectrum in the range 3-40 keV, which resembles very closely a thermal bremsstrahlung model with a temperature of $\sim 40~keV$ (Marshall et al., 1980) led the way to an interpretation in terms of a hot, diffuse intergalactic medium. Such a truly diffuse hot

245

M. Kafatos and Y. Kondo (eds.), Examining the Big Bang and Diffuse Background Radiations, 245–262.
© 1996 IAU.

plasma would, however, produce a severe Compton distortion on the CMB spectrum, which has not been observed by the COBE satellite (Mather et al., 1990). This puts stringent constraints on the fraction of the XRB originating from hot gas and leaves the alternative interpretation of the XRB in terms of discrete sources the only feasible one. Nevertheless, the existence of cooler and/or significantly clumped hot plasma has not been ruled out by the COBE measurements.

At higher X-ray and soft gamma-ray energies (above 4 keV) measurements until recently have been performed only using collimated X-ray detectors with relatively coarse angular resolution (degrees). Consequently, while these measurements yielded very reliable estimates of the intensity and shape of the **total** X-ray background, they were only able to resolve a small fraction ($\sim 3\%$) of the XRB into discrete sources. The situation in the soft X-ray band (0.1-3 keV), where grazing incidence focussing optics can be used, is diametrically opposite. Due to the high sensitivity and angular resolution (below 1 arcmin) a substantial fraction of the X-ray background could already be resolved into discrete sources here. Deep surveys with the ROSAT satellite were able to resolve about 60% of the 1-2 keV background into discrete sources amounting to a surface density of $> 400 \ deg^{-2}$ at a flux of $2.5 \cdot 10^{-15} \ erg \ cm^{-2} \ s^{-1}$ (Hasinger et al., 1993). Direct optical follow-up studies could identify a substantial fraction of these objects as active galactic nuclei (QSOs and Seyfert galaxies, see e.g. Georgantopoulos et al., 1995 and references therein). However, for several reasons the total extragalactic X-ray background in this energy range is very hard to measure, so that its detailed shape and intensity are still a matter of debate (see below).

While the explanation of the X-ray background in terms of the summed X-ray emission of discrete objects (e.g. AGN), integrated in Olber's sense over cosmic distance and time, is a very attractive one (see eg. Setti & Woltjer, 1989), there were two puzzles in recent years which made this interpretation questionable, in particular in the hard X-ray band: the first one is the "spectral paradox", i.e. the fact that no single class of objects known so far has a spectrum resembling that of the XRB (e.g. Boldt, 1987). The second one is the "logN-logS" paradox, i.e. the fact that fluctuation analyses in the "hard" (2-10 keV) band indicate a surface density of objects a factor of 2-3 higher than that in the "soft" (0.5-2 keV) band (e.g. Mushotzky 1992). Another complication is, that the number counts of the faintest objects in the ROSAT deep surveys exceed the predictions from the most recent determination of the AGN X-ray luminosity function (Boyle et al., 1993) by about a factor of two (Hasinger et al., 1993), so that the possibility of a "new class of sources" had to be invoked.

However, a better understanding of the various classes of active galactic nuclei in terms of the "Unified Model", where differences between the classes are mainly due to orientation effects (e.g. Antonucci 1993) has shed new light on these questions. Recent detailed spectral observations of bright, nearby AGN in the X-ray and soft gamma-ray band, as well as ROSAT deep surveys have now obtained new ingredients that led the way to a solution of the two puzzles above in a complete and self-consistent way, assuming only known objects with measured properties (see e.g. Matt, 1994; Comastri et al, 1995; Setti, this volume)

In this *review* I summarize the most recent observations of the X-ray background and what we know about its constituents. I start with a discussion of the spectrum and angular correlation function of the XRB, then I summarize the major soft X-ray surveys and follow-up optical identifications. Finally I shortly discuss the new theoretical models for the X-ray background (this is the topic of the accompanying paper by Setti) and the role of galaxies which seem to become more and more important at fainter X-ray fluxes. At this location I want to thank my collaborators in quite a number of projects for the fruitful cooperation over many years and the permission to show some new material in advance of publication (representative for many more): R. Bower, R. Burg, R. Ellis, R. Giacconi, K. Mason, R. McMahon, M. Schmidt, A. Soltan, J. Trümper, G. Zamorani.

2. The Spectrum of the X-ray Background

2.1. THE HARD X-RAY BACKGROUND

The spectrum of the XRB actually extends over an energy range of more than 4 decades, from soft X-rays to low-energy gamma rays as can be seen in figure 1. In the 3-10 keV band it is largely isotropic, indicating an extragalactic origin. Above 3 keV, the large (and sometimes quite discrepant) variety of "historical" balloon- and satellite-based measurements with collimated proportional counters and scintillators as well as Compton telescopes has been summarized by Gruber (1992) by fitting a simple analytical form to the best available data. In the 2-10 keV band (HEAO-1 A2 data) this spectrum can be well approximated by a power law with an energy index of -0.4 and a normalization of 8 $keV\ cm^{-2}\ s^{-1}\ sr^{-1}\ kev^{-1}$ (Marshall et al, 1980). I would, however, like to caution that an independent measurement of the 2-6 keV background by the Wisconsin group yields a normalization which is almost 40% higher (11 in the above units; McCammon & Sanders 1990).

2.2. THE SOFT X-RAY BACKGROUND

In soft X-rays the situation is much more complex. Here the first reliable measurements of the celestial distribution, intensity and spectral shape of

248

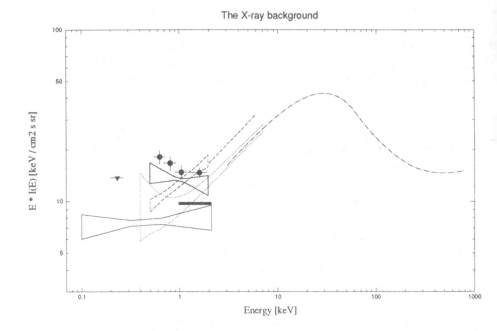

Figure 1. Measurements of the X-ray background spectrum in the energy range 0.1 keV to 1 MeV. The dash-dotted line at high energies gives the analytical representation by Gruber (1992). The trumpet-shaped dotted region refers to the ASCA measurement (Gendreau et al., 1994). The solid and dashed bow-tie shapes in the range 0.5-2 keV represent the ROSAT total background measurements (extragalactic power law component only) by Hasinger (1992) and Georgantopoulos et al. (1994). Filled symbols and the dashed line are from the Wisconsin data (McCammon & Sanders 1990 (see text). The long bow-tie shape in the 0.1-2 keV band is the spectrum of the resolved sources in the deepest ROSAT field; the thick solid line refers to the minimum resolved flux including fluctuations (Hasinger et al., 1993)

the X-ray background have been obtained by the University of Wisconsin in a series of sounding rocket flights using scanning collimated proportional counters (see McCammon & Sanders 1990). The Wisconsin results have been basically confirmed by the ROSAT all-sky survey both in celestial distribution and intensity (to about 10%), however, at dramatically improved angular resolution (Snowden et al., 1994). In all-sky maps below 2 keV there is a substantial variation in galactic coordinates with several discrete diffuse emission features clearly visible (mostly supernova remnants, including the dominant emission from Loop-I). Interstellar hydrogen absorption is also very important in this band, so that the combination of galactic emission and absorption severely complicates the determination of a true extragalactic XRB spectrum. Figure 1 shows a comparison of various X-ray background measurements.

2.2.1. *The C-band background: a "soft thermal component"*

The softest X-ray energy band, the "C-band" below the carbon edge at 0.28 keV, shows the emission of the million-degree hot local bubble (most likely the remnant of the supernova explosion from which the Geminga pulsar originated, see Gehrels & Chen, 1993). This "soft thermal" component is completely dominating the extragalactic light, which is in addition severely attenuated by interstellar absorption. The only chance to get a handle on the extragalactic XRB is by shadowing experiments using extragalactic absorbers. Shadow experiments have been performed by the Wisconsin group using the Small Magellanic Clouds (McCammon & Sanders 1990) and with ROSAT by Barber & Warwick (1994), using other nearby galaxies. The upper limits on the extragalactic background light derived from a lack of any detectable shadow are consistent between the two groups (see upper limit in fig. 1). On the other hand, a substantial fraction of the C-band background ($\sim$ 12%) has already been resolved into - mainly extragalactic - discrete sources, so that a firm lower limit for the extragalactic background can be established (see Hasinger et al., 1993, Snowden et al., 1994, fig. 1).

2.2.2. *The 0.5–2 keV background: a "hard thermal component"*

Between 0.5 and 2 keV, at galactic latitudes above 20° and outside major diffuse and point-like emission features, the XRB is largely isotropic. The "error bars" for the Wisconsin data above 0.5 keV (see figure 1) show the extremely small variation with galactic latitude (of order 10%), indicating that this background might be dominated by extragalactic emission.

Significant line emission by ionized oxygen (OVII at 0.574 keV and OVIII at 0.650 keV), indicating the existence of 2-3 million degree plasma even outside the region of Loop-I has been inferred from early rocket flights (Inoue et al., 1980, Rocchia et al., 1984). This "hard thermal" component on top of any reasonable extrapolation of the high-energy extragalactic power law (see above) has been confirmed by the analysis of many high-latitude ROSAT PSPC pointings across the sky (Hasinger 1992, Wang & McCray 1993, Georgantopoulos et al., 1995). Fitting an optically thin Raymond and Smith spectral model to the ROSAT data, the temperature of this component can be determined surprisingly accurately to 0.15-0.20 keV, despite the relatively coarse PSPC energy resolution and despite known systematic uncertainties of ±3% in the wavelength calibration of the PSPC. (However, one has to carefully select time intervals which are not contaminated by the monochromatic geocoronal oxygen emission line at 0.54 keV on the dayside of the satellite orbit!).

This hard thermal component may represent the hot intracluster gas in the local group of galaxies (Wang & McCray, 1993), similar to the intrgalactic medium that has been detected using ROSAT in several other nearby groups of galaxies, e.g. NGC 2300 (Mulchaey et al., 1993) or HCG 62 (Ponman et al., 1993). Recent ASCA SIS observations of the XRB with high energy resolution confirm the presence and the temperature of the hard thermal component (Gendreau et al., 1994). Unfortunately the dominant oxygen emission lines are not directly visible in the data, otherwise constraints could be put on the redshift of this plasma.

2.2.3. The 0.5-2 keV background: the power law component"

Folding the hard thermal component through the ROSAT PSPC response matrix one finds, that it becomes negligible compared to the extrapolation of the high-energy power law for pulseheight channels above $\sim 0.9 keV$. This means that at 1 keV, where the effective area of ROSAT is highest, the flux of the extragalactic power law component can be determined quite accurately (within the systematic error of the absolute normalization of $\sim 15\%$). This is confirmed by a direct comparison with the Wisconsin data (see figure 1).

At the highest energies accessible to ROSAT ($\sim 2.5 keV$), on the other hand, the quality of the PSPC spectra becomes comparatively poor, due to statistical and systematic errors. Because of the exponentially diminishing mirror reflectivity at higher energies the sky spectrum drops radically above 1.5 keV and the particle background becomes important. Systematic uncertainties in the subtraction of the particle background (Plucinsky et al., 1993) and in the wavelength calibration transform into a flux error of about 30% (a rough estimate) at 2 keV. Because of the very small spectral lever arm (0.9-2 keV), the slope of the extragalactic power law component cannot be determined to better than 0.2. Early estimates of the extragalactic soft X-ray background spectrum using the IPC aboard the Einstein observatory (Wu et al., 1991), with much higher particle background and less energy resolution, which did not take the galactic forground into account, have therefore to be taken with scepticism.

The slope of the extragalactic power law component underlying the hard thermal component in the 0.5-2 keV band has been determined by various authors using ROSAT data (Hasinger 1992, Wang & Mc Cray 1993, Georgantopoulos et al., 1995); the energy index typically lies between -0.5 and -1.1 (see figure 1), a spread which is expected from the large systematic error.

The spectrum of the resolved sources, on the other hand, does not suffer from uncertainties due to the subtraction of diffuse galactic and instrumental backgrounds, because for the discrete sources any diffuse background component can be easily subtracted from source-free regions. The slope of the average source spectrum is therefore usually determined quite accurately, in particular in directions with low interstellar absorption, because of the very long spectral lever arm from 0.1-2 keV (see figure 1). It lies in the range -0.9 to -1.3, depending on the survey depth. Figure 1 shows the resolved source spectrum measured in the deepest ROSAT pointing on the Lockman Hole in the 0.1-2 keV band (Hasinger et al., 1993). Also indicated is the minimum flux due to discrete sources in the 1-2 keV band, which is estimated from a fluctuation analysis in the Lockman Hole (Hasinger et al., 1993).

2.2.4. *The contribution of bright sources*

An important systematic error in background measurements originates from the treatment of bright sources. When comparing background fluxes quoted by different authors which are obtained by measurements with different angular resolution one has to carefully take into account the contribution by resolved sources brighter than the upper flux threshold in a particular observation (see Hasinger et al., 1993, Snowden et al., 1994). While e.g. the $\sim 10^o$ beam used for the Wisconsin observations averages the contribution of all but the brightest extragalactic sources into a "true total background" measurement, some investigators of imaging X-ray observations quote only results for a "residual background", i.e. after removal of all detected discrete sources (e.g. Wang & McCray 1993), which clearly depends on the depth of the survey. Even if "total background" measurements are given, i.e. the sum of all observed photons in a particular sky region (e.g. Georgantopoulos et al., 1995; Gendreau et al., 1994) there is a bias in the resulting flux (roughly 10% for typical ROSAT PSPC fields, maybe more for ASCA fields), because obviously background measurements are taken far away from known bright X-ray sources, which nevertheless contribute to the "true total background" (after all, a large fraction of the background is made of sources). To avoid the poisson error due to the statistical presence or absence of individual bright sources in the field it is therefore advisable to remove all sources brighter than a fixed threshold and to later add their contribution to the results, e.g. estimated from a known logN-logS function (see e.g. Hasinger et al., 1993), in order to predict the "true total background". An estimated contribution of 10% due to brighter sources has therefore been added in figure 1 to the bow-tie shaped 90%-confidence contours for the average extragalactic power law determined by Georgantopoulos et al., 1995. Their average energy slope from 5 medium-deep PSPC fields is -0.56 ± 0.12 (without systematic errors) and the average normalization corrected for brighter sources is $12.8 \pm 0.6 \; keV \; cm^{-2} \; s^{-1} \; sr^{-1} \; keV^{-1}$

at 1 keV. The currently best estimate from my own analysis of night-sky data averaged from 3 deep PSPC pointings yields an energy slope of -1.12 ± 0.12 and a normalization 13.0 ± 0.2 in the above units (Hasinger 1993). The systematic error on the slope determination is 0.2 and has been assumed in figure 1. Again, at 1 keV the two independent flux determinations are consistent with each other and with the Wisconsin data, while the slopes are consistent within the systematic errors. An estimate of the background flux at 1 keV from the combined ROSAT and Wisconsin data is $13.7 \pm 0.6 \; keV \; cm^{-2} \; s^{-1} \; sr^{-1} \; keV^{-1}$. Comparing this with the extrapolation of the HEAO-1 background spectrum from higher energies ($8 \cdot E^{-0.4}$), one would have to conclude that the background spectrum has to steepen somewhere between 2 and 3 keV.

2.2.5. *The ASCA data*

The Japanese X-ray satellite ASCA carries the first imaging X-ray telescope which is sensitive between 0.5 and 10 keV and therefore the first to cover the transition region in the 1-3 keV band with one instrument continuously. It also carries the first X-ray CCD detector with a very high energy resolution compared to proportional counters. Initial results on the X-ray background spectrum have been presented by Gendreau et al. (1994). At higher energies the measurd spectrum agrees very well with the previous observations. Surprisingly however, the ASCA spectrum does not show any trace of the expected steepening in the 2-3 keV band. On the contrary, it remains compatible with the HEAO-1 extrapolation down to energies as low as 1 keV (see figure 1). At lower energies the ASCA data confirm the existence of the hard thermal component (see above), but cannot distinguish between an additional steep power law component and/or non-solar abundances in the hot plasma, so that the estimate of the true extragalactic contribution becomes quite uncertain.

The new measurement disagrees significantly with previous data in the absolute flux level at 1 keV, where ASCA gets a value of 8.9 ± 0.4 versus the above 13.7 ± 0.6 (in the usual units). If one would take the ASCA estimate at face value one would have to conclude that the deepest ROSAT observations have already resolved a much larger fraction of the extragalactic XRB at 1 keV than previously assumed. Including the results of the ROSAT fluctuation analysis we have the paradoxical situation that practically all or even more of the ASCA background at 1 keV have been resolved. What are the possibilities to understand and "cure" this discrepancy?

(1) Might there be a general mismatch of about 50% between the ROSAT/Wisconsin and ASCA absolute calibrations? Although both ASCA and ROSAT/PSPC teams are still in the process of understanding the subtleties in the systematics of their respective calibrations, this seems for-

tunately not to be the case. On the contrary, a flux agreement to better than 10% between the two instruments is generally found when comparing near-simultaneous observations or observations of putatively constant targets with simple spectra (P. Serlemitsos, B. Warwick, N. White, priv. comm., 1994 November).

(2) Could the discrepancy be due to the different energy resolution of the instruments? ASCA is the first X-ray instrument with appreciable spectroscopic capacities and it is clear that it can handle complex spectra much better than the proportional counters with only moderate energy resolution (ROSAT PSPC and Wisconsin). One would therefore tend to believe better in the ASCA results, in particular with respect to the spectral shape. On the other hand, ASCA and ROSAT agree in the existence and the temperature of the hard thermal component and it can be demonstrated that this has a negligible influence for PSPC pulseheight spectra at and above 1 keV. Also, the average spectrum of the resolved sources in PSPC fields is simple and featureless, so that the discrepancy that ASCA finds too little background compared to the resolved sources still remains.

(3) Are there special problems of diffuse background measurements? I have discussed above a number of systematic errors in the treatment of the diffuse XRB. A specific of absolute background measurements is the relatively complicated calculation of the effective area × solid angle product. In the case of ASCA this is particularly complex, because a large and energy-dependent fraction of the diffuse background photons in the field-of-view actually originates from directions outside the field-of-view due to single reflections on one of the two X-ray mirror sections. (For the ROSAT telescope this is not a problem due to a complex system of baffles and field stops, which is not possible for highly nested mirror systems.) However, Gendreau et al. have demonstrated convincingly that they have corrected for this effect using ray tracing simulations. Another, more likely systmatic difference is the bright source bias, as discussed above. According to my understanding the authors of the ASCA paper have not corrected for this effect. For the total 1-10 keV band the correction procedure might actually be quite complicated, because the logN-logS functions are apparently different in the 0.5-2 keV and the 2-10 keV band, and are much less well known in the latter. A rough estimate of the correction that has to be applied to the ASCA 1-2 keV data can be derived as follows: The brightest sources that ASCA is expected to find in its 20×20 $arcmin$ field-of-view has a flux corresponding to a source density $N(> S) = 9$ deg^{-2} (i.e. one source per FOV). Checking with the soft X-ray logN-logS function (Hasinger et al., 1993,1994) this yields a source flux of $5 \cdot 10^{-14}$ erg cm^{-2} s^{-1} $(0.5 - 2\ keV)$. Integrating the logN-logS function above this flux to infinity yields a background intensity of 2.65 kev cm^{-2} s^{-1} sr^{-1} from sources brighter than

the upper flux threshold. Assuming a power law spectrum with an energy slope of −1 in the 0.5-2 keV band for the brighter sources, the spectrum of their contribution to the "true total background" has a normalization of 1.9 $keV\ cm^{-2}\ s^{-1}\ keV^{-1}\ sr^{-1}$ at 1 keV, which has to be added to the spectrum given by Gendreau et al., (1994), such that the total normalization has a value of 10.8. Surprisingly this value is consistent with the Wisconsin data (norm 11) and approaches the ROSAT data. While the estimate of the bright source correction to the soft X-ray background ($\sim$ 21%) is relatively straightforward, the best guess for the correction factor in the 2-10 keV band depends on the relatively uncertain logN-logS function and the average bright source spectrum in this band and could be in the range 10 − 30%. The bright-source contribution might therefore resolve the apparent discrepancies in the observations at 1 keV at the expense of the nice match between ASCA and HEAO-1. The conclusion would then be, that indeed the background spectrum remains flat down to 1 keV with a normalization of 11 (like McCammon & Sanders have told us all along) and that the steepening observed by ROSAT occurs only below 1 keV. A formal fit to the combined ASCA/ROSAT data, after applying the same corrections for bright sources might give the definitive answer.

3. The angular correlation function

Since the X-ray background is made up largely from discrete sources one would expect some variance in the background due to those sources. A signal in the angular correlation function (ACF) can give strong constraints on the clustering properties of the sources contributing to the X-ray background. However, the XRB is remarkably smooth. Until recently no signal could be detected in the XRB ACF neither at soft X-ray nor at hard X-ray energies (see Fabian & Barcons, 1992 for a review and figure 2). A first signal could be found in a 1-2 keV analysis of 50 deep ROSAT pointed observations in about 10% of the fields (Soltan & Hasinger, 1994). This signal could be clearly associated with a few extended, very-low X-ray surface brightness clusters or groups of galaxies at moderate redshift. These objects are now termed "blotches" (Hasinger et al., 1991). In trying to obtain an upper limit on structure in the background due to the clustering of sources producing the bulk of the emission, Soltan & Hasinger excluded the fields with significant cluster emission. Indeed, once those fields were excluded, only upper limits for the ACF could be obtained, however, those limits strongly constrain the nature and clustering properties of the sources contributing to the residual X-ray background. According to this analysis, less than 35% of the residual background can be due to objects with clustering properties similar to QSOs. The objects which make up the remainder of

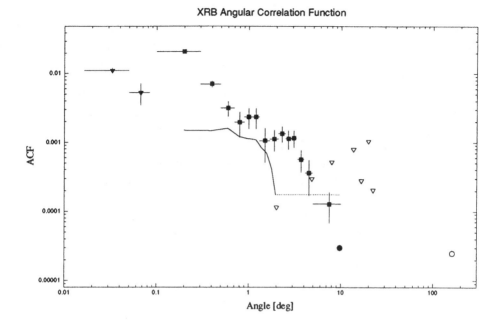

Figure 2. Measurements of the angular correlation function of the X-ray background from different instruments (after Fabian & Barcons, 1992). The open circle corresponds to the measurement of the XRB dipole moment (Shafer & Fabian 1983). The filled circle refers to the upper limit for large-scale surface brightness variations measured by the HEAO-1 A2 experiment (Jahoda 1993). The open triangles are upper limits from the Ginga pointed observations (Carrera et al., 1991). The continuous line shows the 2-sigma upper limits obtained from Ginga scan data (Carrera et al., 1993). The filled squares refer to the first detection of a significant signal in the ACF of the XRB in the ROSAT all-sky survey (Soltan et al., 1994). The filled triangles are upper limits to the ACF obtained by summing up about 50 ROSAT pointed observations (Soltan & Hasinger 1994).

the background must have a clustering length smaller than normal galaxies and/or show very strong cosmologic evolution of their clustering (Soltan & Hasinger, 1994).

In figure 2 the angular correlation function at 2 and 4 arcmin is shown, determined from the average of all 50 ROSAT pointings, i.e. including the "blotchy" fields. A significant signal is detected there. In a recent analysis of the 0.9-1.3 keV background from a "clean" region of 1 sr in the ROSAT all-sky survey, Soltan et al. (1995) detected a very significant signal in the ACF of the X-ray background, extending out to about 10° (see figure 2). The angular resolution of this measurement is 12'. At angles of $0.5 - 5^\circ$ this signal corresponds to roughly 3% fluctuations of the X-ray background. The authors could show convincingly that a large fraction of this signal must be extragalactic, actually galactic contributions to the fluctuations could be largely removed using the angular correlation function in a softer energy band (0.7-0.9 keV). Therefore this measurement represents the first discovery of the long-sought signal in the extragalactic X-ray background flux. Soltan et al. correlated the X-ray background fluctuations with the Abell catalogue of clusters of galaxies and find a significant crosscorrelation signal, not only with the direct cluster emission (within 20') but also an extended component reaching out to $\sim 5^{\circ}$. The magnitude of this effect, which corresponds roughly to one third of the total background fluctuations, excludes the possibility that this signal is due to the cluster-cluster or the galaxy-cluster correlation and the tentative assumption put forward by the authors is, that typical Abell clusters are surrounded by large ($\sim 10\ Mpc$) haloes of diffuse hot gas with an average luminosity of $\sim 10^{43.4}\ erg\ s^{-1}$. This could be the first detection of hot diffuse supercluster gas emission.

Similarly, the ROSAT data are being correlated to catalogues of other object classes (e.g. nearby groups of galaxies), work which will presented elsewhere. A crosscorrelation between the ROSAT all-sky survey map and the 2nd-year COBE DMR map (Smoot et al., 1992) is in progress, but the results are still inconclusive.

4. X-ray surveys

X-ray surveys are important for our understanding of the object classes contributing to the X-ray background as well as their cosmological evolution. Because of the requirement to obtain optical spectroscopic identifications for large, statistically complete samples of X-ray selected objects, X-ray survey work is quite tedious and time consuming. For many years the two X-ray surveys from HEAO-1 (Picinnotti et al., 1982) and the Einstein observatory Extended Medium Sensitivity Survey (EMSS; Gioia et al., 1990) were the only available workhorses. With the advent of the ROSAT X-ray observatory substantial progress has been made both in depth and in coverage of X-ray surveys. The optical follow-up work is, however, still a problem. This is illustrated in figure 3, where various completed and on-going surveys are compared. The filled symbols indicate surveys which are already largely optically complete (typically less than 10% incompleteness), including the EMSS and the Piccinnotti Sample. Open circles refer to ROSAT X-ray sorce catalogues with largely incomplete optical identifications (the all-sky survey, the ROSAT ecliptic pole survey and the deepest ROSAT PSPC survey in the Lockman Hole).

Filled circles indicate a number of independent, relatively complete

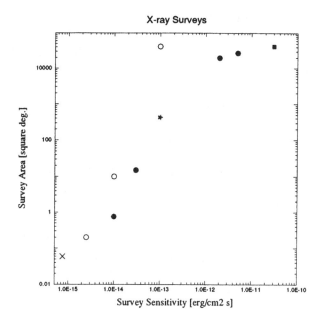

Figure 3. Graphical display of the solid angle versus limiting sensitivity for various X-ray surveys. Filled symbols correspond to surveys which are largely spectroscopically identified, square: HEAO-1 survey (Piccinnotti et al., 1982), star: EMSS (Gioia et al., 1990), circles: ROSAT surveys (see text). The cross corresponds to the ROSAT Ultradeep HRI Survey, which is currently in progress.

ROSAT high galactic latitude survey projects, which contain typically between 100 and 1000 sources. There are two almost complete ROSAT surveys substantially fainter than the EMSS, which contribute most to our understanding of the sources of the XRB: the RIXOS project and the ROSAT Deep Survey project. RIXOS, the ROSAT International X-ray Optical Survey is large program to completely optically identify serendipitous sources from PSPC pointings longer than 8 ksec down to a flux limit of $3 \cdot 10^{-14}$ erg cm^{-2} s^{-1}. For the purpose of optical identifiations 1.5 years of the CCI international time on the Canary Island telescopes was awarded to a large consortium (PI Keith Mason) in the years 1993-1994. In a solid angle of ~ 10 deg^2 (40 fields) the identifications are 100% complete, the remaining ~ 40 fields should be completed by the end of 1994.

A small number of PSPC pointings reach sensitivities substantially below 10^{-14} erg cm^{-2} s^{-1}, the so called ROSAT deep surveys. Optical counterparts in these pointings typically have magnitudes in the range $m_R = 19 - 24$, so that optical identifications become very time consuming. In this review data from a total of 5 ROSAT PSPC fields with optical identifications largely complete down to an X-ray flux of 10^{-14} erg cm^{-2} s^{-1} have been combined (see also Branduardi-Raymont et al., 1994): the Lockman Hole (Hasinger et al., 1993; deRuiter et al., 1995), the Marano Field (Zamorani et al., 1995), the North-ecliptic pole field (Henry et al., 1994; Bower et al., 1995) and the QSF3 and GSP4 fields (Shanks et al., 1991; Boyle et al., 1993).

At fainter X-ray fluxes the typical magnitude of optical counterparts increases correspondingly and the surface density of faint galaxies rises dramatically. While ROSAT PSPC position errors can be as small as 2-3" for brighter sources, the X-ray error boxes at detection threshold are still quite large (10-15" radius), so that the likelihood of spurious associations of faint galaxies with faint X-ray sources increases substantially. In order to overcome these difficulties and to be able to push the limit for secure optical identifications an order of magnitude deeper we have started observations for an Ultradeep ROSAT HRI survey inside the PSPC survey of the Lockman Hole. A total of 1Msec of HRI observations and a new "X-ray speckle" technique to correct for aspect errors are forseen to obtain arcsecond positions for about 70 objects down to a flux limit of $7 \cdot 10^{-16}$ erg cm^{-2} s^{-1}. At the extremely faint optical magnitudes expected, probably only new telescopes of the 8-10m class with multislit spectroscopy capabilities over a large field can achieve completeness in a reasonable exposure time. Nevertheless this is the only possibility to obtain secure information about the role of faint galaxies and clusters for the X-ray background (see below).

TABLE 1. Summary of X-ray Surveys

Survey	S_lim $[erg/cm^2 s]$	Area $[deg^2]$	Total [%]	Unid. [%]	AGN [%]	Gal. [%]	Clus. [%]	Stars
EMSS	10^{-13}	430	835	4	55	2	13	26
RIXOS	$3 \cdot 10^{-13}$	14.9	285	14	51	4	11	20
DEEP	10^{-14}	0.76	61	8	62	7	5	15
ULTRA	$7 \cdot 10^{-16}$	0.06	70?					

5. Optical identifications and the role of galaxies

Table 1 gives a summary of the survey papameters and optical identification content for the EMSS, RIXOS and ROSAT Deep Surveys, which span

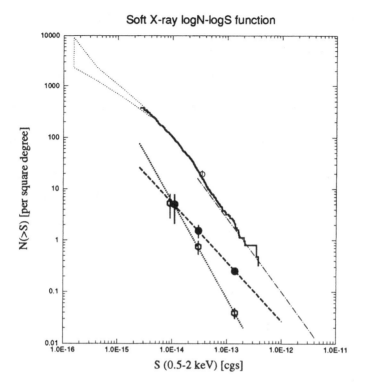

Soft X-ray logN-logS function

Figure 4. Integral soft X-ray source counts (from Hasinger et al., 1993). The filled circles give the contribution of spectroscopically identified clusters of galaxies, the open hexagons that of single galaxies. For clarity, galaxies and clusters for the ROSAT deep surveys are plotted with a small shift in flux.

roughly a decade in limiting flux. Active galactic nuclei, mainly QSOs, represent the majority in all three surveys and are expected to contribute a large fraction of the X-ray background. Foreground stars, galaxy clusters and apparently normal or weak emission line galaxies contribute only 5 − 25% each. Of particular interest is the relative behaviour of galaxy clusters on one hand and apparently normal or weak emission line galaxies on the other hand. Figure 4 shows the source counts for clusters and galaxies in the three samples, compared to the total logN-logS function in the soft X-ray band (from Hasinger et al., 1994). The sub-Euclidean slope for clusters of galaxies found in the Einstein Medium Survey (Gioia et al., 1984) seems to continue to much fainter X-ray fluxes: for the three samples discussed here a power law slope of -1.1 is found for the integral source counts. This is consistent with the strong negative evolution found for the

260

X-ray luminosity of galaxy clusters (Edge et al., 1990; Henry et al., 1992). An independent estimate of the cluster evolution could be obtained from the RIXOS sample (Castander et al., 1994).

Galaxies, on the other hand, seem to have a much steeper logN-logS function than clusters. A slope of -1.9, steeper than Euclidean, has been determined for the cumulative source counts of galaxies in the three samples, indicating a strong positive evolution for this population, which eventually could become the dominant contributor to the X-ray background at the faintest X-ray fluxes. One has to be careful with this interpretation, however, for several reasons: First, as discussed above, the number of spurious identifications of galaxies with faint X-ray sources increases sharply, both because the galaxy number counts and because the increasing X-ray error boxes with decreasing flux. Some fraction of the galaxies in the ROSAT deep surveys could be spurious. Secondly, there may be some distant clusters of galaxies misidentified as single galaxies, e.g. objects which are dominated by a single CD-galaxy in the optical and which are too faint to measure an X-ray extent. Finally a number of galaxies, in particular those with faint emission lines might actually host hidden AGN and might therefore be misclassified. A number of these objects has been detected through a ROSAT/IRAS correlation (Boller et al., 1992; see also Moran et al., 1994). In one particular case an object classified as a normal HII galaxy showed strong and fast time variability (Boller et al., 1993). Interesting is also one of Boller's IRAS/ROSAT correlations which turned out to be a narrow line (type-II) QSO (Elizalde & Steiner, 1994). According to the new unified AGN-background models which can give a satisfactory fit to the background spectrum and many other observational constraints (Comastri et al., 1995; Matt, 1994) we would actually expect type-II AGN, i.e. low-luminosity absorbed Seyfert galaxies and type-II QSOs as one of the major contributors to the X-ray background. This fits well with the fact that the faintest X-ray sources in the Lockman Hole seem to have a substantially hard, intrinsically absorbed spectrum.

References

Antonucci R.R. *Ann. Rev. Astr. Astrophys. Astrophys.*, **31**, 473 (1993).

Barber C.R. & Warwick R.S. *Mon. Not. R. astr. Soc.*, **267**, 270 (1994).

Barcons X. & Fabian A.C., *Mon. Not. R. astr. Soc.*, **243**, 366 (1990).

Boldt E. *Phys. Rep.*, **146**, 215 (1987).

Boller T., Meurs E.J.A., Brinkmann W., Fink H., Zimmermann U. & Adorf H.M. *Astr. Astroph.*, **261**, 57 (1992).

Boller T., Trümper J., Molendi S., Fink H., Schaeidt S., Caulet A. & Dennefeld M. *Astr. Astroph.*, **279**, 53 (1993).

Branduardi-Raymont G., Mason K.O., Warwick R.S. et al. *Mon. Not. R. astr. Soc.*, **270**, 947 (1994).

Bower R. et al. (in preparation) (1994).

Boyle B.J., Griffiths R.E., Shanks T., Stewart G.C. & Georgantopoulos I., *Mon. Not. R. astr. Soc.*, **260**, 49 (1993).

Carrera F.J., Barcons X., Butcher J., Fabian A.C., Stewart G.C., Warwick R.S., Hayashida K. & Kii T. *Mon. Not. R. astr. Soc.*, **249**, 698 (1991).

Carrera F.J., Barcons X., Butcher J.A., Fabian A.C., Stewart G.C., Toffolatti L., Warwick R.S., Hayashida K., Inoue H. & Kondo H. *Mon. Not. R. astr. Soc.*, **260**, 376 (1993).

Castander F.J., Bower R.G., Ellis R.S., Aragon-Salamanca A., Mason K.O., Hasinger G., McMahon R.G., Carerra F.J., Mittaz J.P.D., Perez-Fournon I. & Barcons X. *Nature*, (subm.) (1994).

Comastri A., Setti G., Zamorani G. & Hasinger G. *Astr. Astroph.*, (in press) (1995).

Edge A.C., Stewart G.C., Fabian A.C. & Arnaud K.A. *Mon. Not. R. astr. Soc.*, **245**, 559 (1990).

Elizalde F. & Steiner J.E. *Mon. Not. R. astr. Soc.*, **268**, L47 (1994).

Fabian A.C. & Barcons X. *Ann. Rev. Astr. Astrophys. Astrophys.*, **30**, 429 (1992).

Giacconi R., Gursky H., Paolini F.R. & Rossi B.B., *Phys.Rev.Letters*, **9**, 439 (1962).

Gehrels N. & Chen W. *Nature*, **361**, 706 (1993).

Gendreau K. et al. (preprint) (1994).

Georgantopoulos I., Stewart G.C., Shanks T., Boyle B.J. & Griffiths R.E. *Mon. Not. R. astr. Soc.*, (submitted) (1994).

Gioia I.M., Maccacaro T., Schild R.E., Stocke J.T., Liebert J.W., Danziger I.J., Kunth D. & Lub J. *Astrophys. J.*, **283**, 495 (1984).

Gioia I.M., Maccacaro T., Schild R.E., et al., *Astrophys. J. Suppl.*, **72**, 567 (1990).

Gruber D.E., in *The X-ray Background* X. Barcons & A.C. Fabian eds, (Cambridge University Press), p.45 (1992).

Hasinger G., Schmidt M. & Trümper *Astr. Astroph.*, **246**, L2 (1991).

Hasinger G., in *The X-ray Background* X. Barcons & A.C. Fabian eds, (Cambridge University Press), p.229 (1992).

Hasinger G., Burg R., Giacconi R., Hartner G., Schmidt M., Trümper J. & Zamorani G., *Astr. Astroph.*, **275**, 1 (1993).

Hasinger G. in: *Observational Cosmology*, G.Chincarini, A.Iovino, T.Maccacaro, D.Maccagni, eds., A.S.P. Conf. Ser. Vol. 51, p.439 (1993).

Hasinger G., Burg R., Giacconi R., Hartner G., Schmidt M., Trümper J. & Zamorani G., *Astr. Astroph.*, **291**, 348 (1994).

Henry J.P., Gioia I.M., Maccacaro T., Morris S.L., Stocke J.T. & Wolter A. *Astrophys. J.*, **386**, 408 (1992).

Henry J.P., Gioia I.M., Boehringer H., Bower R.G., Briel U.G., Hasinger G., Aragon-Salamanca A., Castander F.J., Ellis R.S., Huchra J.P., Burg R. & McLean B. *Astron. J.*, **107**, 1270 (1994).

Inoue H. et al. *Astrophys. J.*, **238**, 886 (1980).

Jahoda K. *Adv. Space. Res.* Vol. **13**, No. 12, p.(12)231 (1993).

Marshall F.E. et al., *Astrophys. J.*, **235**, 4 (1980).

Mather J. et al. *Astrophys. J. (Letters)*, **354**, L37 (1990).

Matt G. (preprint), (1994).

McCammon D. & Sanders W.T., *Ann. Rev. Astr. Astrophys. Astrophys.*, **28**, 657 (1990).

Moran E.C., Halpern J.P. & Helfand D.J. *Astrophys. J.*, **433**, L65 (1994).

Mulchaey J.S., Mushotzky R.F. & Weaver K.A. *Astrophys. J.*, **390**, L69 (1992).

Mushotzky R. in: *Frontiers of X-ray Astronomy*, Y.Tanaka, K.Koyama, eds., Universal Academy Press:Tokyo, p.657 (1992).

Piccinotti G., Mushotzky R.F., Boldt E.A., et al. *Astrophys. J.*, **253**, 485 (1982).

Plucinsky P.P., Snowden S.L., Briel U.G., Hasinger G & Pfeffermann E. *Astrophys. J.*, **418**, 519 (1993).

Ponman T.J. & Bertram D. *Nature*, **363**, 51 (1993)

Rocchia R. et al., *Astr. Astroph.*, **130**, 53 (1984).

deRuiter H. (in preparation) (1994).

Setti G. & Woltjer L., *Astr. Astroph.*, **224**, L21 (1989).

Shafer R.A. & Fabian A.C. in: *Early Evolution of the Universe and its Present Structure*, G.O. Abell, G. Chincarini, eds., Reidel:Dordrecht, p.333 (1983).

Shanks T., Georgantopoulos I., Stewart G.C., Pounds K.A., Boyle B.J. & Griffiths R.E. *Nature*, **353**, 315 (1991).

Smoot G.F. et al. *Astrophys. J.*, **396**, L1 (1992).

Snowden S.L., Hasinger G., Jahoda K., Lockman F.J. McCammon D. & Sanders W.T. *Astrophys. J.*, **430**, 601 (1994).

Snowden S.L. et al. (in preparation) 1995

Soltan A. & Hasinger G. *Astr. Astroph.*, **288**, 77 (1994).

Soltan A., Hasinger G., Egger R., Snowden S. & Trümper J. *Astr. Astroph.*, (submitted) (1994).

Wang Q.D. & McCray R. *Astrophys. J.*, **409**, L37 (1993).

Wu X., Hamilton T.T., Helfand D.J. & Wang Q. *Astrophys. J.*, **379**, 564 (1991).

Zamorani G. et al., (in preparation) (1994).

THE SOURCES OF THE HARD X-RAY BACKGROUND

GIANCARLO SETTI AND ANDREA COMASTRI
Dipartimento di Astronomia, Università di Bologna
Osservatorio Astronomico di Bologna
Istituto di Radioastronomia del CNR
Via P. Gobetti, 101 - 40129 Bologna - Italy

1. Introduction

The hard component (3 keV $-$ $\sim$ MeV) of the X-ray background (XRB) comprises the largest portion, $\sim$ 90%, of the overall XRB intensity. The observed isotropy (the entire Galaxy is transparent above 3 keV) provides a *prima facie* evidence of its prevailing extragalactic nature. A large fraction ($\sim$ 75%) of the energy flux falls in the $3 - 100$ keV band, the corresponding energy density being $\simeq 5 \times 10^{-5}$ eV cm^{-3}, of which 50% is confined to the narrower $3 - 20$ keV band. Although the energy flux carried by the XRB is relatively small compared to other extragalactic backgrounds, it was soon realized that it cannot be accounted for in terms of sources and processes confined to the present epoch. An analysis of the combined observed spectra (Gruber 1992) concludes that, while a thermal bremsstrahlung with an e-folding energy $= 41.13$ keV accurately fits the data up to 60 keV, above this energy the sum of two power laws is required with normalizations such that at 60 keV the spectral index is $\sim$ 1.6, gradually flattening to $\sim$ 0.7 at MeV energies. It should also be noted that below 10 keV the XRB energy spectrum is well represented by a power law of index $\alpha = 0.4$ ($I \propto E^{-\alpha}$).

The 40 keV thermal bremsstrahlung has been cosidered by some as evidence of a hot ($> 10^8$ K) diffuse intergalactic gas (IGG). Subsequent investigations (Taylor & Wright 1989 and ref. therein) have shown that a viable model, in the framework of the big-bang cosmology, can only be obtained by assuming that the IGG has been suddenly reheated at a redshift $z = 3 - 5$ to a temperature T $\sim$ $(1 + z_*)40$ keV and its density $\Omega_B > 0.2$, even larger than that allowed by the primordial nucleosynthesis. The ensuing Compton distortion of the CMB is by far inconsistent with COBE

263

M. Kafatos and Y. Kondo (eds.), Examining the Big Bang and Diffuse Background Radiations, 263–270.
© 1996 IAU.

FIRAS results: the contribution of a hot diffuse IGG to the XRB cannot be more than 10^{-4} (Wright et al. 1994). Strong clumping can help in circumventing these difficulties. Nevertheless, consistency with the upper limits on the temperature fluctuations of the CMB requires a large number of unrealistically small size clumps (< few tens kpc; Barcons & Fabian 1988). Indeed, several authors (e.g., Giacconi & Zamorani 1987) have pointed out that the close resemblance of the XRB spectrum to that of a 40 keV thermal bremsstrahlung provides in itself a strong argument against any interpretation involving a hot IGG as the main contributor to the XRB intensity. This is because any reasonable subtraction of the integrated contributions from known classes of extragalactic X-ray sources would destroy the almost perfect thermal shape which, therefore, must be considered as accidental.

Presently, there are no other models which may satisfactorely explain the hard XRB in terms of diffuse processes taking place in the intergalactic medium. As a consequence, the only alternative possibility is that of the summed contribution from extragalactic sources. Among these the AGN, in particular Seyfert (Sy) galaxies and quasars, are known for some time (Setti & Woltjer 1973) to be the most likely candidates. However, unlike the situation in the soft X-rays, only a small fraction of the hard XRB has been actually resolved into sources: the HEAO-1 A2 all-sky survey has resolved $\sim 1\%$ of the $2 - 10$ keV XRB down to a flux limit $\simeq 3 \times 10^{-11}$ erg cm^{-2} s^{-1} and determined a Euclidean Log N $-$ Log S (Piccinotti et al. 1982). At these bright fluxes AGNs and clusters of galaxies are the two dominant classes of sources. P(D) fitting techniques applied to the fluctuations in the XRB measured by GINGA (Warwick & Stewart 1989; Hayashida 1990) and to those of the HEAO-1 A2 experiment (Shafer 1983) are consistent with a Euclidean slope of the source number counts extrapolated from the Piccinotti et al. Log N $-$ Log S down to a flux $\sim 10^{-13}$ erg cm^{-2} s^{-1}, at which flux about 15% of the $2 - 10$ keV XRB is resolved.

The main difficulty with the idea that the AGNs could supply the bulk of the XRB has been that their hard X-ray spectra ($2 - 20$ keV) are well represented by a power law of mean energy spectral index $< \alpha > = 0.7$ (Turner & Pounds 1989, and ref. therein), much steeper than that of the XRB. Recent developments have permitted to overcome this ostacle.

2. Solving the spectral problem

One of the achievements of the GINGA mission has been the discovery that the X-ray spectra of a large fraction of a sample of Sy 1 galaxies flatten beyond ~ 10 keV, the mean spectral index in the $10 - 18$ keV interval being $< \alpha > = 0.28$ with $\sigma = 0.15$ (Nandra & Pounds 1994). This fact has been interpreted either as photoelectric absorption (partial coverage) of an underlying power law continuum source or as a hump produced by reprocessed X-rays (reflection) from thick cold ($< 10^6$ K) matter surrounding the central source, possibly in an accretion disk around a massive black hole (Matsuoka et al. 1990; Pounds et al. 1990). The Compton reflection model is generally thought to be more representative of the physical conditions prevailing in the AGNs.

As conjectured by Schwartz & Tucker (1988), a flattening of the average AGN spectrum above ~ 10 keV may provide a solution to the AGN vs. XRB spectral problem, simply because the redshifted contributions from the flat portion of the source spectra would dominate the $3 - 30$ keV XRB spectral shape, if the AGN volume emissivity adequately increases as a function of look-back time out to a redshift cut-off $z = 3$. The rather sharp decline of the XRB spectrum above 30 keV could be modeled by introducing a high energy cut-off at ~ 100 keV. Following this recipe Morisawa et al. (1990), by adopting a partial coverage absorption model on a power law continuum ($\alpha = 0.7$), obtained a good fit of the $3 - 100$ keV XRB spectrum with parameters typical of the Sy 1 spectra measured by GINGA, assumed to be exponentially cut-off at a computed energy of ~ 115 keV, and with a reasonable cosmological evolution of the local AGN X-ray luminosity function (XLF) from the HEAO-1 bright sample of Piccinotti et al. (1982).

Fabian et al. (1990) emphasized the physical nature of the relatively sharp break of the XRB spectrum at ~ 30 keV and noted that this feature can be adequately reproduced by the redshifted Compton reflection spectrum of sources at $z \sim 2$, thus avoiding the need of introducing artificial cut-offs. It was found that the fit of the XRB spectrum from several keV to ~ 1 MeV requires a new class of strong ($\leq 10^{45}$ erg s^{-1}) sources with more than 90% of the emitted flux in the reflected component, and a strong cosmological evolution such that the comoving volume emissivity of the sources increases as $(1 + z)^4$ out to a redshift $z \sim 5$. 'Reflection' models of this type have been further investigated by Rogers & Field (1991) and Terasawa (1991) with similar conclusions on the strength of the reflected component. While the new class of AGNs postulated by Fabian et al. could not have been hitherto detected, Terasawa's model assumes the local XLF of Piccinotti et al. and, therefore, the conclusion that $\sim 80\%$ of the source flux is channeled in the reflected component exceeds the average value found in the Sy 1 spectra. A general difficulty with these models is that the predicted source counts and/or spectra in the soft X-ray band are inconsistent with those of the EMSS and ROSAT surveys (Setti 1992). It has also been pointed out (Zdziarski et al. 1993a) that these models do not adequately fit the position and the width of the peak of the XRB intensity.

Recently, another 'reflection' model has been proposed by Zdziarski et al. (1993b). Here the typical Sy 1 spectrum is made by Compton reflection of $\sim 50\%$ of a primary X-ray flux produced by thermal Comptonization of seed photons in a hot plasma cloud with a Thompson depth of a few and a temperature $kT = 40$ keV, which means that the spectrum is exponentially cut-off at ~ 100 keV. By adopting the cosmological evolution derived by Boyle et al. (1993) from the analysis of the combined AGN sample from the EMSS and ROSAT surveys, it is found that a very good fit of the $3 - 100$ keV XRB spectrum can be obtained with the evolution extended up to $z = 4$ and with a local volume emissivity about 2 times that of Boyle et al. (1993). No details are given about the expected source counts. However, the adopted source spectral shape, in combination with the high redshifts involved in the model, is likely to produce too many sources with too flat spectra in the soft X-rays.

In general, it appears that the 'reflection' models are rather inefficient in producing the XRB, unless a completely new class of AGNs is postulated.

Alternative models for the synthesis of the XRB assume the existence of a large population of heavily absorbed AGNs. In particular, Setti & Woltjer (1989) have discussed a model based on the X-ray properties of AGN unified schemes, introduced by Antonucci & Miller (1985) for the Seyfert galaxies, in which the central source can be hidden (depending on the viewing angle) by a thick torus of surrounding absorbing matter. Radio-loud quasars and strong radio galaxies can be similarly unified, the quasar phenomenon showing up whenever the source axis is aimed toward us within a specified angle (Barthel 1989), while the existence of a hidden population of radio-quiet quasars is still uncertain, although the luminous IR galaxies have been proposed as likely candidates. Accordingly, it has been shown that the X-rays can be absorbed up to > 20 keV whenever the lines of sight are intercepted by the tori, while above the absorption cut-offs the X-ray properties should be the same as those of the 'unabsorbed' AGNs, unless the X-ray emission itself is largely beamed. X-ray observations of Sy 2 galaxies and strong radio galaxies support this hypothesis (Awaki et al. 1991; Allen & Fabian 1992). By adopting source spectra with the canonical $2 - 10$ keV slope ($\alpha = 0.7$), very simple fractional distributions of the absorption cut-offs and an internally consistent cosmic evolution of the AGN XLF out to $z = 3$, Setti & Woltjer demonstrated that the $3 - 30$ keV XRB spectral intensity can be accurately reproduced with a number of 'absorbed' AGNs about equal to that of the 'unabsorbed' ones. Very good fits of the XRB have been obtained by more detailed investigations, thus confirming the basic validity of this scenario (Madau et al. 1993; Matt and Fabian 1994).

We have briefly reviewed a variety of models meant to explain the hard XRB as a superposition of the emission from AGNs. We have also under-lined that it is critically important to check these models against other observational constraints, in particular the predicted source counts and spectra in the soft X-rays. One important feature of Setti & Woltjer's proposal is that the 'absorbed' AGNs, essential to explain the XRB above several keV, may not show up significantly in the soft X-rays, so that these two spectral regions are to some extent indepedent. Therefore, the next step is the construction of a model consistent with the main statistical properties of AGN X-ray samples. An attempt in this direction by Comastri et al. (1994) will be briefly described in the following section (see also Madau et al. 1994).

3. A baseline model of the XRB from AGNs

The basic assumptions of the model, aimed at reproducing the XRB spectrum in the $3 - 100$ keV interval, can be summarized as follows:

a) A typical, primary AGN continuum spectrum with spectral indices $\alpha = 1.3$ below 1.5 keV and $\alpha = 0.9$ above this energy, exponentially cut-off at an e-folding energy of $\simeq 320$ keV. A Compton reflection component (50% of the primary flux) has been added to the continuum of the low luminosity AGNs ($< 7 \times 10^{43}$ erg s^{-1} in the 0.3-3.5 keV interval). This is consistent with the broad spectral characteristics observed in Sy 1 and quasars. b) The AGNs are surrounded by tori of absorbing material (solar composition) with 45° half opening angle, such that hydrogen column densities N_H of up to 10^{25} cm^{-2} can be intercepted, depending on the source characteristcs and orientation with respect to the line of sight to the central source. c) The adopted ($z = 0$) XLF is that determined by Boyle et al. (1993) in the 0.3-3.5 keV band, evolved in luminosity by $(1 + z)^{2.6}$ up to $z = 2.25$, thereafter constant and cut-off at $z = 4$. It covers a wide luminosity range from 10^{42} to 10^{47} erg s^{-1}.

The number distribution of 'absorbed' sources as a function of N_H ($> 10^{21}$ cm^{-2}) has been parameterized by dividing the N_H range into four equal intervals on a log scale and by assigning to the absorbed objects the mean N_H value of the corresponding interval. It is found that the observational constraints are met by a number distribution of 'absorbed' AGNs, normalized to the 'unabsorbed' ones, given by 0.35,1.10,2.30,1.65, in order of increasing N_H.

Specifically, the following constraints have been identified and checked for consistency: 1) The XRB from several keV to about 100 keV is well approximated (Fig. 1) to better than 5% accuracy with respect to the best fit curve of Gruber (1992). Below ~ 5 keV the fit underestimates the background intensity by $5 - 10$ % to allow for possible contributions from other classes of objects. 2) The source counts in the $2 - 10$ keV band (Fig. 2) are consistent with the AGN surface density of the bright sample from the

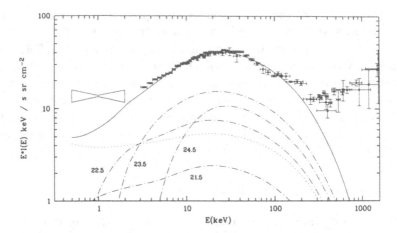

Figure 1. The XRB spectrum: 0.5-2.0 keV data from ROSAT (Hasinger et al. 1993) and > 3 keV from a compilation of Gruber (1992). Represented: the fit of the AGN baseline model (solid line), the contributions from unabsorbed sources (dotted line) and absorbed ones (dot-dashed lines; labels = Log N_H).

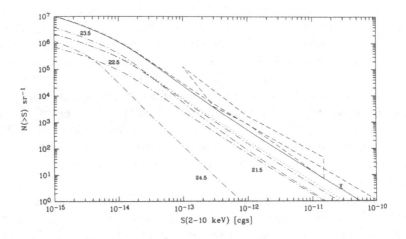

Figure 2. Baseline model predicted counts, AGN (solid) and AGN + galaxy clusters (upper dashed) compared with AGN surface density (• ;Piccinotti et al. 1982) and GINGA P(D) counts (dashed area). Represented: counts from unabsorbed sources (dotted) and from absorbed ones (dot-dashed; labels = Log N_H).

HEAO-1 A2 all-sky survey and with the constraints from the fluctuation analysis of GINGA fields. The local X-ray volume emissivity (3.8×10^{38} erg s^{-1} Mpc^{-3}) is fully consistent with that obtained by Miyaji et al. (1994) from a cross-correlation of the HEAO-1 A2 maps with the IRAS galaxies. 3) The redshift and absorption (N_H) distributions of bright AGNs are in good agreement with those of the Piccinotti et al. sample. 4) The predicted source

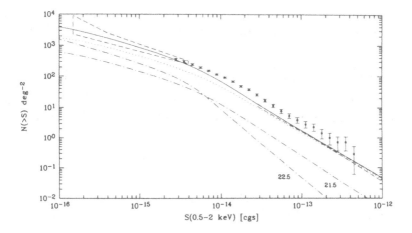

Figure 3. AGN baseline model predicted counts (solid) compared with EMSS AGN counts (dashed; Della Ceca et al. 1992), with ROSAT total counts and P(D) extension (dashed area; Hasinger et al. 1993). Represented: counts from unabsorbed sources (dotted line) and from absorbed ones (dot-dashed; labels = Log N_H).

counts in the soft X-rays (Fig. 3) are in good agreement with the EMSS Log N - Log S (Della Ceca et al. 1992) and with the ROSAT deep survey counts (Hasinger et al. 1993). An extrapolation of the predicted counts down to a zero flux accounts for $\sim$ 74% of the $1 - 2$ keV ROSAT XRB. Moreover, and most important, the source spectra at different flux levels have mean slopes in agreement with those found in the EMSS and ROSAT surveys. 5) The redshift distributions of the EMSS AGN sample and of the AGNs so far identified in five deep ROSAT fields are well reproduced by the model.

The ratios between absorbed and unabsorbed objects of the baseline model cannot be immediately associated with the Sy 2 / Sy 1 ratio. This is because objects with intrinsic absorptions corresponding to N_H values of up to several times 10^{22} cm^{-2} have been frequently classified as Sy 1 galaxies and, viceversa, Sy 2 galaxies sometimes have $N_H < 10^{22}$ cm^{-2}. If one takes column densities in the range $10^{22} - 10^{22.5}$ cm^{-2} as the dividing line between the two types, then the model predicts a ratio Sy 2 / Sy 1 = $2.4 - 3.7$, in good agreement with that found by Huchra & Burg (1992) from a complete sample of optically selected Sy galaxies.

4. Conclusions

The main conclusion to be drawn is that it is possible, in the framework of the simplest version of the AGN unified schemes, to construct a baseline model which accurately explain the $5 - 100$ keV XRB spectral intensity and is consistent with the available data both in the soft and hard X-rays. The 'absorbed' AGNs are an essential feature of the model. The contribution of the AGNs to the soft XRB measured by ROSAT is only about 75%. The spectrum above 100 keV, not discussed here, could be due to a sub-class of AGNs with flat X-ray spectra up to high energies, such as is the case of the flat spectrum quasars and BL Lacs detected by CGRO/EGRET possibly contributing the γ-ray background above 100 MeV (Setti & Woltjer 1994). The ongoing observations with the Japanese satellite ASCA should be able to further check the validity of this model.

References

Allen S.W. & Fabian A.C. 1992, *Mon. Not. R. astr. Soc.*, **258**, 29P

Antonucci R.R.J. & Miller J.S. 1985, *Astrophys. J.*, **297**, 621

Awaki H., Koyama K., Inoue H., Halpern J.P. 1991, *Pub. Astr. Soc. Japan*, **43**, 195

Barcons, X. & Fabian, A.C. 1988, *Mon. Not. R. astr. Soc.*, **230**, 189

Barthel P.D. 1989, *Astrophys. J.*, **336**, 606

Boyle B.J., et al. 1993, *Mon. Not. R. astr. Soc.*, **260**, 49

Comastri A., Setti G., Zamorani G., Hasinger G. 1994, *Astron. Astrophys.*, in press

Della Ceca R., Maccacaro T., et al. 1992, *Astrophys. J.*, **389**, 491

Fabian A.C., George I.M., Miyoshi S., Rees M.J. 1990, *Mon.Not.R.astr. Soc.*, **242**, 14P

Giacconi R. & Zamorani G. 1987, *Astrophys. J.*, **313**, 20

Gruber D.E. 1992, in *The X-ray background*, eds. X. Barcons and A.C. Fabian (Cambridge Univ. Press, Cambridge), p.44

Hasinger G., et al. 1993, *Astron. Astrophys.*, **275**, 1

Hayashida K. 1990, Ph.D. Thesis, Tokyo University, ISAS RN 466

Huchra J., & Burg R. 1992, *Astrophys. J.*, **393**, 90

Madau P., Ghisellini G., Fabian A.C. 1993, *Astrophys. J.*, **410**, L7

Madau P., Ghisellini G., Fabian A.C. 1994, *Mon. Not. R. astr. Soc.*, **270**, L17

Matsuoka M., Piro L., Yamauchi M., Murakami T. 1990, *Astrophys. J.*, **361**, 440

Matt G. & Fabian A.C. 1994, *Mon. Not. R. astr. Soc.*, **267**, 187

Miyaji T., Lahav O., Jahoda K., Boldt E. 1994, *Astrophys. J.*, in press

Morisawa K., Matsuoka M., Takahara F., Piro L. 1990, *Astron. Astrophys.*, **236**, 299

Nandra K. & Pounds K.A. 1994, *Mon. Not. R. astr. Soc.*, **268**, 405

Piccinotti G., et al. 1982, *Astrophys. J.*, **253**, 485

Pounds K.A., et al. 1990, *Nature*, **344**, 132

Rogers R.D. & Field G.B. 1991, *Astrophys. J.*, **378**, 117

Schwartz D.A. & Tucker W.H. 1988, *Astrophys. J.*, **332**, 157

Setti G. & Woltjer L. 1973, in IAU Symposium No. 55, *X- and Gamma-Ray Astronomy*, eds. H. Bradt and R. Giacconi (Reidel, Dordrecht), p.208

Setti G. 1992, in *The X-ray background*, eds. X. Barcons and A.C. Fabian (Cambridge Univ. Press, Cambridge), p.187

Setti G. & Woltjer L. 1989, *Astron. Astrophys.*, **224**, L21

Setti G. & Woltjer L. 1994, *Astrophys. J. Suppl.*, **92**, 629

Shafer R.A. 1983, Ph.D. Thesis, University of Maryland, NASA Tech.Mem. 85029

Taylor G.B. & Wright E.L. 1989, *Astrophys. J.*, **339**, 619

Terasawa N. 1991, *Astrophys. J.*, **378**, L11

Turner T.J. & Pounds K.A. 1989, *Mon. Not. R. astr. Soc.*, **240**, 833

Warwick R.S. & Stewart G.C. 1989, in 23rd ESLAB Symposium, in *Two-Topics in X-Ray Astronomy*, ESA SP-296, Vol.2, p.727

Wright E.L., et al. 1994, *Astrophys. J.*, **420**, 450

Zdziarski A.A., Zycki P.T., Svensson R., Boldt E. 1993a, *Astrophys. J.*, **405**, 125

Zdziarski A.A., Zycki P.T., Krolik J.H. 1993b, *Astrophys. J.*, **414**, L81

THE COSMOLOGICAL DIFFUSE γ-RAY BACKGROUND: MYTH OR REALITY ?

P. LAURENT
CEA/DSM/DAPNIA/Sap
CE Saclay, 91191 Gif sur Yvette Cedex, FRANCE

1. Introduction

This paper is by no means a critical review of what have been thought, calculated or written about the cosmological diffuse γ-ray background. More simply, it is intended to tell its story, in order to answer a may be more historical than astronomical question:

"How ideas about the cosmological diffuse γ-ray background has evolved since the beginning of the space era ?"

As the γ-ray energy range is quite large, this story will be divided in two parts. The first one, which is given here, will address only the low energy part of the γ-ray domain, that is γ-rays whose energy is less than a few MeV. The high energy domain will be described by Prof. Bignami (this issue).

2. First observations of the "MeV bump"

So, let's return to the sixties, at the beginning of our story ... An "interstellar" γ-ray emission has been observed indeed for the first time by the experiments Ranger 3 and 5 in 1964 (Metzger *et al.*, 1964). These experiments consisted of a phoswitch system composed of a caesium iodide scintillation crystal and a plastic scintillator. Most of the instruments we will consider later on were built on the same principle. This is an important point to keep in mind as we will see below.

At that moment, it was only known that this emission did not come from the Solar System, so they called it "interstellar". The "diffuse γ-ray

M. Kafatos and Y. Kondo (eds.), Examining the Big Bang and Diffuse Background Radiations, 271–277.
© 1996 IAU.

272

background" was not yet really born ... However, as we can see in Figure 1, the spectrum of this emission shows a excess over a single power law around one MeV, excess which will be very important further on in our story.

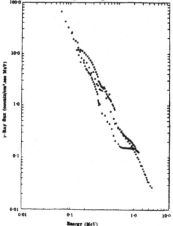

Figure 1. First spectrum obtained from space (Ranger 3 and 5, Metzger et al. 1964)

This excess, which will be called later on the "MeV bump", was detected afterwards by several experiments such as the omnidirectional NaI detector onboard the ERS 18 satellite (Vette *et al.*, 1970), or the 7×7 cm NaI detector placed on a boom 7.6 m away from the Apollo 16 spacecraft (Trombka *et al.*, 1973). The bump is clearly seen in the ERS 18 data , but is less prominent in the Apollo 15 spectrum (see Figure 2), a difference that we will explain in the next part.

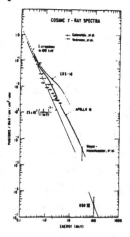

Figure 2. Appolo 15 spectrum (Trombka et al. 1973)

3. The MeV bump: myth or reality ?

In fact, an isotropical diffuse γ-ray background is intrinsically difficult to separate from a local background. As the emission is everywhere in the sky, there is no possibility of "on-off" (one observation on sources followed by one observation off sources in order to measure the background). One has to rely on other criteria, such as the study of the shape of the spectrum, in order to determine the origin of the detected emission.

If difficult to observe, the MeV sky is nevertheless interesting for astrophysicists, as it is the domain where nuclear reactions occurs. It is then the key range of energy for the nucleosynthesis studies. Unfortunately, it is also, of course, the energy range where nuclear reactions occurs *in* the detector itself and *in* its environment, reactions induced by the high energy protons crossing the whole experiment.

So, these reasons make the determination of the origin of the "MeV bump" difficult. A solution of these problems consisted to use a boom in order to remove the detector from the spacecraft, supposed to be the locally induced background source. This has been used for the Ranger 3/5 experiment but the boom was only 2 meters long. A longer one (7.60 m) has been used by the astronauts on Apollo 15/16, which has enabled them to obtain a better discrimination between the local and the cosmic background (see Figure 3). This explain the discrepancies between the results of Ranger 3/5, ERS 18 on one side and Apollo 15/16 on another side.

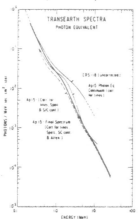

Figure 3. Appolo 15: spectra at different stages of data correction (Trombka et al. 1973)

This importance of the local background at MeV energies, has been pointed out as early as 1972 by Fishman et al., who have shown that they can reproduce the MeV bump seen by ERS 18, with Monte-Carlo simulation of the crossing of protons in the detector, and also by measures in particles accelerator (see Figure 4). As we have noticed earlier, as the detectors onboard Ranger 3/5, ERS 18 and Apollo 15/16 are of the same type, these calculation applied to all of them in the same way.

As it can be seen in Figure 2, taking into account the local background in the Apollo 15/16 case reduce seriously the MeV bump intensity, but there is still an excess over a single power law. This excess has been detected later on by numbers of satellites and balloon (Schönfelder, 1980), so even if the possibility of a local emission cannot be rejected, we can have some confidence in its existence. So, if we suppose that the cosmic MeV bump really exists, where does it come from ? This will be the leading question of our story from now ...

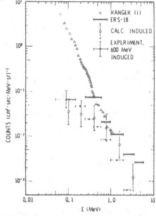

Figure 4. Ranger III and ERS 18 data points compared to simulations and experimental measures of the local background (Fishman 1972)

4. Is the MeV bump cosmological in origin ?

As soon as the discovery of the MeV bump was known, a lot of theoretical works have been done in order to explain it. Most of these theories was cosmological, as γ-ray coming from cosmological distances are practically not absorbed. So, "extragalactic γ-ray astronomy must be considered with cosmological questions" (Pinkau, 1979). It will be too long to describe all these models in details here, so we give below only a list of the most frequently used ones:

- Compton scattering of electrons leaking from radio-galaxies (Brecher *et al.*, 1969).
- Redshifted π^0 annihilation (Stecker *et al.*, 1971).
- Nuclear γ-ray in supernovae from distant galaxies (Clayton *et al.*, 1969).
- Intergalactic electron bremsstrahlung (Arons *et al.*, 1971).
- Radiation from exceptional galaxies (Bignami *et al.*, 1979).

One of the cosmological model the most used in the seventies explained the "MeV bump" in term of annihilation of cosmologically redshifted π^0 (Stecker 1971, see Figure 5). As π^0 annihilation gives rise to two 70 MeV photons, a redshift of 70, and thus a z of nearly 100 is needed to shift these photons down to one MeV. So, this annihilation must occur in the early phases of the universe. According to Stecker (1971), this emission arises from matter-antimatter transition zones, zones which should be created naturally in a baryon-symmetric big bang model (Omnes, 1969).

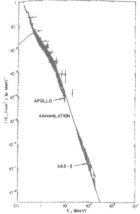

Figure 5. Redshifted π^0 annihilation model (Stecker et al. 1971)

5. Is the MeV bump extragalactic in origin ?

The situation changed completely at the beginning of the eighties, with the observation of a hard tail up to one MeV from the Seyfert galaxy NGC 4151 (see Figure 6), by the MISO experiment in 1979 and 1980 (Perotti *et al.*, 1981). Indeed, it appears that the MeV bump observed by Apollo and several other experiments could result from the sum of MeV emissions from several AGN. This was detailed in a subsequent paper by Bignami et al. (1979):

"With reasonable value of luminosities and present space density, Seyferts, BL Lacs objects and quasars may account for a major portion of the observed isotropic diffuse gamma-ray emission above 1 MeV ..." (Bignami *et al.*, 1979).

This "extragalactic" explanation took then for some years the precedence of the "cosmological" one, as it can be seen from the quotation below:

"In view of this discussion, it is perhaps not surprising that no single power law dependence is observed over the entire X and gamma-ray range, since different types of galaxies may contribute and dominate at different energies; the question of a really diffuse component like the one from

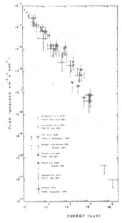

Figure 6. MISO spectrum of NGC 4151 (Perotti et al. 1980)

matter- antimatter annihilation in a baryon symmetric universe can only be answered if much more information on the X and gamma-ray emission of active galaxies and quasars is available. Only then will it be possible to derive that part of the background that cannot be explained by unresolved sources." (Schönfelder, 1985)

6. Back to cosmology ?

But, in 1984 and 1992 , the need for more information about the X and γ-ray emission of AGN became indeed more and more evident with the observation of a spectral break around 50 keV in the NGC 4151 spectrum, by the HEAO 1 (Baity *et al.*, 1984) and SIGMA (Jourdain *et al.*, 1992) experiments. This leads to the existence of at least two spectral states in Seyfert galaxies, a hard one as the one detected by MISO in 1979 and a soft one observed by HEAO and SIGMA (see Figure 7). So the knowledge of the origin of the MeV bump became more and more related to the knowledge of the occurrence of these spectral states. The spectral break around 50 keV was confirmed later on by the *GRO*/OSSE calculation of the Seyfert mean spectrum (see Figure 8), obtained between 1991 and 1993 (Kurfess *et al.*, 1994). The soft states seems then to occurs much more often than the hard one, which strengthen a cosmological origin of the "MeV bump".

7. What is the situation now ?

So, a definitive explanation of the origin of the MeV bump, if this bump really exists, is not yet available, and must await a good knowledge of the AGN spectral states. This will be precised by the future observations done by the *Compton* observatory, and by the *INTEGRAL* observations available at the beginning of the next millennium. Only then shall we know

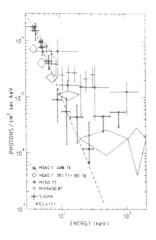 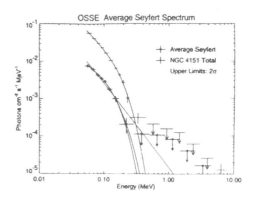

Figure 7. (a) SIGMA spectrum of NGC 4151 compared to the MISO and the HEAO 1 observations (Jourdain et al. 1992) (b) Seyfert mean spectrum (*GRO*/OSSE, Kurfess 1994)

if it originates from active galaxies or from regions of high cosmological redshift, or both.

References

Arons J., et al. (1971) *ApJ*, **170**, 431
Baity W.A., et al. (1984) *ApJ*, **279**, 555
Bignami G.F., et al. (1979) *ApJ*, **232**, 649
Brecher K. & Morrison P. (1969) *Phys. Rev. Letters*, **23**, 802
Clayton D.D. & Silk J. (1969) *ApJ*, **158**, L43
Fishman G.J., (1972) *ApJ*, **171**, 163
Jourdain E. et al. (1992) *A&A*, **256**, L38
Kurfess J., et al. (1994) Proc. of Les Houches school, eds. M. Signore P. Salati and G. Vedrenne, NATO:ASI, in press
Metzger A.E., et al. (1964) *Nature*, **204**, 766
Omnes R. (1969) *Phys. Rev. Letters*, **23**, 38
Perotti F., et al. (1981) *ApJ*, **247**, L63
Pinkau K., (1979) *Nature*, **277**, 17
Schönfelder V., (1980) *ApJ*, **240**, 350
Schönfelder V., (1985) Proc. of 19th ICRC La Jolla, Rapporteur Talk, **9**, 93
Stecker , et al. (1971) *Phys. Rev. Letters*, **27**, 1469
Trombka J.I., et al. (1973) *ApJ*, **181**, 737
Vette J.I., et al. (1970) *ApJ*, **160**, L161

DIFFUSE GAMMA RAYS OF GALACTIC AND EXTRAGALACTIC ORIGIN:PRELIMINARY RESULTS FROM EGRET

P. SREEKUMAR

NASA/Goddard Space Flight Center (USRA)
Greenbelt, MD 20771, USA

AND

D.A. KNIFFEN

Hampden-Sydney College
Hampden-Sydney, VA 23943, USA

Abstract. The all-sky survey in high energy gamma rays (E>30 MeV) carried out by the Energetic Gamma Ray Experiment Telescope (EGRET) aboard the Compton Gamma Ray Observatory provides for the first time an opportunity to examine in detail diffuse gamma-ray emission of extragalactic origin. The observed diffuse emission at high galactic latitudes is generally assumed to have a galactic component arising from cosmic-ray interactions with the local interstellar gas and radiation, in addition to an isotropic component presumably of extragalactic origin. The galactic component can be estimated from a model of the interstellar medium and cosmic-ray distribution. Since the derived extragalactic spectrum depends very much on the success of our galactic model, the consistency of the galactic diffuse emission model is examined both spectrally and spatially with existing EGRET observations. In conjunction with this model, EGRET observations of the high latitude emission are used to examine the flux and spectrum of the residual extragalactic emission. This residual emission could be either truly diffuse in origin or could arise from accumulated emission from unresolved sources particularly in the light of EGRET observations showing the presence of numerous gamma-ray bright active galactic nuclei.

M. Kafatos and Y. Kondo (eds.), Examining the Big Bang and Diffuse Background Radiations, 279–288.
© *1996 IAU.*

1. Introduction

Historically, the first indication of an extended diffuse component of the high energy gamma radiation was obtained by Kraushaar et al.(1972) with the OSO-3 satellite. Later observations by the second Small Astronomy Satellite (SAS-2) clearly showed for the first time a strong correlation of the observed diffuse emission with column density of interstellar gas (Fichtel, Simpson & Thompson 1978; Thompson & Fichtel 1982). In addition, this correlation also indicated the presence of a residual emission at zero column density which was interpreted as not being associated with the Galaxy and hence of extragalactic origin. SAS-2 was unable to truly examine the isotropic nature of this residual emission due to the premature loss of the satellite after about 6 months of operation. The subsequent COS-B mission did not result in any additional understanding of the isotropic emission due to the high instrumental background. Neither the placement of material surrounding the instrument nor the orbit were optimized to allow observations of the weak extragalactic diffuse radiation. With more than an order of magnitude increase in sensitivity over previous experiments and a low instrumental background, EGRET, on board the Compton Gamma Ray Observatory (CGRO) provides the first opportunity to study in detail the spectrum and distribution of the extragalactic gamma-ray emission. During the early phase of the CGRO mission (April 1991 - Nov 1992), EGRET carried out the first all sky survey in high energy gamma rays (30 MeV$\leq E \leq$ 30 GeV). The most striking aspect of the all sky survey is the dominant diffuse emission from the plane of the Galaxy. In addition, numerous point sources are also observed, many of them have been identified with pulsars, molecular clouds, active galactic nuclei and a nearby normal galaxy (Fichtel et al.1994). Along the galactic plane, point source detection is difficult since they are embedded in the extremely strong galactic diffuse emission. Currently, the unambiguously identified point sources along the plane include, five sources identified as pulsars (PSR B0531+21 (Crab) (Nolan et al.1993), PSR B0833-45 (Vela) (Kanbach et al.1994), PSR 0630+178 (Geminga) (Bertsch et al.1992; Mayer-Hasselwander et al.1994), PSR B1706-44 (Thompson et al.1992) & PSR B1055-52 (Fierro et al.1993) on the basis of their characteristic timing signature. At high galactic latitudes point source detection capability is enhanced by the much weaker galactic emission. A total of 36 sources have been identified with a class of active galactic nuclei called blazars. In general, blazars are radio bright AGNs with flat radio spectra, strong optical polarization, often exhibit superluminal motion and show significant time variability at most wavelengths. EGRET observations have shown that for a majority of the gamma-ray luminous blazars, the gamma-ray emission dominates the bolometric luminosity. This important finding further strengthens the theoretical attempts to explain the origin of the extragalactic emission as a superposition of sources. Other high latitude as well as galactic point sources have also been detected by EGRET and these have yet to be identified with any *likely* candidates.

This paper briefly discusses the approach used to study the extragalactic gamma-ray background by carefully accounting for the foreground galactic emission. The level of our understanding of the galactic model is demonstrated by showing the consistency both spatially and spectrally between the prediction and EGRET observations. The observational results presented here are based on the all sky survey phase of the CGRO mission (April 1991 to Nov 1992). Details on the instrument and its capabilities are discussed by Thompson *et al.*(1993). Preliminary conclusions derived on the extragalactic gamma-ray background are also discussed.

2. Diffuse Gamma Rays of Galactic Origin

Beyond the point sources discussed above, the emission that appears diffuse in nature in the all sky survey is dominated by a strong galactic component. Diffuse emission of galactic origin strongly traces out the plane of the Milky Way including the warp in the atomic hydrogen gas disk around longitude of 90°. Apart from a possible contribution from unresolved sources such as pulsars (Bailes & Kniffen 1992), this emission is generally understood to arise from cosmic rays interacting with the interstellar gas and radiation (Bertsch *et al.*1993). Away from the plane, the galactic emission gets weaker but is non-negligible even at the poles. The detailed comparison of the model against observational data from EGRET will be discussed by Hunter *et al.*(1995) for the galactic plane and by Sreekumar *et al.*(1995) for regions at high latitudes. In order to understand and characterize diffuse gamma rays of extragalactic origin, it is important to remove the galactic component from the observed diffuse emission . Consequently, for the study of the extragalactic radiation, it is more appropriate to examine regions away from the galactic plane where the galactic contribution is weaker.

Gamma-ray observations also provide a direct way to examine the cosmic-ray distributions in our Galaxy as well as in other external galaxies. Thus in addition to the reason given above to model the galactic emission, another important goal is to use gamma-ray observations to probe the spectrum and spatial density distribution of cosmic rays in the Galaxy. The three principal gamma-ray production processes are bremsstrahlung by energetic cosmic-ray electrons, decay of neutral pions resulting from nucleon-nucleon interactions and by inverse Compton scattering of cosmic-ray electrons on the low energy 2.7K, infra-red, optical and UV photons in the Galaxy.

Electron bremsstrahlung dominates the observed diffuse emission at energies below 100 MeV. However, above a few 100 MeV, diffuse gamma-ray emission is dominated by photons from the decay of neutral pions. While relativistic cosmic-ray electrons can be studied using radio observations, cosmic-ray protons which carry most of the energy in cosmic rays are best studied using high energy gamma-ray observations. Beyond the Milky Way, we have detected for the first time gamma-ray emission from an external galaxy the Large Magellanic Cloud (Sreekumar *et al.*1992). In addition, our observation of the Small Magellanic Cloud has provided clear evidence for the galactic origin of the bulk of cosmic rays (Sreekumar *et al.*1993). This result provides the fundamental basis to model galactic cosmic rays as being coupled to the interstellar gas (see Bertsch *et al.*,1993, for further discussion). Since the observed gamma-ray emission depends also on the density distribution of the interstellar gas, the model can be used to limit the CO to H_2 normalization factor (X-factor) used to convert the observed CO emission into molecular hydrogen column density. Prior to the launch of the Compton Observatory, the modeling of the COS-B data was carried out using a multiparameter fit to the observed emission. The analysis showed indications of a variation of cosmic-ray density with galactocentric radius as well as an upper limit of 2.3×10^{20} K km s^{-1} for the X-factor (Strong *et al.*1988). Conclusions from our analysis are discussed later on in this paper.

2.1. MODEL

The primary input parameters to calculate the galactic diffuse gamma-ray intensity include a model of the interstellar gas (neutral and ionized), photons and cosmic rays. The neutral atomic hydrogen distribution is obtained from 21-cm observations while the molecular hydrogen distribution is derived indirectly using the strength of $^{12}C^{16}O$ emission at 2.6 mm. At present, diffuse gamma-ray observations provide one of the better tools to constrain the conversion factor (X) used to derive the molecular hydrogen column density from the CO. The model uses the all sky 21-cm map of Dickey & Lockman (1990) for regions at high latitudes (above $\pm 10°$) and data from the Maryland-Parkes survey(Kerr *et al.*1986), Weaver & Williams (1974) and Leiden-Green Banks survey (Burton & Liszt 1983) for the galactic plane. Along the plane, the model deconvolves the observed column density of neutral hydrogen along the line of sight using the galactic rotation curve. The molecular hydrogen density was derived from the CO data of Dame *et al.*(1987). The ionized interstellar medium is not as well determined and here we use the model of Taylor and Cordes (1993) which is based on results from pulsar observations. The cosmic rays are assumed to be closely coupled to the interstellar gas, the coupling length

scale along the galactic plane is adjusted to best fit the EGRET observations. Additional details on the model are discussed in Bertsch *et al.*(1993).
Since the interstellar neutral gas distributions have a narrow scale height
(110 pc for atomic H; 60 pc for molecular H), at high latitudes, we assume
the neutral gas is all local and hence the cosmic-ray density to be similar
to that in our galactic neighborhood. The ionized gas is generally believed
to have a larger scale height of several hundred parsecs (Reynolds 1993).
Radio continuum observations have indicated the cosmic ray electron scale
height to be significantly larger than that of the gas. Our model uses a
cosmic ray scale height of 1 kpc and assumes a ratio of electrons to protons
of 1/100, which is the generally accepted value in our galactic neighborhood. The larger scale heights lead to larger relative contributions from the
ionized medium and the inverse Compton process at high latitudes. The
structure and scale height of the ionized medium and the spatial distribution of the soft photons in the Galaxy (2.7K, NIR, FIR, Optical and UV)
that participate in the inverse Compton process, are not well known. This
introduces uncertainties in the predicted high latitude gamma-ray emission.
Thus, our current results on the extragalactic background emission are also
subject to these uncertainties. Efforts to improve the model in this regard
are underway.

2.2. RESULTS

The observed diffuse gamma-ray emission from the plane can be compared
with the model calculations of Bertsch *et al.*(1993). Figure 1 shows the longitude dependence of the observed intensity averaged over $\pm$ 10° in latitude.
The solid line represents the model of Bertsch *et al.*(1993) together with
an isotropic extragalactic component of 1.5×10^{-5} photonscm^{-2}s^{-2}ster^{-1}.
Considering the model has only two adjustable parameters and does not
force fit the observations rigorously, it is remarkable that it reproduces
many of the spatial features at various longitudes which correspond to spiral arm tangent points. The assumption by Bertsch *et al.*(1993) that cosmic
rays are closely tied to the interstellar gas distribution appears to have a
strong basis as seen from figure 1. The latitude distribution averaged over
the longitude range (90°-150°), is shown in figure 2 and shows good consistency with the model predictions. The model requires an X-factor (one of
the adjustable parameters of the model) of 1.8×10^{20} K km s^{-1} or less to be
consistent with the EGRET observations. The analysis also indicated that
the cosmic-ray - matter coupling length scale is about (2 ± 1) kpc in the
plane of the Galaxy. The corresponding length scale derived using a typical
diffuse coefficient (Ginzburg *et al.*1980) and a cosmic-ray mean lifetime of
2×10^7 years is about 3 kpc which is within the errors of our observation-

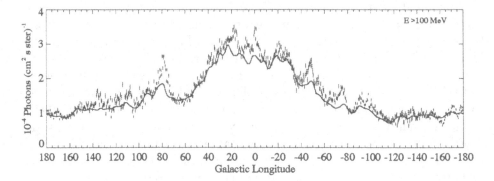

Figure 1. EGRET observations of the Galactic diffuse emission: distribution in galactic longitudes, averaged over ±10°in latitude. The solid line represents the model prediction along with an assumed isotropic component of $1.5 \times 10^{-5} photons cm^{-2} s^{-1} ster^{-1}$. Only the clearly identified sources along the galactic plane viz., the five gamma-ray pulsars are subtracted (Hunter *et al.*1995).

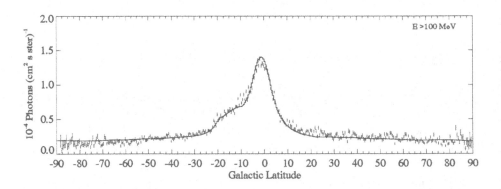

Figure 2. EGRET observations of the Galactic diffuse emission: distribution in galactic latitudes, averaged over the longitude range of 190°- 240°.The solid line represents the model prediction along with an assumed isotropic component of $1.5 \times 10^{-5} photons cm^{-2} s^{-1} ster^{-1}$. EGRET detected point sources at high latitudes have been removed. Along the galactic plane, only the five gamma-ray pulsars are subtracted from the data (Hunter *et al.*1995; Sreekumar *et al.*1995).

ally derived value. The point source analysis using a maximum likelihood method (Mattox *et al.*1995) indicates < 15% of the total observed diffuse emission from the Milky Way arise from resolvable point sources. There may

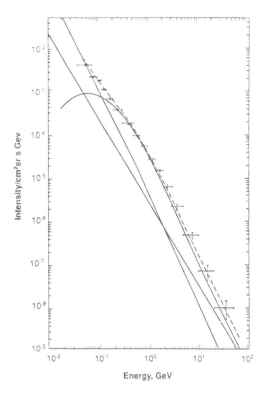

Figure 3. Galactic diffuse emission spectrum near the galactic center (from Fichtel *et al.*1993). The dashed line indicates the sum of the three components: π° decay, electron bremmstrahlung and inverse Compton process.

still be unresolved point sources below the detection threshold of EGRET. The spectrum from the region near the galactic center is shown in figure 3. It shows for the first time the predicted feature from gamma rays arising from π° decay. This is an important signature of cosmic-ray protons in the Galaxy. Thus the galactic model shows consistency both spectrally and spatially with the observed diffuse emission. Additional results on the diffuse emission from the galactic plane will be discussed by Hunter *et al.*(1995).

3. Diffuse Gamma Rays of Extragalactic Origin

The measurement of the extragalactic diffuse radiation is one of the most difficult of all high-energy measurements. This is because the intensity is low and there is no unique signature such as spatial or temporal profile to separate the cosmic signal from the various backgrounds. Not the least of these is the instrumental background produced by the interactions of ambient cosmic rays with the material in and around the detector. So two separate factors must be minimized, the cosmic-ray intensity and instrument and/or spacecraft material outside of the anticoincidence scintillator.

Only two instruments have been placed in orbit with these two factors designed to give a background well below the expected diffuse background intensity. These are the spark chamber telescopes flown aboard the SAS-2, and the EGRET instrument on the Compton Observatory. The SAS-2 instrument was launched into a near equatorial orbit and EGRET into the standard 28° inclination orbit for a shuttle launch. Because of the higher inclination, the Compton Observatory is kept below a 450 kilometer altitude to minimize the effects of the trapped radiation. The design of the Compton Observatory was complicated also by the discovery during the repair of the Solar Maximum Mission (SMM) satellite that space debris was more intense than expected. A penetration of the light shields surrounding the instruments could cause light leaks which could degrade or damage the photomultiplier tubes and adversely affect instrument performance. For this reason the mass in the detector aperture on EGRET, ~ 0.19 g cm^{-2}, was slightly higher than the ~ 0.16 g cm^{-2} of SAS-2. Nevertheless, extensive preflight calibration at the Brookhaven Laboratory have indicated that the expected instrumental background of EGRET is an order of magnitude less than the intensity of extragalactic radiation derived from the SAS-2 data. The small disadvantages of EGRET are more than overcome by the much improved instrument sensitivity of well over an order of magnitude, the long exposure to high latitude regions of the sky, and broad spectral coverage (30 MeV - 30 GeV).

3.1. EXPOSURE AT HIGH GALACTIC LATITUDES

The most extensive exposure off the galactic plane to date is to the Virgo region. Over 20 weeks of exposure have been obtained in phase 1 to 3 (April 1991 to Jan 1994) and another 5 more weeks are allocated for phase 4. Furthermore, there are a number of other high latitude (+ and -) exposures, shorter in duration, that will add to the study. In order to complete the analysis of these data great care must be exercised to ensure proper normalization between different exposures and other even more subtle corrections, including subtraction of background due to discrete source and galactic diffuse contamination. This work is nearing completion, but a normalized spectrum is not yet available. Preliminary results indicate a good power law fit to the extragalactic spectrum with a slope of -(2.2±0.2) (Sreekumar *et al.*1994) .

3.2. WORK IN PROGRESS

Current work is aimed toward optimizing the subtraction of backgrounds since these and an accurate exposure calculation are crucial to obtaining a good normalized spectrum. In addition, several approaches are being taken to investigate the origins of this radiation. In particular, fluctuation analysis using a logN-logS technique as well as topological and wave packet analysis of spatial fluctuations on all scales are being explored.

4. Conclusions

We have successfully modeled the diffuse gamma ray emission from the Galaxy. The model predicted diffuse emission spectrum and the spatial distribution in galactic longitude and latitude are in good agreement with the EGRET observations. This makes it possible to remove the galactic component from the observed high latitude diffuse emission. The determination of the diffuse spectrum of the extragalactic high-energy gamma radiation is very difficult and must be done with extreme care. Instrumental response must be thoroughly understood and folded into the analysis. This work is well underway, but not yet complete. The results which currently exists suggests a power law slope of -(2.2±0.2). This is consistent with the slope of $-2.34^{+0.4}_{-0.3}$ reported by SAS-2 (Thompson and Fichtel 1982). This is also consistent with the average spectra reported for AGNs detected by EGRET and hence the spectral shape of the extragalactic emission could be explained by an accumulation of unresolved AGN emission. Future work on fluctuation analysis may shed some light on this question.

5. References

Bailes, M., & Kniffen. D.A., 1992, ApJ, 391, 659

Bertsch, D. L., Dame, T. M., Fichtel, C. E., Hunter, S. D., Sreekumar, P., Stacy, J. G., & Thaddeus, P., 1993, ApJ, 416, 587

Bertsch, D.L., et al.1992, Nature, 357, 306

Burton, W.B., & Liszt, H.S., 1983, A&AS, 52,63

Dame, T.M., et al.1987, ApJ, 322, 706

Dickey, J.M., & Lockman, F.J., 1990, Ann. Rev. Astron. Astrophys., 28, 215

Fichtel, C.E. et al.1994, ApJS, 94, 551

Fichtel, C.E., et al.1993, A&AS, 97, 13

Fichtel, C. E., Özel, M., Stone, R., & Sreekumar, P., 1991, ApJ, 374, 134

Fichtel, C.E., Simpson, G.A., & Thompson, D.J., 1978, ApJ, 222, 833

Fierro, J.M., et al.1993, ApJ, 413, L27

Ginzburg, V. L., Khazan, Ya. M., Ptuskin, V.S., 1980, Astrophys. Space Sci., 68, 295

Hunter, S.D. *et al.*1995, (in preparation)

Kanbach, G., *et al.*1994, A&A (in press)

Kerr, F.J., Bowers, P.F., Jackson, P.D., & Kerr, M., 1986, A&AS, 66,373

Kraushaar, W.L., *et al.*1972, ApJ, 177, 341

Mattox, J.R., *et al.*1995, (in preparation)

Mayer-Hasselwander, H.A., *et al.*1994, ApJ, 421, 276

Nolan, P.L., *et al.*1993, ApJ, 409, 697

Reynolds, R.J., 1993, *Back to the Galaxy* , AIP conf. Proc. 278, Ed. S.Holt & F. Verter

Sreekumar, P. *et al.*1992, ApJ, 400, L67

Sreekumar, P. *et al.*1993, PRL, 70, 127

Sreekumar, P., *et al.*1994, Bull. Amer. Astron. Soc., 25, 1293

Sreekumar, P. *et al.*1995, (in preparation)

Strong, A., *et al.*1988. A&A, 207, 1

Taylor, J.H., & Cordes, J.M., 1993, ApJ, 411, 674

Thompson, D. J., & Fichtel, C. E., 1982, A&A, 109, 352

Thompson, D.J., *et al.*1993, ApJS, 86, 629

Thompson, D.J., *et al.*1992, Nature, 359, 615

Weaver, H., & Williams, D.R.W., 1974, A&AS, 17,1

Big Bang Scenario and Nature of Dark Matter

I. Novikov[1,2*,3,4]

1) Astro Space Center of P.N.Lebedev Physical Institute

Profsoyuznaya 84/32, Moscow, 117810, Russia

2) University Observatory, Øster Voldgade 3, DK-1350 Copenhagen K, Denmark

3) NORDITA, Blegdamsvej 17, DK-2100 Copenhagen, Ø Denmark

4) Theoretical Astrophysics Center, Blegdamsvej 17, DK-2100,

Copenhagen Ø, Denmark

---

*) Permanent address

Abstract

We give a brief review of a problem of dark matter in the Universe. The key questions regarding dark matter are discussed: what is the distribution of dark matter at different scales? What is the nature of dark matter? Is there baryonic dark matter? We discuss a few new techniques of the investigation of dark matter in the Universe. We discuss the peculiar periodic dependence of the initial spectrum of the baryonic matter distribution on wavelength at the moment of recombination (Sakharov oscillations). Sakharov oscillations should manifest themselves by the specific anomalies in the angular correlation function of the microwave background anisotropy on scales $10' - 2°$, practically in any cosmological model. We discuss special methods for filtering of this effect in the observational data.

1 Introduction

First International Astronomical Union Symposium devoted to the problem of Dark Matter in the Universe was held in 1985, but the history of the problem began much earlier. F. Zwicky (1933) was the first to point out that the sum of masses of visible galaxies in Coma Clusters is essentially less than the total mass of this cluster producing the gravitational field, assuming that cluster is a gravitationally bound and well relaxed system, and the virial theory can be

M. Kafatos and Y. Kondo (eds.), Examining the Big Bang and Diffuse Background Radiations, 289–300.

applied. He concluded that either the cluster is short lived or visible matter is not a good guide for the mass. Thus invisible mass must exist in this cluster.

Another evidence about dark matter came in the 1970's from observations of rotation curves (dependence of the rotational Kepler's velocity on radius r) of spirals.

Nowadays the problem of dark matter in the Universe is more than 60 years old and yet is at the heart of all astrophysics and physics. This is not a surprise. The mass of the dark matter dominates in the Universe, determines the dynamics of the Universe and the quervature of the 3-dimensional space. On the other hand the dark matter plays a crucial role in the process of growth of small fluctuations in the matter distribution (progenitors of galaxies), in the processes of the origin of the large scale structure in the Universe, in the process of galaxy formation and evolution. The last but not list, dark matter determines the physical contents of the Universe and carries the information about the physics of the Early Universe.

In the Preface of the Proceedings of the 117 IAU Symposium, J.Kormendy and G.R.Knapp wrote: "This is the first time that the International Astronomical Union has held a symposium on objects of totally unknown nature" (Kormendy and Knapp, 1987). Unfortunately, today we have to repeat that the nature of dark matter is still unknown.

2   Dark matter on different scales

What can one say about the amount of various types of matter in the Universe? Surveys of visible matter (visible parts of galaxies, hot gas etc.) get (in dimensionless units $\Omega \equiv \langle \rho \rangle / \rho_{crit}$)

$$\Omega_{vis} \approx 0.003 - 0.006, \tag{1}$$

see for example Persic and Salucci (1992), Steigman (1994). On larger and larger mass/length scales the value Ω increases. If one considers only the visible parts of galaxies then the value of Ω is given by (1). However, on the scale of 1-3 hundred Kpc (scales of spiral halos and small groups of galaxies), the value of Ω has increased and $\Omega_{Halo} \approx 0.01 - 0.09$ (Griest (1994), Steigman (1994)). This trend continues up to the scale of rich clusters and beyond. On the scales of clusters $\Omega_{cl} \approx 0.1 - 0.3$ (Steigman (1994), Bahcall (1994)). The velocity distribution of galaxies on very large scales (large scale flows) provide the probes of the largest scales. A. Dekel (1993), (1994) concludes that the velocity data indicates a high value for Ω_{LSF} near unity, with $\Omega_{LSF} \leq 0.3$ strongly ruled out.

It is worth noting the following three new techniques of the investigation of missing mass (dark matter) in the Universe.

First of them is the amplification of light from distant stars as they crossed very near to the line-of-sight due to the relativistic microlensing effect. This method was proposed by Paczynski (1986) to detect Massive Astrophysical Compact Halo Objects (MACHOs). A few events were detected (Alcook et al. (1993), Aubourg et al. (1993)) from photometric measurements of millions of stars in the Large Magellanic Cloud. Probably MACHOs have masses of about 0.1 solar masses. The nature of them is not known. It is still unclear if the total number of MACHOs is enough to be halo dark matter.

Second technique is mapping cosmic dark matter in clusters of galaxies via the gravitational lens technique (Tyson (1994)). Foreground mass concentrations in a cluster of galaxies can be detected and mapped directly by the way they distort images of background sources. One can use as sources QSOs or far galaxies. QSOs are not so good due to the relative scarcity of QSOs compared with galaxies at similar redshifts. Thus they can give a little information on the mass distribution in the lens. Tyson et al. (1990) and Tyson (1994) have pointed out that faint blue galaxies could be excellent background sources permitting the tomographic mapping of the projected dark matter distribution.

The third technique is so called POTENT reconstruction (see Dekel (1994)). The POTENT method recovers the smoothed dynamical fluctuation fields of potential velocity and mass density from observed radial peculiar velocities of galaxies under quasilinear gravitational instability.

3 The nature of dark matter

What can one say about nature of dark matter? There are strong restrictions on the possible average baryon density, Ω_b, from the comparison of the predictions of the theory of Big Bang nucleosynthesis with the observational abundances of the light elements. Big Bang nucleosynthesis calculations predict the primordial abundances of the light nucleides. The predictions depend on the baryon abundance and, consistency with observation is found for

$$0.010 \leq \Omega_b h^2 \leq 0.016, \tag{2}$$

where $H_0 = 100 h \text{kms}^{-1} \text{Mpc}^{-1}$, see Olive et al. (1990), Pagel (1990), see also Steigman and Tosi (1992), the analysis of new data see Pagel (1994 a,b,c) analysis of the new observation of D/H ratio by Songalia et al. (1994) and Carswell et al. (1994) see Pagel (1994, a). Thus we get

$$\Omega_b > \Omega_{vis}, \tag{3}$$

suggesting that perhaps most baryons in the universe are dark. Part of this dark baryons could be in the form of hot gas in the rich clusters of galaxies. This gas is optically dark but visible in x-rays. Another part may be hiding in the halos of galaxies.

The data on clusters and large scale flows suggests Ω_{Halo}, $\Omega_{LSF} \geq 0.1 - 0.3$. Thus we get

$$\Omega_{Halo}, \Omega_{LSF} > \Omega_b. \tag{4}$$

The conclusions that most of the mass in the Universe must be non-baryonic. What is it then?

The nature of the non-baryonic dark matter is not known. The formation of the large scale structure in the Universe depends on some general properties of dark matter candidates. The analysis of the large scale structure formation is the main probe for the investigation of the nature of the dark matter. According to the convenient classification the dark matter candidates are divided into two categories: "Hot Dark Matter" (HDM) and "Cold Dark Matter" (CDM). At present various types of the large scale formation models are discussed: HDM, CDM, "Mixed" (i.e. with the following content: baryons $\Omega_b = 0.05$, HDM $\Omega_{HDM} = 0.25$ and CDM $\Omega_{CDM} = 0.70$). Other possibilities are to include in the consideration the cosmological constant Λ or topological defects such as cosmic strings.

What could be the suspect weakly-interacting massive particles left over from earliest moments of the Universe and which now form the dark matter? Massive neutrinos are the favored HDM candidates, a favored CDM candidate is the axion. The modern physics provides many excellent other candidates (see Kolb and Turner (1990)), but I have to repeat that we know practically nothing about the physical nature of dark matter.

4   Sakharov oscillations

In the subsequent part of the paper we will discuss one more technique of the investigation of the nature of dark matter. This technique was discussed in our papers Naselsky and Novikov (1993a,b) and Jørgensen *et al.* (1994). I'll follow the last of these papers.

Let us consider the transitional phenomena which take place at the epoch of recombination of cosmic plasma in the expanding Universe. Sakharov (1965) has been the first to realize that the transitional phenomena in the growing perturbations lead to a peculiar periodic dependence of the perturbation amplitudes on wavelength. He considered the cold model of the Universe, though, qualitatively speaking, the same phenomena occur in the hot model as a result of recombination. These Sakharov oscillations have been discussed, for example, by Peebles and Yu (1970), Zeldovich and Novikov (1983) in the framework of the baryonic model with adiabatic initial perturbations.

It should be emphasized that in any cosmological model at the time of recombination ($z_{rec} \simeq 1100$) there existed acoustic modes of perturbations in the cosmic plasma due to the interaction between the baryonic matter and radiation. The wavelengths of the modes correspond to the comoving linear scales between the Silk damping scale, $l_s \sim 10h^{-1}\mathrm{Mpc}$, and the acoustic horizon $R = r_{rec}\frac{v_{acoust}}{c} \approx 100 - 200h^{-1}\mathrm{Mpc}$, r_{rec} is the event horizon at $z_{rec} \simeq 1100$, all being scaled to the present epoch, $v_{acoust} = 3^{-1/2}/\sqrt{1 + 3\rho_b/4\rho_r}$ where ρ_b/ρ_r is the ratio of baryon and radiation densities at z_{rec}. After recombination the motion of baryonic matter is independent of the background radiation.

In non-baryonic dark matter models (for example - CDM models) the baryonic density fluctuations follow that of the dark matter, due to the peculiar gravitational field of the dark matter, whereas the distribution of the microwave background photons gets frozen on the celestial sphere thus preserving the information about the inhomogeneities at the last scattering surface. Therefore, in the present spatial distribution of matter, which is determined mainly by that of the dark matter, acoustic modes are practically erased, while the angular distribution of CMB preserves the signature of the acoustic motions of matter.

Now many groups observe the anisotropy of CMB on the scales of the order $1°$. Because the upper limit on the scale of acoustic perturbations, R, is determined only by r_{rec} and $\rho_b/\rho_r(z_{rec})$, it is clear that detecting the effects of these modes in $\Delta T/T(\theta)$ would give unique information on the amount of baryons in the Universe as well as the spectrum of density perturbations on scales around $r \simeq 100h^{-1}\mathrm{Mpc}$. The analysis of the manifestation of the Sakharov's oscillations see, for example, in the papers Holtzman (1989), Muciaccia *et al.* (1993). Authors use another names for this effect: "Doppler peak" for example.

5 Filtering of Sakharov effect in the anisotropy of CMB.

In the papers by Naselsky and Novikov (1993a,b) and Jørgensen *et al.* (1994) we investigated the manifestation of Sakharov oscillations in the anisotropy of Cosmic Microwave Background (CMB). We called the corresponding peculiarities of the angular correlation function of $\Delta T/T$ Long Distance Correlations (LDC). The LDC are thus a specific signature of the acoustic modes.

In the paper Naselsky and Novikov (1993a,b) it was demonstrated that at the angular scales $\theta \approx \theta_R$ the Sakharov oscillations lead to an appearance of a specific anomaly (a "resonance" type) which has a half-width of order θ_c, where θ_R is equal to the ratio of the acoustic horizon R at the epoch of recombination (scaled to the present epoch) to the present-day particle horizon h, and $\theta_c = l_s/h$, l_s is the Silk damping scale.. An analogous feature occurs at $\theta \to 2\theta_R$. The existence of the feature at $\theta \approx 2\theta_R$ for the limit case $\Omega_b \to 0$ was mentioned in the paper of Starobinskii (1988).

We would like to emphasize that the anomalies at $\theta \approx \theta_R$ and $\theta \approx 2\theta_R$ have similar nature. The reason for both of them is the existence of acoustic modes evolving in the cosmic plasma at the epoch of hydrogen recombination.

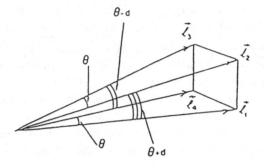

Figure 1: Space orientation of the unit vectors for the 4-beams experiment.

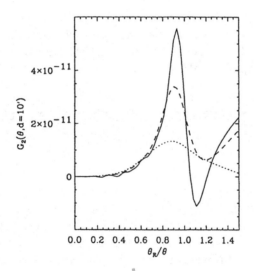

Figure 2: Function $G_2(\theta, d, \theta_A)$ for the model $\Omega = 1$; $\Omega_b = 0.03$; $h = 0.5$; $d = 10'$ and the Harrison -Zel'dovich spectrum. Solid line for $\theta_A = 0$, dashed line for $\theta_A = 0.1\theta_R$, dotted line for $\theta_A = 0.3\theta_R$.

These acoustic modes left traces on the correlation function of the CMB angular fluctuations due to Silk (1968) and Doppler effects.

It is important to emphasize the following fact. In the "resonance" $\theta \sim \theta_R$ the Silk effect only gives the contribution, while in the "resonance" $\theta \approx 2\theta_R$ both effects (Silk and Doppler) make their contributions. Thus we have an unique possibility of an experimental test of these effects, which are important both for the clarification of the character of the initial perturbations and for the investigation of the ionization history of the cosmic plasma.

What could be the method of filtering of the microwave background anisotropies which 1) cancels the regular component of the correlations and 2) emphasizes/preserves the effect of LDC? One method is to investigate the multipole moments representation of the angular correlation function

$$C(\theta) = \langle T(\alpha)T(\alpha + \theta)\rangle_\alpha = \frac{1}{4\pi}\sum_l (2l+1)C_l P_l(\cos\theta), \tag{5}$$

where C_l are multipole moments, see, for example Smoot (1994).

We proposed another method (see Naselsky and Novikov (1993a,b)) which is an analogy of the method used by Watson and Gutierrez de la Cruz (1993): the observer should measure the difference of intensity of CMB in l_1 and l_3 directions, and in l_2 and l_4 as it is shown on Fig.4. We introduce an auxiliary function of two variables θ, d for given θ_A as:

$$G_2(\theta, d, \theta_A) = \frac{\langle [T(l_1) - T(l_3)][T(l_2) - T(l_4)]\rangle}{T^2} =$$

$$= C(\theta + d, \theta_A) + C(\theta - d, \theta_A) - 2C(\theta, \theta_A), \tag{6}$$

where $l_1 \cdot l_4 = l_2 \cdot l_3 = \cos\theta$, $l_1 \cdot l_2 = \cos(\theta + d)$, $l_3 \cdot l_4 = \cos(\theta - d)$, T is the average temperature, θ_A is the half-width of the antenna-beam. The value of θ should vary somewhat more than the range $\theta_R \leq \theta \leq 2\theta_R$, and $d \leq \theta_c$. The method can be used both for the direct measurement with the 4-beam experiment as well as the specific data-reduction technique.

One can see from (6) that for $d \ll \theta$ the level of anisotropy is related to the second derivative of the correlation function $C''(\theta, \theta_A)$ via:

$$G_2(\theta, d, \theta_A) \simeq C''(\theta, \theta_A)d^2. \tag{7}$$

We show that the quantity $G_2(\theta, d, \theta_A)$ introduced above allows one to measure in an optimal way the contribution of acoustic modes to the anisotropy of CMB.

On Fig.2 and Fig.3 we plot the result of the numerical calculations of $G_2(\theta, d, \theta_A)$ for the Harrison-Zeldovich and "tilted" spectra of the initial perturbations of the metric for different values of θ_A. The results of the computations of $G_2(\theta, d, \theta_A)$ in the neighborhood of the extrema $\theta \approx \theta_R$ can easily be interpreted using the

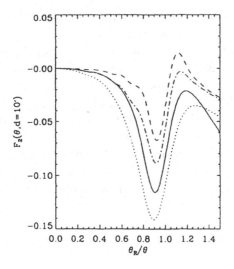

Figure 3: Function $F_2(\theta, d, \theta_A) = G_2(\theta, d, \theta_A)/G_2(0, d, \theta_A)$ for the same model as on Fig.2 ($d = 10'$). We consider the following values of the exponent of the spectrum of primordial perturbations n. Solid line for $n = 3$, $\theta_A = 0.1\theta_R$, Dashed line for $n = 3$, $\theta_A = 0$, Dotted line for $n = 3.5, \theta_A = 0.1\theta_R$, Dashed-dotted line for $n = 3.5$, $\theta_A = 0$.

properties of $C(\theta)$ which we have been found in the paper of Jørgensen $et\ al.$ (1994). We have demonstrated that the LDC effect manifests itself analytically most clearly at the range $\theta \approx \theta_R$

$$G_2(\theta \approx \theta_R, d, \theta_A) \approx \frac{9\varepsilon\ d^2}{a\ (1+\varepsilon)^{1/4}} \frac{1}{\theta_R\theta_c} C_0(0, \theta_{ef}) x \left(1 - \frac{5}{18}\beta x^2\right), \qquad (8)$$

where $x = \frac{\theta - \theta_R}{\theta_c}$, $\epsilon = (1 + \nu)\frac{3\rho_b}{4\rho_r}|_{z_{rec}}$; $\nu = \frac{2r_{eq}}{(\sqrt{2}-1)r_{rec}}$; R is an acoustic horizon at $z = z_{rec}$, r_{eq} is the Lagrangian radial coordinate of the particle horizon at the redshift of equality of radiation and dark matter densities (note that $\Omega \gg \Omega_b$), r_{rec} is the same at the moment of recombination.

One can see that $G_2(\theta \approx \theta_R, d, \theta_A)$ equal to zero at $\theta = \theta_R$ and looks like an almost linear function of x up to the extrema. Further the factor $1 - \frac{5}{18}\beta\ x^2$ plays a significant role (see Figures 2 and 3). The analogous phenomenon takes place at $\theta \approx 2\theta_R$, however in this range the amplitude of $G_2(\theta \approx 2\theta_R, d, \theta_A)$ is very small.

Let us now consider the dependence of the function $G_2(\theta, d, \theta_A)$ on the value of the half-width of the antenna-beam θ_A. For this purpose we calculated $G_2(\theta, d, \theta_A)$ for $(\theta_A/\theta_R)^2 = 0.1; 0.01$ (see Fig.2). As it is seen on the Fig.2, the peaks are smoothed out for a significant θ_A. Thus for the filtering of the effect under consideration we must have both θ_A, $d < \theta_c$.

On Fig.3 we plot the function $F_2(\theta, d, \theta_A) = G_2(\theta, d, \theta_A)/G_2(0, d, \theta_A)$ for the "tilted" models of the initial power spectrum (see, for example Muciaccia *et al.*, 1993). One can see that the level of modulation depends on the exponent of the spectrum and θ_{ef}, $\theta_{ef}^2 = \theta_A^2 + \theta_c^2$. Besides that, if we have a fixed value of the antenna beam θ_A, the depth of the negative peak at $\theta \approx \theta_R$ increases when we go from a Harrison-Zeldovich spectrum to the "tilted" one with $n = 3.5$. For fixed n we have an analogous feature for increasing θ_A. At the same time the strength of the peak of $G_2(\theta, d, \theta_A)$ at $\theta \approx \theta_R$ decreases (see Fig.2).

Concluding this section we note, that the 4-beam experiment possesses additional possibilities for an investigation of the field of the CMB. As it follows from Fig.1 in the 4-beam scheme the directions l_1, l_2 and l_3, l_4 form two bi-ray schemes. Using this fact we introduce the following function

$$G_1(\theta, d, \theta_A) = \left\langle \left[\frac{T(l_1) - T(l_2)}{T} \right]^2 \right\rangle - \left\langle \left[\frac{T(l_3) - T(l_4)}{T} \right]^2 \right\rangle =$$

$$= 2[C(\theta - d, \theta_A) - C(\theta + d, \theta_A)]. \tag{9}$$

For the angles $\theta \approx \theta_R$ and $\theta \approx 2\theta_R$ at $d \ll \theta_R$ the function $G_1(\theta, d, \theta_A)$ is strictly connected with the value of the first derivative of the correlation function $C(\theta, \theta_A)$ with respect to the angle θ:

$$G_1(\theta, d, \theta_A) \approx -4C'(\theta, \theta_A)d, \tag{10}$$

where $C'(\theta, \theta_A) \equiv \frac{dC}{d\theta}$.

One can evaluate the character of the function $G_1(\theta, d, \theta_A)$ in the neighborhood of the point $\theta \approx \theta_R$. In the paper Jørgensen *et al.* (1994) we have demonstrated that

$$G_1(\theta \approx \theta_R, d, \theta_A) \approx \frac{12\varepsilon}{a(1+\varepsilon)^{1/4}} \frac{d}{\theta_R} C_0(0, \theta_A) \left(1 - \frac{3}{2}x^2 + \frac{5}{24}\beta\, x^4 \right), \tag{11}$$

where $a = 9\epsilon^2 + \frac{2+\epsilon}{2(1+\epsilon)^{3/2}}$ and $\beta = \frac{3(5-n)}{40} \left(\frac{\theta_c}{\theta_{ef}} \right)^2$. As one can see from (11), the function $G_1(\theta = \theta_R, d, \theta_A)$ reaches an extremum at the point $\theta = \theta_R$ and has a form of a pronounced peak. This result is confirmed by numerical computations of $G_1(\theta, d, \theta_A)$ which are shown on Fig.4. It is seen that in this case the formation of peak-like anomalies takes place at the points $\theta \approx \theta_R$ and $\theta \approx 2\theta_R$ where the function $G_2(\theta, d, \theta_A)$ has the assymptotics $G_2(\theta \to \theta_R, d, \theta_A) \propto (\theta - \theta_R)$ and $G_2(\theta \to 2\theta_R, d, \theta_A) \propto (\theta - 2\theta_R)$.

We have to mention another important property of the functions $G_1(\theta, d, \theta_A)$ and $G_2(\theta, d, \theta_A)$. As we has shown above, the half-width of the peaks of the function $G_2(\theta, d, \theta_A)$ at $\theta \approx \theta_R$ is determined by the coefficient β (see (8)). For the function $G_1(\theta, d, \theta_A)$ the half-width of the peaks is strictly determined by the angle θ_c (see (11)). Thus, using the properties of the functions $G_1(\theta, d, \theta_A)$ and $G_2(\theta, d, \theta_A)$ at $\theta \approx \theta_R$ we can, in principle, get θ_c and β after estimation of

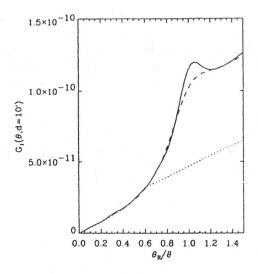

Figure 4: Function $G_1(\theta, d = 10', \theta_A)$ for the model $\Omega = 1$; $\Omega_b = 0.03$; $h = 0.5$; $d = 10'$ and the Harrison-Zel'dovich spectrum. Solid line for $\theta_A = 0$, dashed line for $\theta_A = 0.1\theta_R$, pointed line for $\theta_A = 0.3\theta_R$

the half-width of the peaks. Both these parameters play an important role in the determination of the statistical properties of the random temperature fluctuations of CMB on the celestial sphere (Zabotin and Naselsky (1985); Sazhin (1985); Bond and Efstathiou (1987)).

6   Conclusion

We gave a brief review of a problem of dark matter in the Universe and described a few new techniques of the investigation of the dark matter.

We have considered the peculiarities of the angular correlation function for $\frac{\Delta T}{T}$ at the most interesting angular scales $\theta \simeq 20' - 2°$. We emphasized that the Sakharov oscillations lead (in a CDM model) to appearance of a specific anomalies ("resonances") at the angles θ_R and $\theta_{2R} \approx 2\theta_R$.

The dependence of the "resonances" on the parameters of the cosmological model and the width of an antenna was investigated. We pointed out the special importance of the first "resonance" at θ_R for the comparison with observational data. This first "resonance" is specific for the model with non-baryonic dark matter and it is absent in the model with baryonic dark matter.

We propose a specific method of filtering of the "resonances" in the correlation function, which is reduced to the direct measure of the second derivative of the correlation function in the vicinity of the "resonance".

7 Acknowledgements

This investigation was supported in part by the Danish Natural Science Research Council through grant 11-9640-1 and in part by Danmarks Grundforskningsfond through its support for an establishment of the Theoretical Astrophysics Center.

References.

Alcock, C. *et al.* 1993, Nature 365, 621.

Aubourg, E. *et al.* 1993, Nature 365, 623.

Bahcall, N.A., 1994, Princeton Observatory Preprint, POP-570.

Bond, Y.R. & G. Efstathiou: 1987. MNRAS, **226**, 655.

Carswell, R.F., Rauch, M., Weymann, R.J., Cooke, A.J., and Webb, J.K., 1994, MNRAS, 268, L1.

Dekel, A. 1993, in the book: 2nd Course, current topics in astrofundamental physics, ed. by . N. Sanchez and A. Zichichi, World Scientific, p. 297.

Dekel, A. 1994, Annual Reviews of Astronomy and Astrophysics **32**.

Griest, K. 1994, Particle- and Astro-Physics of Dark Matter, APS Summer Study 94, Snowmass.

Holtzman, J.A., 1989, Ap.J. Suppl., **71**

Jørgensen H.E., E.Kotok, P.Naselsky, I.Novikov, 1994, Astronomy and Astrophysics, in press.

Kormendy, J. and G.R.Knapp ; 1987, in the book: Dark Matter in the Universe, Proceedings of the 117-th IAU symposium, ed. by J.Kormendy and G.Knapp, p.XV, D.Reidel Publishing Company

Kolb, E.W., Turner, M.S. 1990, The Early Universe, Addison-Wesley Publishing Company.

Muciaccia, P.F., S. Mei, G. de Gaspers, and N. Vittorio: 1993, Ap.J. **410**, L61.

Naselsky, P.D., and I.D. Novikov: 1993a, Ap.J.**413**,14.

Naselsky, P.D. and I.D. Novikov: 1993b, Proc. of Int.Conf (Erice).

Olive, K.A., Schramm, D.N., Steigman, G., walker, T.P. 1990, Phys. Lett. B, 236, 254.

Paczynski, B. 1986, ApJ., 304, 1.

Pagel, B.E.J. 1990, in the book: Baryonic Dark Matter, eds. Lynden-Bell, D., Gilmore, G., Kluwer, Dardrecht, p. 237.

Pagel, B. 1994a, Light Elements: From Big Bang to Stars, Invited review at Third Intl. Symp. Nuclei in the Cosmos, Gran Sasso, 8-13 July, 1994, M. Busso, R. Gallino & C. Raiteri (eds.), Amer. Inst. Phys. Publ.

Pagel, B. 1994b, helium in Hii Gegions and Stars, Invited review at ESO/EIPC Workshop on Light Elements, Elba, 23-28 May 1994.

Pagel, B. 1994c, Fine Structure in The main Sequence: Primordial Helium and $\Delta Y/\Delta X$ paper contributed to IAU Symposium 166: Astronomical and Astrophysical Objectives of Sub-milliarcsecond Astronometry, The Hague

Peebles. P.J.E., I.T. Yu: 1970, *Ap.J.* **162**, 815.

Persic, M., Salucci, P. 1992, MNRAS, 258, p. 14.

Sazhin, M.V.: 1985, MNRAS, **216**, 25.

Sakharov, A.D.: 1965, JETP, **49**, 345.

Silk, J.: 1968 *Ap.J.* **151**, 459.

Smoot, G.F., 1994, in the book: Present and Future of the Cosmic Microwave Background, ed. by J.L.Sanz, E.Martinez-Gonsalez, L.Cayon. p.67, Springer-Verlag.

Songaila, A., Cowie, L.L., Hogan, C.J., and Rugers, M. 1994, Nature, 368, 599.

Starobinskii, A.A.: Sov.Astronomy Lett. **14** (3), 166, 1988.

Steigman, G., 1994, An Introduction to Cosmological Dark Matter, Preprint, The Ohio State University, Theoretical Astrophysics.

Steigman, G., Tosi, M. 1992, ApJ, 401, 150.

Tyson, J.A. 1994, in the book: Cosmology and Large Scale Structure, Les Houches, August 1993, ed. R. Schaeffer, Elsevier Science Publishers B.V.

Tyson, J.A., Valders, F., and Wenk, R.A. 1990, ApJ, 349, L1.

Watson, R.A., Gutierrez de la Cruz, C.M.: Ap.J. **413**, L5, 1993.

Zabotin, N. and P. Naselsky: 1985: Sov. Astron **29**, 614.

Zel'dovich, Ya.B. and I.D. Novikov: 1983, *The Structure and Evolution of the Universe.* (The University of Chicago Press.1983) v.2.

Zwicky, F. 1933, helv. Phys. Acta., 6, 110.

The Hot Big Bang and Beyond

Michael S. Turner

Departments of Physics and of Astronomy & Astrophysics
Enrico Fermi Institute, The University of Chicago, Chicago, IL 60637-1433

NASA/Fermilab Astrophysics Center
Fermi National Accelerator Laboratory, Batavia, IL 60510-0500

ABSTRACT

The hot big-bang cosmology provides a reliable accounting of the Universe from about 10^{-2} sec after the bang until the present, as well as a robust framework for speculating back to times as early as 10^{-43} sec. Cosmology faces a number of important challenges; foremost among them are determining the quantity and composition of matter in the Universe and developing a detailed and coherent picture of how structure (galaxies, clusters of galaxies, superclusters, voids, great walls, and so on) developed. At present there is a working hypothesis—cold dark matter—which is based upon inflation and which, if correct, would extend the big bang model back to 10^{-32} sec and cast important light on the unification of the forces. Many experiments and observations, from CBR anisotropy experiments to Hubble Space Telescope observations to experiments at Fermilab and CERN, are now putting the cold dark matter theory to the test. At present it appears that the theory is viable only if the Hubble constant is smaller than current measurements indicate (around $30 \, \mathrm{km \, s^{-1} \, Mpc^{-1}}$), or if the theory is modified slightly, e.g., by the addition of a cosmological constant, a small admixture of hot dark matter ($5 \, \mathrm{eV}$ "worth of neutrinos"), more relativistic particles, or a tilted spectrum of density perturbations.

M. Kafatos and Y. Kondo (eds.), Examining the Big Bang and Diffuse Background Radiations, 301–320.

1 Successes

The success of the hot big-bang cosmology (or standard cosmology as it is known) is simple to describe: It provides a reliable and tested accounting of the Universe from a fraction of a second after the bang (temperatures of order a few MeV) until the present 15 Billion years later (temperature 2.726 K). When supplemented by the standard model of particle physics and various ideas about physics at higher energies (e.g., supersymmetry, grand unification, and superstrings) it provides a sound foundation for speculations about the Universe back to 10^{-43} sec after the bang (temperatures of 10^{19} GeV) and perhaps even earlier [1].

The fundamental observational data that support the standard cosmology are: the universal expansion (Hubble flow of galaxies); the cosmic background radiation (CBR); and the abundance of the light elements D, ^{3}He, ^{4}He, and ^{7}Li. The Hubble law ($z \simeq v/c \simeq H_0 d$) has been tested to a redshift $z \sim 0.05$ [2] and the highest redshift object is a QSO with $z = 4.90$. (One plus redshift is the size of the Universe today relative to its size at the time of emission, $1 + z = R_0/R_E$; R is the cosmic scale factor, or relative size of the Universe).

The surface of last scattering for the CBR is the Universe at an age of a few hundred thousand years ($T \sim 0.3$ eV and redshift $z \sim 1100$). COBE has determined its temperature to be 2.726 ± 0.005 K and constrains any deviations from a black-body spectrum to be less than 0.03% [3]. The CBR temperature is very uniform: the difference between two points separated by angles from arcminutes to 90° is less than $300 \mu K$, indicating that the Universe had a very smooth beginning. There is a dipole anisotropy in the CBR temperature of about 3 mK, due to our motion with respect to the cosmic rest frame (the "peculiar velocity" of the Local Group is 620 km s^{-1} toward the constellation Leo), and temperature differences on angular scales from 0.5° to 90° have been detected by about ten experiments at the level of about $30 \mu K$ [4].

The abundance of the light elements, which range from about 24% for ^{4}He to 10^{-5} for D and ^{3}He and 10^{-10} for ^{7}Li are consistent with the predictions of big-bang nucleosynthesis. The synthesis of the light elements occurred when the Universe was of order seconds old and the temperature was of order MeV. Big-bang nucleosynthesis is the earliest test of the standard cosmology, and it passes it with flying colors [5].

Finally, the standard cosmology provides a general framework for understanding how the Universe evolved from a very smooth beginning to the abundance of structure observed today—galaxies, clusters of galaxies, superclusters, voids, great walls and so on. Small (primeval) variations in the matter density ($\delta\rho/\rho \sim 10^{-5}$) were amplified by gravity (the Jeans' instability in the expanding Universe) eventually resulting in the structure seen today [6]. The CBR temperature fluctuations detected on angular scales from $0.5°$ to $90°$ are strong evidence for the existence of primeval density fluctuations.

In addition to accounting for the evolution of the Universe from 0.01 sec onward, the standard cosmology provides a sound framework for speculating about even earlier times—at least back to the Planck time (10^{-43} sec and temperature of order 10^{19} GeV). Of course, advances in particle theory have played an important role here.

According to the standard model of particle physics the fundamental particles are point-like quarks and leptons whose interactions are weak enough to treat perturbatively. The cosmological implications of this are profound: The Universe at temperatures greater than about 150 MeV (times earlier than 10^{-5} sec) consisted of a hot, dilute gas of quarks, leptons, and gauge bosons (photons, gluons, and at high enough temperatures W and Z bosons, the carriers of the electromagnetic, strong and weak forces). The standard model of particle physics, which has been tested up to energies of several hundred GeV, provides the microphysics needed to discuss times as early as 10^{-11} sec. In addition, it provides a firm platform for speculations about the unification of forces and particles (e.g., supersymmetry and grand unification), and in turn, the necessary microphysics for extending cosmological speculations back to the Planck epoch. Earlier than the Planck time a quantum description of gravity is needed, and superstring theory is a good candidate for such.

While the hot big-bang cosmology and modern particle theory allow "sensible"—and very interesting—speculations about the early Universe, there is no evidence *yet* that any of these speculations is correct. However, contrast this with the situation before the early 1970s. The count of "elementary particles" (baryons and mesons) had exceeded 100 and was growing exponentially with mass; this, the strength of their interactions and their finite sizes precluded any sensible speculation about the Universe at times earlier than about 10^{-5} sec [7].

2 Challenges

Cosmology is not without its challenges. In its success, the hot big-bang model has allowed cosmologists to ask even deeper questions. They include: What is the quantity and composition of the ubiquitous dark matter in the Universe? What is the origin and nature of the primeval density perturbations that seeded structure and precisely how did the structure form? What is the origin of the cosmic asymmetry between matter and antimatter? Why is the observed portion of the Universe so smooth and flat? And does this mean that the entire Universe is the same? Are there observational consequences of the phase transitions that the Universe has undergone (transition from quarks to nucleons and related particles, electroweak symmetry breaking, and possibly others) during its earliest moments? Are there observable consequences of the quantum-gravity epoch? Why does the Universe have four dimensions? What caused the expansion in the first place?

The first two of these challenges, the nature of the dark matter and the details of structure formation, are in my opinion the most pressing and may well be resolved soon. Thus they offer an excellent opportunity for extending the big-bang cosmology back to much earlier times.

That is not to say that the other challenges are not important or do not have potential for advancing our understanding. In addition, there more practical challenges that I have not listed; for example, a precise determination of the three traditional parameters used to describe "our world model," the Hubble constant, the deceleration parameter, and the cosmological constant, or an explanation for the primeval magnetic fields required to seed the magnetic fields seen throughout the Universe today.

2.1 Discard the big bang?

There are few who believe the big bang faces challenges of such enormity that they will led to its downfall [8]. For example, if the Hubble constant is as large as some determinations indicate, say around $80 \, \mathrm{km \, s^{-1} \, Mpc^{-1}}$ and the oldest stars are as old as some determinations indicate, say around $16 \, \mathrm{Gyr}$, then a real dilemma exists because without recourse to a cosmological constant the time back to the back is less than $12 \, \mathrm{Gyr}$ [9].

Before the COBE discovery of CBR anisotropy in 1992 [10], some argued that the absence of anisotropy precluded inhomogeneity of a size large enough to seed all the structure seen today. The big-bang has weathered that storm: Fluctuations in the CBR temperature have been detected and are now seen on scales from 0.5° to 90°. In fact, careful calculations indicate that if anything the level of temperature fluctuations seen is slightly larger than is expected, given the structure seen today [4, 11].

The only two competitors to the big bang are the quasi steady-state model [12] and the plasma universe model. At the moment the problems that these models face seem far more daunting: themalization of starlight to produce 2.726 K black body background with no spectral distortion (quasi steady-state) and the formulation of a model definite enough to be tested (plasma universe). Until these models (or another model) can account for the cosmological data that have been firmly established (expansion, CBR, light elements, and structure formation), the standard cosmology is without a serious competitor.

2.2 Dark Matters

An accurate inventory of matter in the Universe still eludes cosmologists. What we do know is: (i) luminous matter (i.e., matter closely associated with bright stars) contributes a fraction of the critical density that is about $0.003h^{-1}$ [13]; (ii) based upon big-bang nucleosynthesis baryons contributions a fraction of critical density between $0.009h^{-2}$ and $0.022h^{-2}$ [5], which for a generous range of the Hubble constant corresponds to between about 0.01 and 0.15 of the critical density; (iii) there are indications that the fraction of critical density contributed by all forms of matter is *at least* $0.1 - 0.3$ [14]—flat rotation curves of spiral galaxies, virial mass determinations of rich clusters—and perhaps around the critical density—the peculiar motions of galaxies, cluster mass determinations based upon gravitational lensing and x-ray measurements [14, 15]. (Here the Hubble constant $H_0 = 100h \, \mathrm{km \, s^{-1} \, Mpc^{-1}}$ and the critical density $\rho_{\mathrm{crit}} = 3H_0^2/8\pi G = 1.88h^2 \times 10^{-29} \, \mathrm{g \, cm^{-3}} \simeq 1.05h^2 \times 10^4 \, \mathrm{eV \, cm^{-3}}$.)

From this one concludes that: (i) most of the matter in the Universe is dark; (ii) most of the baryons are dark; (iii) the dark matter is not closely associated with bright stars, i.e., it is more diffusely distributed, e.g., in the extended halos of spiral galaxies; and (iv) if the total mass density is greater than about 20% of the critical density, then there must be another form of matter since baryons can at most account for 15% of the critical density (and only for a low value of the Hubble constant).

The case for $\Omega_0 \gtrsim 0.2$ and nonbaryonic dark matter receives additional support, albeit indirectly, from other lines of reasoning. First, it is difficult to reconcile all the data concerning the formation of structure in the Universe with a theory that has no nonbaryonic dark matter (the one model that *may* be able to do so is Peebles' primeval baryon isocurvature model or PBI [16]). Second, the most compelling and comprehensive theory of the early Universe, inflation [17], predicts a flat Universe (total energy density equal to the critical density) and thus requires something other than baryons. Third, since the deviation of Ω from unity grows with time, if Ω_0 is not equal to unity, the epoch when Ω_0 just begins to deviate significantly from one is a special epoch and is today(!) (this is often called the Dicke-Peebles timing argument).

Last but not least, there are three compelling candidates for the nonbaryonic dark matter: an axion of mass between 10^{-6} eV and 10^{-4} eV; a neutralino of mass between 10 GeV and 1000 GeV; and a light neutrino species of mass between 10 eV and about 50 eV [18]. By compelling, I mean these particles arose out of efforts to unify the forces of Nature, and the fact that a particle was predicted whose relic mass density is close to critical is a bonus. This may be the "Grand Hint" or the "Great Misdirection."

For the axion, the underlying particle physics is Peccei-Quinn symmetry which is the most attractive solution to the so-called strong-CP problem (the fact that standard model of particle physics predicts the electric dipole moment of the neutron to be almost ten orders of magnitude larger than the current upper limit). For the neutralino, it is supersymmetry, the symmetry that relates fermions and bosons and which helps to explain the large discrepancy between the weak scale (300 GeV) and the Planck scale and may hold the key to unifying gravity with the other forces. Unlike the axion or the neutralino, neutrinos are known to exist, come in three varieties, and have a relic abundance known to three significant figures (113 cm^{-3} per species); the only issue is their mass. Almost all attempts to unify the forces and particles of nature lead to the prediction that neutrinos have mass, often in the "eV range" (meaning anywhere from 10^{-6} eV or smaller to keV).

The axion and neutralino are referred to as "cold dark matter" because they move very slowly (neutralinos because they are heavy and axions because they were produced in the early Universe with very small momenta). Neutrinos on the other hand are referred to as "hot dark matter" because they move rapidly (due to their small mass). The distinction between the two is crucial for structure formation: at early times neutrinos can "run out" of overdense regions and into underdense regions, damping density perturbations on scales smaller than those corresponding to superclusters. This means that in the absence of additional seed perturbations that don't involve neutrinos (e.g., cosmic string) the sequence of structure formation in a hot dark matter universe proceeds from the "top down:" objects like superclusters form first and then fragment into smaller objects (galaxies and the like). Because there is now much evidence that "small objects" (galaxies, quasars, neutral hydrogen clouds, and clusters) were ubiquitous at redshifts from 1 to 4, and "large objects" are just forming today, hot dark matter (without additional perturbations) is not viable.

To end on a sober note, at present the data can neither prove nor disprove either: (i) $\Omega_0 = \Omega_B \simeq 0.15$; (ii) $\Omega_0 = 1$ with $\Omega_B \sim 0.05$ and $\Omega_{CDM} \sim 0.95$. (In the first case the Hubble constant must be near its lower extreme since the nucleosynthesis measurement of $\Omega_B = 0.009h^{-2} - 0.022h^{-2}$.) In any case, I will devote the rest of this paper to the second, more radical possibility.

2.3   Coherent picture of structure formation

Because the energy densities of matter (baryons + CDM?) and radiation (photons, light neutrinos, and at early times all the other particles in the thermal plasma) evolve differently, R^{-3} for matter and R^{-4} for radiation, the energy density in radiation exceeded that in matter earlier at early times, $t \lesssim t_{EQ} \sim 10^4 \, \mathrm{yr}$ ($T \gtrsim T_{EQ} \sim 5 \, \mathrm{eV}$ and $R \lesssim R_{EQ} \sim 3 \times 10^{-5} R_{today}$). Moreover, matter density perturbations do not grow during the radiation-dominated era, and thus the formation of structure did not begin in earnest until the epoch of matter-radiation equality. After that, (linear) perturbations in the matter grow as the scale factor, for a total (linear) growth factor of around 30,000. This factor sets the characteristic amplitude of density perturbations, about $few \times 10^{-5}$ (nonlinear structures have formed by the present) and thus the expected size of temperature fluctuations in the CBR (density perturbations lead to comparable sized fluctuations in the CBR temperature).

The detection of CBR anisotropy at the level of about 10^{-5} validates the gravitational instability picture of structure formation. This success should be viewed in the same way that the evidence for a large primeval mass fraction of ^{4}He validated the the basic idea of primordial nucleosynthesis in the late 1960s. From this early success, big-bang nucleosynthesis developed into a coherent and detailed explanation for the abundances of D, ^{3}He, ^{4}He and ^{7}Li, and now provides the earliest test of the big bang, the most reliable determination of the baryon density, and an important probe of particle physics. It is not unreasonable to hope that a detailed and coherent picture of structure formation develop and will lead to similar advances in our understanding of the Universe.

The two crucial elements must underlay any detailed picture: specification of the quantity and composition of the dark matter and the nature of the density perturbations. With regard to the latter, what is wanted is a mathematical description of the spectrum of density perturbations. For example, the Fourier components δ_k of the density field and their statistical properties.

At present there are three viable theories: cold dark matter models; topological-defect models [19]; and the primeval baryon isocurvature model (PBI) [16]. The effort being brought to bear on this problem—both experimental and theoretical—is great, and I am confident that *at least* two of these models, if not all three (!), will be falsified soon. It is my view that only cold dark matter will survive the next cut, but of course others may hold a different opinion.

Topological defect models, where the seeds are cosmic string, monopoles or textures produced in an early Universe phase transition and the dark matter is either neutrinos (cosmic string) or cold dark matter (textures), seem to predict CBR anisotropy on the degree scale that is significantly less than that measured. In addition, when normalized to the COBE measurements of anisotropy, they require a high level of "bias;" bias refers to the discrepancy between the light and mass distributions, $b \simeq (\delta n_{\mathrm{GAL}}/n_{\mathrm{GAL}})/(\delta\rho/\rho)$, which is generally believe to be of order $1-2$. Much of the difficulty in assessing the defect models is on the theoretical side; density perturbations are constantly being produced as the defect network evolves and thus cannot easily be described by Fourier components whose evolution is simple.

The basic philosophy behind the PBI model is to explain the formation of structure by using "what is here," rather then what early-Universe theorists (like myself) hope is here! The parameters for PBI are: $\Omega_0 = \Omega_B \sim 0.2$ and $H_0 \sim 70\,\mathrm{km\,s^{-1}\,Mpc^{-1}}$. An arbitrary power-law spectrum of fluctuations in the local baryon number (cut off at small scales to avoid difficulties with primordial nucleosynthesis) is postulated and its parameters (slope and normalization) are determined by the data (CBR fluctuations and large-scale structure). PBI has some serious problems: the baryon density violates the nucleosynthesis bound by a wide margin ($\Omega_B h^2 \sim 0.1 \gg 0.02$; it is difficult to make PBI consistent with the measurements of CBR anisotropy [20]. To wit, Peebles has considered variations on the basic theme [21] (e.g., adding a cosmological constant, or even cold dark matter). At the very least PBI provides a useful model against which scenarios that postulate exotic dark matter can be compared; at best, it may represent our Universe.

3   Inflation and Cold Dark Matter

Inflation represents a bold attempt to extend the standard big-bang cosmology to times as early as 10^{-32} sec and to resolve some of the most fundamental questions in cosmology. In particular, inflation addresses squarely both the dark matter and structure formation problems, as well as providing an explanation for the flatness and smoothness of the Universe. If successful, inflation would be a truly remarkable addition to the standard cosmology.

At present there is no standard model of inflation; however, there are many viable models, all based on well defined speculations about physics at energy scales of around 10^{14} GeV and higher [22]. Inflation makes three robust predictions: (1) spatially flat Universe [23]; (2) nearly scale-invariant spectrum of density (scalar metric) perturbations [24]; (3) nearly scale-invariant spectrum of gravity waves (tensor metric perturbations) [25].

With regard to metric perturbations; they are imprinted during inflation, arising from quantum mechanical fluctuations excited on extremely small scales ($\lesssim 10^{-23}$ cm), which are stretched to astrophysical scales ($\gtrsim 10^{25}$ cm) by the tremendous growth in the scale factor during inflation ($\gtrsim 10^{25}$). In almost all models of inflation the statistics of the perturbations are gaussian, and the Fourier power spectrum, $P(k) \equiv |\delta_k|^2$, completely specifies the statistical properties of the density field.

While the metric perturbations are predicted to nearly scale invariant, the small deviations that can occur encode much about the underlying inflationary model. Likewise, the amplitudes of the metric perturbations are model dependent and hold equally important information. (Scale-invariant density perturbations means fluctuations in the gravitational potential that are equal on all scales at early times; for the gravitational waves, scale invariant means that all gravity waves cross the horizon with equal amplitude.)

The first prediction means the total energy density (including matter, radiation, and the vacuum energy density associated with a cosmological constant) is equal to the critical density, that is $\Omega_0 = 1$. Coupled with our knowledge of the baryon density, this implies that the bulk of matter in the Universe (95% or so) must be nonbaryonic. The two simplest possibilities are hot dark matter or cold dark matter. Structure formation with hot dark matter has been studied, and, sadly, does not work; thus we are led to to cold dark matter.

For cold dark matter there is no damping of perturbations on small scales, and structure is built from the "bottom up:" Clumps of dark matter and baryons continuously merge to form larger objects. "Typical galaxies" are formed at redshifts $z \sim 1-2$; "rare objects" such as quasars and radio galaxies can form earlier from regions where the density perturbations have larger than average amplitude. Clusters form in the very recent past (redshifts less than order unity), and superclusters are just forming today. Voids naturally arise as regions of space are evacuated to form objects [26].

3.1 Almost, but is something missing?

Broadly speaking, testing the cold dark matter scenario involves measuring the quantity, composition, and distribution of dark matter and determining the spectrum of density perturbations. I have already discussed the current state of our knowledge of dark matter. While a host of observations provide information about the primeval spectrum of density perturbations, measurements of the anisotropy of the CBR and mapping the distribution of matter today (as traced by bright galaxies) are perhaps most crucial. (For reference, perturbations on scales of about 1 Mpc correspond to galactic sized perturbations, on 10 Mpc to cluster size perturbations, on 30 Mpc to the large voids, and 100 Mpc to the great walls.)

CBR anisotropy probes the power spectrum on large scales. The CBR temperature difference measured on a given angular scale is related to the power spectrum on a given length scale: $\lambda \sim (\theta/\deg)100h^{-1}\,\mathrm{Mpc}$. Since the COBE detection, a host of ground-based and balloon-borne experiments have also detected CBR anisotropy, on scales from about $0.5°$ to $90°$, at the level of around $30\mu K$ ($\delta T/T \sim 10^{-5}$). The measurements are consistent with the predictions of cold dark matter, though there are still large statistical uncertainties as well as concerns about contamination by foreground sources [4]. There is a great deal of experimental activity (more than ten groups), and measurements in the near future should improve the present situation significantly. The CBR contains important information on angular scales down to about 0.1 deg (anisotropy on smaller angular scales is washed out due to the finite thickness of the last scattering surface). A follow-on to COBE, being studied in both Europe and the US, and a variety of earth-based and balloon-based experiments should hopefully map CBR anisotropy on scales from 0.1 deg to 90 deg in the next decade.

The COBE detection of CBR anisotropy not only provided the first evidence for the existence of primeval density perturbations, but also an unambiguous way to normalize the spectrum of density perturbations: Given the shape of the power spectrum (for cold dark matter, approximately scale invariant) the COBE measurement (on a scale of around $10^3 h^{-1}$ Mpc) ties down the spectrum on all scales. This leads to definite predictions that can be tested by other CBR measurements and observations of large-scale structure.

The comparison of predictions for structure formation with present-day observations of the distribution of galaxies is very important, but fraught with difficulties. Theory most accurately predicts "where the mass is" (in a statistical sense) and the observations determine where the light is. Redshift surveys probe present-day inhomogeneity on scales from around one Mpc to a few hundred Mpc, scales where the Universe is nonlinear ($\delta n_{GAL}/n_{GAL} \gtrsim 1$ on scales $\lesssim 8h^{-1}$ Mpc) and where astrophysical processes undoubtedly play an important role (e.g., star formation determines where and when "mass lights up," the explosive release of energy in supernovae can move matter around and influence subsequent star formation, and so on). The distance to a galaxy is determined through Hubble's law ($d = H_0^{-1} z$) by measuring a redshift; peculiar velocities induced by the lumpy distribution of matter are significant and prevent a direct determination of the actual distance. There are the intrinsic limitations of the surveys themselves: they are flux not volume limited (brighter objects are seen to greater distances and vice versa) and relatively small (e.g., the CfA slices of the Universe survey contains only about 10^4 galaxies and extends to a redshift of about $z \sim 0.03$). Last but not least are the numerical simulations which bridge theory and observation; they are limited dynamical range (about a factor of 100 in length scale) and in microphysics (in the largest simulations only gravity, and in others only a gross approximation to the effects of hydrodynamics/thermodynamics).

This being said, redshift surveys do provide an important probe of the power spectrum on small scales ($\lambda \sim 1 - 300\,\mathrm{Mpc}$). Even with their limitations redshift surveys (as well as other data) indicate that while the simplest version of COBE-normalized cold dark matter is in broad agreement with the data, the shape of the power spectrum as well as its amplitude on small scales is not quite right [11, 27]. At least three possibilities come to mind: (i) the comparison of numerical simulations and the observations is still too primitive to draw firm conclusions; (ii) cold dark matter has much, but not all, of the "truth;" or (iii) cold dark matter has been falsified.

For three reasons I believe that it is worthwhile exploring possibility (ii), namely that something needs to be added to cold dark matter. First, cold dark matter is such an attractive theory and part of a bold attempt to extend greatly the standard cosmology. Second, many observations seem to point to the same problem (e.g., the abundance of x-ray clusters and the cluster-cluster correlation function). Third, there are other reasons to believe that the Universe is more complicated than the simplest model of cold dark matter.

3.2   Five cold dark matter models

Somewhat arbitrarily, standard cold dark matter has come to mean: precisely scale-invariant density perturbations; baryons + CDM only; and Hubble constant of $50\,\mathrm{km\,s^{-1}\,Mpc^{-1}}$ (to ensure a sufficiently aged Universe with a Hubble constant still within the range of observations). This is the vanilla or default model, which, when normalized to COBE has too much power on small scales and the wrong spectral shape on larger scales.

The spectrum of density perturbations today depends not only upon the primeval spectrum (and the normalization on large scales provided by COBE), but also upon the energy content of the Universe. While the fluctuations in the gravitational potential were initially approximately scale invariant, the fact that the Universe evolved from an early radiation-dominated phase to a matter-dominated phase imposes a characteristic scale on the spectrum of density perturbations seen today; that scale is determined by the energy content of the Universe (the characteristic scale $\lambda_{\mathrm{EQ}} \sim 10h^{-1}\,\mathrm{Mpc}\,g_*^{1/2}/\Omega_{\mathrm{mat}}$ where g_* counts the relativistic degrees of freedom and $\Omega_{\mathrm{matter}} = \Omega_B + \Omega_{\mathrm{CDM}}$). In addition, if some of the nonbaryonic dark matter is neutrinos, they will inevitably suppress power on small scales through freestreaming. With this in mind, let me discuss the small modifications of cold dark matter that improve its agreement with the observations.

(1) Low Hubble Constant + cold dark matter (LHC CDM). Remarkably, simply lowering the Hubble constant to around $30 \, \text{km s}^{-1} \, \text{Mpc}^{-1}$ solves all the problems of cold dark matter. Recall, the critical density $\rho_{\text{crit}} \propto H_0^2$; lowering H_0 lowers the matter density and postpones matter-radiation equality, which has precisely the desired effect on the spectrum of perturbations. It has two other added benefits: it makes the expansion age of the Universe comfortably consistent with the ages of the oldest stars and raises the baryon fraction of critical density to a value that is consistent with that measured in x-ray clusters (see below). Needless to say, such a small value for the Hubble constant flies in the face of current observations; further, it illustrates the fact that the problems of cold dark matter get even worse for the larger values of H_0 that have been determined by recent observations [9].

(2) Hot + cold dark matter (νCDM). Adding a small amount of hot dark matter can suppress density perturbations on small scales; of course, too much leads back to the longstanding problems of hot dark matter. The amount required is about 20%, corresponding to about "5 eV worth of neutrinos" (i.e., one species of mass 5 eV, or two species of mass 2.5 eV, and so on). This admixture of hot dark matter rejuvenates cold dark matter provided the Hubble constant is not too large, $H_0 \lesssim 55 \, \text{km s}^{-1} \, \text{Mpc}^{-1}$.

(3) Cosmological constant + cold dark matter (ΛCDM). (A cosmological constant corresponds to a uniform energy density, or vacuum energy.) Shifting 60% to 80% of the critical density to a cosmological constant lowers the matter density and has the same beneficial effect as a low Hubble constant. In fact, a Hubble constant as large as $80 \, \text{km s}^{-1} \, \text{Mpc}^{-1}$ can be tolerated. In addition, the cosmological constant allows the age problem to solved even if the Hubble constant is large, addresses the fact that few measurements of the mean mass density give a value as large as the critical density (most measurements of the mass density are insensitive to a uniform component), and allows the fraction of matter in baryons to be large (see below). Not everything is rosy; cosmologists have invoked a cosmological constant twice before to solve their problems (Einstein to obtain a static universe and Bondi, Gold, and Hoyle to solve the earlier age crisis when H_0 was thought to be $250 \, \text{km s}^{-1} \, \text{Mpc}^{-1}$). Further, particle physicists can still not explain why the energy of the vacuum is not at least 50 (if not 120) orders of magnitude larger

than the present critical density.

(4) Extra relativistic particles + cold dark matter (τCDM). The epoch of matter-radiation equality can also be delayed by raising the level of radiation. In the standard cosmology the radiation content today consists of photons + three (undetected) cosmic seas of neutrinos (corresponding to $g_* \simeq 3.36$). While we have no direct determination of the radiation beyond that in the CBR, there are at least two problems: (i) what are the additional relativistic particles?; and (ii) can additional radiation be added without upsetting the successful predictions of primordial nucleosynthesis which depend critically upon the energy density of relativistic particles. The simplest way around these problems is an unstable tau neutrino (mass anywhere between a few keV and a few MeV) whose decays produce the radiation. This fix can tolerate a larger Hubble constant, though at the expense of more radiation.

(5) Tilted cold dark matter (TCDM). While the spectrum of density perturbations in most models of inflation is very nearly scale invariant, there are models where the deviations are significant and lead to smaller fluctuations on small scales. Further, not only do density perturbations produce CBR anisotropy, but so do the gravitational waves; if gravity waves account for a significant part of the CBR anisotropy, the level of density perturbations must be lowered. A combination of tilt and gravity waves can solve the problem of too much power on small scales, but does not seem to address the shape problem as well as the other fixes.

In evaluating these better fit models, one should keep the words of Francis Crick in mind (loosely paraphrased): A model that fits all the data at a given time is necessarily wrong, because at any given time not all the data are correct(!). ΛCDM provides an interesting example; when I discussed it in 1990, I called it the best-fit Universe, but not the best motivated and was certain it would fall by the wayside [28]. In 1995, it is still probably the best-fit model.

Let me end by defending the other point of view, namely, that to add something to cold dark matter is not unreasonable, or even as some have said, a last gasp effort to saving a dying theory. Standard cold dark matter was a starting point, similar to early calculations of big-bang nucleosynthesis. It was always appreciated that the inflationary spectrum of density perturbations was not exactly scale invariant [29] and that the Hubble constant was unlikely to be exactly $50 \, \mathrm{km \, s^{-1} \, Mpc}$. As the quality and quantity of data improve, it is only sensible to refine the model, just as has been done with big-bang nucleosynthesis. Cold dark matter seems to embody much of the "truth." The modifications suggested all seem quite reasonable (as opposed to contrived). Neutrinos exist; they are expected to have mass; there is even some experimental data that indicates they do have mass. It is still within the realm of possibility that the Hubble constant is less than $50\,\mathrm{km\,s^{-1}\,Mpc^{-1}}$, and if it is as large as $70\,\mathrm{km\,s^{-1}\,Mpc^{-1}}$ to $80\,\mathrm{km\,s^{-1}\,Mpc^{-1}}$ a cosmological constant seems inescapable based upon the age problem. There is no data that can preclude more radiation than in the standard cosmology and deviations from scale invariance were always expected.

4 The Future

4.1 Testing and discriminating

The stakes for cosmology are high: if correct, inflation/cold dark matter represents a major extension of the big bang and our understanding of the Universe, which can't help but shed light on the fundamental physics at energies of order $10^{14}\,\mathrm{GeV}$ or higher.

How and when we will have definitive tests of cold dark matter? Because of the large number of measurements that are being carried out and can have significant impact, I believe sooner rather than later. The list is long: CBR anisotropy; larger redshift surveys (e.g., the Sloan Digital Sky Survey will have 10^6 redshifts); direct searches for the nonbaryonic in our neighborhood (e.g., axion and neutralino searches) and baryonic dark matter (microlensing); x-ray studies of galaxy clusters; the use of back-lit gas clouds (quasar absorption line systems) to study the Universe at high redshift; galactic evolution (as revealed by deep images of the sky taken by the Hubble Space Telescope and Keck 10 meter telescope); a variety of measurements of H_0 and q_0; mapping of the peculiar velocity field at large redshifts through the Sunyaev-Zel'dovich effect; dynamical estimates of the mass density (weak gravitational lensing, large-scale velocity fields, and so on); age determinations of the Universe; gravitational lensing; searches for supersymmetric particles (at accelerators) and neutrino oscillations (at accelerators, solar-neutrino detectors, and other large underground detectors); searches for high-energy neutrinos from neutralino annihilates in the sun using large underground detectors; and on and on. Consider the possible impact of a few specific examples.

A definitive determination that H_0 is greater than $55 \, \mathrm{km \, s^{-1} \, Mpc^{-1}}$ would falsify LHC CDM and νCDM. Likewise, a definitive determination that H_0 is $75 \, \mathrm{km \, s^{-1} \, Mpc^{-1}}$ or larger would necessitate a cosmological constant. A flat Universe with a cosmological constant has a very different deceleration parameter than one dominated by matter, $q_0 = -1.5\Omega_\Lambda + 0.5 \sim -(0.4 - 0.7)$ compared to $q_0 = 0.5$, and this could be settled by galaxy number counts or numbers of lensed quasars. The level of CBR anisotropy in τCDM and LHC CDM on the 0.5° scale is about 50% larger than the other models, which should be easily measurable. If neutrino-oscillation experiments were to provide evidence for a neutrino of mass $5 \, \mathrm{eV}$ (two of mass $2.5 \, \mathrm{eV}$) νCDM would seem almost inescapable.

A map of the CBR with $0.5° - 1°$ resolution could separate the gravity-wave from density perturbation contribution to the CBR anisotropy and provide evidence for the third robust prediction of inflation. Further, mapping CBR anisotropy on these scales or slightly smaller offers the possibility determining the geometry of the Universe (the position of the "Doppler" peak scales as $0.5°/\sqrt{\Omega_0}$ [30]).

X-ray observations of rich clusters are able to determine the ratio of hot gas (baryons) to total cluster mass (baryons + CDM) (by a wide margin, most of the baryons "seen" in cluster are in the hot gas). To be sure there are assumptions and uncertainties; the data at the moment indicate that this ratio is $0.07h^{-3/2}$ [31]. If clusters provide a fair sample of the universal mix of matter, then this ratio should equal $\Omega_B/(\Omega_B + \Omega_{CDM}) \simeq (0.009 - 0.022)h^{-2}/(\Omega_B + \Omega_{CDM})$. Since clusters are large objects they should provide an approximately fair sample. Taking the numbers at face value, cold dark matter is consistent with the cluster gas fraction provided either: $\Omega_B + \Omega_{CDM} = 1$ and $h \sim 0.3$ or $\Omega_B + \Omega_{CDM} \sim 0.3$ and $h \sim 0.7$, favoring LHC CDM or ΛCDM.

If cold dark matter is correct, then a significant, if not dominant, fraction of the dark halo of our galaxy should be cold dark matter (the halos of spiral galaxies are not large enough to guarantee that they represent a fair sample). Direct searches for faint stars have failed to turn up enough to account for the halo [32]. Over the past few years, microlensing has been used to search for dark stars (stars below the $0.08M_\odot$ limit for hydrogen burning). Five stars in the LMC have been observed to change brightness in a way consistent with their being microlensed by dark halo objects passing along the line of sight.

While the statistics are small, and there are uncertainties concerning the size of dark halo, these results indicate that only a small fraction (5% to 30%) of the dark halo is in the form of dark stars [33].

4.2 Reconstruction

If cold dark matter is shown to be correct, then a window to the very early Universe ($t \sim 10^{-34}$ sec) will have been opened. While it is certainly premature to jump to this conclusion, I would like to illustrate one example of what one could hope to learn. As mentioned earlier, the spectra and amplitudes of the the tensor and scalar metric perturbations predicted by inflation depend upon the underlying model, to be specific, the shape of the inflationary scalar-field potential. (Inflation involves the classical evolution of a scalar field ϕ rolling down its potential energy curve $V(\phi)$.) If one can measure the power-law index of the scalar spectrum and the amplitudes of the scalar and tensor spectra, one can recover the value of the potential and its first two derivatives around the point on the potential where inflation took place [34]. (Measuring the power-law index of the tensor perturbations in addition, allows an important consistency check of inflation.) Reconstruction of the inflationary scalar potential would shed light both on inflation as well as physics at energies of the order of 10^{14} GeV.

4.3 Concluding remarks

We live in exciting times. We have a cosmological model that provides a reliable accounting of the Universe from 0.01 sec until the present. Together with the standard model of particle physics it provides a framework for both asking deeper questions about the Universe and making sensible speculations. With inflation and cold dark matter we may be on the verge of a very significant extension of the standard cosmology. Most importantly, the data needed to test the cold dark matter theory is coming in at a rapid rate. At the very least we should soon know whether we are on the right track or if it's back to the drawing board.

This work was supported in part by the DOE (at Chicago and Fermilab) and by the NASA through grant NAGW-2381 (at Fermilab).

References

[1] For textbooks on modern cosmology see e.g., E.W. Kolb and M.S. Turner, *The Early Universe* (Addison-Wesley, Redwood City, CA, 1990) or P.J.E. Peebles, *Principles of Physical Cosmology* (Princeton University Press, Princeton, NJ, 1993).

[2] See e.g., J. Mould et al., *Astrophys. J.* **383**, 467 (1991).

[3] J. Mather et al., *Astrophys. J.* **420**, 439 (1994).

[4] See e.g., M. White, D. Scott, and J. Silk, *Ann. Rev. Astron. Astrophys.* **32**, 319 (1994).

[5] C. Copi, D.N. Schramm and M.S. Turner, *Science* **267**, 192 (1995).

[6] See e.g., G. Efstathiou, in *The Physics of the Early Universe*, eds. J.A. Peacock, A.F. Heavens and A.T. Davies (Adam Higler, Bristol, 1992).

[7] See e.g., S. Weinberg, *Gravitation and Cosmology* (J. Wiley, New York, 1972).

[8] See e.g., H. Arp et al., *Nature* **346**, 807 (1990).

[9] G.H. Jacoby et al., *Proc. Astron. Soc. Pacific* **104**, 599 (1992); M. Fukugita, C.J. Hogan, and P.J.E. Peebles, *Nature* **366**, 309 (1993); W. Freedman et al, *Nature* **371**, 757 (1994).

[10] G. Smoot et al., *Astrophys. J.* **396**, L1 (1992).

[11] J.P. Ostriker, *Ann. Rev. Astron. Astrophys.* **31**, 689 (1993).

[12] F. Hoyle, G. Burbidge, and J.V. Narlikar, *Astrophys. J.* **410**, 437 (1993); *Mon. Not. R. astron. Soc.* **267**, 1007 (1994); *Astron. Astrophys.* **289**, 729 (1994).

[13] S. Faber and J. Gallagher, *Ann. Rev. Astron. Astrophys.* **17**, 135 (1979).

[14] V. Trimble, *Ann. Rev. Astron. Astrophys.* **25**, 425 (1987).

[15] See e.g., N. Kaiser et al., *Mon. Not. R. astr. S.* **252**, 1 (1991); M. A. Strauss et al., *Astrophys. J.* **397**, 395 (1992); A. Dekel, *Ann. Rev. Astron. Astrophys.* **32**, 371 (1994).

[16] P.J.E. Peebles, *Nature* **327**, 210 (1987); *Astrophys. J.* **315**, L73 (1987).

[17] A. Guth, *Phys. Rev. D* **23**, 347 (1981).

[18] M.S. Turner, *Proc. Natl. Acad. Sci. (USA)* **90**, 4822 (1993).

[19] See e.g., N. Turok, *Physica Scripta* **T36**, 135 (1991).

[20] W. Hu and N. Sugiyama, astro-ph/9403031.

[21] P.J.E. Peebles, *Astrophys. J.* **432**, L1 (1994).

[22] See e.g., A.D. Linde, *Particle Physics and Inflationary Cosmology* (Harwood, Chur, 1990), or M.S. Turner, in *Proceedings of TASI-92*, eds. J. Harvey and J. Polchinski (World Scientific, Singapore, 1993).

[23] Inflationary models have been constructed with $\Omega_0 < 1$; see e.g., P.J. Steinhardt, *Nature* **345**, 47 (1990); M. Bucher et al., hep-th/9411206.

[24] A. H. Guth and S.-Y. Pi, *Phys. Rev. Lett.* **49**, 1110 (1982); S. W. Hawking, *Phys. Lett. B* **115**, 295 (1982); A. A. Starobinskii, *ibid* **117**, 175 (1982); J. M. Bardeen, P. J. Steinhardt, and M. S. Turner, *Phys. Rev. D* **28**, 697 (1983).

[25] V.A. Rubakov, M. Sazhin, and A. Veryaskin, *Phys. Lett. B* **115**, 189 (1982); R. Fabbri and M. Pollock, *ibid* **125**, 445 (1983); A.A. Starobinskii *Sov. Astron. Lett.* **9**, 302 (1983); L. Abbott and M. Wise, *Nucl. Phys. B* **244**, 541 (1984).

[26] See e.g., G.R. Blumenthal et al., *Nature* **311**, 517 (1984).

[27] A.D. Liddle and D. Lyth, *Phys. Repts.* **231**, 1 (1993).

[28] M.S. Turner, *Physica Scripta* **T36**, 177 (1991).

[29] P.J. Steinhardt and M.S. Turner, *Phys. Rev. D* **29**, 2162 (1984).

[30] M. Kamionkowski et al., *Astrophys. J.* **426**, L57 (1994).

[31] See e.g., U.G. Briel et al., *Astron. Astrophys.* **259**, L31 (1992); S.D.M. White et al., *Nature* **366**, 429 (1993); D.A. White and A.C. Fabian, *Mon. Not. R. astron. Soc.*, in press (1995).

[32] See e.g., J. Bahcall et al., *Astrophys. J.* **435**, L51 (1994) and References therein.

[33] E. Gates, G. Gyuk, and M.S. Turner, astro-ph/9411073; C. Alcock et al., astro-ph/9501091.

[34] E.J. Copeland, E.W. Kolb, A.R. Liddle, and J.E. Lidsey, *Phys. Rev. Lett.* **71**, 219 (1993); *Phys. Rev. D* **48**, 2529 (1993); M.S. Turner, *ibid*, 3502 (1993); M.S. Turner, *ibid* **48**, 5539 (1993); L. Knox and M.S. Turner, *Phys. Rev. Lett.* **73**, 3347 (1994); L. Knox, *Phys. Rev. D*, in press (1995).

INFLATION, MICROWAVE BACKGROUND ANISOTROPY, AND OPEN UNIVERSE MODELS

J.A. FRIEMAN

NASA/Fermilab Astrophysics Center
Fermi National Accelerator Laboratory
Batavia, IL 60510

1. Introduction

The inflationary scenario for the very early universe has proven very attractive, because it can simultaneously solve a number of cosmological puzzles, such as the homogeneity of the Universe on scales exceeding the particle horizon at early times, the flatness or entropy problem, and the origin of density fluctuations for large-scale structure [1]. In this scenario, the observed Universe (roughly, the present Hubble volume) represents part of a homogeneous inflated region embedded in an inhomogeneous space-time. On scales beyond the size of this homogeneous patch, the initially inhomogeneous distribution of energy-momentum that existed prior to inflation is preserved, the scale of the inhomogeneities merely being stretched by the expansion.

In its conventional form, inflation predicts a nearly scale-invariant spectrum of density perturbations produced by the inflaton field, and that the Universe is observationally indistinguishable from being spatially flat ($k = 0$). In the absence of a cosmological constant or exotic forms of matter, this implies that the present matter density parameter $\Omega_0 \equiv 8\pi G \rho_m(t_0)/3H_0^2$ is very close to unity. However, it is not clear that such an Einstein-de Sitter Universe jibes with astronomical observations. As is well known, dynamical estimates of mass-to-light ratios from galaxy rotation curves and cluster dynamics [2] typically indicate $\Omega_0 \simeq 0.1 - 0.2$. Similar conclusions have recently been reached from the consistency of the ROSAT observations of X-ray emission from the Coma cluster and Big Bang nucleosynthesis constraints on the baryon density Ω_B [3].

M. Kafatos and Y. Kondo (eds.), Examining the Big Bang and Diffuse Background Radiations, 321–327.
© 1996 IAU.

Moreover, if $\Omega_0 = 1$ the age of the Universe is $t_0 = (2/3H_0) = 6.7 \times 10^9 h^{-1}$yrs (where the present Hubble parameter is $H_0 = 100h$ km/sec/Mpc). This is less than globular cluster age estimates of $t_{gc} \simeq 13 - 15 \times 10^9$ yr if $h \geq 0.5$, and a number of extragalactic distance indicators suggest $h \simeq 0.8$. A large age is also indicated by the colors of stellar populations of radio galaxies at high redshift, $z \simeq 4$ [4]. The presence of galaxies and perhaps even protoclusters at $z \geq 3.5$ is also easier to explain in a low-density Universe, where structures should have collapsed by $z \simeq \Omega_0^{-1} - 1$ [5]. On larger scales, the situation is still uncertain: several analyses of large-scale peculiar motions suggest higher values of Ω_0, consistent with unity [6], while other methods are consistent with low values of Ω_0 [7].

In sum, the current observational status of Ω_0 is at best inconclusive, with much of the data pointing to a low-density Universe. In the context of inflation, the simplest way to accomodate $\Omega_0 < 1$ is to incorporate a cosmological constant $\Lambda = 3H_0^2 \Omega_\Lambda$, retaining spatial flatness by imposing $\Omega_0 + \Omega_\Lambda = 1$. However, initial studies of observed gravitational lens statistics indicate the bound $\Omega_\Lambda \lesssim 0.7$ [8], marginally disfavoring the spatially flat, low-density model.

The other logical possibility is an open, negatively curved universe, and various suggestions have been made to try to accommodate an open, low-Ω_0 Universe within inflation [9]. While the models differ in the mechanisms that drive inflation, their common feature is that the homogeneous patch that encompasses the presently observable Universe was inflated by just the right number of e-foldings to ensure that $1 - \Omega_0 \simeq 1$; generally, this implies that the present size L_0 of the inflated patch is comparable to the current Hubble distance, H_0^{-1}.

Points separated by distances larger than the scale of the inflated homogeneous patch have never been in causal contact, and one thus expects large density fluctuations, $(\delta\rho/\rho)_L \sim 1$, on scales $L \gtrsim L_0$. However, if the size of the homogeneous region is close to the present Hubble radius, such non-linear inhomogeneities on large scales will induce significant microwave background anisotropy via the Grischuk-Zel'dovich (GZ) effect [10]. In order of magnitude, the quadrupole anisotropy induced by superhorizon-size fluctuations of lengthscale L is $Q_L \simeq (\delta\rho/\rho)_L (LH_0)^{-2}$. The COBE DMR has measured a quadrupole anisotropy of $Q_{COBE} = (4.8 \pm 1.5) \times 10^{-6}$ from the first year of data and $Q_{COBE} = (2.2 \pm 1.1) \times 10^{-6}$ from the first two years of data [11]. Consequently, assuming order unity density fluctuations on scales $L \gtrsim L_0$, the size of the inflated patch must be significantly larger than the present Hubble radius, $L_0 > 500 H_0^{-1}$ [12, 13]. However, the Grischuk-Zel'dovich analysis was performed for a spatially flat ($k = 0$) universe; to self-consistently exclude an open model, it must be extended to the case of negative curvature.

This talk summarizes an investigation of the Grischuk-Zel'dovich effect in an open universe, done in collaboration with Alexander Kashlinsky and Igor Tkachev [14]. We found that the constraint on L_0 generally becomes even tighter when $\Omega_0 < 1$. If the Universe began from inhomogeneous initial conditions, the comoving size of the quasi-homogeneous patch that encompasses our observable universe must extend to at least $500 - 2000$ times the present Hubble radius. Thus, the required large size of the inflated patch is very improbable in low-Ω inflationary models.

To relate the size of the inflated patch to the local value of Ω_0 we write the Friedmann equation as $-K = [1 - \Omega(t)]H^2(t)a^2(t)$, where $a(t)$ is the global expansion factor and $K = +1, -1$, or 0 is the spatial curvature constant. Note that the global topology of the inflationary Universe could be quite complex, with e.g., locally Friedmann universes of both positive and negative spatial curvature connected by wormhole throats. We will focus on the open, negatively curved ($K = -1$) model, since it is the open model that attracts attention as an alternative to the flat Universe on observational grounds. Thus, we can relate the present scale L_0 of the homogeneous patch to its size L_s at the start of inflation, $(1 - \Omega_0)H_0^2 L_0^2 = (H_s^2 L_s^2)(1 - \Omega_s)$ (where subscript 's' denotes quantitites at the onset of inflation). By the onset of inflation, we expect that causal microphysical processes could have smoothed out initial inhomogeneities only on scales up to the Hubble radius, so that $H_s L_s \sim 1$. This is also a sufficient condition for spatial gradients to be subdominant compared to the vacuum energy density driving accelerated expansion [15]. Inflation was proposed in part to allow $1 - \Omega_s \sim 1$ as an initial condition, but in any case $1 - \Omega_s \leq 1$. Consequently, we expect the present size of the homogeneous patch to satisfy $L_0^2 \lesssim H_0^{-2}/(1 - \Omega_0) \equiv R_{curv}^2$, i.e., the present size of the inflated patch is at most comparable to the present curvature radius R_{curv}. If $1 - \Omega_0 \ll 1$, the Universe is nearly spatially flat, and the present curvature radius is much larger than the Hubble radius. On the other hand, if $1 - \Omega \simeq 1$, then $R_{curv} \sim H_0^{-1}$, implying $L_0 \lesssim H_0^{-1}$, and in particular $L_0 \ll 500 H_0^{-1}$. This simple argument shows that the GZ effect is only naturally suppressed in the limit $\Omega_0 \to 1$, and that the required large size of the homogeneous domain of our observable Universe implied by the microwave background measurements is difficult to produce in $\Omega_0 \ll 1$ inflationary models. However, as noted above, the effect of spatial curvature on the GZ anisotropy can be significant, and this calculation should be done self-consistently in an open universe.

Microwave background anisotropies in an open universe have been studied by a number of authors [16]. We write the background metric of the open universe in the form $ds^2 = a^2(\eta)[d\eta^2 - d\chi^2 - \sinh^2(\chi)(d\theta^2 + \sin^2\theta d\phi^2)]$, where $\eta = \int dt/a(t)$ is conformal time, and χ is the comoving radial distance in units of the curvature scale (i.e., the physical distance $\chi_{phys} =$

$R_{\mathrm{curv}}\chi$). For the matter-dominated universe, the scale factor is given by $a(\eta) = a_m (\cosh \eta - 1)$, where a_m is a constant and $\eta = 0$ corresponds to the initial singularity. At a given conformal time, the density parameter is given by $\Omega(\eta) = 2(\cosh \eta - 1)/\sinh^2 \eta$. Thus, at early times, $\eta \ll 1$, the universe is effectively flat, $\Omega(\eta \ll 1) \simeq 1$, and at late times, $\eta \gtrsim 1$, it is curvature-dominated.

To describe the propagation of waves in curved space, we expand them in terms of eigenfunctions of the Helmholtz equation $(\nabla^2 + k^2 + 1)f(\chi, \theta, \phi) = 0$, where ∇^2 is the Laplace operator on the three-surface of constant negative curvature. The solutions are of the form $X_l(k; \chi)Y_l^m(\theta, \phi)$, where Y_l^m are the spherical harmonics and the radial eigenfunctions are given by

$$X_l(k; \chi) = (-1)^{l+1}\frac{(k^2 + 1)^{l/2}}{N_l^{-1}(k)} \sinh^l \chi \frac{d^{l+1}(\cos k\chi)}{d(\cosh \chi)^{l+1}} . \tag{1}$$

Here $N_l^k = k^2(k^2 + 1)...(k^2 + l^2)$. The normalization is chosen such that in the limit $\Omega \to 1$ the radial eigenfunctions become spherical Bessel functions. For a perturbation of comoving wavelength λ, the comoving wavenumber $k = 2\pi/\lambda = k_{\mathrm{phys}}R_{\mathrm{curv}}$. Using this relation and the Friedmann equation, the comoving wavenumber corresponding to the size of the inflated patch is $k_0 \simeq R_{\mathrm{curv}}/L_0 = 1/L_s H_s \sqrt{1 - \Omega_s} \gtrsim 1$.

Similarly, the microwave background temperature can be expanded in spherical harmonics on the sky, $\delta T/T = \sum a_{lm}Y_{lm}(\theta, \phi)$, and the multipole moments of the anisotropy are then given by the Sachs-Wolfe relation

$$\langle |a_l|^2 \rangle = \frac{2}{\pi} \int |\Phi_k(\eta = 0)|^2 |\tilde{\theta}_l(k)|^2 \frac{N_l(k)}{(k^2 + 1)^l}dk , \tag{2}$$

where

$$\tilde{\theta}_l(k) = \frac{F(\eta_{ls})}{3}X_k^l(\eta_0 - \eta_{ls}) + 2\int_{ls}^{\eta_0} \frac{dF}{d\eta}X_k^l(\eta_0 - \eta)d\eta , \tag{3}$$

η_{ls} denotes the epoch of last scattering, and the gravitational potential fluctuation satisfies $\Phi_k(\eta) = \Phi_k(\eta = 0)F(\eta)$, with (ignoring the decaying mode)[17]

$$F(\eta) = 5\frac{\sinh^2 \eta - 3\eta \sinh \eta + 4\cosh \eta - 4}{(\cosh \eta - 1)^3} . \tag{4}$$

Note that $F(\eta) = 1$ for $\Omega_0 = 1$; in an open universe, $F(\eta) \simeq 1$ for $\eta \lesssim 1$ and decays as $1/a(\eta)$ for $\eta \gtrsim 1$. Eqs. (1) - (4) allow one to estimate the anisotropy due to superhorizon-size perturbations, with wavelengths $\lambda \gg H_0^{-1}$. The potential Φ is a gauge-invariant measure of the spatial curvature perturbation, related to the density fluctuation by the relativistic curved-space analogue of the Poisson equation [18]. For perturbations on scales larger than the Hubble radius, $k\eta \lesssim 1$, it satisfies $\Phi_k \simeq -\delta_k/2 \simeq$ constant, where δ is a gauge-invariant measure of the density perturbation amplitude, equal to the density fluctuation in the longitudinal (conformal Newtonian) gauge [17]. For such long wavelengths, the dominant anisotropy is generally the quadrupole $l = 2$ (for some values of Ω_0, the quadrupole is accidentally suppressed, and the main contribution would be the $l = 3$ octupole moment, as we discuss below). The quadrupole anisotropy due to such superhorizon-scale modes is thus

$$\langle|a_2|^2\rangle \simeq \frac{1}{2\pi} \int_0^{k_0} \frac{k^2+4}{k^2+1} |\tilde{\theta}_2(k)|^2 \langle|\delta_k|^2\rangle k^2 dk \ . \tag{5}$$

For superhorizon-size modes, the quadrupole mode contribution $|\tilde{\theta}_2(k)|^2$ quadrupole depends on Ω_0: for example, for $\Omega_0 = 0.1$, we find $|\tilde{\theta}_2(k)|^2 \simeq 0.1$ for these modes, while for $\Omega_0 = 0.7$, $|\tilde{\theta}_2(k)|^2 \simeq 0.02$. For $\Omega_0 = 0.4$, the mode contribution is strongly suppressed, due to a near cancellation of the line-of-sight contribution (the second term on the RHS of eq. (3)) with the last scattering term (see below). We emphasize that for modes outside the scale of the homogeneous patch, $k \leq k_0$, the pre-inflation perturbation amplitude δ_k is preserved and expected to be of order unity. To study the implications of this result, we consider two limits: Ω_0 close to unity $(1 - \Omega_0 \ll 1)$ and low-density models with $1 - \Omega_0 \sim 1$.

Ω_0 **close to 1**: Using the relation $\cosh\eta - 1 = 2(1 - \Omega)/\Omega$, the limit $\Omega_0 \to 1$ corresponds to taking $\eta_0^2 \simeq 4(1 - \Omega_0) \to 0$. Taking this limit in Eq. (1) while keeping $k_{\rm phys}$ fixed, we find $X_2(k; \eta \to 0) \simeq (1+k^2)\eta^2/15 \simeq 4(1+k^2)(1 - \Omega_0)/15$. In this limit, the line-of-sight integral in Eq. (3) becomes $\int (dF/d\eta) X_k^2 d\eta \simeq -(1 + k^2)\eta_0^4/630$, which can be neglected compared to the last scattering term. As a result, the quadrupole arising from modes with $k\eta_0 \lesssim 1$ can be expressed as

$$\langle|a_2|^2\rangle \simeq \frac{8}{225\pi} \frac{(1 - \Omega)^2}{9} \int_0^{k_0} dk\ k^2(k^2 + 1)(k^2 + 4)\langle|\delta_k|^2\rangle \ . \tag{6}$$

The usual flat-space result can be recovered from Eq. (6) by taking the limit $k \gg 1$ and keeping $k_{\rm phys}$ fixed in the relation $k_{\rm phys} = k H_0 \sqrt{1 - \Omega_0}$. Eq. (6) can be used to constrain Ω_0 with any given pre-inflation power spectrum $\langle|\delta_k|^2\rangle$ on scales $k \leq k_0$. A plausible assumption is that $\langle|\delta_k|^2\rangle \sim k^n$ with $n \geq 0$, i.e., random Poisson fluctuations (or less). For example, such a spectrum would arise if one imagines that prior to inflation the universe consisted of uncorrelated, quasi-homogeneous regions of size k_0^{-1}. However, quantitatively the result does not depend strongly upon the shape of the power spectrum. With the assumption of no fine tuning prior to inflation, i.e., $\langle|\delta(k_0)|^2\rangle \simeq 1$, and since in inflationary models $k_0 \gtrsim 1$, the

COBE measurement of the quadrupole moment translates eq. (6) into the constraint

$$\Omega_0 > 1 - a_2(\text{COBE}) \simeq 1 - 10^{-6} \ . \tag{7}$$

Thus, *if* an epoch of inflationary expansion was responsible for the homogeneity of our observable Universe, the density parameter Ω_0 cannot differ from 1 by more than one part in $Q^{-1} \sim 10^6$.

Low Ω_0: We now consider the case of low Ω_0 and estimate the scale out to which the Universe must be homogeneous in light of the COBE results, independent of considerations of inflation, namely we allow $k_0 \ll 1$. If Ω_0 is not very close to 0.4 or 1, $\tilde{\theta}_2(k)$ is nearly independent of k for small k, and we can set $\tilde{\theta}_2(k) \simeq \tilde{\theta}_2(0)$ to good approximation for $k < 1$. (This is very different from the spatially flat model, where $|\tilde{\theta}_2(k)| = j_2(2k)$ and goes to zero as k^2 at small k). The zero-mode contribution $|\tilde{\theta}_2(0)|$ as a function of Ω_0 is shown in Fig. 2. In this case the quadrupole becomes:

$$\langle |a_2|^2 \rangle \simeq |\tilde{\theta}_2(0)|^2 \frac{1}{2\pi} \int_0^{k_0} \frac{k^2 + 4}{k^2 + 1} \langle |\delta_k|^2 \rangle k^2 dk \ . \tag{8}$$

Again conservatively assuming an initial spectrum that falls at least as white noise ($n \geq 0$), eq. (8) yields a lower bound on the scale $k_0^{-1} \sim L_0 H_0 \sqrt{1 - \Omega_0}$ over which the Universe must be homogeneous if $\Omega_0 \lesssim 1$,

$$k_0^{-1} > \left(\frac{|\tilde{\theta}_2(0)|}{a_{2,COBE}} \right)^{2/3} \simeq 10^4 |\tilde{\theta}_2(0)|^{2/3} \ . \tag{9}$$

Eq. (9) implies that the Universe must to be homogeneous over scales $k_0^{-1} \gtrsim 2000$ for $\Omega_0 \lesssim 0.1$ and over scales $k_0^{-1} \gtrsim 500$ for $\Omega \simeq 0.5 - 0.8$. In inflation models these bounds on $k_0 \ll 1$ require superhorizon-sized correlations prior to inflation. Note that for a given constant value of $|\delta_k|^2$ the quadrupole anisotropy for $k_0 \ll 1$ scales as $\sim k_0^{7/2}$ for $\Omega = 1$ and only as $\sim k_0^{3/2}$ for $\Omega \ll 1$.

$\Omega_0 \simeq 0.4$: The quadrupole due to long wavelength modes is suppressed not only at $\Omega_0 \to 1$ but also accidentally for $\Omega_0 \sim 0.4$, due to cancellation between the last scattering term and the line-of-sight integral. (The positive last scattering term dominates at $\Omega_0 \to 1$, while the negative line-of-sight term dominates at $\Omega_0 \to 0$.) As Ω_0 is varied over a small interval around 0.4, the wavenumber where the two terms cancel varies over the interval $(0, \eta_0^{-1})$. While interesting, this suppression cannot make inflation and low-Ω_0 compatible, for in this case the contribution to the octupole ($l = 3$) mode will be dominant and lead to similarly severe constraints on L_0.

We have arrived at two results of significance for inflation and open universe models. (1) Inflation can produce a homogeneous patch encompassing the observable Universe (the present Hubble volume) and be consistent with the microwave background observations only if the present density parameter Ω_0 differs from unity by no more than 1 part in $Q_{COBE}^{-1} \sim 10^6$. (2) On the other hand, if Ω_0 is significantly below 1, the Universe must be homogeneous on scales $k_0^{-1} > (500 - 2000)$. If this is the case, inflation does not by itself solve the horizon problem. Indeed, if we assume that the distribution of quasi-homogeneous regions satisfies Poisson statistics, the probability of finding one such region per volume k_0^{-3} in curvature units is $P \simeq k_0^{-3} \times \exp(-k_0^{-3})$, which is negligibly small for the k_0^{-1} values above. If it turns out that the universe is open, $\Omega_0 < 1$, this implies that our Hubble volume occupies a very special place in the space of initial conditions, which is precisely the condition inflation was meant to alleviate.

I thank my collaborators Sasha Kashlinsky and Igor Tkachev. This work was supported by the DOE and NASA grant NAGW-2381 at Fermilab.

References

1. For a review see K. A. Olive, Phys. Rep. C **190**, 307 (1990).
2. See, *e.g.*, S. Kent and J. Gunn, Astron. J. **87**, 945 (1982).
3. S. D. M. White *et al.*, Nature **366**, 429 (1993).
4. S. Lilly, Ap. J **333**, 161 (1988); K. C. Chambers and S. Charlot, Ap.J. **348**, L1 (1990); K. C. Chambers, *et al.*, Ap. J. **363**, 21 (1990).
5. J. Uson *et al.*, Phys. Rev. Lett. **67**, 3328 (1992); M. Giavalisco *et al.*, Ap. J. **425**, L5 (1994); A. Kashlinsky, Ap. J. **406**, L1 (1993).
6. A. Dekel, Ann. Rev. Astr. Astrophys., in press (1994).
7. A. Kashlinsky, Ap. J. **386**, L37 (1992), and in *"Evolution of the Universe and its observational quest"*, ed. K. Sato, in press (1994); R. Scaramella *et al.*, Ap. J. **422**, 1 (1994).
8. M. Fukugita and E. L. Turner, MNRAS **253**, 99 (1991); D. Maoz and H. W. Rix, Ap. J. **416**, 425 (1993); C. Kochanek, Ap. J. **419**, 12 (1993).
9. J. R. Gott, Nature **295**, 304 (1982); G. F. R. Ellis, Class. Quantum Grav. **5**, 891 (1988); P. Steinhardt, Nature **345**, 47 (1989); D. H. Lyth and E. D. Stewart, Phys. Lett. **B252**, 336 (1990); B. Ratra and P. J. E. Peebles, preprints PUPT-1444, PUPT-1445 (1994); M. Bucher, A. S. Goldhaber, and N. Turok, preprint PUPT-94-1507 (1994).
10. L. Grischuk and Ya. B. Zeldovich, Sov. Astron. **22**, 125 (1978).
11. G. Smoot *et al* Ap. J. **396**, L1 (1992); C. Bennett *et al.* Ap. J. (1994), submitted.
12. M. Turner, Phys. Rev. D **44**, 3737 (1991).
13. L. Grischuk, Phys. Rev. D **45**, 4717 (1992).
14. A. Kashlinsky, I. Tkachev, and J. Frieman, Phys. Rev. Lett. **73**, 1582 (1994).
15. D.S. Goldwirth and T. Piran, Phys. Rev. Lett. **64**, 2852 (1990).
16. See, *e.g.*, M. Wilson, Ap. J. **253**, L53 (1983); L. F. Abbott and R. K. Schaefer, Ap. J. **308**, 546 (1986); J. Traschen and D. Eardley, Phys. Rev. D **34**, 1665 (1986), and the last two references in [9]; M. Kamionkowski and D. Spergel, preprint IASSNS-HEP-93/73 (1993).
17. V. F. Mukhanov, H. A. Feldman and R. H. Brandenberger, Phys. Rep. C **215**, 203 (1992).
18. J. Bardeen, Phys. Rev. D **22**, 1882 (1980).

THE QUASI-STEADY STATE COSMOLOGY

F. HOYLE
102 Admirals Walk, West Cliff Road, Dorset, BH2 5HF Bournemouth, United Kingdom

G. BURBIDGE
University of California, San Diego, Department of Physics and Center for Astrophysics & Space Sciences, 9500 Gilman Drive, La Jolla, California 92093-0111

AND

J.V. NARLIKAR
Inter-University Centre for Astronomy & Astrophysics, Poona University Campus, Pune 411 007, India

1. Introduction

At IAU Symposium No. 168 the presentation was divided into two parts:
I.   Theoretical Foundations and
II. Observational Facts and Consequences

Nearly all of the work presented here is contained in four papers published by Hoyle, Burbidge and Narlikar (1993; 1994a,b,c) which will be abbreviated in the text to HBN 1993, HBN 1994a, HBN 1994b and HBN 1994c.

In this presentation Section 2 is devoted to the basic theory and in Section 3 we describe the observations and the way that we interpret them using the theory. Also in Section 3 we discuss various predictions relating to the theory.

2. Theoretical Foundations

To begin with we show with the help of a model how the problems of space-time singularity and violation of the energy momentum conservation law that are present in the standard cosmology can be avoided by introducing

M. Kafatos and Y. Kondo (eds.), Examining the Big Bang and Diffuse Background Radiations, 329–368.

a scalar field minimally coupled to gravity and having its sources in events where matter is created.

We than show that matter creation preferentially occurs near collapsed massive objects and the scalar field created at such mini-creation events has a feedback on spacetime geometry causing the universe to have a steady expansion as in the de Sitter model but with periodic phases of expansion and contraction superposed on it.

The parameters of our model can be empirically fixed in relation to the cosmological observations thus providing tests of the theory. In the second part the observational aspects are dealt with.

Next we argue that the model arises from a deeper theory which is Machian in origin with the inertia of a particle determined by the rest of the particles in the universe in a long range conformably invariant scalar interaction. The characteristic mass of a particle created is then the Planck mass. The Planck particle decays quickly to baryons. The inertial effects produced by the Planck particles during their brief existence generate the scalar field of the toy model while the inertial effects of the stable baryonic particles give the more familiar Einstein equations of relativity.

Finally we show that extending the theory to the most general conformably invariant form automatically leads to the cosmological constant whose sign and magnitude are of the right cosmological order.

We begin with the tentative definition that cosmology refers to a study of those aspects of the universe for which spatial isotropy and homogeneity can be used, with the spacetime metric taking the form

$$ds^2 = dt^2 - S^2(t)\left[\frac{dr^2}{1 - kr^2} + r^2(d\theta^2 + sin^2\theta d\phi^2)\right] \tag{1}$$

in terms of coordinates t, r, θ, ϕ with $r = 0$ at the observer. The topological constant k in this so-called Robertson–Walker form can be shown to be 0 or ± 1. The "particles" to which (1) applies are thought of as galaxies or clusters of galaxies, each "particle" having spatial coordinates r, θ, ϕ independent of the universal time t. They form what is often referred to as the Hubble flow.

Big-Bang cosmology in all its forms is obtained from the equations of general relativity,

$$R_{ik} - \frac{1}{2}g_{ik}R + \lambda g_{ik} = -8\pi G T_{ik}, \tag{2}$$

which follow from the variation of an action formula

$$\mathcal{A} \;=\; \frac{1}{16\pi G}\int_V (R+2\lambda)\sqrt{-g}\,d^4x + \int_V \mathcal{L}_{phys}(X)\sqrt{-g}\,d^4x \qquad (3)$$

with respect to a general Riemannian metric

$$ds^2 \;=\; g_{ik}dx^i dx^k \qquad (4)$$

within a general spacetime volume V. The physical Lagrangian $\mathcal{L}_{phys}(X)$ generates the energy-momentum tensor T_{ik} in this variation of g_{ik}. In the standard big-bang cosmology the physical Lagrangian includes only particles and the electromagnetic field, whereas in inflationary forms of big-bang cosmology a scalar field is also considered to be added to $\mathcal{L}_{phys}$. This is done in various ways, being severally advocated by different authors (see Narlikar and Padmanabhan 1991 for a review).

The initial conditions assumed in the standard model are :

(i) The universe was sufficiently homogeneous and isotropic at the outset for the metric (1) to be used immediately over a range of the r-coordinate of relevance to presentday observation,

(ii) $k = 0$,

(iii) $\lambda = 0$,

(iv) The initial balance of radiation and baryonic matter was such that the light elements D, ^{3}He, ^{4}He, ^{7}Li were synthesised in the early universe in the following relative abundances to hydrogen

$$\frac{D}{H} \;\simeq\; \frac{^3He}{H} \simeq 2\times 10^{-5},\frac{^4He}{H}\simeq 0.235,\frac{^7Li}{H}\simeq 10^{-10}.$$

From detailed calculations these abundances can be shown to require

$$\rho_{\text{baryon}} \;\simeq\; 10^{-32}T^3 \ \text{gcm}^{-3}, \qquad (5)$$

the radiation temperature being in degrees kelvin. (Gamow 1946, Alpher, et al 1950, 1953, Hoyle and Tayler 1964).

There is a fundamental problem here. The instant t = 0 is the so-called spacetime singularity at which the field equations (2) break down. This is identified with the big bang epoch. All matter that we see in the universe (as well as radiation) is supposed to be given as an initial condition at $t = \epsilon > 0$. The initial instant ϵ can be taken arbitrarily close to $t = 0$ but not identified with it. Thus the action principle (3) itself gets restricted in validity since the singular epoch must be excluded from it too.

Conceptually this is an exceptional step to take. In theoretical physics the basic laws or principles like the action principle are considered superior to the specific solutions based on them. Yet here we seek to restrict the validity of (2) and (3) because the solution so warrants it! There is thus a clear indication here of an inconsistency of the overall framework.

The other problems of the standard big bang model often referred to as the horizon and flatness problems also relate to the above initial conditions assumed at $t = \epsilon > 0$. While the need for such far reaching assumptions as (i) to (iv) has always prompted a measure of unease they were widely accepted for a decade and a half, and are indeed still fully accepted by the more orthodox supporters of the standard model. Others, however, welcomed the inflationary idea of including a scalar field in the physical lagrangian that initially dominated both matter and other fields and which varied adiabatically in such a way as to give

$$\frac{\dot{S}^2}{S^2} = C, \quad S(t) = S(0) \exp \sqrt{C} t, \tag{6}$$

with C a constant. The solution (6) is considered to apply from $t = \epsilon > 0$, where $\epsilon \sim 10^{-36}$s to a value of t large compared to $1/\sqrt{C}$. It greatly reduces the range of the r-coordinate over which (i) is needed and it effectively removes the k-term from (1). It also removes any initial contributions from matter and radiation, but these are considered to be reasserted through a physical transition of the scalar field, which jumps the solution (6) to

$$\dot{S}^2 = \frac{A}{S}, S \simeq \left(\frac{9}{4}A\right)^{\frac{1}{3}} t^{\frac{2}{3}}, \tag{7}$$

which is the so-called closure model with matter just having sufficient expansion to reach a state of infinite dispersal, a condition that is considered most favorable for the eventual formation of stars and galaxies.

A major problem associated with inflation is how to effectively eliminate the cosmological constant. The value of this constant which gave the exponential solution (6) above must reduce to zero or, if the cosmological observations so demand, become as small as 10^{-108} of its initial value.

Any theoretical trick invoked to achieve this has a contrived appearance (Weinberg 1989).

The papers referred to earlier (HBN 1993, 1994abc) show how we have developed an alternative scenario.

The action principle (3) has a second term which is supposed to include physical contributions other than gravity. A close parallel exists between the scalar field used for inflation and the scalar field used earlier by Hoyle and Narlikar (1963) for obtaining the steady state model from Einstein's field equations. To begin with we will use the 1963 formalism as a "toy model" for describing creation of matter without violating the law of conservation of energy-momentum and without encountering spacetime singularity.

Thus the classical Hilbert action leading to the Einstein equations is modified by the inclusion of a scalar field C whose derivatives with respect to the spacetime coordinates x^i are denoted by C_i. The action is given by

$$
\begin{aligned}
\mathcal{A} = & -\sum_a \int_{\Gamma_a} m_a ds_a + \int_V \frac{1}{16\pi G} R\sqrt{-g}d^4x - \frac{1}{2}f \int_V C_i C^i \sqrt{-g}d^4x \\
& + \sum_a \int_{\Gamma_a} C_i da^i
\end{aligned} \tag{8}
$$

where C is a scalar field and $C_i = \partial C/\partial x^i$. f is a coupling constant. The last term of (8) is manifestly path-independent and so, at first sight it appears to contribute no new physics. The first impression, however, turns out to be false if we admit the existence of broken worldlines. For, if particles $a, b, \ldots$ are created at world points $A_0, B_0, \ldots$ respectively, then the last term of (8) contributes a non-trivial sum

$$
-\{C(A_0) + C(B_0) + \ldots\}
$$

to $\mathcal{A}$.

Thus, if the worldline of particle a begins at point A_0, then the variation of $\mathcal{A}$ with respect to that worldline gives

$$
m_a \frac{da^i}{ds_a} = g^{ik}C_k \tag{9}
$$

at A_0. In other words, the C-field balances the energy-momentum of the created particle.

The field equations likewise get modified to

$$R_{ik} - \frac{1}{2}g_{ik}R \;=\; -8\pi G\left[\overset{T_{ik}}{m} + \overset{T_{ik}}{c}\right] \tag{10}$$

where

$$\overset{T_{ik}}{c} \;=\; -f\left\{C_iC_k - \frac{1}{2}g_{ik}C^lC_l\right\}. \tag{11}$$

Thus the energy conservation law is

$$\overset{T^{ik}}{m}{}_{;k} \;=\; -\overset{T^{ik}}{c}{}_{;k} = fC^iC^k{}_{;k}. \tag{12}$$

That is, matter creation via a nonzero left hand side of (12) is possible while conserving the overall energy and momentum. The C-field tensor has negative stresses which lead to the expansion of spacetime, as in the case of inflation.

From (9) we therefore get a necessary condition for creation as

$$C_iC^i \;=\; m_a^2; \tag{13}$$

this is the 'creation threshold' which must be crossed for particle creation. How this can happen near a massive object, can be seen from the following simple example.

The Schwarzschild solution for a massive object M of radius $R > 2GM/c^2$ is

$$ds^2 = dt^2\left(1 - \frac{2GM}{r}\right) - \frac{dr^2}{1 - \frac{2GM}{r}} - r^2(d\theta^2 + \sin^2\theta d\phi^2), \tag{14}$$

for $r \geq R$. Now if the C-field does not seriously change the geometry, we would have at $r \gg R$,

$$\dot{C} \approx m, \qquad C' \equiv \frac{\partial C}{\partial r} \cong 0. \tag{15}$$

If we continue this solution closer to $r \approx R$, we find that

$$C^i C_i \equiv \left(1 - \frac{2GM}{r}\right)^{-1} m^2. \tag{16}$$

In other words $C_i C^i$ increases towards the object and can become arbitrarily large if $r \approx 2GM$. So it is possible for the creation threshold to be reached *near* a massive collapsed object even if $C_i C^i$ is *below* the threshold far away from the object. In this way massive collapsed objects can provide new sites for matter creation. Further, because of the negative stresses the created matter is expelled outwards from the site while the C-field quanta escape with the speed of light. Thus, instead of a single big bang event of creation, we have mini-creation events near collapsed massive objects.

Since the C-field is a global cosmological field, we expect the creation phenomenon to be globally cophased. Thus, there will be phases when the creation activity is large, leading to the generation of the C-field strength in large quantities. However, the C-field growth because of its large negative stresses leads to a rapid expansion of the universe and a consequent drop in its background strength. When that happens creation is reduced and takes place only near the most collapsed massive objects thus leading to a drop in the intensity of the C-field. The reduction in C-field slows down the expansion, even leading to local contraction and so to a build-up of the C-field strength. And so on!

We can describe this up and down type of activity as an oscillatory solution superposed on a steadily expanding de Sitter type solution of the field equations as follows. For the Robertson–Walker line element the equations (10)–(12) give

$$3\frac{\dot{S}^2 + kc^2}{S^2} = 8\pi G(\rho - \frac{1}{2}f\dot{C}^2), \tag{17}$$

$$2\frac{\ddot{S}}{S} + \frac{\dot{S}^2 + kc^2}{S^2} = 4\pi Gf\dot{C}^2, \tag{18}$$

where $S(t)$ is the scale factor and k the curvature parameter ($= 0, \pm 1$). The cosmic time is given by t. These equations have a deSitter type solution given by

$$S \propto \exp(t/P), k = 0, \qquad \dot{C} = \text{constant}, \qquad \rho = \text{constant} \qquad (19)$$

The oscillatory solution is given by

$$k = +1, \qquad \dot{C} \propto 1/S^3, \qquad \rho \propto 1/S^3. \qquad (20)$$

Thus (17) becomes, in the latter case

$$\dot{S}^2 = -c^2 + \frac{A}{S} - \frac{B}{S^4}, \quad A, B = \text{constant.} \qquad (21)$$

Here the oscillatory cycle will typically have a period $Q \ll P$.

Although the exact solution of (21) will be difficult to obtain, we can use the following approximate solution of (19) and (20) to describe the short-term and long-term cosmological behaviour :

$$S(t) = \exp\left(\frac{t}{P}\right)\left\{1 + \alpha\cos\frac{2\pi t}{Q}\right\}. \qquad (22)$$

Note that the universe has a long term secular expanding trend, but because $|\alpha| < 1$, it also executes non-singular oscillations around it. For this reason this model has been called "quasi-steady state cosmology". We can determine α and our present epoch $t = t_0$ by the observations of the present state of the universe. Thus an acceptable set of parameters is

$$\alpha = 0.75, \ t_0 = 0.85Q, \ Q = 4 \times 10^{10} yr., \ P = 20Q. \qquad (23)$$

We shall show how their values relate to the observed parameters in Part II.

But now we discuss the theory underlying the physics of matter creation.

An important property of physical theories is scale invariance or conformal invariance. Maxwell's equations and the Dirac equation for a massless particle are conformably invariant but general relativity is not. If, however, the inertial mass transforms inversely as the length scale in conformal transformation then the Dirac equation for a massive fermion as well as classical and quantum electrodynamics will become conformably invariant. Can general relativity be suitably reformulated to be conformably invariant? We indicate the steps towards this goal since they naturally lead to a comprehensive theory of matter creation that encompasses our model.

It is necessary to begin by finding an action A that is unaffected in its value by a scale transformation. The second term on the right-hand side of (3) can be made to satisfy this requirement. For a set of particles $a, b, \ldots$ of masses $m_a, m_b, \ldots$ the form of $\mathcal{L}_{\text{phys}}$ usually considered in gravitational theory is

$$\sum_{a,b,\ldots} \int \frac{\delta_4(X, A)}{\sqrt{-g(A)}} m_a(A) da, \qquad (24)$$

where the possibility of the particle masses varying with the spacetime position requires the mass $m_a(A)$ of particle a to vary with the point A on its path, and similarly for the other particles. Hence the second term on the right-hand side of (3) is $-\sum_a \int m_a(A) da$.
With $da^* = \Omega da$ and $m_a^* = \Omega^{-1} m_a$ it is clear that (25) is invariant with respect to a conformal (scale) transformation.

$$\Box_X M(X) + \frac{1}{6} R M(X) = \sum_a \int \frac{\delta_4(X, A)}{\sqrt{-g(A)}} da. \qquad (25)$$

The possibility of particle masses varying with spacetime coordinates arises most naturally in a Machian approach. Here the property of inertia is not entirely intrinsic to a particle but is also related to its presence in a non-empty universe. A quantitative description of this idea that we will follow here is based on an early work by two of us (Hoyle and Narlikar 1964). In this inertia is expressed as a scalar conformably invariant long range interaction between particles.

To begin with choose a scalar mass field $M(X)$ to satisfy

$$-\sum_a \int m_a(A) da. \qquad (26)$$

Equation (26) has both advanced and retarded solutions. We particularize an advanced solution $M^{\text{adv}}(X)$ and a retarded solution $M^{\text{ret}}(X)$ in the following way. $M^{\text{ret}}(X)$ is to be the so-called fundamental solution in the flat spacetime limit (Courant and Hilbert, 1962). This removes for $M^{\text{ret}}(X)$

the ambiguity that would obviously arise from the homogeneous wave equation. The corresponding ambiguity for $M^{adv}(X)$ is removed by the physical requirement that fields without sources are to be zero. Since

$$\left[M^{adv} - M^{ret}\right] + \frac{1}{6}R[M^{adv} - M^{ret}] = 0, \tag{27}$$

the immediate consequences of this boundary condition is that $M^{adv} - M^{ret}$, being without sources, must be zero, so that

$$M^{adv}(X) = M^{ret}(X) = M(X) \text{ say.} \tag{28}$$

The gravitational equations are now obtained by putting

$$m_a(A) = M(A), \quad m_b(B) = M(B), \ldots \tag{29}$$

It can also be shown that in a conformal transformation the mass field $M(X)$ transforms as

$$M^*(X) = \Omega^{-1}(X)M(X), \tag{30}$$

a result that follows from the form of the wave equation (10) (c.f. Hoyle and Narlikar, 1974, 111). The outcome (loc. cit., 112 et seq) is

$$K\left(R_{ik} - \frac{1}{2}g_{ik}R\right) = -T_{ik} + M_iM_k - \frac{1}{2}g_{ik}g^{pq}M_pM_q$$
$$+G_{ik}K - K_{;ik}, \tag{31}$$

where

$$K = \frac{1}{6}M^2. \tag{32}$$

These gravitational equations are scale invariant. It may seem curious that from a similar beginning, (24) for the action rather than (3), the outcome is more complicated, but this seems to be a characteristic of the physical laws. As the laws are improved they become simpler and more elegant in their initial statement but more complicated in their consequences.

Now make the scale change

$$\Omega(X) = M(X)/\tilde{m}_0, \tag{33}$$

where $\tilde{m}_0$ is a constant with the dimensionality of $M(X)$. After the scale change the particle masses simply become $\tilde{m}_0$ everywhere and in terms of transformed masses the derivative terms drop out of the gravitational equations. And defining the gravitational constant G by

$$G = \frac{3}{4\pi \tilde{m}_0^2}, \tag{34}$$

the equations (31) take the form of general relativity

$$R_{ik} - \frac{1}{2}g_{ik}R = -8\pi G T_{ik}. \tag{35}$$

It now becomes clear why the equations of general relativity are not scale invariant. These are the special form to which the scale invariant equations (31) reduce with respect to a particular scale, namely that in which particle masses are everywhere the same.

It is also clear that the transition from (31) to (35) is justified provided $\Omega(X) \neq 0$ or $\Omega(X) \neq \infty$. For example, if $M(X) = 0$ on a spacelike hypersurface the above conformal transformation breaks down. It is because of the existence of such time sections that the use of (35) leads to the (unphysical) conclusion of a spacetime singularity. It was shown (Hoyle and Narlikar 1974, Kembhavi 1979) that the various spacetime singularities like that in the big bang or in a black hole collapse arise because of the illegitimate use of (35) in place of (31).

It is easily seen from the wave equation (26) that $M(X)$ has dimensionality $(\text{length})^{-1}$, and so has $\tilde{m}_0$. Units are frequently used in particle physics for which both the speed of light c and Planck's constant $\hbar$ are unity and in these units mass has dimensionality $(\text{length})^{-1}$. If we suppose these units apply to the above discussion then from (34)

$$\tilde{m}_0 = (3/4\pi G)^{1/2}, \tag{36}$$

which with $c = \hbar = 1$ is the mass of the Planck particle. This suggests that in a gravitational theory without other physical interactions the particles must be of mass (36), which in ordinary practical units is about 10^{-5} gram, the empirically determined value of G being used. This conclusion can be supported at greater length [See HBN 1994c]. We next consider what happens when the Planck mass decays into the much more stable baryons.

A typical Planck particle a exists from A_0 to $A_0 + \delta A_0$, in the neighborhood of which it decays into n stable baryonic secondaries, $n \simeq 6.10^{18}$, denoted by $a_1, a_2, \ldots a_n$. Each such secondary contributes a mass field $m^{(a_r)}(X)$, say, which is the fundamental solution of the wave equation

$$m^{(a_r)} + \frac{1}{6} R m^{(a_r)} = \frac{1}{n} \int_{\sim A_0 + \delta A_0} \frac{\delta_4(X, A)}{\sqrt{-g(A)}} da, \tag{37}$$

while the brief existence of a contributes $c^{(a)}(X)$, say, which satisfies

$$c^{(a)} + \frac{1}{6} R c^{(a)} = \int_{A_0}^{A_0 + \delta A_0} \frac{\delta_4(X, A)}{\sqrt{-g(A)}} da. \tag{38}$$

Summing $c^{(a)}$ with respect to $a, b, \ldots$ gives

$$c(X) = \sum_a c^{(a)}(X), \tag{39}$$

the contribution to the total mass $M(X)$ from the Planck particles during their brief existence, while

$$\sum_a \sum_{r=1}^{n} m^{(a_r)}(X) = m(X) \tag{40}$$

gives the contribution of the stable particles.

Although $c(X)$ makes a contribution to the total mass function

$$M(X) \;=\; c(X) + m(X) \tag{41}$$

that is generally small compared to $M(X)$, there is the difference that, whereas $m(X)$ is an essentially smooth field, $c(X)$ contains small exceedingly rapid fluctuations and so can contribute significantly to the derivatives of $c(X)$. The contribution to $c(X)$ from Planck particles a, for example, is largely contained between two light cones, one from A_0, the other from $A_0 + \delta A_0$. Along a timelike line cutting these two cones the contribution to $c(X)$ rises from zero as the line crosses the light cone from A_0, attains some maximum value and then falls back effectively to zero as the line crosses the second light cone from $A_0 + \delta A_0$. The time derivative of $c^{(a)}(X)$ therefore involves the reciprocal of the time difference between the two light cones. This reciprocal cancels the short duration of the source term on the right-hand side of (40). The factor in question is of the order of the decay time τ of the Planck particles, $\sim 10^{-43}$ seconds. No matter how small τ may be the reduction in the source strength of $c^{(a)}(X)$ is recovered in the derivatives of $c^{(a)}(X)$, which therefore cannot be omitted from the gravitational equations.

The derivatives of $c^{(a)}(X), c^{(b)}(X), \ldots$ can as well be negative as positive, so that in averaging many Planck particles, linear terms in the derivatives do disappear. It is therefore not hard to show that after such an averaging the gravitational equations become

$$R_{ik} - \frac{1}{2} g_{ik} R = \frac{6}{m^2} \left[-T_{ik} + \frac{1}{6}(g_{ik} m^2 - m^2_{;ik}) + (m_i m_k - \frac{1}{2} g_{ik} m_l m^l) \right.$$
$$\left. + \frac{2}{3}\left(c_i c_k - \frac{1}{4} g_{ik} c_l c^l \right) \right]. \tag{42}$$

Since the same wave equation is being used for $c(X)$ as for $m(X)$, the theory remains scale invariant. A scale change can therefore be introduced that reduces $M(X) = m(X) + c(X)$ to a constant, or one that reduces $m(X)$ to a constant. Only that which reduces $m(X)$ to a constant, viz

$$\Omega \;=\; \frac{m(X)}{m_0} \tag{43}$$

has the virtue of not introducing small very rapidly varying ripples into the metric tensor. Although small in amplitude such ripples produce non-negligible contributions to the derivatives of the metric tensor, causing difficulties in the evaluation of the Riemann tensor, and so are better avoided. Simplifying with (43) does not bring in this difficulty, which is why separating of the main smooth part of $M(X)$ in (41) now proves an advantage, with the gravitational equations simplifying to

$$8\pi G = \frac{6}{m_0^2}, \quad m_0 \text{ a constant,} \tag{44}$$

$$R_{ik} - \frac{1}{2}g_{ik}R = -8\pi G[T_{ik} - \frac{2}{3}(c_i c_k - \frac{1}{4}g_{ik}c_l c^l)]. \tag{45}$$

Using the metric (1) with $k = 0$ the dynamical equations for the scale factor $S(t)$ are

$$\frac{2\ddot{S}}{S} + \frac{\dot{S}^2}{S^2} = \frac{4\pi}{3}G\bar{c}^2, \tag{46}$$

$$\frac{3\dot{S}^2}{S^2} = 8\pi G\left(\bar{\rho} - \frac{1}{2}\bar{c}^2\right), \tag{47}$$

with $\bar{\rho}$ the average particle mass density and $\bar{c}^2$ being the average value of c^2, the average value of terms linear in c and of $\ddot{c}$ being zero. It is easily shown from (46) and (47) that

$$\frac{\partial\bar{\rho}}{\partial t} + \frac{3\dot{S}}{S}\bar{\rho} = \frac{1}{2}\left(\frac{\partial\bar{c}^2}{\partial t} + \frac{4\dot{S}}{S}\bar{c}^2\right). \tag{48}$$

If at a particular time there is no creation of matter then at that time the left-hand side of (48) is zero with $\bar{\rho} \propto S^{-3}$. And with the right-hand side also zero at that time $\bar{c}^2 \propto S^{-4}$. The sign of the $\bar{c}^2$ term in (46) is that of a negative pressure, a characteristic of the fields introduced into inflationary cosmological models. The concept of Planck particles forces the appearance of a negative pressure. In effect the positive energy of created particles is compensated by the sign of the $\bar{c}^2$ terms, which in (46) increases

$\ddot{S}/S$ and so causes the universe to expand. One can say that the universe expands because of the creation of matter. The two are connected because the divergence of the right-hand side of the gravitational equations (45) is zero.

As would be expected from this conservation property the sign of the $\bar{c}^2$ term in (47) is that of a negative energy field. Such fields have generally been avoided in physics because in flat spacetime they would produce catastrophic instabilities – creation of matter with positive energy producing a negative energy $\bar{c}^2$ term, producing more matter, producing a still larger $\bar{c}^2$ term, and so on. Here the effect is to produce explosive outbursts from regions where any such instability takes hold, through the $\bar{c}^2$ term in (46) generating a sharp increase of $\ddot{S}$. The sites of the creation of matter are thus potentially explosive. The explosive expansion of space serves to control the creation process and avoids the catastrophic cascading down the negative energy levels.

As will be discussed in II, this is in agreement with observational astrophysics which in respect to high energy activity is all of explosive outbursts, without evidence for the ingoing motions required by the popular accretion-disk theory for which there is no direct observational support. The profusion of sites where X-ray and γ-ray activity is occurring are on the present theory sites where the creation of matter is currently taking place.

A connection with our model can now be given. Writing

$$C(X) \; = \; \tau c(X), \qquad (49)$$

where τ is the decay lifetime of the Planck particle, the action contributed by Planck particles $a, b, \ldots$,

$$-\sum_a \int_{A_0}^{A_0+\delta A_0} c(A)da \qquad (50)$$

can be approximated as

$$-C(A_0) - C(B_0) - \ldots, \qquad (51)$$

which form was used in the model. And the wave-equation for $C(X)$, using the same approximation, is

$$C + \frac{1}{6}RC \;=\; \tau^{-2}\sum_a \frac{\delta_4(X, A_0)}{\sqrt{-g(A_0)}}, \tag{52}$$

which was also used in the model, except that the $1/6\,RC$ term is included in the wave equation and previously an unknown constant f appeared in place of τ^2.

Writing $M^{(a)}(X), M^{(b)}(X), \ldots$ as the mass fields produced by the individual Planck particles $a, b, \ldots$, the total mass field

$$M(X) \;=\; \sum_a M^{(a)}(X) \tag{53}$$

satisfies the wave equation (26) when $M^{(a)}, M^{(b)}, \ldots$ satisfy

$$M^{(a)} + \frac{1}{6}RM^{(a)} \;=\; \int \frac{\delta_4(X, A)}{\sqrt{-g(A)}} da, \ldots \tag{54}$$

Scale invariance throughout requires all the mass fields to transform as

$$M^{*(a)} \;=\; M^{(a)}\Omega^{-1} \tag{55}$$

with respect to the scale change Ω, when both the left and right hand sides of every wave equation transform to its starred form multiplied by Ω^{-3}, i.e. the left hand side of (54) goes to $(M^{*(a)} + \frac{1}{6}R^* M^{*(a)})\Omega^{-3}$ and the right hand side to

$$\Omega^{-3}\int \frac{\delta_4(X, A)}{\sqrt{-g^*(A)}} da^*. \tag{56}$$

Then the factor Ω^{-3} cancels to give the appropriate invariant equation. This cancellation is evidently unaffected if, instead of (54) for the wave equation satisfied by M^a, we have

$$M^{(a)} + \frac{1}{6}RM^{(a)} + M^{(a)^3} \;=\; \int \frac{\delta_4(X, A)}{\sqrt{-g(A)}} da. \tag{57}$$

Since the cube term transforms to $M^{*(a)^3}\Omega^{-3}$ with respect to Ω changing (54) to (57) preserves scale invariance in what appears to be its widest form. Since in other respects the laws of physics always seem to take on the widest ranging properties that are consistent with the relevant forms of invariance we might think it should also be here, in which case (57) rather than (54) is the correct wave equation for $M^{(a)}$, and similarly for $M^{(b)}, \ldots$, the mass fields of the other Planck particles.

But this departure from linearity in the wave equations for the individual particles prevents a similar equation being obtained for $M = \sum_a M^{(a)}$. Nevertheless, the addition of the individual equations can be considered in a homogeneous universe to lead to an approximate wave equation for M of the form

$$M + \frac{1}{6}RM + \Lambda M^3 = \sum_a \int \frac{\delta_4(X, A)}{\sqrt{-g(A)}} da, \tag{58}$$

$$\Lambda = N^{-2}, \tag{59}$$

where N is the effective number of particles contributing to the sum $\sum_a M^{(a)}$. The latter can be considered to be determined by an Olbers-like cut-off, contributed by the portion of the universe surrounding the point X in $M(X)$ to a redshift of order unity. In the observed universe this total mass $\sim 10^{22} M_\odot$, sufficient for $\sim 2.10^{60}$ Planck particles. The actual particles are of course nucleons of which there are $\sim 10^{79}$ But if suitably aggregated they would give $\sim 2.10^{60}$ Planck particles and with this value for N

$$\Lambda \simeq 2.5 \times 10^{-121}. \tag{60}$$

The next step is to notice that the wave-equation (58) would be obtained in usual field theory from $\delta\mathcal{A} = 0$ for $M \to M + \delta M$ applied to

$$\mathcal{A} = -\frac{1}{2} \int (M_i M^i - \frac{1}{6}RM^2)\sqrt{-g}d^4x + \frac{1}{4}\Lambda \int M^4 \sqrt{-g}d^4x$$
$$- \sum_a \int \frac{\delta_4(X, A)}{\sqrt{-g(A)}} M(X)da. \tag{61}$$

In the scale in which M is m_0 everywhere the derivative term in (61) vanishes and since $G = 3/4\pi m_0^2$ the term in R is the same as in (3), as are also the line integrals, requiring the remaining term to be the same gives

$$\lambda = -3\Lambda m_0^2. \tag{62}$$

Thus we have obtained not only a cosmological constant but also its magnitude, something that lies beyond the scope of the usual theory. With 2.5×10^{-121} for Λ as in (60) and with m_0 the inverse of the Compton wavelength of the Planck particle, $\sim 3.10^{32}$ cm^{-1}, (62) gives

$$\lambda \simeq -2.10^{-56} \text{ cm}^{-2}, \tag{63}$$

agreeing closely with the magnitude that has previously been assumed for λ. In the classical big bang cosmology there is no dynamical theory to relate the magnitude of λ to the density or other physical properties of matter. For observational consistency it is assumed that λ is of order (63). A dynamical derivation is possible if one goes into the very early inflationary epochs. However, the values of λ deduced from those calculations are embarrassingly large, being $10^{108} - 10^{120}$ times the value given by (63). The problem then is, how to reduce λ from such high values to the presently acceptable range (Weinberg 1989). By contrast, the present derivation leads to the acceptable range of values with very few theoretical assumptions.

The theory developed in this paper differs from big-bang cosmology in what we believe to be an important aspect, that the gravitational equations are scale invariant. The gravitational equations including both the creation terms and the cosmological constant then reduce in the constant mass frame to

$$R_{ik} - \frac{1}{2}g_{ik}R + \lambda g_{ik} = -8\pi G \left[T_{ik} - \frac{2}{3}(c_i c_k - g_{ik}\frac{1}{4}c_l c^l) \right]. \tag{64}$$

The immediate successes of the theory are :

(i) The circumstance that G determined by (34) is necessarily positive requires gravitation to act as an attractive force. Aggregates of matter must tend to pull together. This is unlike general relativity where gravitation can as well be centrifugal, with aggregates of matter blowing always apart, as follows if G in the action (3) of general relativity is chosen to be negative.

(ii) In the cosmological case with homogeneity and isotropy the pressure contributed by the c-field term in the gravitational equations is negative, explaining the expansion of the universe.

(iii) Also in the cosmological case, the energy contribution of the c-field is negative, which ensures that when the creation conditions (9) are satisfied the creation process tends to cascade with explosive consequences.

(iv) The magnitude of the constant λ is shown to be of the order needed for cosmology. Unlike big-bang cosmology this is a deduction not an assumption.

These ideas therefore generate hopes for a more comprehensive framework for relating the property of inertia of matter and the phenomena of matter creation to cosmology. It is not claimed that what is outlined here is the final product; rather it should be looked upon as a preliminary attempt. The successes claimed above have to be followed up by a quantum version of the Machian theory and the empirical values of the parameters of the quasi-steady state model have to be related to the fundamental constants of the theory as well as to cosmological boundary conditions. This is the direction in which our future theoretical work will go.

3. Observational Facts and Consequences

Earlier we showed that the approximate solution for the scale factor $S(t)$ is given by equation

$$S(t) = \exp\left(\frac{t}{P}\right)\left\{1 + \alpha\cos\frac{2\pi t}{Q}\right\}.$$
where $P \gg Q$

(22)

We have chosen values of Q and P as follows.

$$Q = 4 \times 10^{10} yr, \ P = 20Q$$

(23)

In (22) α, P, Q are constants determining the model, with $P \gg Q$ a consequence of creation being slow. We now examine the astrophysical consequences of the scale factor $S(t)$ being given by (22).

We are thus concerned with an oscillatory model in which some matter creation occurs, especially near the minimum in each cycle, as was already visualized in HBN 1993. At each oscillation the universe experiences an expansive push. To give a framework for discussion, we suppose creation occurs so that the ratio S_1/S_2 stays fixed, as (22) requires it to do, with S_1

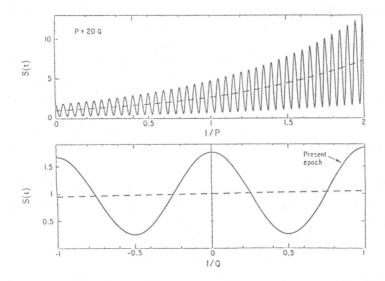

Figure 1. S(t) plotted against t/P (upper panel) and against t/Q (lower panel).

and S_2 both increasing as the slow exponential factor $exp\ t/P$. Thus the timescale for the universe to expand irreversibly by e is $P \gg Q$, which is to say in each exponentiation there are many oscillations. The situation is analogous to the classical steady-state model but with each exponentiation of the scale factor broken into many oscillations.

In Fig 1 we show $S(t)$ plotted against t/P and against t/Q for the assumed value of $P = 20Q$. We also put $\alpha = 0.75$. The time in Fig 1 is measured in units of Q. In order to relate this model to the current state of the observed universe we also need to assign a value for t_0, the present epoch, in relation to the phase of the oscillatory cycles. We choose $t_0 = 0.85$ being 85 per cent of the way through the current cycle, cycles being reckoned maximum to maximum.

The parameter Q is related to the observed values of the Hubble constant H_0 and the deceleration term q_0 respectively. For $P \gg Q$, the effect on H_0, q_0 of the overall expansion will hardly be noticed. The time dependent quantities H, q defined as

$$H = \frac{\dot{S}}{S}\ ,\ q = -\frac{S\ddot{S}}{\dot{S}^2} \tag{65}$$

have the following properties. Starting from the minimum phase of an oscillation, H begins at zero, rises to a maximum and then falls back to zero at maximum phase, while q starts sharply negative and grows to zero, and then

TABLE 1.

Q(years)	$H_0(km \ sec^{-1} \ Mpc^{-1})$
30×10^9	86.2
40×10^9	64.7
50×10^9	51.7

goes to markedly positive values as maximum phase is approached. The observed value of H, H_0, lies between $\sim 50 km \ sec^{-1} \ Mpc^{-1}$ (cf Sandage 1993) and $\sim 80 km \ sec^{-1} \ Mpc^{-1}$ (cf Tully 1993) while Kristian, Sandage and Westphal (1978) gave $q_0 \simeq 1.5$ but with considerable uncertainty. These values are generally indicative of a phase in the current oscillation approaching maximum. Without knowing the precise present-day phase, the time that has elapsed since the last minimum is somewhat uncertain but is probably close to $\frac{1}{2}H_0^{-1}$.

This leads us to the following numerical values relating Q to H_0.

In HBN 1994a we chose the value $Q = 40 \times 10^9$ years with $H_0 = 64.7 km \ sec^{-1} Mpc^{-1}$ as a compromise between the high and low values of H_0 which are a subject of continuous debate. The parameter q_0 is determined from (22) to be 1.725, again close to the value given above. With the choices of α and t_0 the maximum redshift of objects in the present cycle is $z = 4.86$. This is not a limit, however, since the corresponding redshift from the previous cycle will be $z = 5.166$ and so on step by step. However the Hubble diagram for this model shows that the objects in each cycle will be fainter than those in the most recent cycle by about 3 magnitudes. Thus by specifying α, t_0 and the parameter Q based on the observed value of H_0, we obtain reasonable values for q_0, the maximum redshift in this cycle, and a Hubble diagram.

In this model we also have to explain two other properties of the observed universe which have previously been thought to provide strong evidence for the so-called standard model.

3.1. THE COSMIC MICROWAVE BACKGROUND

It has been known for many years that the energy density of the microwave background is almost exactly equal to the energy released in the conversion of hydrogen to helium in the visible baryonic matter in the universe (cf Hoyle 1968). This density is $\rho \simeq 3 \times 10^{-31} gm \ cm^{-3}$ and we suppose that about $7.5 \times 10^{-32} gm \ cm^{-3}$ is H_e. Thus the energy released in the production of this H_e through the conversion $H \to He$ is $4.5 \times 10^{-13} erg \ cm^{-3}$, which if thermalized gives a radiation field of $2.78K$.

In the standard Big-Bang cosmology this agreement with the observed value is considered to be purely fortuitous, but within the framework of QSSC it is a clear indication that the microwave background was generated ultimately by the burning of hydrogen into helium in stars, through many creation cycles each of length Q. The optical and ultraviolet light must have progressively been degraded and scattered by dust, much of it in the form of iron needles so that it now forms a smooth black body form, as discussed at length in HBN 1994, where we predict a temperature of $2.68K$, very close to the observed temperature of $2.735 \pm 0.06K$ (Mather et al. 1990).

How many cycles are required, i.e. what is the value of P/Q? We have shown in HBN 1994 that the ratio P/Q can be obtained from the observed $\log N - \log S$ curves for radio sources. This is because radio sources from earlier cycles are contained in the counts. The reason for this is that while there will be optical obscuration near oscillatory minima and optical sources from earlier cycles will not be easily detectable, this will not apply at long enough radio wavelengths. For a simple model in which it is assumed that radio sources appear at a uniform rate per unit proper volume, we find that $P/Q \simeq 20$. With this value we not only can understand the origin of the microwave background but also the shape of the $\log N - \log S$ radio observations to very faint levels. Thus we have shown that in this model we have two timescales $Q = 40 \times 10^9$ years and $P = 8 \times 10^{11}$ years.

3.2. PRODUCTION OF THE LIGHT ISOTOPES

In the standard model the production of the light elements is attributed to nuclear reactions early in the explosion. In HBN 1993 (Section 6 and Appendix) in Hoyle (1992), and most recently in Hoyle, Burbidge and Narlikar (1995),(HBN 1995), a detailed analysis has been given of a similar process in QSSC in which the light elements are synthesized in a creation process starting with a Planck fireball. In view of the fact that remarkable results are obtained using this approach, we describe it again in some detail here.

Because the early stages in the development of a Planck fireball belong to the realm of unknown physics, it is necessary to begin with a specification of initial conditions. Fermions of familiar types are necessarily excluded by degeneracy conditions at early stages when the fireball dimension is only $\sim 10^{-33}cm$. Indeed, fermions of familiar types cannot appear until the interparticle spacing within the expanding fireball has increased to $\sim 10^{-13}cm$.

We take the view in specifying the model to be investigated that energy considerations discriminate against charmed, bottom and top quarks. We also take the view that degeneracy considerations, together with the need

TABLE 2. Densities and Temperatures at $1 < r < 4$ in the expansion of a Planck Fireball

r	1	1.25	1.5	1.75	2	2.5	3	3.5	4
log N	36.08	35.79	35.55	35.35	35.18	34.89	34.65	34.45	34.27
T_9	0	13.9	19.3	21.2	21.7	20.8	19.3	17.7	16.3

for electrical neutrality, prevent the strange quark from being discriminated against. When the up, down and strange quarks combine to baryons, equal numbers of N, P, Λ $\Sigma^{\pm}$, Σ^0, Ξ^0, Ξ^- are thus formed, with only a negligible amount of Ω^-. Because of the long lifetimes, $\sim 10^{-10}$ seconds, of Λ, $\Sigma^{\pm}$, Ξ^0, Ξ^-, the strange quark survives the effective stages in the expansion of the fireballs, although Σ^0 goes to Λ plus a γ-ray at a stage proceeding the synthesis of the light elements. Finally, we consider that baryons containing the strange quark do not form stable nuclei. Ultimately they decay into N and P, but only after the particle density has fallen so far that the production of light elements has stopped. With N going on a much longer time scale (10 minutes) into P, six of the baryons of the octet go at last into hydrogen. Thus we see immediately that the fraction by mass of helium, Y, to emerge from Planck fireballs is given by

$$Y = 0.25(1 - y), \qquad (66)$$

where $1 - y$ is the fraction of the original N and P to go to ^4He. Anticipating that y will be shown in the next section to be ~ 0.085, equation (66) gives $Y = 0.229$, somewhat lower than the value of ~ 0.237 obtained previously (Hoyle, 1992).

The numerical values used in the detailed calculations of later sections are given in Table 2.

Here N is the number per cm^3 of each baryon type, the values in the table being such that N declines with increasing r as r^{-3}. The unit of r depends on a specification of the total number of baryons in the fireball. Thus for a total of 5.10^{18} the unit of r is $5.10^{-7}cm$. However, since this total is uncertain, because the Planck mass, usually given as $(3\hbar c/4\pi G)^{1/2}$, is theoretically uncertain to within factors such as 4π, we prefer to leave the unit of r unspecified – we shall not need it in the calculations. Suffice it that there will always be a unit for r such that N has the values in the table.

Taking the expansion of the fireball to occur at a uniform speed v, the time t of the expansion to radius r is proportional to r, $t \propto r$. In specifying the model we take the factor of proportionality here to be 10^{-16} seconds. With the unit of r chosen as $5.10^{-7}cm$ this requires $v = 5.10^9 cm \ s^{-1}$, a rather low speed. But for a Planck mass increased by 4π above $(3\hbar c/4\pi G)^{1/2}$ the expansion speed is raised by $(4\pi)^{1/3}$ to $1.16 \times 10^{10} cm \ s^{-1}$. Thus

$$t = 10^{-16} r \text{ seconds,} \tag{67}$$

thereby relating t to N and T_9 through the values in Table 2. The numerical coefficient of equation (2) can be regarded as a parameter of the theory, but it is not a parameter that can be varied by more than a small factor, unlike the parameter η in Big-Bang nucleosynthesis which could be varied by many orders of magnitude for all one knows from the theory.

The temperature values in Table 2 are calculated from a heating source which comes into play at $r = 1$, i.e. at $t = 10^{-16} s$. The heating source is from the decay of π^0 mesons with a mean life of $8.4 \times 10^{-17} s$. The temperature values in Table 2 correspond to a π^0 meson concentration of $2/3N \ cm^{-3}$, which is to say one π meson to each neutron and each proton, with π^0, $\pi^{\pm}$ in equal numbers.

The decay of a π^0 meson into two 75 Mev γ-rays does not immediately deposit energy into the temperature T_9 of the heavy particles. It does not even lead to more than a limited production of $e^{\pm}$ pairs, because at these densities this is prevented by electron degeneracy. Thus the energy of π^0 decay is at first stored in the form of relativistic particles, quanta and some $e^{\pm}$, the latter being adequate, however, to prevent the γ-rays from escaping out of the fireball.

As the fireball expands, confined relativistic particles lose energy proportional to $1/r$, the energy loss going to the heavy particles, for each type of which there is a conservation equation of the form $dQ = dE + PdV$, viz

$$-\alpha d(1/r) \ = \ 3/2kdT + 3kTdV/V, \tag{68}$$

an equation that integrates to give

$$T_9 \ = \ \frac{2\alpha}{3k} \frac{r-1}{r^2}, \tag{69}$$

the constant of integration being chosen to given $T_9 = 0$ at $r = 1$. The constant α in is easily determined from the energy yield of the π^0 mesons. Sharing the energy communicated to the heavy particles equally among all of them, leads to the values of T_9 in Table 2.

The energy is considered to have all gone to the heavy particles by the stage of the expansion when r reaches 4, after which T_9 declines as r^{-2}, i.e. adiabaticlly, the heavy particles being non-relativistic in their thermal motions. Thus for $r > 4$ we have

$$T_9 = 16.3 \cdot \left(\frac{4}{r}\right)^2 = \frac{260.8}{r^2}, \tag{70}$$

$$t = 10^{-16} r = \frac{1.62 \times 10^{-15}}{T_9^{1/2}} \text{seconds}, \tag{71}$$

while the particle densities decline as r^{-3}.

(a) The Abundance of ^4He

It will be shown that neutrons and protons are in statistical equilibrium with ^4He up to $r = 3$ in Table 2, but not for $r > 3$. Defining a parameter ζ by

$$\log \zeta = \log N - 34.07 - \frac{3}{2} \log T_9 \tag{72}$$

it was shown by Hoyle (1992) that the fraction y of neutrons and protons remaining free at temperature T_9 and particle density N for each nucleon type is given in statistical equilibrium by

$$\log \frac{1-y}{y^4} = 0.90 + 3 \log \zeta + \frac{142.6}{T_9}, \tag{73}$$

the values of T_9 and N in Table 2 at $r = 3$ giving $y = 0.085$, leading to the value $Y = 0.229$ given above. A similar calculation at $r = 2.5$ yields $y = 0.083$, much the same as at $r = 3$. For $r < 2.5$ the values of y fall away to ~ 0.06. Thus in moving to the right in the table the values of y increase towards $r = 3$, where the falling value of T_9 eventually freezes the equilibrium.

The condition for freezing is that the break-up of ^4He by $^4He(2N,T)T$, followed by the break-up of T and D into neutrons and protons should just be capable of supplying the densities of P and N, $n(P) = n(N) \simeq 5.10^{33} cm^{-3}$ for the range of r from 2.5 to 3 and $y \simeq 0.085$. The time available for this break-up of ^4He is that for r to increase from ~ 2.5 to ~ 3, i.e. 5.10^{-17} seconds. In this time the break-up of $n(A) \simeq 2.9 \times 10^{34} cm^{-3}$, Using the reaction rates of Fowler, Coughlan and Zimmerman (1975) we verify that.

$$\frac{1.67 \times 10^9}{T_9} \cdot \frac{3.28 \times 10^{-10}}{T_9^{2/3}} \exp -\frac{4.872}{T_9^{1/3}} \cdot \exp -\frac{131.51}{T_9}$$
$$(1 + 0.086 T_9^{1/3} - 0.455 T_9^{2/3} - 0.271 T_9 + 0.108 T_9^{4/3} + 0.225 T_9^{5/3}$$
$$\left(\frac{n(N)}{6.022 \times 10^{23}}\right)^2 n(A). \, 5 \times 10^{-17} \qquad (74)$$

Here we put $T_9 \simeq 20$ for the range of r from 2.5 to 3, and putting $n(N) = 5.10^{33} cm^{-3}$, $n(A) = 2.9 \times 10^{34} cm^{-2}$, the value of (74) is $2.85 \times 10^{33} cm^{-3}$, This is close enough to the required value of 2.5×10^{33}.

This is already an astonishing result. That so complicated an expression as (74) should combine so exactly to produce such an outcome is not a consequence of the parametric choice of the model. The freedom of choice of the numerical coefficient in (74) is entirely dwarfed by the factors 10^{34}, 10^{33}, 10^9, 10^{-10}, 10^{-17} in (74), while even some variation in the parameter α in (4), as it affects the value of the factor $\exp -131.51/T_9 \approx 2.5 \times 10^{-3}$, is also dwarfed by the much larger powers in (74). The most license that can be permitted to a critic would be to accept the above result as model-dependent to the extent that it already consumes essentially all the available degrees of freedom of the model, leaving all further results to be judged as effectively parameter independent.

(b) The Abundances of D and ^3He

We have omitted the analysis which leads to the values

$$D/H =^3 He/H \simeq 5 \times 10^{-5} \qquad (75)$$

which are given in the summary Table 3.

(c) The Abundance of ^7Li

Writing $n(P)$, $n(A)$ for the densities of protons and alpha particles we have

$$n(P) = 1.58 \times 10^{33} \left(\frac{T_9}{16.3}\right)^{3/2} cm^{-3}, n(A) = 8.5 \times 10^{33} \left(\frac{T_9}{16.3}\right)^{3/2} \qquad (76)$$

The ratio of the abundance of ^7Li to ^8Be established in statistical equilibrium at temperature T_9 is given by

$$
\begin{aligned}
\log \frac{^7Li}{^8Be} &= \frac{3}{2}\log\frac{7}{8} + \log 4 - \log n(P) + 34.07 \\
&\quad + \frac{3}{2}\log T_9 - \frac{5.04}{T_9} \times 17.35, \\
&= 3.20 - \frac{87.44}{T_9},
\end{aligned}
\tag{77}
$$

with

$$
\log \frac{^8Be}{^4He} = \frac{3}{2}\log 2 + \log n(A) - 34.67 - \frac{3}{2}\log T_9 = -2.11 \tag{78}
$$

also given by statistical equilibrium.

The abundance of ^7Li established at T_9 according to (23) will, however, be subject to attenuation as the temperature declines from T_9, according to an attenuation factor

$$
A \int_0^{T_9} \exp -\frac{30.443}{T_9} dt, \tag{79}
$$

with

$$
dt = \frac{8.1 \times 10^{-16}}{T_9^{3/2}} dT_9 \tag{80}
$$

as before and A a numerical coefficient obtained from the reaction rate for $^7Li(P, A)^4He$ given by Fowler et al. (1975), viz

$$
A = 1.7 \times \frac{1.05 \times 10^{10}}{T^{3/2}} \cdot \frac{2.40 \times 10^{31}}{6.022 \times 10^{23}} T^{3/2} = 7.25 \times 10^{17} s^{-1} \tag{81}
$$

The factor 1.7 here arises from an estimate of the combined effect of various terms adding to the rate of $^7Li(P, A)^4He$, the rest of A being the main term. Evaluating (79) leads to

$$\sim 19.3 T_9^{1/2} \exp -\frac{30.443}{T_9} \tag{82}$$

as the attenuation factor to be applied to the abundance of ^7Li. With $\log \frac{^4He}{H} = -1.08$ we thus have

$$
\begin{aligned}
\log \frac{^7Li}{H} &= 3.20 - 2.11 - 1.08 \\
&\quad -19.3 \times 0.4343 T_9^{1/2} \exp -\frac{30.443}{T_9} \\
&= 0.01 - 8.38 T_9^{1/2} \exp -\frac{30.443}{T_9}
\end{aligned}
\tag{83}
$$

which has a maximum of -9.60 at $T_9 \simeq 12$. Thus the surviving lithium abundance is

$$\frac{^7Li}{H} \simeq 2.50 \times 10^{-10}, \tag{84}$$

a result in good agreement with the observational requirement, again calculated from highly complicated formulae, again without any model adjustment.

(d) The Abundance of ^{11}B

A similar calculation for ^{11}B leads to $^{11}B/H \simeq 10^{-18}$, below the observational detection limit. This is significantly lower than the value calculated by Hoyle (1992) who used an attenuation factor that was not sufficient. From an observational point of view the model therefore predicts that there is effectively no 'plateau' under boron. Such boron as exists is required to come from cosmic-ray spallation on ^{12}C, ^{16}O.

(e) The Abundance of ^9Be

As noted in Hoyle (1992), the nucleus of ^9Be is exceptionally fragile, leading to a particularly low freezing temperature. Statistical equilibrium at higher temperatures establishes

$$\log \frac{^9Be}{H} = \frac{3}{2} \log \frac{9}{8} + \log \frac{4}{3} - 0.15 + \log D/H$$

$$+ \log \frac{^8Be}{^4He} + \log \frac{^4He}{n(P)} - \frac{3.28}{T_9} \tag{85}$$

with respect to the reaction $^9Be(P,D)2^4He$. Using $\log D/H = -4.30$ already calculated, $\log {}^8Be/^4He = -2.11$, $\log^4 He/n(P) = 0.73$, gives $-5.63 - 3.28/T_9$ for the right hand side of (?). Because $^9Be(P,A)^6Li$ contributes equally with $^9Be(P,D)2^4He$ to the destruction of ^9Be, whereas at $T_9 \simeq 1$ it contributes essentially nothing to the production of ^9Be, the equilibrium concentration of ^9Be is lowered by a further factor 2, so that

$$\log \frac{^9Be}{H} = -5.93 - \frac{3.28}{T_9}. \tag{86}$$

Freezing of the equilibrium condition at $T_9 = 0.50$ for ^9Be would thus give

$$\log \frac{^9Be}{H} = -12.5 \tag{87}$$

in satisfactory agreement with the apparent observed plateau under ^9Be (Boesgaard, 1994).

The estimated freezing temperature according to the model can be obtained by requiring that the product of the expansion time scale, $1.62 \times 10^{-15}/T_9^{1/2}$ seconds at temperature T_9 and the sum of the reaction rates terms for $^9Be(P,D)2\,^4He$ and of those for $^9Be(P,A)^6Li$ be unity, viz

$$2.\frac{1.03 \times 10^9}{T_9} \cdot \frac{2.40 \times 10^{31}}{6.033 \times 10^{23}} \cdot \frac{1.62 \times 10^{-15}}{T_9^{1/2}} T_9^{3/2} . \exp -\frac{3.046}{T_9}$$
$$= 1. \tag{88}$$

The factor 2 on the left of this formula comes from the circumstance that at the values of T_9 in question the highly complicated non-resonant contribution given by FCZ about doubles the resonant reaction rates. Equation (88) determines a freezing temperature $T_9 = 0.623$, reasonably close to the required value of 0.5.

(f) The Abundances of ^{12}C and ^{16}O

The reaction rate on ^9Be from $^9Be(A,N)^{12}C$ as given by FCZ is

$$\sim \frac{2.40 \times 10^8}{T_9^{3/2}} \frac{n(A)}{6.023 \times 10^{23}} . \exp -\frac{12.732}{T_9}. \tag{89}$$

Using (74) for $n(A)$ and putting $T_9 \simeq 10$, at which temperature most of the production of ^{12}C takes place, gives 1.44×10^{16} for (89). Multiplying by the time-scale $1.62 \times 10^{-15}/T_9^{1/2}$ then gives ~ 7.4, implying that an abundance $^9Be/H \simeq 5.5 \times 10^{-7}$ given by (20) is converted 7.4 times over to ^{12}C, leading to

$$\frac{^{12}C}{H} \simeq 4.1 \times 10^{-6} \tag{90}$$

The value of $^{16}O/H$ is of a similar order.

(g) The External Medium

All of the above followed from just the N and P members of the baryon octet. The other six baryons are considered not to form stable nuclei. They decay in $\sim 10^{-10}$ seconds, by which time a Planck fireball has effectively expanded into its surroundings, which according to the QSSC model (HBN 1993,1994a,b) is necessarily a strong gravitational field in which the decay products of Λ, $\Sigma^\pm$, Ξ^0 and Ξ^- may be expected rapidly to lose energy. The Ξ^0 baryon decays to Λ and π^0 in a mean life of $3.0 \times 10^{-10}s$, Σ^+ which decays in a mean life of $8.0 \times 10^{-11}s$, gives a π^0 meson in about a half of the cases, so that together with Λ, which decays in a mean life of $2.5 \times 10^{-10}s$, there is a late production of about $2.5\pi^0$ per baryon octet, yielding ~ 5 late γ-rays per octet, typically with energies $\sim 100~Mev$. It is these γ-rays and their products that are expected to be subjected to energy loss in strong gravitational fields.

The production of Planck particles near large masses of the order of galactic clusters occurs typically in an environmental density $\sim 10^{-16}g~cm^{-3}$ at which density γ-rays of $100~Mev$ have path lengths of $\sim 10^{18}cm$, ample for considerable redshifting effects to occur, when quanta and particles in the $1 - 100~kev$ range would arise. Although such particles and quanta are readily shielded against, it is an interesting speculation that pathways into the external universe may be briefly opened and that the mysterious γ-ray bursts arise in such situations.

(h) Summary of Abundances and Conclusions

The calculations more accurate than those described earlier in Hoyle (1992) and in HBN (1993). They lead to the abundances and results shown in the following table.

To obtain a ratio $^9Be/H \simeq 3.10^{-13}$ requires a freezing temperature $T_9 \simeq 0.5$ which is close but not equal to the calculated freezing temperature $T_9 \simeq 0.62$.

TABLE 3. Summary of Results

$$
\begin{array}{rcl}
{}^{4}He/H & = & Y \simeq 0.229 \\
D/H \simeq {}^{3}He/H & \simeq & 5.10^{-5} \\
{}^{7}Li/H & = & 2.5 \times 10^{-10} \\
{}^{11}B/H & & \text{very small} \\
{}^{12}C/H & \simeq & 4.1 \times 10^{-6}
\end{array}
$$

We conclude that a certain model of the decay of Planck particles leads to interesting values for the abundances of the light elements. The work is deductive, and in this sense the model used is not subject to negotiation, any more than the axioms on which a mathematical theorem is proved are subject to negotiation. Or any more than supporters of Big-Bang nucleosynthesis regard the choice of their parameter η as a matter of negotiation. Thus the only basis for judging the situation is to assess how good, or bad, are the results. Our assessment is the following:

(i) Our result for $^{4}He/H$ is very good.

(ii) The ratios D/H, $^{3}He/H$ are too high by factor ~ 2. A more detailed calculation might well lower $^{3}He/H$ to its observational value. But at the expense of a further increase in D/H, necessitating an epicycle for the theory in which the observed $D/H \simeq 1.5 \times 10^{-5}$ is due to environmental effects.

(iii) The ratio $^{7}Li/H$ is very good.

(iv) The prediction of essentially no 'floor' under ^{11}B is subject to test. The 'floor' under ^{9}Be requires a freezing temperature $T_9 \simeq 0.50$, whereas the calculated freezing temperature was $T_9 \simeq 0.62$. Considering the very complicated expressions of FCZ, especially that involved in a cut-off procedure for non-resonant contributions, this correspondence is adequately close.

Finally we may ask how this situation for the synthesis of light elements from Planck particles compares with the situation in Big-Bang nucleosynthesis. In that case

(a) The classic choice $\eta = 3.10^{-10}$ for the baryon to photon ratio is good for $^{3}He/H$ but is too low for $^{7}Li/H$ and too high for Y and D/H.

(b) While reducing η brings Y and $^{7}Li/H$ into good agreement with observation the value of D/H becomes so large that the theory requires an astration epicycle to save itself.

(c) Raising η to $\sim 6.10^{-10}$ gives good results for D/H, $^{3}He/H$ and $^{7}Li/H$ but the resulting value $Y = 0.25$ is too high, and hardly savable by any epicycle or combination of epicycles.

(d) The theory predicts no plateau under ^9Be, which seems wrong. A recourse to inhomogeneous cosmological models would be to make the theory wildly epicyclic.

(e) Big-bang nucleosynthesis, but not the present model, predicts a present-day average baryon density in the universe much below the cosmological closure value, either forcing a change to a so-called open model (when galaxy formation is made difficult or impossible) or leading to the proposal that most of the material in the universe must be dark and non-baryonic. It is this argument that has led to the proposal that non-baryonic matter dominates the universe. None has so far been found.

3.3. THE VALUE OF ρ_0

To determine ρ_0 (the mean density at this epoch) we have no simple relation like $\rho_0 = 3H_0^2/8\pi G$ in the closure model of Friedmann cosmology. However from the analysis in Part I we have that

$$\frac{\dot{S}^2}{S^2} = \frac{8\pi G\rho_0}{3}\left(\frac{S_0}{S}\right)^3 + \frac{1}{3}\lambda. \tag{91}$$

Also neglecting the slight variation of $exp\ t/P$ over the current half-cycle,

$$\frac{S_0}{S} = \frac{1 + 0.75\cos 1.7\pi}{1 + 0.75\cos 2\pi t} \tag{92}$$

in which $\alpha = 0.75$ and $t_0 = 0.85$ are used. Applying (91) and (92) at $t = 1$, the next maximum when $\dot{S} = 0$, the value of ρ_0 is related to λ by

$$\lambda = -0.558.\ 8\pi G\rho_0. \tag{93}$$

Substituting λ given by (93) in (91) now gives

$$\frac{\dot{S}^2}{S^2} = \frac{8\pi G}{3}\rho_0\left[\left(\frac{S_0}{S}\right)^3 - 0.558\right]. \tag{94}$$

Since $\dot{S}^2/S^2$ at $t = t_0$ is H_0^2 we therefore get

$$0.442\rho_0 = \frac{3H_0^2}{8\pi G}, \tag{95}$$

the coefficient 0.442 being appropriate only for the present moment $t_o = 0.85$. Putting $H_o = 64.7 km\ sec^{-1} mpc^{-1}$, $(Q = 40 \times 10^9)$, determines the present-day average cosmological density as

$$\rho_o = 1.79 \times 10^{-29} g\ cm^{-3} . \tag{96}$$

Thus, since the density of visible matter is $\approx 3 \times 10^{-31} gm\ cm^{-3}$ in this theory as in the standard big bang the bulk of the matter is dark. However the major difference between the QSSC and the big bang is that the dark matter in our theory is made up of baryons. This points up the fact that it is only in the big bang scenario that there is any reason at all to suppose that dark matter is non-baryonic.

3.4. THE COMPOSITION OF THE DARK MATTER

Since a typical galaxy goes through many cycles Q in our model, stellar evolution is not limited by the value of $2/3H_0 \simeq 15 \times 10^9$ years as it is in the big bang.

In our Galaxy we then will expect to have the following stellar components:

1. The known stellar population with stars with ages ranging from $\sim 10^7$ years for those most recently formed to $\sim 15 \times 10^9$ years for the oldest globular clusters.
2. Stars which were formed from matter created in a previous cycles, and are now $\sim 50 \times 10^9$ years old, and remnants from even earlier cycles.
3. Failed stars – the so-called MACHOS.

The distribution of all of this baryonic matter in disk and halo is determined by the formation process and its dynamical behavior. Creation events on a scale sufficient to augment the masses of galaxies on a significant scale will produce violent gravitational disturbances, mostly disrupting and expelling into a halo all previously existing stars. Thus the present day galactic disk (Population I) is only the most recent of a number of major star-forming episodes. Previous episodes occurring at ages $(15 + 50P) \times 10^9$ years, P–1,2,3,..., are now in the halo. Indeed they form the halo, and the process just described gives a picture of how the halos of galaxies are built up. All galaxies have a largest value of P, corresponding to the epoch of their formation, the larger P_{max} corresponding to massive ellipticals and the smaller P_{max} corresponding to 'late' type spirals and irregulars. We estimate $P_{max} \sim 10$ for a galaxy such as our own and $P_{max} \sim 20$ for a typical massive elliptical.

The main-sequence of a stellar population of age 15×10^9 years is burnt out down to masses of solar order and is usually taken to have mass to light ratio of ~ 3, while a population of age $(15 + 50P) \times 10^9$ years will have its main-sequence burnt-out down to a mass $\sim M_\odot(\frac{1+10P}{3})^{\frac{1}{4}}$ and will have a mass to light ratio $\sim 3 \times (1 + 10\ P/3)$ which for $P = 10$ have the values $0.41 M_\odot$ and ~ 100 respectively.

The stellar mass function determined in the Galaxy from the solar neighborhood has a 'missing mass' factor of about 2. While this is uncertain, it can be explained by the addition of so-called brown dwarfs to the mass function. If the stellar mass function is taken to be everywhere and always the same, high mass-to-light ratios for halo populations and for elliptical galaxies cannot be explained wholly through a brown dwarf component. But much or most of such high mass-to-light ratios can be explained if such populations have ages of $(15 + 50P) \times 10^9$ years with $P \sim 10$ or more.

It is a prediction of our theory that the main-sequence of halo stars should thicken for red dwarfs with masses $\sim 0.5 M_\odot$. Attempts to confirm or deny this prediction would need to be confined, because of the low luminosities of such dwarfs, to the solar neighborhood, say to with $\sim 50pc$ of the sun. Being halo stars they would pass through the solar neighborhood with high velocities, $\sim 200km\ s^{-1}$. An accurate color-magnitude for the high velocity stars of the solar neighborhood would therefore be of great cosmogonic interest. To which should be added the need for accurate theoretical studies of the evolution of very old stars with masses $\sim 0.5 M_\odot$, especially with regard to the possibility of the long-term mixing of the products of nucleosynthesis.

3.5. FAINT GALAXIES

The apparent luminosity of a galaxy of radial co-ordinate r and intrinsic luminosity L observed at a redshift z is given by

$$\sim \frac{L}{4\pi r^2} \frac{1}{(1+z)^2} \frac{1}{S^2(t_o)} \qquad (97)$$

Putting $rS(t_o) \simeq cH_o^{-1} \simeq 2 \times 10^{28} cm, z = z_1 = 5$, a galaxy of absolute magnitude -21 would thus be observed with an apparent bolometric magnitude of about $+27$. Since this is within the range of observation it follows that galaxies with still larger values of r should also be observable, such galaxies having an emission time t that occurred in the previous universal oscillation. For those where emission occurred at the last oscillatory maximum there would be a blueshift with $S(t) > S(t_o)$. If we suppose that at the present we are not far from maximum phase, the blueshift will be comparatively small, and the second factor in (97) will be greater than unity but not greatly so. With $rS(t_o) \simeq 5 \times 10^{28} cm$ in such a case, the apparent bolometric magnitude of a galaxy of absolute magnitude -21 would be somewhat fainter than +26 if absorption of a magnitude at the last oscillatory minimum is included. The theory thus predicts that a multitude of blue galaxies should be observed at about this brightness level, some indeed with spectrum lines that are blueshifted rather than redshifted. The blueness is not an intrinsic property of the galaxies themselves but arises from the oscillatory character of the scale function $S(t)$ together with there being many oscillations occurring in the characteristic expansion time $P \simeq 10^{12}$ years of the universe, and with there being little change of the universe from one oscillation to the next.

Thus the prediction is that faint blue galaxies will appear in profusion at faint magnitude levels. Other explanations of this *observed* phenomenon have been proposed (cf Koo and Kron 1992) but here we have an explanation which comes naturally out of the cosmological model.

Also in this model the universe changes much more slowly than had hitherto been supposed since the appropriate timescale is determined by P. Some 5-10 exponentiations of the scale factor $S(t)$ are required to expand an initially local situation with a dimension of a few megaparsecs to dimensions $\sim 3000 Mpc$. In terms of the galaxies we observe, the average age is $P/3 \sim 3 \times 10^{11}$ years, while the ages of the oldest objects at the limit of observation are $5 - 10P \approx 10^{13}$ years. Clusters of ellipticals like the Coma cluster may well have ages intermediate between these values, i.e. $\sim 2.10^{12}$ years. On this basis we would expect that a large part of the mass will be in the form of evolved stars, not only white dwarfs, neutron stars or black holes but also dead stars with $M > 0.5 M_\odot$, including brown dwarfs. Some part of this may be in the form of completely evolved galaxies.

3.6. THE COSMOGONY ASSOCIATED WITH THE QSSC

The long time scale associated with this cosmology means that galaxies can form at all epochs and that different components in the same galaxy will have very different ages. Thus stellar components with ages in the range $10^7 - 10^{12}$ years will be present.

Since the mass creation events will be at maximum intensity and frequency in the minima of the cycles and decrease in intensity and numbers at the peaks of the cycles young objects will be comparatively rare among the galaxies at comparatively small redshifts.

Observational evidence of a range of ages is present. The evidence is of several different kinds.

(a) A class of faint galaxies often called H II galaxies which were originally investigated by Sargent and Searle (1970) and Kunth and Sargent (1983) are found among the faint galaxies.. Their spectra have very weak continua and line emission characteristics of H II regions together with O and B stars. All of the strong features in the spectra arise from stars with ages $\leq 10^8$ yr. There is some ambiguity about the continua. In one case, Sandage (1963) argued that the colors of the continuum in NGC 2444-2445 were similar to those in the Large Magellanic Cloud and that there could be an underlying old system. However, in general there is no strong evidence that any old stellar population is present.

b) Highly luminous *IRAS* galaxies: Far-infrared observations ($\sim 25 - 200\mu m$) made originally in a few cases from the ground or high-flying aircraft, and then much more extensively from *IRAS*, have shown that there is a large population of spiral and irregular galaxies which emit powerfully in the far-infrared (Soifer, Houck, & Neugebauer 1988). It is generally believed that the far-infrared radiation is thermal emission from dust heated by main-sequence stars, and that the powerful sources are regions in which star formation is dominant – so-called starburst regions. The most extreme examples are the high-luminosity *IRAS* galaxies with luminosities in excess of $10^{12}L_\odot$, practically all in the far-infrared. Nearby examples of such galaxies are Arp 220 and NGC 6240. Such systems are all very irregular. While it has often been argued that such objects are a result of mergers between previously separate galaxies (cf. monograph edited by Wielen 1989), one of us (Burbidge 1986; Burbidge & Hewitt 1988) has made the case that these objects are genuinely young ($\ll 10^9$ years old) systems made up of successive generations starting with fairly massive stars ($20M_\odot \leq M \leq 50M_\odot$) which have themselves in the early generations made the dust which is now being heated by ultraviolet radiation from later generations. These galaxies fulfill all of the tests for young systems. There are no stars detected in them older than A-type systems, they contain huge masses of molecules and we predict that the dust in them will not be of typical galactic form, since it will have condensed from metallic oxides and other compositions typical of material ejected from massive stars. According to our proposal made here, galaxies of the types found in categories *a* and *b* have arisen from recent mass-creation events. Another unexpected observational discovery has been the finding of very young stellar systems in old galaxies. There are many examples. We mention here two recent discoveries.

c) In our own Galaxy, Krabbe et al. (1991) have shown that within 0.5 pc of the center there is a cluster of young massive stars with random motions of $\sim 200km\ s^{-1}$. These have been found by high-resolution imaging of a $2.06\mu m$ recombination line of He I. The stars are broad emission line objects which must have ages less than 10^6 yr. The radio source Sgr A lies in this cluster. This can readily be explained by mass creation very close to the center.

d) Observations using the HST have shown what appear to be very young globular clusters in the central region of the well-known radio galaxy

NGC 1275 (Holtzman et al. 1992). This galaxy gives every indication that parts of it at least have an age $\sim 10^{10}$ yr. Both of these examples suggest that mass creation may well be continuing at a low level in the nuclear regions of old galaxies.

A more general question is to what extent there is good evidence that the majority of galaxies have ages $\sim 10^{10}$ yr. In fact we can only determine ages accurately when we can observe the color-magnitude diagrams of clusters of coeval stars and detect their turn-off on the main sequence or the positions of their horizontal branches. This restricts us so far to our Galaxy and objects like the LMC. The arguments based on color measurements which have been made in general (cf. Larson & Tinsley 1978) involve assumptions about the mix of stellar populations that are not very secure, though they are often implicitly assumed to support the claim that the majority of galaxies are old. If we look at dynamical motions, clusters of galaxies like the Coma Cluster appear from their forms to be totally relaxed, and this means that they are very well mixed. Since in a typical case it takes about 10^9 yr for a galaxy to traverse a cluster diameter, such cluster galaxies should have ages $\gtrsim 10^{10}$ yr. On the other hand, there are many clusters which are clearly not relaxed (e.g., the Hercules Cluster) which contain many bright spirals. The dynamical argument would then suggest that the age is no more than $\sim 10^9$ yr. This was the position taken by Ambartsumian (1958, 1965) when he first discussed expanding associations of galaxies.

So how are galaxies formed in this model? It appears likely that there are two routes. Given a small seed mass, creation in the vicinity of the center will add to that mass which will be driven out but will still remain bound to the overall system. In this way, galaxies will increase in mass as a function of time.

On the other hand, creation at the center will lead to matter which is ejected from the galaxy and can form seeds for the formation of new galaxies.

Explosive ejections of this type is what we see in radio galaxies, QSOs and other active nuclei.

The classical explanation of the latter phenomena is that they arise in a rather mysterious way after some matter from an accretion disk falls into a black hole. A discussion of the many ingenious arguments which have been made has recently been summarized by Blandford & Rees (1992). How are these phenomena to be alternatively explained in a mass-creation event?

The creation units described in this paper have an early stage in which gravitational fields are strong, with creation taking place close to an event horizon, which introduces a time dilatation factor large compared with unity, a factor upward of 10^6 for large creation units. This influences the time scale as measured by an external observer in which the creation unit bursts away from its state near the event horizon, an effect of a strong negative pressure term in the dynamical equations similar to that in the inflationary model (Guth 1981). Now it seems unlikely, especially for a rotating object, that the time dilatation factor will be everywhere the same. Using spherical polar coordinates, the dilatation factor will not be precisely of the form $(1 - 2GM/R)^{-1/2}$ but will also have some dependence on the angular coordinates as in the Kerr metric. To an external observer the moment of breakout from the strong gravitation field will therefore appear dependent on θ and φ. Even though to a comoving observer the times of breakout may not be greatly variable with respect to θ, φ, to an external observer the situation will appear otherwise. That is to say instead of the object expanding after the creation phase as a uniform object, it is likely to emerge in a series of blobs or jerks, every blob appearing as a distinct object in its own right. This type of model may be important in attempts to understand the properties of radio galaxies and other active objects in which matter and high-energy particles are ejected in jets.

Let us consider a few examples starting with M87, the classical radio galaxy with a jet interpreted in this way. The well-known synchrotron jet in the large galaxy M87 lies in position angle 290°. It has been known since 1960 that the position angle of the line joining M87 to the radio galaxy M84 is coincident with the position angle of the jet (Wade 1960). This is as direct evidence as one can have of a changing gravitating object at the center of M87, a blob that later becomes the galaxy which we know as M84. Arguing from the angular coincidence of the position angles, the probability of this being so was already about 100 to 1 in favor, while now we have at least the beginning of a theoretical explanation of how such situations arise. Indeed, taking place repeatedly from the breakup of an initial large mass, the product would be a cluster of galaxies. It is an interesting possibility that the Virgo Cluster was formed in this way or, at any rate, the elliptical galaxies in the Virgo Cluster. The spirals are more likely to have been formed as gas from the object, created as M87, impinging on gas from other neighboring creation units. The circumstances that we still see the jet of M87 suggests that the process of forming M84 occurred fairly recently and that here we have good evidence of a young galaxy.

In addition to this, Arp (1987) has shown that a number of X-ray-emitting QSOs are also aligned in the position angle of the jet, and this also is very suggestive of ejection.

We suggest that it is this type of event – creation process in the center of a massive object – which is largely responsible for the generation of powerful radio sources associated with elliptical galaxies.

There is nearly always a preferred axis of ejection in a powerful source, and many optical and near-IR observations have shown that there is a great deal of optical emission along the major axis of the radio emission. Most of these galaxies are very faint, and many have $z > 1$. In some cases blobs are seen, but most of them are far enough away so that structure of the kind seen in M87 will not be detected.

The recent studies (McCarthy et al. 1991: McCarthy, Elston, & Eisenhardt 1992a; McCarthy, Persson, & West 1992b; McCarthy 1991) show that the optical emission is correlated with the radio emission, and everything suggests that the activity takes place from the inside out.

We would also like to explain the existence of QSOs ejected from galaxies as a variation on this same process, but here much more work needs to be done before we have a satisfactory model. Certainly the ejection process may be understood within the framework of the creation in an active nucleus, and there are some very well aligned ejection cases (cf. Arp 1987; Burbidge et al. 1990). While we believe that the large number of associations between bright galaxies (often spirals) and QSOs with large redshifts provide overwhelming evidence for non-cosmological redshifts, we have not yet found any way of explaining these redshifts using the theory outlined in this paper. We believe that this aspect of the problem of violent activity remains a major challenge.

References

Alpher, R.A., Follin, J.W. and Herman, R.C. (1950) Rev. Mod. Phys., 22, 153

Alpher, R.A., Follin, J.W. and Herman, R.C. (1953) Phys. Rev., 92, 1347

Ambartsumian, V.A. (1958) in: Stoops, R. (ed.) Proc Solvay Conf on Structure of the Universe (Brussels), 241

Ambartsumian, V.A. (1965) in: Proc Solvay Conf on Structure and Evolution of the Galaxies (Brussels), Wiley-Interscience, New York, 1

Arp, H.C. (1987) Quasars, Redshifts, and Controversies (Berkeley: Interstellar Media)

Blandford, R. and Rees, M. (1992) in AIP Conf Proc 254, Black Hole-Accretion Disk Paradigm (New York: AIP), 3

Boesgaard, A. (1995) Proc. Workshop on Light Elements, ESO, held on Elba, May 1994

Burbidge, G. (1986) PASP, 98, 1252

Burbidge, G. and Hewitt, A. (1988) in Comets to Cosmology, ed. A. Lawrence (Lecture Notes in Physics; Berlin: Springer-Verlag), 320

Courant, J. and Hilbert, D. (1962) Methods of Mathematical Physics, Vol. II, (Interscience, New York), p. 727-744

Fowler, W.A., Coughlan, G. and Zimmerman, B. (1975) ARA&A, 13, 69

Gamow, G. (1946) Phys. Rev., 70, 572

Guth, A. (1981) Phys. Rev. D., 23, 347

Holtzman, J.A., Faber, S.M., Shaya, E.J., et al. (1992) AJ, 103, 691

Hoyle, F. (1968) Proc. Roy. Soc. A., 308, 1

Hoyle, F. (1992) AP&SS, 198, 177

Hoyle, F., Burbidge, G. and Narlikar, J.V. (1993) ApJ, 410, 437

Hoyle, F., Burbidge, G. and Narlikar, J.V. (1994a) MNRAS, 267, 1007

Hoyle, F., Burbidge, G. and Narlikar, J.V. (1994b) A& A, 289, 721

Hoyle, F., Burbidge, G. and Narlikar, J.V. (1994c) Proc. Roy. Soc. A, December, 1994

Hoyle, F., Burbidge, G. and Narlikar, J.V. (1995) Proc. Workshop on Light Elements, ESO, held in Elba, May 1994

Hoyle, F. and Tayler, R. (1964) Nature, 203, 1108

Kembhavi, A.K. (1979) MNRAS, 185, 807

Koo, D. and Kron, R. (1992) ARA&A, 17, 135

Krabbe, A., Genzel, R., Drapatz, S., Rotaciuc, V., (1991) ApJ, 382, L19

Kristian, J., Sandage, A. and Westphal, J. (1978), ApJ, 221, 383

Kunth and Sargent, W.L.W. (1983) ApJ, 273, 81

Larson, R.B. and Tinsley, B. M. (1978) ApJ, 219, 46

Mather, J.C. Cheng, E.S., Eplee, R.E., et al. (1990) ApJ, 354, L37

McCarthy, P. (1991) AJ, 102, 518

McCarthy, P., van Breugel, W., Kapahi, V.V., and Subramanya, C.R. (1991) AJ, 102, 522

McCarthy, P., Elston, R. and Eisenhardt, P. (1992a) ApJ, 387, L29

McCarthy, P., Persson, S.E. and West, S.C. (1992b) ApJ, 386, 52

Sandage, A. (1963) ApJ, 138, 863

Sandage, A. (1993) in Chincarini, G., Iovino, A., Maccacaro, T., Meccagni, D. (eds.) Proc Conf Observational Cosmology, Milan, ASP Conf. Series 51, p. 3

Sargent, W.L.W. and Searle, L. (1970) ApJ, 162, L155

Soifer, B.T., Houck, J.R. and Neugebauer, G. (1987) ARA&A, 25, 231

Tully, R.B.,(1993) in Chincarini, G., Iovino, A., Maccacaro, T., Meccagni, D. (eds.) Proc Conf Observational Cosmology, Milan, ASP Conf Series 51, p. 18

Wade, C. (1960) Observatory, 80, 235

Wielen (1989) in: Proc Int Conf on Dynamics and Interactions of Galaxies (Heidelberg: Springer-Verlag)

Weinberg, S. (1989) REv. Mod Phys., 61, 1

X-Ray Observations of Galaxy-Quasar Associations

H. Arp

Max-Planck-Institut für Astrophysik
D-85740 Garching, Germany

Abstract Five examples of close associations of quasars with bright, low redshift galaxies have been observed in X-ray wavelengths with ROSAT. In three cases where the galaxies are detected strongly, the nuclei of the galaxies have X-ray extensions in the direction of the adjacent quasars.

In all cases the active galaxies and quasars are located at the origin of apparent lines or pairs of X-ray sources, some involving filamentary X-ray connections to fainter quasars or candidate blue stellar objects. Brighter X-ray sources in these fields are found to be in excess of average survey values. The filaments and connections have measured fluxes of $1 \lesssim F_X \lesssim 60 \times 10^{-13} erg\ cm^{-2}s^{-1}$ and they tend to radiate more strongly in the harder end of the 0.1-2.4 keV energy band.

The most surprising result is the evidence that X-ray sources of optically diverse character are linked together in lines and extended filaments. Because of the rapid decay time of the high energy X-ray radiation, it is implied that we are observing some ongoing process possibly related to matter creation or emergence.

Introduction

From 1966 on a number of galaxy-quasar pairs were discovered. In addition to the very low accidental probability of these quasars falling so close to bright galaxies, the galaxies themselves tended to be unusually active and disturbed. With the capability of pointed observations by the ROSAT X-ray telescope these associations could be observed in wavelengths where the possibility of observing interactions and connections between the high energy components of the quasars and galaxies would be maximized.

We consider the X-ray extensions from the nuclei of the three active galaxies observed here to be the result with the most immediate import. The apparent lines and filaments of X-ray sources are more surprising but potentially also of great importance. The mutually supporting results in the five fields presented here are particularly important for establishing the phenomenon of lines of sources and filaments which characteristically emerge from the strongest X-ray objects in the field.

M. Kafatos and Y. Kondo (eds.), Examining the Big Bang and Diffuse Background Radiations, 369–380.
© 1996 IAU.

1. Mark474 and nearby objects This field is particularly interesting because it illustrates the tendency for companions to large galaxies to be bluer, more active, have slightly higher redshifts and to be associated with more active objects such as Seyfert galaxies and quasars. (Arp 1987; 1994a). A prototype configuration of the Sa galaxy NGC5689, the companion S_d, NGC5682, and the Seyfert galaxy Mark474 was discovered by Arp and Khachikian (1973). A quasar within 95 arc sec of NGC5682 was reported by Arp, Baldwin and Wampler (1975). Observations of a large sample of companion galaxies have shown associations with quasars at an over density of a factor of about 20 (Arp 1983; 1987). In this particular case the probability of accidentally finding an unrelated quasar this close to NGC5682 was only of the order of 5×10^{-3}. However, the probability that the nearby galaxy would be a later type companion was of an order of magnitude less and the probability that such an active Seyfert galaxy would fall so closeby was several orders of magnitude less. Statistically this group appears to represent a highly significant physical association.

Many of the important results are introduced in Fig.1 where it is evident from the 12.8ks, ROSAT PSPC exposure that the Seyfert 1 galaxy, Mark474, is a very strong X-ray source. In these observations Mark474 registered a total of 10,830 counts which enabled an accurate spectrum to be derived. It was fitted to a power law spectrum of photon index $\Gamma = -2.31\pm0.06$ with an HI absorption between $3.6\pm.23$ and $5.5\pm1.0\times10^{20}cm^{-2}$ (depending on the fitting of an additional black body component below 0.5 keV, see Bi 1994). It is also evident that the large Sa galaxy in the field, NGC5689, is erupting X-ray material from its nucleus just along its northern minor axis. Finally there is an apparent extension of X-rays between Mark474 and the quasar of $z = 1.94$ immediately to the northwest.

Fig.2 shows a larger scale view of the X-ray connection between Mark474 and the quasar to a somewhat lower isophote level in the same energy band, 0.5–2.4 keV (softer energy bands do not contain much relative signal in this field). The extension from Mark474 toward the quasar is about $2'$ in length and about $0\!.7$ in width. The width is, however, hard to judge because between the galaxy and the quasar is a low isophote. The most certain result of this feature, however, as shown in Figs.1 and 2, is that the isophotes around the quasar are extended along a line directly away from the nucleus of Mark474.There are about 30 counts associated with this quasar in the 0.5-2.0 keV bands. This is in a region not affected by extended emission arround Mark474 or NGC5682 and represents the clearest evidence for the association of the quasar with the galaxy.

Fig. 1(left) A12.8ks exposure overlayed on a 200-inch photograph shows the strong X-ray Seyfert Mark474, the S_d companion just to the NW, the S_a with X-rays emerging along the minor axis, and the quasar of $z = 1.94$ connected back to the Seyfert. Fig.2 (right) enlargement with lowest X-ray isophotes at about 3, 4, 5 sigma.

In Fig.1 the sources are: a) $z = 1.94$ quasar; b) compact blue galaxy identified in halo of Mark474 by application of deconvolution program (Bi 1994); c) extended source, grouping of faint objects some of which may be blue; d) S_a galaxy NGC5689; e) seven reddish galaxies distributed linearly in shape of a "V". The X-ray source, however, coincides with a 20 mag BSO. Other X-ray sources which coincide with blue stellar objects are (BSO); 7) 21 mag. BSO 15) 20 mag. BSO; 6, 10 and 11) 20 mag. BSO.

2. NGC4651 and 3C275.1 Cross correlation of a complete survey of radio (3CR) quasars with bright galaxies (Burbidge et al. 1971) showed the chance of the observed association between the two samples being accidental was only 5×10^{-3}. This result was later confirmed by Kippenhahn and de Vries (1974). For this particular pair the chance of accidental association was 3×10^{-3}. What was not included in this improbability, however, was the fact that the galaxy showed more striking evidence for physical ejection than almost any other galaxy in the Reference Catalog.

The picture of the luminous jet and counterjet emerging from NGC4651 (Sandage et al. 1956) revealed, however, that the jet from the galaxy did not point directly at the quasar but instead was at a position angle about 20° different. This was taken by many as a reason for not

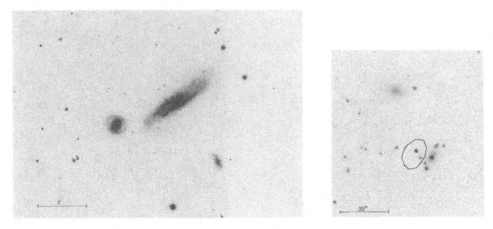

Fig.3 (left) Quasar (a) and compact blue galaxy (b) shown aligned across Mark474. Fig.4 (right) BSO X-ray source in a non-equilibrium configuration of galaxies (source e).

believing that the quasar was related to the galaxy. But, close inspection of the Sandage picture showed evidence for ejection from NGC4651 not only in the direction of the long filament but evidence also for ejection at greater position angles; down to p.a. $\simeq 94°$, much closer to the position angle of the quasar at p.a. $= 100°$. The recently exposed, Palomar, deep IIIa-J survey plate shown here in Fig.5 confirms the optical evidence for material at greater position angles than the main filament and hence closer to the position angle of the quasar. As it turns out in the X-ray observations discussed in the following section as well as in a growing number of optical cases (see for egs. M82 and NGC1097 in Arp 1987) there is evidence that active galaxies tend to eject in cones as well as lines and also in more than one direction. (See also Wilson and Tsvetanov 1994.)

The most important results from the 10.5 ks PSPC exposure are presented in Fig.5 and 6 where it is seen that the galaxy NGC4651, which we have seen is so optically active, has an X-ray jet, or extended X-ray emission, from its nucleus pointing directly at the quasar 3C275.1.

In Fig.6 the broad energy band (0.1–2.0 keV) has been deliberately undersmoothed (FWHM= 24″) to maximize the resolution and the lowest contour at 2σ plotted in order to emphasize the close approach of the X-ray extension from the nucleus of NGC4651 to the quasar. The overall length of the X-ray extension from the nucleus of NGC4651 is $\sim 2.'4$ at the 3 sigma

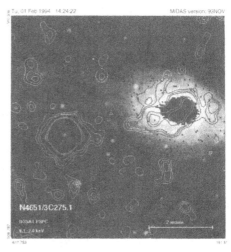

Fig.5. (left) X-ray sources are shown by contours superposed on a IIIa-J photograph from 2nd Palomar Sky Survey. Symbols DG identify very low surface brightness dwarf galaxies. Fig.6. (right) Enlarged view of galaxy and quasar with X-ray isophote levels of 2, 3, 4 and 8 sigma.

level. The extension from the nucleus toward 3C275.1 is 112″ and in the direction away from 3C275.1 it is only 32″. In this 0.1-2.0 keV broad band there are about 90 photons giving a flux of $2 \times 10^{-13} erg\ cm^{-2} s^{-1}$. In the hard energy range, 0.7-2.0 keV, the extension is also pointed directly at the quasar but does not reach so far ($\sim 100″$).

The quasar itself, however, is also extended, in the broad band about 76″ toward the nucleus of the galaxy (the radius orthogonal to this is only 51″ on average). In the hardest (0.9-2.0 keV) band, however, the quasar is only slightly extended toward the galaxy and strongly ($\sim 83″$) extended away from the galaxy.

Therefore the maximum extension between the quasar and the galaxy is in the broad band and the maximum extension away from the galaxy is in the hardest band. Both NGC4651 and 3C275.1, integrated, are among the hardest sources in the field. The galaxy hardness ratio (HR1=.78) is even harder than the quasar (HR1=.54), however, perhaps lending support to the idea that we are seeing a fairly young jet.

As Fig.6 shows there are two apparent condensations on either side of the nucleus of NGC4651. As far as can be determined they are equally spaced across the nucleus (57″ either side). It is an intriguing question as to whether, in the whole extension from the galaxy nucleus, it is more like knots in a continuous filament or separate condensations which are blended instrumentally to resemble a jet.

3. NGC3067 and 3C232. This is another galaxy-quasar pair that Burbidge et al. (1971) found to fall improbably close together in the sky. Later the surprising discovery was reported that low redshift absorption lines of the galaxy appeared in the spectrum of the quasar. The investigators who modeled the galaxy under the assumption it was a normal equilibrium object (Boksenberg and Sargent 1978; Rubin et al. 1982), however, ignored the obvious fact that NGC3067 was a very unusual, high surface brightness, star burst galaxy. In continuum light the galaxy is shattered and chaotic and in the light of H alpha emission there are conspicuous ejection filaments. (Arp 1989) Most recently HI observations showed an extraordinary hydrogen filament leading from the galaxy directly to the quasar. (Carilli, et al. 1989) Therefore, the probability of association of this quasar with the galaxy was extremely high.

The field was observed for 5.6ks with the ROSAT PSPC. The primary result on the galaxy-quasar pair is shown in Fig.7. There it is seen that X-ray material is emerging in a bipolar direction from the nuclear regions of the disturbed, active galaxy, NGC3067. In a high resolution photograph with the KPNO, 4 meter telescope, there appears to be more obscuration on the North side of the galaxy suggesting that is the near side. If this is so we are looking at the underside of the galaxy and the X-ray ejection would be more obscured by the disk of the galaxy on the north side.

Fig.7. Hard X-ray isophotes are plotted around the quasar 3C232 and the starburst galaxy NGC3067. The isophotes suggest a bipolar ejection from the nucleus of the galaxy. Absorption in the galaxy suggests the stronger X-ray lobe is on the nearer, less obscured side and that the longer one is curving over slightly toward the quasar as it extends.

The overall extension measured for the jet is 1.'4. Measured from the optical center of the galaxy it is 46″ in extent, i.e. about 21% longer in the direction toward the quasar than in the direction away from the quasar. The northern X-ray extension starts out at a position angle $\sim 35°$ away from the quasar but then appears to start curving gently over in the direction of the quasar. If the relatively short exposure of 5.6ks had been as long as the 10.5ks of the preceding NGC4651 field, it is possible that the NGC3067 extension would have reached relatively closer to the quasar. In the 0.7-2.0 keV band there are 20 net photons yielding a $F_X = 0.5 \times 10^{-13} erg cm^{-2} s^{-1}$ for the whole nuclear feature in NGC3067. The quasar 3C232 (like 3C275.1) is very hard (again HR1 = .54). But again the galaxy, NGC3067 is the hardest source in the field at HR1 = .63.

4. X-ray Observations with HRI of NGC4319/Mark205.

The bright apparent magnitude, quasar-Seyfert galaxy Markarian 205 falls only 44″ south of the extremely disrupted spiral galaxy NGC4319. Extensive optical observations have established that a luminous filament connects these two objects of much different redshift. (Arp 1971; Sulentic 1983) Since the optical connection is only 2/3 arc min long, the maximum resolution of the ROSAT HRI was required to investigate the area of interest. The immediately apparent result of the 12.4 ks exposure, was that detectable X-ray radiation filled a large part of the central field out to about 8′ radius. The compact, active Mark205 was detected strongly, but also NGC4291, a morphologically undistinguished companion galaxy at 6.4 distance, as well as a number of point sources which are shown in Figs.8 and 9.

X-ray Connections from Mark205. In order to study the lower surface brightness features in the field, the results were smoothed with a special filter that smoothed the 8″ image pixels in the field which had one photon with a $16 \times 8″ = 128″$ gaussian filter, those pixels with two photons with an $11.3 \times 8″ = 90.4$ gaussian filter and then successively $n/\sqrt{2}$ where n goes from 8 to .5. Image pixels with more than 12 photons were not smoothed. Finally all smoothing levels were added into one final picture. This program gives the greatest detectivity for faint surface brightness features by smoothing over large areas and at the same time retains good resolution on the strongest features.

Fig.8 shows the startling result that a number of the point X-ray sources are connected back to the central Mark 205 by luminous X-ray filaments. Of particular importance is the fact that the two brightest quasars to the northwest and south are connected by long, continuous filaments leading from the central region of Mark 205 in roughly opposite directions to each quasar (16.3 from Mark205 to the northern X-ray source and 13.8 to the southern source). The third identified quasar of $z = 1.259$ lies along the filament extending to the southernmost quasar. (These quasars were measured as part of the Einstein, Medium Sensitivity Survey, Stocke et al. 1991).

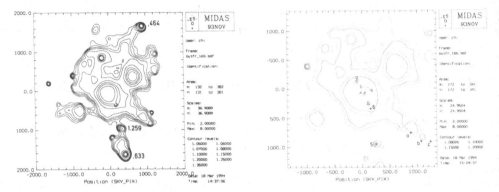

Fig.8 (left) HRI observations show Mark205 to be the strongest X-ray source. Smoothed with a variable filter which gives maximum detectivity to low surface brightness features, the X-ray filaments lead to quasars with the marked redshifts. Fig.9 (right) An enlarged view of higher surface brightness jet emerging 13′ from Mark205. Radio ejections from NGC4319 (Sulentic 1986) are shown inside central isophote.

The fact that Mark 205 is actually ejecting is shown by Fig. 9 where a broad, strong jet emerges at a position angle of $p.a. = 235°$. Tracing this 13′ jet of Fig.9 back shows a pair of sources only 30″ either side of the nucleus at $p.a. = 236°$. Apparently there is a strong, perhaps more recent, ejection from Mark205 at $p.a. = 236°$ and a fainter, thinner ejection at $p.a. = 195°$. The latter extends further and may be somewhat older. This is the ejection that now contains the bright quasars identified in Fig. 8.

There seems to be no trace of an X-ray nucleus in NGC4319. A possible explanation is that violent events have disrupted the morphology of the galaxy and this could account for the now minimal X-ray nucleus. In support of this, optical observations by Sulentic and Arp (1987) showed that some process has removed the hydrogen gas from NGC4319.

5. X-ray Observations of NGC5832 and 3C309.1

This galaxy-quasar pair is another of the high probability associations from the complete 3C radio survey but the galaxy is optically not very disturbed and its separation from the quasar is the largest in the group at 6′2. The PSPC observations reported here for the relatively short exposure time of 4.3 ks.

None of the reductions of this observation showed any trace of X-ray emission at the positions of NGC 5832. Unless longer exposures showed some at the position of the galaxy nothing can be said about any X-ray relation to the quasar. Contour maps of this field are shown in the medium hard band in Fig.10. The apparent alignment of sources there serves to introduce the final result that there is a surprising but ubiquitous tendency for bright apparent magnitude, active X-ray sources to have lines of fainter X-ray sources emerging in at least one, sometimes two directions.

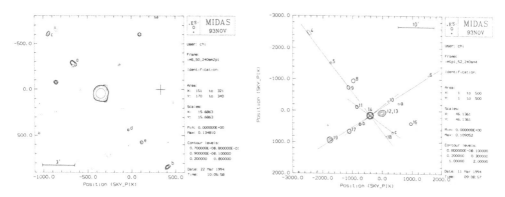

Fig.10. (left) The X-ray source is 3C309.1. The optical position of NGC5832 is indicated by a cross. In the .5-2.4 keV band shown here the sources are 1-1.5 cts ks^{-1}. In the 0.1-2.0 keV band a and b are 10 and 14.5 cts ks^{-1}. Fig.11. (right) All sources in the 0.5=2.4 keV energy range in the NGC4651/3C275.1 field. Lowest contour = .08 photons per 5$''$ image pixel. Lines of sources originating from the quasar are indicated. No. 4 is catalogued quasar of $z = 1.477$.

6. Alignments of X-ray Sources.
The clearest lines of X-ray sources are seen in Fig.5 and particularly Fig.11 emanating from the quasar 3C275.1. Three strong sources extend in a line SE (b, 17 and 19). On the other side are sources 10 and 6. Even more conspicuous in Fig.11, however, is another line of X-ray sources running NE to SW. This includes sources 4, 5, 8, 9 and, on the other side, 18 and c. A strong source at the end of the NE line is no.4 which is a catalogued quasar of $z = 1.477$. Although its outer isophotes are extended due to its 29$'$ distance from the field center, its inner isophotes are conspicuously extended back toward 3C275.1, a characteristic not shared by other sources at this distance from the field center.

378

As independent support for the significance of these alignments we show Fig.12 where the individual photons are recorded between 3C275.1 and source 17. It is apparent that source b is actually a linear distribution of photons stretching from the quasar along the line to the SE.

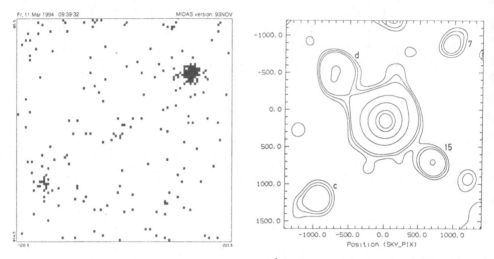

Fig.12 (left) Individual hard photons show source b is elongated by a ratio of about 4 to 1 along the SE line from 3C275.1. Fig.13 (right) Low surface brightness contours in the field of Mark474 showing sources 7, 15 and c are extended away from the Seyfert along previously defined lines.

Initially, in Fig.1 we saw alignments of sources across Mark474. Sources a and b were aligned across the Seyfert at $p.a. = 314°$ and $132°$. Sources 7 and c were aligned at $p.a. = 319°$ and $138°$ and paired at $11.'9$ and $11.'6$ distance. Then almost orthogonal to this is the alignment of NGC5689 nucleus (d) across the Seyfert nucleus with 15 ($p.a. = 50° \pm 1$ and spacing $7.'9$ and $7.'6$). Also roughly along this latter line are sources 6 (BSO) and e (BSO). Now Fig.13 shows the low surface brightness contours of these sources around Mark474 extend away from the Seyfert along the previously identified lines of sources–another independent check of the physical significance of the source alignments.

Of course we saw luminous filaments connecting the quasars back to Mark205 in Fig.8 and alignments of faint sources from 3C309.1 in Fig.10. For the fifth galaxy quasar pair, NGC3067 and 3C232 we show low surface brightness contours in Fig.14.

Discussion. For some the "quasar" Mark205 is technically a Seyfert Galaxy. This is because on the cosmological hypothesis a quasar is arbitrarily defined as $M \approx -23$ mag. and Mark205, at $z = .07$, is only

Fig.14. An area $10' \times 10'$ around NGC3067/3C232 shown in broad energy band, 0.1-2.0 keV. The line of emission extending NE and SW from 3C232 is primarily "soft" while the extension N-S is most conspiouous in hard bands. Smoothing is $35''$ FWHM. Continuation of the northern X-ray material, $17'$ from NGC3067 leads to a $7'$ diameter ring of 5-6 BSO's, all strong quasar candidates.

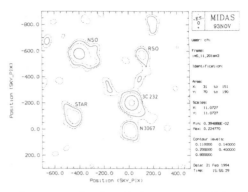

$M = -22.5$ mag. The important point, however, is that it is connected by luminous X-ray filaments to three quasars of much higher redshift. Two of these quasars are optically $m \approx 18 mag$ and statistically they have only a 10^{-3} chance of being accidentally associated with Mark205 at their angular distances. If the X-ray fluxes are plotted on the log N, log S relation for average fields (Hasinger et al. 1993) it is seen that the Mark205 field sources are more than an order of magnitude in excess of normal. This would seem to add up to inescapable evidence that these quasars of high redshift are physically associated with this active object of low redshift.

In three additional cases of the five examined here, active galaxies which had compelling statistical probability of association, were found to have jets or extensions pointing at or near the adjacent, high redshift quasars. The implication of these four cases is that active galaxies eject smaller, higher redshift objects. This process is supported by the fact that the active objects, both galaxies and quasars, have lines of X-ray sources emerging from them. Even in the fifth case observed here, which showed no adjacent active galaxy, the quasar 3C309.1 ($z = .904$) exhibited lines of small X-ray sources emerging from it. These lines from the quasars themselves (such as 3C275.1) contained some confirmed, additional higher redshift quasars as well as numerous blue stellar objects (BSO's) which could easily be confirmed as additional quasars.

The ejection interpretation seems reasonable because radio ejections are well known to emerge in opposite directions from active galaxies, X-ray jets are often seen inside radio jets, and the rotation with time of ejection would naturally explain the slightly different alignment of inner sources with outer sources as seen in Mark474 and 3C275.1.

There is some difficulty with plus and minus ejection velocities, however, and the apparent varying ages of the objects along the ejection path. Also X-rays confined to a filament will decay more quickly than optical

380

or radio photons, counter to what is observed. All this might cause us to consider the possibility of continuous creation along "white lines". This modification would avoid ejection of material bodies and instead suggest emergence of objects at nodes along a "string" at different times. Their intrinsic redshifts would be explained as a function of their age since creation (Narlikar and Arp 1993). In this case the "jet" could represent the energized material in active galaxy nuclei guided out along a creation line.

Regardless of the ultimate theoretical interpretation, however, the ROSAT X-ray observations of galaxies and quasars appear to require a change in basic assumptions which will lead to entirely new distances, masses and ages of objects which comprise the universe.

Arp, H. 1971, Astrophys.Lett. 9, 1.

Arp, H. and Khachikian, E. 1973, Astrofizika 9, 509.

Arp, H., Baldwin, J.A. and Wampler, J. 1975, Ap.J. 198, L3.

Arp, H. 1983, Ap.J. 271, 479.

Arp, H. 1987, "Quasars, Redshifts and Controversies" Interstellar Media, Berkeley.

Arp, H. 1989, ESO Workshop on Extranuclear Activity in Galaxies, ed. E.J.A. Meurs and R.A.E. Fosbury, ESO, May 1989, p.90.

Arp, H. 1994a, Companion Galaxies - A Test of the Assumption that Velocities can be Inferred from Redshifts, Ap.J. in press (July 20).

Bi, H. 1994, I.A.U. Symp. 159, ed. Courvoisier and Blecha, Kluwer, p.366.

Boksenberg, A. and Sargent, W.L.W. 1978, Ap.J. 220, 42

Burbidge, E.M., Burbidge, G.R., Solomon, P.M. and Strittmatter, P.A. 1971, Ap.J. 170, 233.

Carilli, C.L., van Gorkom, J.H. and Stocke, J.T. 1989, Ap.J. 338, 31.

Hasinger, G., Burg, R., Giacconi, R., Hartner, G., Schmidt, M., Trümper, J. and Zamorani, G. 1993, A&A 275, 1.

Kippenhahn, R. and deVries, H.L. 1974, Astrophys. Space.Sci. 26, 131.

Narlikar, J. and Arp, H. 1993, Ap.J. 405, 51.

Rubin, V.C., Thonnard, N. and Ford, W.K. 1982, A.J. 87, 477.

Sandage, A.R., Véron, P. and Wyndham, J., 1965, Ap.J. 142, 1307.

Stocke, J., Burns, J.O. and Ch ristiansen, W.A. 1985, Ap.J. 299, 799.

Stocke, J., Morris, S.L. Gioia, I.M., Maccacaro, T., Schild, R., Wolter, A., Fleming, T.A. and Henry, J.P. 1991, Ap.J.Suppl. 76, 813.

Sulentic, J.W. 1983, Ap.J. 265, L49.

Sulentic, J.W. and Arp, H. 1987, Ap.J. 319, 693.

Wilson, A.S. and Tsvetanov, Z.I. 1994, Ap.J. April issue and STScI Preprint no. 814.

THE PRESENT STATUS OF THE DECAYING NEUTRINO THEORY

D.W. SCIAMA
International School for Advanced Studies (SISSA), Trieste,
International Centre for Theoretical Physics, Trieste, Italy
Department of Physics, Oxford, UK

A discussion is given of recent observations relevant to the decaying neutrino theory. The HST spectrum of the halo star HD 93521 obtained and analysed by Spitzer and Fitzpatrick supports our predictions that warm opaque clouds in the Galaxy should possess substantial partial ionisation, and that the electron density in such clouds near the sun should be constant.

A recent stringent upper limit on the intergalactic ionising flux at $z \sim 0$ obtained by Vogel et al. requires us to reduce the energy of decay photons to less than 13.8eV. The decay flux at $z \geq 2$ could still be mainly responsible for the ionisation of Lyman α clouds and of the intergalactic medium at such red shifts. A further consequence of the Vogel et al. result is that decay photons can no longer ionise nitrogen.

Recent estimates of the isotropic cosmic background at $1500\mathring{A}$ by Henry and Murthy and by Witt and Petersohn agree well with the predictions of the decaying neutrino theory.

A satellite experiment will be carried out in 1995 to search for the predicted decay line from neutrinos within a parsec of the sun. The detectors are being built by Bowyer and his colleagues at the Center for EUV Astrophysics in Berkeley and will be launched in a Spanish minisatellite (principal investigator C. Morales).

1. Introduction

I am very grateful to the organisers of this Symposium for the opportunity to bring my decaying neutrino theory up to date. This theory was originally proposed to explain the widespread ionisation of hydrogen in the Milky Way, which at the time was puzzling astronomers (Sciama 1990, 1993a). I suggested that if most of the dark matter in our Galaxy consists

M. Kafatos and Y. Kondo (eds.), Examining the Big Bang and Diffuse Background Radiations, 381–388.
© 1996 IAU.

of neutrinos with non-zero rest mass, then decay photons from these neutrinos might be the main ionising source for the interstellar medium. This hypothesis is relevant to our present Symposium because it turns out that the decay lifetime required could make the cosmological distribution of neutrinos the main ionising source for the pregalactic medium, the intergalactic medium and Lyman α clouds (Sciama 1993a). Thus the hypothesis, while speculative, is strongly unifying. In addition it makes a number of rather precise predictions, and is therefore very vulnerable to observational disproof. Nevertheless it remains a viable theory at the present time.

In this talk I shall describe the impact on the theory of a number of recent or still unpublished observations. They are the following:

a) The HST spectrum of the halo star HD 93521 (Spitzer and Fitzpatrick 1993).

b) H_α observations of the intergalactic HI cloud 1225+01 (Vogel et al. 1994).

c) A probable Gunn-Peterson absorption trough in HeII (Jakobsen et al. 1994).

d) The isotropic extragalactic background at 1500 Å (Henry and Murthy 1993, Witt and Petersohn 1994).

e) A future satellite experiment to search for the predicted decay line from neutrinos within ~ 1 parsec of the sun.

2. The basic idea

Particle physicists tell us that if neutrinos have a non-zero rest-mass, then a neutrino of type 1 (say τ) would be expected to decay into a photon and a neutrino of type 2 (say μ) of lower mass. The decay lifetime cannot yet be predicted, since it depends strongly on unknown details of particle physics models. We would thus have the decay

$$\nu_1 \to \gamma + \nu_2. \tag{1}$$

Conservation of energy and momentum tell us that the energy E_γ of the decay photon in the rest frame of the parent neutrino is given by

$$E_\gamma = \frac{1}{2}m_1 \left(1 - \frac{m_2^2}{m_1^2}\right). \tag{2}$$

Hence the process produces a *line*, a fact of great importance for the subsequent analysis. It is likely that $m_1 \gg m_2$, in which case

$$E_\gamma \approx \frac{1}{2}m_1, \tag{3}$$

which we shall assume from now on. In order to ionise hydrogen we need

$$E_\gamma \geq 13.6 \text{ eV}. \tag{4}$$

Hence

$$m_1 \geq 27.2 \text{ eV}. \tag{5}$$

We may compare this condition with the Tremaine-Gunn phase space constraint for neutrinos of type 1 to dominate the mass of the Galaxy. One obtains (Sciama 1993a)

$$m_1 \geq 27.6 \text{ eV}, \tag{6}$$

which is essentially the same condition. One also notices that one is immediately in the cosmologically interesting range for m_1.

We now consider the required value of the decay lifetime τ. This was evaluated in Sciama (1990) from the condition that the decay photons should be the main source of the interstellar density of free electrons. One obtains

$$\tau \approx 2 - 3 \times 10^{23} \text{ secs}, \tag{7}$$

which is about a million times longer than the age of the universe.

This estimate for τ would conflict with the analysis by Davidsen at al. (1991) of their HUT observations of the rich cluster of galaxies A665, which led to $\tau > 2 \times 10^{24}$ secs. However a re-analysis by Melott et al, (1994) using an n-body simulation of cluster formation in the presence of neutrinos, modified this constraint to $\tau > 2 \times 10^{23}$ secs, which is just compatible with our required value.

3. The HST observations of HD 93521

In order to prepare for these observations we first consider two basic predictions of the decaying neutrino theory. These predictions apply to an interstellar cloud which is opaque to photons with energy E_γ. Essentially every decay photon produced in the cloud ionises an H atom in the cloud which then recombines. We thus have in a steady state

$$\frac{n_\nu}{\tau} = \alpha n_e^2, \tag{8}$$

where n_ν is the neutrino density in the cloud, n_e is the electron density and α is the recombination coefficient. For clouds within 1 or 2 kpc of the sun n_ν would be nearly constant (halo population). For warm clouds with $T \sim 10^4$ K, α would be nearly constant. Hence n_e is nearly constant, that is, is independent of the total gas density in the cloud. This is our first prediction.

The second prediction follows immediately from the fact that we have a strong source of ionisation uniformly distributed throughout an opaque cloud. We thus expect to find substantial *partial ionisation* within such an opaque cloud. This would not be expected in the prevailing theory (Miller and Cox 1993, Domgorgen and Mathis 1994, Dove and Shull 1994) in which UV photons from distant O stars are regarded as the main source of ionisation. In this case only the outer skins of opaque clouds could be ionised (apart from a small contribution to the inner ionisation from carbon atoms and from cosmic rays). In addition the electron density would then be governed by the total density of the skins, which need not be nearly constant.

We now consider the observations of Spitzer and Fitzpatrick (1993). Their HST spectrum of HD 93521 revealed four opaque slowly moving clouds which are unlikely to be disturbed by shock waves. They derived the electron density n_e in each cloud from the observed column density of CII ions in the upper excited fine-structure level of the ground state.

They found that

a) n_e is the same in each cloud (with only a 10% scatter about the mean)
b) the ionised gas in each cloud is mixed up with the opaque neutral gas, rather than being confined to its surface. Thus the clouds seem to be partially ionised in their interiors. The agreement of these results with the predictions of the neutrino decay theory was pointed out and analysed by Sciama (1993b).

4. The intergalactic ionising flux

A new more stringent upper limit on the intergalactic ionising flux F at $z \sim 0$ can be derived from recent H_α observations of the intergalactic HI cloud 1225+01 (Vogel et al. 1994). This limit is

$$F_{obs} \leq 1.6 \times 10^5 \text{ photons cm}^{-2} \text{ sec}^{-1}. \tag{9}$$

This result imposes a new stringent upper limit on the energy E_γ of a decay photon. To see this we compute the contribution F_ν to F from the cosmological distribution of decaying neutrinos. To allow for the effect of the red shift, which limits the volume of integration, we put

$$E_\gamma = 13.6 + \epsilon \text{ eV}. \tag{10}$$

Then

$$F = \frac{n_\nu}{\tau} \frac{c}{H_0} \frac{\epsilon}{13.6}, \tag{11}$$

where n_ν is here the cosmological neutrino density at $z \sim 0$ and c/H_0 is the radius of the universe. With $n_\nu \approx 100$ cm^{-3}, $\tau \approx 2 \times 10^{23}$ sec and $c/H_0 \approx 2 \times 10^{28}$ cm (see later) we deduce that

$$\epsilon < 0.2 \text{ eV}. \tag{12}$$

Hence

$$E_\gamma < 13.8 \text{ eV}, \tag{13}$$

and

$$m_1 < 27.6 \text{ eV}. \tag{14}$$

Accordingly

$$m_1 = 27.4 \pm 0.2 \text{ eV}, \tag{15}$$

which constrains m_1 to better than 1 percent.

If we allow for the contribution to the density of the universe from baryons, according to the standard model of big bang nucleosynthesis, we find that

$$\Omega h^2 = 0.28 \pm 0.003 \tag{16}$$

where $H_0 = 100$ h km sec^{-1}Mpc^{-1}. If the cosmological constant λ is zero, then arguments concerning the age of the universe (Sciama 1993a) show that

$$\Omega \approx 1 \tag{17}$$

and

$$H_0 \approx 54 \text{ km sec}^{-1}\text{Mpc}^{-1}. \tag{18}$$

This value for H_0 is close to that advocated by Tammann at this Symposium. If $\Omega = 1$ exactly, then our uncertainty in H_0 is less than

$$\pm \frac{1}{2} \text{ km sec}^{-1}\text{Mpc}^{-1}. \tag{19}$$

We now consider the implications of this theory for the ionising flux at large z. With $\Omega \sim 1$ we are close to the Einstein-de Sitter model in which

$$F(z) = (1 + z)^{3/2} F(0). \tag{20}$$

Hence, for example,

$$F_\nu(2) < 8 \times 10^5 \text{ cm}^{-2}\text{sec}^{-1}. \tag{21}$$

According to the proximity effect for Lyman α clouds, as discussed at this Symposium by Bechtold

$$F_\nu(2) \sim 5 \times 10^5 \text{ cm}^{-2}\text{sec}^{-1}. \tag{22}$$

The contribution to $F(2)$ from quasars is still uncertain, but some estimates lead to a value a few times less than this. It is thus still possible that decay photons make the main contribution to $F(2)$, so long as F_ν is close to its present upper limit.

The absence of a Gunn-Peterson absorption trough in HI could also be mainly due to F_ν, but this depends on the presently uncertain density of the intergalactic medium. This question has recently been illuminated by the important probable discovery of a Gunn-Peterson absorption trough in HeII (Jakobsen et al. 1994) which was reported on at this Symposium by Jakobsen. The implied lower limit on the ratio J_{HI}/J_{HeII} of the effective intergalactic ionising fluxes at the HI and HeII edges can be converted to a lower limit on the ratio Φ_{HI}/Φ_{HeII} of the emissivity of the sources by allowing for absorption in the intergalactic medium and in Lyman α clouds and Lyman limit systems (Madau and Meiksin 1994). If this ratio is found to be too steep to be due to quasars it would indicate that a source which preferentially ionises HI is making a major contribution to J_{HI}. This would fit in nicely with the decaying neutrino theory, but more observational work is needed before this possibility can be assessed.

5. The ionisation of nitrogen

One important consequence of our new upper limit on E_γ is that decay photons can no longer ionise nitrogen (whose threshold is at 14.5eV). We must therefore rediscuss the ionisation of nitrogen observed in the local interstellar medium, throughout the Galaxy, and in NGC 891, which we previously attributed to decay photons (Sciama 1993a). Since this Symposium is devoted to cosmology, we do not discuss this question here in detail, and will return to it elsewhere. We simply remark that the large column density of NII towards βCMa implied by the observations of Gry et al. (1985) can be attributed to radiation from βCMa itself. The reason for this is that the column density of HI towards βCMa is 2×10^{18} cm^{-2} (Drew 1994) whereas that towards the nearby star βCMa is only 9×10^{17} cm^{-2}(Cassinelli et al. 1994, Vallerga and Welsh 1994). This disparity indicates that an additional cloud is present along the line of sight towards βCMa, and since the ionising flux in this direction is about ten times greater than in neighbouring directions, it is natural to attribute this extra flux to the additional cloud lying suitably close to βCMa. This extra flux would then be responsible for most of the ionisation of nitrogen in this direction.

With these few remarks we leave the subject of the ionisation of nitrogen and return to it elsewhere.

6. The cosmic background at 1500Å

The earliest lower limits on the lifetime τ of a decaying neutrino were based on observational estimates of the cosmic background in the far ultra-violet (Stecker 1980, Kimble, Bowyer and Jakobsen 1981). Today the value of the observed cosmic background is somewhat controversial (Jakobsen, this Symposium). The most recent estimates for an isotropic background at 1500Å (Henry and Murthy 1993, Witt and Petersohn 1994) are

$$300 \pm 80 \text{ photons } cm^{-2}sec^{-1}ster^{-1}\mathring{A}^{-1} \text{ (C.U.)} \quad (23)$$

The most recent value (Armand et al. 1994) for the contribution due to galaxies is

$$40 - 120 \text{ C.U.} \quad (24)$$

The red shifted contribution from decay photons for $\tau \sim 2 \times 10^{23}$ secs is

$$200 \text{ C.U.} \quad (25)$$

Thus the decaying neutrino theory is close to being tested by these observations.

7. A future satellite experiment

It is planned to search for the predicted decay line from neutrinos near the sun using detectors on board a satellite. The sun is known to be immersed in a cloud with $n_{HI} \sim 0.1$ cm^{-3}, so that unit optical depth for decay photons would occur at a distance ~ 0.5 pc. The resulting decay flux at the Earth would then be about 600 photons cm^{-2} sec. Bowyer and his colleagues at the Center for EUV Astrophysics in Berkeley are building detectors with sufficient sensitivity and energy resolution to observe such a flux. The equipment will be flown on board a Spanish minisatellite (Principal Investigator Carmen Morales) and launch is planned for the end of 1995. The observation period will be about a year, and we hope to obtain a result in time to report it at the next General Assembly of the IAU.

References

Armand, C., Milliard, B., and Deharveng, J.M. (1994), *A & A*, **284**, 12.
Cassinelli, J.P. et al. (1994), to be published.
Davidsen, A.F. et al. (1991), *Nature* **351**, 128.

388

Domgorgen, H. and Mathis, J.S. (1994), *Ap. J.* **428**, 647.

Dove, J.B. and Shull, J.M. (1994), *Ap. J.* **430**, 222.

Drew, J.E. (1994), private communication.

Gry, C., York, D.G. and Vidal-Majdar, A. (1985), *Ap. J.* **296**, 593.

Henry, R.C. and Murthy, J. (1993), *Ap. J.* **418**, L17.

Jakobsen, P. et al. (1994), *Nature* **370**, 35.

Kimble, R., Bowyer, S. and Jakobsen, P. (1981), *Phys. Rev. Lett.* **46**, 80.

Madau, P. and Meiksin, A. (1994), to be published.

Melott, A.L., Splinter, R.F., Persic, M. and Salucci, P. (1994), *Ap. J.* **416**, 12.

Miller, W.W. and Cox D.P. (1993), *Ap. J.* **417**, 579.

Sciama, D.W. (1990), *Ap. J.* **364**, 54.

Sciama, D.W. (1992), *Int. Journ. of Mod. Phys. D.* **1**, 161.

Sciama, D.W. (1993a), *Modern Cosmology and the Dark Matter Problem*, Cambridge University Press.

Sciama, D.W. (1993b), *Ap. J.*, **409**, L25.

Stecker, F.W. (1980), *Phys. Rev. Lett.*, **45**, 1460.

Spitzer, L. and Fitzpatrick, E.L. (1993), *Ap. J.*, **409**, 299.

Vallerga, J.V. and Welsh, B.Y. (1994), to be published.

Vogel, S.N., Weymann, R., Rauch, M. and Hamilton, T. (1994), to be published.

Witt, A.N. and Petersohn, J.K. (1994), in *The First Symposium on the Infrared Cirrus and Diffuse Interstellar Clouds*, ASP Conference Series Vol. 58, ed. R.M. Cutri and W.B. Latter.

BACKGROUND RADIATION: PROBES AND FUTURE TESTS

MARTIN J. REES
Institute of Astronomy
Madingley Road
Cambridge, CB3 0HA.

1. The 'Standard' Hot Big Bang

The clearest evidence for the 'hot big bang' is of course the microwave background radiation. Its spectrum is now known, from the FIRAS experiment on COBE, to be a very precise black body – indeed, the deviations due to high-z activity, hot intergalactic gas, etc are smaller than many people might have expected. Also the light element abundances have remained concordant with the predictions of big bang nucleosynthesis, thereby giving us confidence in extrapolating back to when the universe was a few seconds old (see Copi, Schramm and Turner 1994 for a recent review). These developments give us grounds for greater confidence in this model than would have been warranted ten years ago. Several things could have happened which would have refuted the picture, but they haven't happened. For instance:

(i) Objects could have been found where the helium abundance was far below 23 per cent.

(ii) The background spectrum at millimetre wavelengths could have been weaker than a black body with temperature chosen to fit the Rayleigh-Jeans part of the spectrum.

(iii) A stable neutrino might have been discovered in the mass range 100eV-1MeV.

The key features that determine the present universe – the baryon/photon ratio, the fluctuations, etc – are legacies of exotic physics at ultra-early eras. These issues are now coming into sharper focus. However any inferences about the first microsecond remain tentative because the basic physics is itself uncertain. It is only when the universe has cooled down below 100 Mev that 'conventional' physics becomes adequate, and we can have confidence in quantitative models. We have, however, heard at this meeting about one detailed alternative to the standard picture of the microwave background and nucleosynthesis. This is the model developed by Burbidge, Hoyle and Narlikar (these proceedings). I'd like to mention how measurements of anisotropies in the microwave background might help to distinguish this model from the 'standard' hot big bang.

M. Kafatos and Y. Kondo (eds.), Examining the Big Bang and Diffuse Background Radiations, 389–398.
© 1996 *IAU.*

It is a generic feature of all models which attribute the microwave background to a dense big bang that the dominant opacity on the last scattering surface would be electron scattering. This means that the last scattering surface is located at the same redshift, and has the same thickness, whatever microwave observing frequency is used. Any angular fluctuations attributed to a 'last scattering surface' at high redshift should be the same at each frequency: if a strip of sky were scanned at two frequencies, the temperature fluctuations would track each other closely. (On the Rayleigh-Jeans part of the spectrum this is still true even when there is a Sunyaev-Zeldovich contribution). But in the model of Burbidge *et al.,* the relevant opacity (due to carbon 'whiskers' etc) depends strongly on frequency. Scans at different frequencies are therefore probing 'surfaces' at different distances. One would therefore not expect the same fluctuations, except maybe on the very largest angular scales.

2. Dark Matter

Intimations of dark matter date back to studies of motions in clusters of galaxies in the 1930s; to analyses of motions within the local group, particularly the classic 1959 paper of Kahn and Woltjer; and to radio and optical studies of rotation in the outer parts of disc galaxies. This is not primarily a historical review, but I would like nevertheless to go back 20 years, to 1974, because it was in that year that a consensus about the existence of dark matter was crystalised, particularly in two important papers. One of the classic papers, by Einasto, Kaasik, and Saar (1974), stated that "the mass of galactic coronae exceeds the mass of populations of known stars by 1 order of magnitude, as do the effective dimensions. The mass luminosity ratio rises to $f = 100$ for spiral and $f = 120$ for elliptical galaxies. With $H = 50$ km/sec/Mpc this ratio for the Coma cluster is 170". In the second paper, by Ostriker, Peebles, and Yahil (1974), it was stated that "currently available observations strongly indicate that the mass of spiral galaxies increases almost linearly with radius to nearly 1 Mpc, and that the ratio of this mass to the light within the Holmberg radius is $200 \ M_\odot/L_\odot$".

These particular inferences have been buttressed enormously by progress in the last 20 years; but it is remarkable that the conclusions have not be drastically changed.

Another indirect constraint on the amount of dark matter in baryonic form comes from the abundances of light elements predicted by cosmic nucleosynthesis. As has been well-known since the late 1960s, these abundances depend on the baryon density when the universe cools through the temperature range from 1 Mev to 100 keV, and therefore (since the present background temperature is known) can be related directly to the present baryon density. The predicted helium abundance increases only slowly with density, but the measurements of helium are now precise enough to provide a significant upper limit. Deuterium, however, is a more sensitive measure of the primordial baryon density. Since it is an intermediate product in the production of helium, more deuterium survives in a universe of low baryon density. Moreover, it is now much more clearer than it was in the 1960s that deuterium is best explained as a relic of the early universe.

1974, plainly a vintage year for this subject, also saw the publication of a review by Gott, Gunn, Schramm, and Tinsley (1974). These authors adopted a synoptic approach, and tried to seek consistent ranges for the density parameter Ω and the Hubble constant. They considered three constraints. The first was the requirement that the universe (whose age depends on the Hubble constant and, in Friedman models, on Ω) should be older than the oldest stars. The second was an upper limit on the baryon density from deuterium. And the third was a lower limit to Ω, of order 0.1, set by the amount of dark matter that was reliably established by dynamical arguments. They claimed that there was a very small window, with Ω of order 0.1 and a low Hubble constant, such that the age constraints could be satisfied and all the reliably-established dark matter could be baryonic.

How have Gott *et al.*'s arguments fared in the last twenty years? First, the uncertainties about the Hubble constant and stellar ages are still with us. So let us confront that problem squarely and pass on.

Much more is known about the amount of dark matter in clusters of galaxies, though the net effect of the newer evidence does not substantially change the old estimates. However, the issue of extra dark matter between clusters, maybe even sufficient to provide the critical density, is now a more lively one, and I shall return to it later.

Estimates of deuterium as a measure of baryon density have improved, particularly through a better understanding of the relationship of deuterium and helium[3]. There has recently been a flurry of interest in cosmic deuterium, stimulated by the claim of a high relative abundance of deuterium to hydrogen, of order 3×10^{-4}, in a high redshift damped Lyman-alpha absorption system along the line of sight to a quasar (Songalia *et al.* 1994). If this result were to stand up, it would push down the permitted baryon density, completely ruling out the possibility that most halo dark matter could be baryonic unless one abandons other standard assumptions. However, it would be wise to suspend judgement on this issue. The alleged deuterium line is a weak satellite of a very strong feature attributed to high-column-density HI. It is indeed displaced by 80 kilometres per second, equivalent to the expected isotopic shift, from the centre of a strong hydrogen feature, and there is only a few per cent chance of finding a random weak line in the Lyman forest in this position. But there may very well be an excess of weak 'satellite' lines close to any damped Lyman alpha system (due to gas associated with the same 'protogalaxy'). Until we are sure that there are more systems displaced by 80 km/s than by, say, 60 or 100 km/s the significance of this claim for high deuterium must remain in doubt. Further data, particularly from the Keck Telescope, ought to settle this question within the next couple of years.

The possibility of baryonic dark matter in stars or stellar remnants was addressed with particular thoroughness by Carr, Bond, and Arnett (1984). These authors showed, through a variety of arguments that are now well-known, that there were two possible mass ranges. Dark matter could exist in black holes in the mass range between a few hundred and 10^6 solar masses, which could be a remnant of a population of early massive stars that ended their lives collapsing via the pair production instability. The other possibility is brown dwarf or planetary mass objects, similar to stars except they are below the threshold of around 0.07 solar masses needed to trigger hydrogen fusion. Some constraints on high mass objects in our Galaxy are set by the lack of evidence for accretion onto those passing through the Disc, and so forth. But the most interesting recent work, involving gravitational microlensing and the search for evidence of lensing by low-mass compact objects in our own galaxy, features strongly in another symposium being held in parallel with ours, so I shall say no more about it here.

Traditionally, the dark matter in clusters has been inferred from application of the virial theorem to galaxy motions. But there are now two other lines of attack. Maps of the X-ray brightness profile and temperature are now good enough to allow estimates of the depth of the gravitational well confining the hot X-ray emitting gas. And the detection of large numbers of very faint background galaxies whose shapes are distorted, often into conspicuous arcs, by the effects of light bending due to the clusters gravitational field, will soon offer very direct information about the total mass distribution, whatever that mass may be. One of the early highlights of the data from the post-refurbishment HST is a superb picture of the cluster Abell 2218, with a redshift 0.18, by Ellis, Kneib, and Smail (1994), which shows very large numbers of obvious background arcs.

It will soon seem natural to discuss the structure and dynamics of cluster masses in an order different from the traditional one. We shall first infer the depth of the potential well directly by reconstructing it from the observed distortion by gravitational lensing of background galaxies. It will then be possible to decide whether the observed spatial distribution of galaxies, and the spread in their velocities is consistent with an isotropic equilibrium in that particular potential; if it isn't, the angular distribution of velocities must otherwise be more complex, or the system must be out of equilibrium. X-ray maps will reveal whether the gas has a temperature and density profile consistent with that potential. If it isn't, we shall be motivated to consider whether the gas is partially supported by rotation, macroscopic bulk motion, magnetic pressure, relativistic particles, etc. (The gas can of course be somewhat inhomogeneous, but the clumping factor is constrained because gas confined in the potential well cannot be on a very much higher adiabat than the gas that dominates the X-ray emission.)

Of course clustering must be seen in the more general context of overall cosmic structure formation. Numerical simulations of this are now a heavy industry, and an increasingly sophisticated one. Most of these simulations are based on the assumption that the dominant gravitating stuff is nonbaryonic. So let us briefly consider this option.

One of the main changes since Gott *et al.* wrote their 1974 paper has been the much greater willingness to invoke nonbaryonic matter. Non-zero neutrino masses are no longer thought theoretically unacceptable, and there is a willingness to invoke new kinds of particles, particularly those predicted by supersymmetric theories. What are the prospects for direct detection of nonbaryonic matter? Neutrinos seem impossible to detect by feasible current techniques, and axions present a very severe experimental challenge. But there has been substantial interest in detecting heavy neutral particles, such as the lightest stable supersymmetric particles. These techniques involve detecting the recoil in the rare event when one of these particles, which would pervade the entire halo moving with speeds of about $10^{-3}c$, interacts with a nucleus in an experimental detector. We should certainly spare a thought, and give every possible encouragement, to those of our colleagues, mainly working down mineshafts, who have accepted the challenge to detect dark matter. Even the optimist cannot predict success with great confidence, but the attainable upper limits are themselves becoming significant, and detection of such particles would tell us what 90 per cent of the Universe is made of, as well as perhaps discovering an entirely new class of particles that cannot be produced terrestrially.

Apart from direct detection, another way of reducing the range of non-baryonic options for the dark matter would be by progress in particle physics. If we knew what particles should exist in the ultra-early universe, together with their masses and annihilation cross-sections, it should be possible to calculate which (if any) survive in sufficient numbers to contribute to the dark matter, with the same confidence that we can now apply to calculations of primordial nucleosynthesis. Even in the absence of such direct knowledge, something can be learnt about non-baryonic dark matter by exploring its implications for cosmogony.

3. The 'Cold Dark Matter' Model

The most intensively studied model for structure formation involves the hypothesis that the dark matter is 'cold', in the sense that its thermal motions are never sufficient to smear out small-scale structure. A particular benchmark for comparison of observations has been the so-called 'standard' CDM model. This model involves a package of five assumptions.

1. The primordial spectrum has the Harrison-Zeldovich form, and the fluctuations are Gaussian.

2. The universe is dynamically dominated by cold nonbaryonic matter which interacts only gravitationally with everything else.

3. The density is taken to be equal to the critical value, in other words, $\Omega = 1$.

4. Galaxies are related to dark matter by a simple biasing prescription.

5. Neutrino masses are taken to be zero.

The outcome of these simulations is tested against the data by comparing the relative amplitude of clustering on different scales with what is actually observed at the current epoch. The z-dependence of the structure offers another test. There is limited evidence on how the observed large-scale structure has evolved, but on galactic scales there are constraints back to redshifts of 5 from quasars, neutral hydrogen clouds, etc.

It is now well-known that this 5-item package, the 'standard' CDM model, runs into some problems with reconciling small- and large-scale structure and the microwave background fluctuation amplitude. However, this doesn't mean that the dark matter cannot be in the form of 'cold' non-baryonic matter, because there are a number of modifications of the other four hypotheses which are physically motivated and by no means simply 'ad hoc'. First, the primordial fluctuation spectrum could be tilted, so that the amplitude increases slowly with scale. Indeed most inflationary models predict that this should occur. Also one could consider models where Ω is different from unity: these are either open or else flat with the extra curvature made up by a non-zero cosmological constant. Another uncertainty concerns the relation between the galaxies and the dark matter. The simple scheme which depicts the biasing in terms of one parameter is certainly oversimplified: the galaxy formation efficiency may depend on environment, etc., in many ways. It may be easier to test the models by directly probing the distribution of the dark matter, either by determining the motions it induces in the galaxies which deviate them from the Hubble flow, or by detecting weak lensing due to inhomogeneities on supercluster scales.

And it may turn out that neutrino masses are not exactly zero. The so-called 'hybrid' or 'mixed' dark matter models, in which a neutrino has a mass of a few eV, surmount some of the difficulties of standard CDM. If experimentalists find such evidence for neutrino masses, believers in CDM would delightedly incorporate it in their existing models, ending up with a better fit.

4. Is $\Omega = 1$?

The other change since Gott *et al.*'s classic paper is that there is now a strong theoretical prejudice in favour of $\Omega = 1$, stemming from the attractiveness of the general concept of an inflationary universe. Such models naturally predict that the universe expands enough to stretch the universe flat, in the sense that the Robertson-Walker curvature radius would become enormously larger than the present Hubble scale. Anything different from a flat universe would, as is well known, involve fine tuning in the expansion factor. This tuning is implausible at the level of a few per cent, even in the more optimistically contrived scenarios. However, most variants of inflation allow an even stronger argument in favour of $\Omega = 1$. In these models, if the universe had inflated only enough to make the present Robertson-Walker curvature of order the Hubble radius, there would be quadrupole or dipole effects in the microwave background of order unity. Some recently developed models, however, manage to avoid this latter constraint.

What, then, is the observational case for or against a critical density? I think everyone would agree that this is still tentative. Some of the classical 'geometrical' methods should soon become more helpful. The Hubble diagram for supernovae may be extended to high enough redshifts to reveal the deceleration parameter; further studies may firm up the earlier tentative evidence from the angular diameters of high redshift sources in favour of a high density. On the other hand, if the Hubble constant error bars are reduced, and the Hubble time becomes less than 15 billion years, this will obviously argue against a critical density in which the time since the big bang is only two thirds of the Hubble time.

I should like to conclude this section by mentioning some rather less direct lines of evidence on the density.

Clusters of galaxies offer two such arguments. The first is an inference from the irregular shapes of most clusters, indicating that they have undergone recent mergers of subcomponents each comprising a substantial fraction of the total mass. In a low density universe, structure forms early and is thereafter frozen in. On the other hand, formation continues if Ω is high. Therefore the prevalence of conspicuous substructure points towards a high Ω, though I think this is not yet quantitative enough to allow us to say that it requires the full critical density.

But a quite distinct argument, again based on clusters, suggests a low Ω. X-ray data show that the baryon fraction in a cluster, mainly in hot gas, is typically between 10 and 20 per cent. (The exact fraction depends, of course, on the Hubble constant). This has been inferred from detailed study of the Coma cluster (White *et al.* 1993), and also, in a recent paper by White and Fabian (1994), for a sample of 19 clusters. When the baryonic fraction of the mass in the cluster is compared with the baryonic fraction in the universe allowed by standard big bang nucleosynthesis, there is a contradiction if Ω is more than about 0.3. If

Ω is indeed high, one has either to abandon standard nucleosynthesis, or understand how baryons can be segregated relative to dark matter (by a factor of about 3) even on scales as large as the turnaround radius of a cluster. It could be that the resolution of this dilemma will come from a combination of small effects and uncertainties, but at the moment it seems a serious argument against Ω of 1.

Another quite different estimate of Ω will soon come from studies of microwave background fluctuations. The COBE data refer to angular scales of $10°$. However, several other groups are now reporting fluctuations on angular scales of order $1°$. These latter scales seem to display a larger amplitude than found by COBE. This is precisely what is expected if one is probing scales smaller than the horizon at recombination, because there is then a contribution from Doppler motions, etc. That angular scale is about $2°$ in a flat model, but scales as $\Omega^{\frac{1}{2}}$. Firm evidence for an upturn in the background fluctuation amplitude on angular scales of one or two degrees would be hard to reconcile with a low Ω, where any Doppler contribution would be restricted to angular scales below one degree.

5. Origin of Magnetic Fields

This symposium has dealt with background radiation in all wavebands. At the risk of 'stretching' the definition of background radiation rather far, I'd like to conclude with some remarks on the zero-frequency (DC) limit – large-scale cosmic magnetic fields – whose origin is a mysterious and under-discussed aspect of cosmogony. Fuller details are given elsewhere (Rees 1994).

Cosmic magnetic fields probably owe their present pervasive strength to dynamo amplification. But there must then have been an initial seed field – otherwise the dynamo process would have had nothing to feed on. It seems to be generally 'taken for granted' that the requisite seed field will be there. In many astrophysical contexts this confidence may be justifiable: if the dynamical (and amplification) timescale is short enough, there can be a huge number of e-foldings; a merely infinitesimal statistical fluctuation might then suffice. But the large-scale fields in disc galaxies seem to pose a less trivial problem. The amplification timescale may be 2.10^8 years; even by the present epoch there has been time for only 50 e-foldings. The galactic field could not, therefore, have built up to its observed strength by the present day, unless the seed were of order $10^{-20}G$ – very weak, but not infinitesimal. Moreover, if it turned out that substantial fields existed even in high-z galaxies whose discs may have only recently formed, the seed would need to have been correspondingly higher.

Star formation would proceed differently (with regard both to its rate, and the shape of the initial mass function) if there were no magnetic field: the field modifies the Jeans mass and contributes to transfer of angular momentum. So we cannot hope to model galactic evolution adequately without knowing when the field builds up to a dynamically-important strength. (Moreover, even a weaker field may be significant through its influence on thermal conductivity, etc). If several galactic rotation periods elapsed before a dynamically-significant field built up, then the oldest stars may well, for this reason alone, have a different luminosity function. There is as much reason to believe that the absence of a magnetic field affects the IMF as to believe that a lack of heavy elements does so (though the quantitative nature of the effect is as uncertain in the one case as the other).

Could a field even if only $10^{-20}G$, have been created in the early stages of the big bang? The ultra-early universe may have undergone a phase transition; and maybe this transition could (as in a cooling ferromagnetic material) spontaneously create a field. Because the relevant physics is exotic and poorly understood, we plainly cannot rule this possibility out. However, the correlation scale would be limited to the scale of the horizon. So, even if the field had a high local energy density, it would be primarily on such small scales that it would quickly decay, and there would be no chance of getting even $10^{-20}G$ on the scale of a protogalaxy.

This is a generic problem with attributing a cosmological origin to the field, even if a convincing microphysical mechanism could be found. (Of course, this problem would be surmounted if there were an overall cosmic anisotropy).

A cosmic 'battery' mechanism would have to await nonlinearities that lead to shock waves or the formation of bound systems that exert tidal torques on each other. Compton drag can then (cf Zeldovich, Rosmaikin and Sokoloff, 1983) gradually build up a current in a rotating protogalaxy. If plasma moves at speed V relative to the frame in which the microwave background is isotropic, its motion would be damped out on a timescale $(m_p/m_e)t_{comp}$, where $t_{comp} = m_e c/\sigma_T(aT^4)$ is the usual Compton cooling timescale for electrons. To couple electrons and ions, an E-field of strength $m_e V/et_{comp}$ must maintain itself in the plasma. A protogalaxy of radius R rotating with speed V would be gradually braked by Compton drag, and the E field within it (with, of course, non-zero curl) would build up a B-field at a rate $(m_e c^2/et_{comp})(V/R)$. For a protogalaxy at redshift $z \simeq 5$, this process yields a field only of order $10^{-21}G$.

It is more promising to consider a later origin. I'd like briefly to mention two options: field generation by the first generation of stars, and also in radio galaxies.

Protostars condensing in the present-day interstellar medium start off with too much magnetic flux rather than too little. But the field in a star at the end of its life may be insensitive to the conditions at its birth: even if a star initially had zero field, the Biermann battery could generate a seed field, on which dynamo amplification (by a huge number of factors of e if necessary) could operate. If such a star exploded as a supernova, then a wind spun off the remnant pulsar could pervade several cubic parsecs with a field of order $10^{-4}G$ (just as in the Crab Nebula). So the first few supernovae could have created a weak field throughout the galactic disc, even if a larger-scale battery hadn't already done so.

Provided that the large-scale modes could be preferentially amplified, these stellar-generated fields would be adequate seeds for a galactic dynamo. For a quantitative estimate, note that each hemisphere of the Crab Nebula contains an (equal and opposite) flux of order $10^{34}G$ cm^2. If N similar remnants formed in, for instance, a young galactic disc, the net flux would then be larger by a factor N^x. The appropriate value for x isn't obvious. The net effect depends on the two hemispheres evolving differently – otherwise the net flux cancels out. To assume that $x = 1/2$ may therefore be over-optimistic. A better guess might be $x = 1/3$. This is appropriate if the remnants are randomly oriented, and the galaxy can be modelled as the interior of a surface which slices a fraction N of the remnants. As an example, if $N = 10^6$, the large-scale component of the field in a protogalactic disc of 10 kpc radius would be $3.10^{-8} - 3.10^{-9}G$, for x in the range 1/3 - 1/2.

The highest-redshift radio galaxies formed when the formation of typical galaxies (especially those with discs) still lay in the future. The fields in the lobes of radio galaxies could have been generated in the active nucleus of the associated galaxy and expelled along collimated jets (resembling a scaled-up and directional version of the relativistic pulsar wind that generates the Crab Nebula's field). In the nucleus itself, the dynamical timescale may be as short as a year, or even a few hours if the relevant processes occur close to a black hole. So we need not worry about what seeded the AGN itself: there is time for a battery process to operate, or even for a dynamo to be seeded by an infinitesimal field. Thus, a radio galaxy's field, like that in a supernova remnant, can be accounted for even if the progenitor central object had zero field when it formed.

Galaxies may acquire their discs at $z < 2$ via collapse of a slowly-rotating cloud with turn-around radius > 50 kpc. If the infalling material had been 'contaminated' by a fraction f of a radio source lobe, the large-scale component of the seed field would be $3.10^{-8}(f/10^{-4})G$. So only a small value of f might suffice. However the seed fields in discs could only be attributed to early radio sources if the lobe material were subsequently mixed into a larger volume. This is because radio galaxies are relatively thinly spread through the universe, being far less common than disc galaxies.

The origin of the seed field for the galactic dynamo is a more challenging question than the seeding of smaller-scale cosmic dynamos because the galactic timescale is so long, and the amplification correspondingly slow. (And I have assumed, of course, that the galactic dynamo mechanism is indeed efficient – the problem is obviously far worse if it isn't.) There are as yet no firm grounds for expecting significant fields in the ultra-early universe — indeed there are good reasons for expecting the large-scale components of any such field to be uninterestingly small. And the galactic-scale batteries where Compton drag provides the emf would be barely enough to yield an adequate seed. More promising, in my view are supernova remnants from early stars, or the lobes of high-z radio galaxies, either of which could yield $\sim 10^{-9}G$.

These mechanisms are not mutually exclusive; and there are clearly strong inter-relations between fields in stars, in AGNs or radio galaxies, and in galactic discs. The build-up of a galactic magnetic field depends on how strong the seed field is and when it was generated. Because of the field's importance in star formation, we have little chance of really understanding what a high-redshift galaxy should look like until these issues have been given a good deal more attention by experts in cosmic magnetism.

References

Carr, B.J., Bond, J.R. and Arnett, W.D. 1984, Ap.J., 277, 455

Copi, E., Schramm, D.N. and Turner, M.S. 1994, Science (in press)

Einasto, J., Kaasik, A. and Saar, E. 1974, Nature, 250, 309

Ellis, R.S., Kneib, J.P. and Smail, I. 1994, Ap.J. (in press)

Gott, J.R., Gunn, J.E., Schramm, D.N. and Tinsley, B. 1974, Ap.J., 194, 543

Kahn, F.D. and Woltjer, L.W. 1959. Ap.J., 130, 703

Ostriker, J.P., Peebles, P.J.E. and Yahil, A. 1994, Ap.J. (Lett.), 193, L1

Rees, M.J. 1994 in "Cosmical Magnetism" ed. D. Lynden-Bell, p155 (Kluwer, Dordrecht)

Songalia, A., Cowie, L.L., Hogan, C.J. and Rugers, M. 1994, Nature, 368, 599

White, D.A. and Fabian, A.C. 1994, MNRAS (in press)

White, S.D.M. *et al.* 1993, Nature, 366, 429

Zeldovich, Y.B., Ruzmaikin, A.A., and Sokolov, D.D. 1983, "Magnetic Fields in Astro-physics", Gordon and Breach.

A PANEL DISCUSSION OF
"MAJOR UNSOLVED PROBLEMS OF COSMOLOGY"

Yoji Kondo
Goddard Space Flight Center
Greenbelt, MD 21044, U.S.A.

I would first like to ask my distinguished co-panelists to introduce themselves briefly. I will then ask each to make a five-minute comment on "major unsolved problems of cosmology". It will be followed by discussions among the panelists and also with the audience.

[The panel consisted of Holton Arp of Max Planck Institute for Astrophysics in Munich, Geoffrey Burbidge of the University of California at La Jolla, Menas Kafatos of the George Mason University, Yoji Kondo (Chairman) and John Mather of the NASA Goddard Space Flight Center, Philip Morrison of the Massachusetts Institute of Technology, Bruce Partridge of the Haverford College, Martin Rees of the University of Cambridge, and Michael Turner of the University of Chicago.]

I will lead off the discussion by stating what, in my view, are major unsolved problems in cosmology today. Since I am the first to speak, I wish to make it clear that this is not an attempt to preempt those topics from further discussions by my co-panelists. (A) Is the universe closed or open? What is the value of omega? The resolution of this question may hinge upon the answer to the next question.
(B) Does dark matter exist? If it does exist, what is it or what are they? Is it true that a substantial fraction of the matter in the universe is dark so that it is not directly observable as luminous substance? The evidence for its existence consists primarily of the following three lines of argument.

(1) A number of galaxies -- that are sufficiently near so that we can determine their rotational velocities that are not consistent with the mass content as estimated from the stellar densities and spectral types in those galaxies. Up to some ninety percent of the mass in those galaxies may be non-luminous. This is perhaps the least model-dependent evidence for the presence of dark matter.

(2) If clusters of galaxies are to remain dynamically bound for a cosmological time scale -- say, for at least several billion years -- more mass than is observable in the form of luminous matter must be keeping those galaxies gravitational bound. More than ninety percent of the matter may be dark in such clusters.

(3) If the inflationary Big Bang Model is correct, omega must be close to unity. In order for that to be true, more than ninety-nine percent of the matter in the universe must be invisible. This is probably the most hypothetical of the three lines of arguments.
(C) What is the value of the Hubble Parameter? Is it truly a linear function of distance

M. Kafatos and Y. Kondo (eds.), Examining the Big Bang and Diffuse Background Radiations, 399–400.
© 1996 IAU.

throughout the universe?

(D) What are the properties of large scale structures, such as the 'Great Wall' and the 'Great Attractor'? Are they real or are they simply artifacts of observation?

Throughout the millennia of human history, we have witnessed the birth, death and evolution of various cosmological models. Answers to those questions may help use select or develop a viable cosmological model.

Fundamental Observational Problems

H. Arp
Max-Planck-Institut für Astrophysik
D-85740 Garching, Germany

Almost all of cosmology and extragalactic astronomy are built on the assumption that redshifts are caused by recession velocities and hence measure distances. Evidence has been accumulating since 1966, however, that high redshift quasars and other active objects violate this assumption and are associated with nearby, relatively low redshift galaxies (For recent reviews see Arp 1987; 1992a, 1994a,b and the present conference).

Another recent confirmation of physical association of quasars with galaxies is shown in Fig.1. Two compact X-ray sources are conspicuously paired across the nucleus of the Seyfert galaxy NGC4258 ($z = .0017$). Since this galaxy is known to be ejecting radio and optical material in opposite directions, the authors of this data (Pietsch et al. 1994) conclude these two X-ray sources "may be bipolar ejecta from the nucleus". They report also that at the center of each X-ray source is a blue stellar object of about $m = 20mag$. Optical spectra of these two objects would

determine their redshifts. Considering the large number of expensive new telescope facilities, it is astonishing that more than a year has passed, still without the completion of these two short, crucially

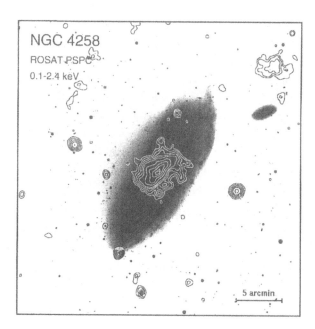

Fig.1. X-ray contours (by Pietsch et al. 1994) overlayed on photograph of NGC4258. Note two strong point-like X-ray sources, each centered on a blue stellar object across the active galaxy nucleus.

M. Kafatos and Y. Kondo (eds.), Examining the Big Bang and Diffuse Background Radiations, 401–406.
© 1996 IAU.

important observations. I think, however, we can conclude that these are two quasars, probably in the redshift range $.3 < z < 1.4$, which have been ejected from this active galaxy.

Another recent result involves the first discovered, most famous quasar, 3C273. Fig.2 shows that this brightest apparent magnitude quasar falls in the largest, brightest cluster of galaxies near us, the Virgo Cluster. There are seven independent sets of evidence that 3C273, or quasars like it, fall in the Virgo Cluster. But the most compelling proof of all is evidenced by the X-ray filament in Fig.2 which connects the center of the Virgo Cluster (M49) in one direction to the powerful radio galaxy 3C274 (M87) and in the other direction to the radio galaxy 3C270 ($z = .007$) a QSO ($z = .334$) and 3C273 ($z = .158$). The luminosity weighted mean of the glaxies conventionally supposed to comprise the Virgo Cluster is $z = .003$.

The importance of this evidence can be judged by the fact that *Nature* published the upper part of the Virgo Cluster X-ray survey but refused to publish the lower part that showed the conspicuous physical connection to the famous quasar. One might draw the conclusion from this that the observations are only photons as a function of x and y – but the conventional theory of distant quasars is knowledge of such higher certainty that is unimpeachable by mere observations.

It is interesting to further note that another famous quasar, 3C279, lies further south and a little east of 3C273. Both 3C273 and 3C279 are strong X-ray sources (0.1 to 2.4 keV) but 3C279 ($z = .538$) becomes stronger than 3C273 in the highest energy gamma rays ($\sim 100 < Mev < \sim 1000$). A picture of 3C279 in gamma rays was published in *Sky and Telescope* (Dec. 1992 p. 634). Closeby, and joined to it by an obvious extension of luminous gamma ray material was 3C273. But by joining a quasar of $z = .158$ to a quasar of $z = .538$ the observations violate the fundamental assumption that redshifts mean distances. Still for practical reasons some observations must be published. The solution: 3C273 was simply not identified in the picture!

Some of these gamma ray pictures are shown and discussed in Arp 1994a. It can be seen there that as one proceeds from the center of the Virgo cluster $\sim$ southward the X-radiation becomes harder, turns into gamma radiation and then ends on the highest gamma radiation of all, going

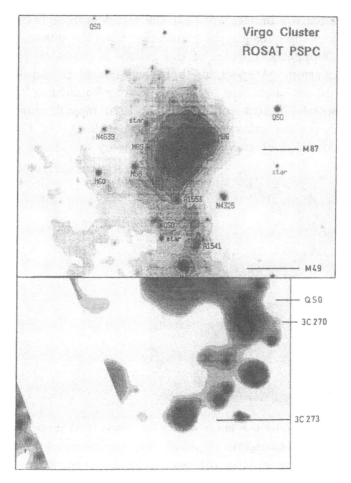

Fig.2 Map of Virgo Cluster in X-rays from ROSAT survey. Upper frame from Böhringer et al. (1994) and lower frame Arp (1994a). Center of Virgo is at M49. In the southern extension 3C273 has $z = .158$ and QSO has $z = .334$.

up to 20,000 Mev at 3C279. The redshifts steadily get higher along this sequence and it is difficult to escape the implication that this is a time sequence of creation of new matter with the youngest object having the highest redshifts (Narlikar and Arp 1993).

Shifting to another case, four quasars ($z = 1.7$) associated so closely ($\lesssim 1''$) with a galaxy ($z = .04$) that no one has dared to call it an accident, we show a photograph of the "Einstein Cross" in Fig. 3. The photograph represents high resolution Hubble Space Telescope images added together and processed with a Lucy image restoration algorithm. Since the images are near the wavelength of Lyman alpha at the redshift of the quasars, the connection from the quasar on the right (D) to the elongated nucleus of the galaxy is indicated to be Lyman alpha in emission. A spectrum between quasars B and A shows the Lyman alpha emission line becoming *narrow* confirming this is a low density gaseous filament connecting the high redshift quasar and the low redshift central galaxy (Arp and Crane 1992). In Fig.3 the theoretical predictions of gravitational lens theory are shown on the right

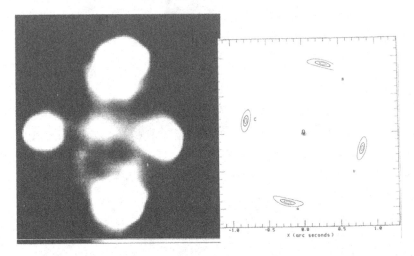

Fig.3 Space Telescope image of Einstein Cross (left) predictions of gravitational lens (right).

and are seen to be exactly orthogonal to the observations. In fact it should have been seen long ago that resolved quasars (as many must be in the conventional picture of a magnified host galaxy) should be drawn into arclets on a circumference in contradiction to the observations. Observations from this, the world's most expensive telescope, are routinely published in Ap.J.Lett. These observations were rejected.

If high red shift quasars and active galaxies have intrinsic redshifts what about more normal galaxies? Smaller companion galaxies in the completely studied nearest galaxy groups *all* show intrinsic redshifts of order of 150 kms^{-1} (Arp 1994b). They also show quantization at 72.4kms^{-1} (Arp 1986). So real velocity components must be less. But now Napier and Guthrie (1993) show that all the most accurately measured redshifts, when differenced, show an enormously significant periodicity of 37.5kms^{-1}. Here real Doppler velocities must be less than about 20kms^{-1} to avoid smearing out the periodicity shown in Fig.4.

Whatever the explanation for this periodicity turns out to be it seems that real motions in the universe must be drastically less than presently supposed. Hence the presently accepted distances, masses and luminosities must be hugely over estimated. Is this compatible with currently accepted physics? Surprisingly the answer is yes. If a solution of the Einstein field equation's more general than the current Friedmann solution is made then particle masses can vary with space and time, $m = m(t)$. This gives all physics as currently observed in the laboratory and in the galaxy for matter of our creation epoch (Narlikar and Arp 1993). It is only when we go extragalactic do we get redshifts as a function of age in an episodicaly creating, non-expanding universe.

But again, regardless of whether this is the correct theory of how the universe works, it is clear that the present observations, of which only a small sam-

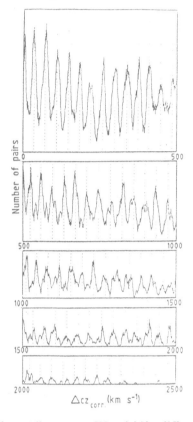

Fig.4 All accurate HI redshifts differenced (Napier and Guthrie 1993)

ple have been shown in this paper, inescapably contradict the current Big Bang hypothesis. In my opinion progress can only be turned from negative to positive in cosmology today by accepting that the current paradigm has been disproved.

At this point I would like to respond briefly to some of the remarks made by Martin Reese during this panel: First he presents the observed superluminal expansions as evidence for the conventional picture. Actually, the situation is the reverse. The observations as they stand violate the Einsteinian upper limit of signal velocity as c. One needs a highly contrived model of jets beamed almost exactly at the observer with quite implausible bulk motions in the 0.99c range. On the other hand if we accept the observations, the quasars are generally 10-100 times closer and their observed velocities of ejection are in a quite precedented range of physical velocity. For example, the "superluminal" 3C273 becomes quite reasonable when moved into the Virgo Cluster.

Secondly, it is particularly disturbing that he should say I do not believe in the age of our own galaxy. I was among the first to do the hard work of measuring globular cluster main sequences which lead to the current best age values. That is the age which requires the observed Hubble constant, $H_o = 50$, even in the non-expanding version of cosmology which I favor. What I do find unwarranted, however, is the extrapolation of this age of our own galaxy to every other galaxy in the universe. Demonstrably, other galaxies are younger and empirically it is this younger age which correlates with their intrinsic redshifts.

Finally, I would like to comment that the efforts to ban the observation of critical objects and to suppress the communication of discordant results is disasterous for science. I feel it is the primary responsibility of a scientist to face, and resolve, discrepant observations.

Arp, H. 1986, Astr. Astrophysics 156, 207.

Arp, H. 1987, "Quasars, Redshifts and Controversies" Interstellar Media, Berkeley.

Arp, H. 1992a, IAU Highlights of Astronomy 9, 43.

Arp, H. 1992b, Phys. Lett. A 168, 6.

Arp, H. 1993, Progress in New Cosmologies, Proceedings of XIII Cracow School of Cosmology, Plenum Press, p2.

Arp, H. 1994a, Frontiers of Fundamental Physics, Proceedings of the Olympia Conference, ed. F. Selleri and M. Barone, Plenum Publ. Corp.

Arp, H. 1994b, Ap.J. 430, 74.

Böhringer, H., Briel, U.G., Schwarz, R.A., Voges, W., Hartner, G. and Trümper, J. 1994, Nature 368, 828.

Napier, W.M. and Guthrie, B.N.G. 1993, Progress in New Cosmologies, Proceedings of XIIIth Cracow School of Cosmology, Plenum Press, p.29.

Narlikar, J.V. and Arp, H.C. 1993, Ap.J. 405, 51.

Pietsch, W., Vogler, A., Kahabka, P., Jain, A. and Klein, U. 1994, Astr. Astrophys. 284, 386.

REDSHIFTS OF UNKNOWN ORIGIN

G. BURBIDGE
University of California, San Diego
Department of Physics and
Center for Astrophysics & Space Sciences
9500 Gilman Drive
La Jolla, California 92093-0111

1. Introduction

Probably the biggest problem in cosmology is one that many people don't even think about or want to think about. It has to do with the nature of the redshifts of astronomical bodies.

In general, we only acknowledge the existence of three ways of explaining redshifts. These are motions, giving rise through Doppler effect to both blueshifts and redshifts, the expansion of the universe, the explanation normally attributed to Hubble but developed by the leading theoreticians in the period 1927-1930 for the redshifts of galaxies, and gravitational redshifts. In the early 1930s the so-called "tired light" hypothesis was invoked as an alternative to expansion by Zwicky, MacMillan and others to explain the redshifts of galaxies, but it was never accepted though the idea was revived by Max Born and others in the 1950s. There are severe difficulties of a fundamental physical nature in this explanation. Comparatively recently Sandage has demonstrated from the surface brightness test that the redshifts of normal galaxies are due predominantly to expansion and not to tired light.

Why then do I consider that the nature of the redshifts is a major problem. It is because, in my view, there is <u>abundant observational evidence</u> that <u>not all</u> of the redshifts of astronomical objects can be explained by expansion, by Doppler effects, or by gravitation.

In general, the shifts due to different physical processes are additive. Let us suppose that the measured redshift of an object is made up of three terms, a term due to the kinetic motions of stars or galaxies $z_r = v_r/c$, a

407

M. Kafatos and Y. Kondo (eds.), Examining the Big Bang and Diffuse Background Radiations, 407–418.
© 1996 *IAU.*

term due to the cosmological expansion z_c, and a term of unknown origin, z_u. For all extended objects (galaxies) the gravitational redshift cz_g must be very small and even for stars, it amounts to no more than $\sim 1 - 2km\ sec^{-1}$.

The observed redshift z_0 is related to the other quantities by the expression

$$(1 + z_0) = (1 + z_r)(1 + z_c)(1 + z_g)(1 + z_u). \tag{1}$$

For stars and nebulae in our own galaxy $z_c = 0$, and the highest kinetic velocities measured are $\overset{\sim}{<} 1000km\ sec, z_r \leq 0.003$. Thus $z_0 = z_r + z_g + z_u$. However in hot stars there is evidence that z_u is not zero. Historically this is known as the K term.

2. The K Term

Where is the K term found and how large is it? The K term originally named by Campbell in 1911, is an excess redshift always seen in the spectra of high luminosity (O and B) stars. It amounts to about $10km\ sec^{-1}$ or $z_u = 0.00003$. While it is very small, the value is well determined, and it is highly significant at the 10 σ level (cf Trümpler 1956). The standard textbooks of the 1930s (cf Russell Dugan & Steward 1938) and distinguished astronomers such as Otto Struve and others all believed the K term to be real. Initially it was thought that it could be explained as a gravitational redshift, but it soon became clear that the gravitational redshift at the surface of a hot star cannot be more than 1-2 $km\ sec^{-1}$ or $z_g \leq 0.000003$. Thus we have a small but measurable redshift term which is *real* but *unexplained*. In a recent study Arp (1992) has extended the earlier work of Trümpler to the A & B supergiants in $h + \chi$ Persei and other young associations, and by using the very accurate redshifts from the 21cm line he has been able to detect the same effect in the most luminous stars in the Magellanic Clouds and in other nearby galaxies.

Arp has also made the very important point that there is a great deal of evidence that the hot stars have strong stellar winds. If this is taken into account the intrinsic redshift term must be equal to the algebraic sum of the observed value of z_u and cv_s where v is the stellar wind velocity at the level at which the lines are formed. Thus the true value of z_u must be considerably in excess of 0.00003, and may be as large as 0.0001.

As we shall show, the pattern of investigation common in astronomy is to ignore a result when it cannot be understood theoretically . In the case of the K term the first explanation was that it was a gravitational redshift. When it was clear that this would not work, it was ignored. *But it remains.* At least in this case many reputable scientists remained aware that there was a problem. In contrast nowadays when a phenomenon cannot be understood, there is not only an attempt made to ignore it, but also to suppress studies of it, and treat very harshly those who persist in working in the field.

3. The Tifft Effect

Starting in 1974, Tifft (1974, 1976) claimed that ordinary galaxies show quantized differential redshifts with a period $c\Delta z_u = 70 - 75km\ sec^{-1}$. He first found this effect in the differences between the redshifts of members of the Coma cluster and later in the redshift differences between physical pairs of galaxies (Tifft 1980). Also Holmberg, and later Arp and Sulentic (1985) showed that in small groups of galaxies dominated by a bright galaxy (e.g. the M81 group) the differences are not distributed at random about the redshift of the main galaxy as would be expected if they were due to satellite motions, or even if they were expanding away from the primary galaxy. It turns out that the mean shift with respect to the central galaxy is displaced to the red. The majority, and some cases <u>all</u> of the differences relative to the central galaxy are redshifts. Not only that but the distribution is quantized with $c\Delta z_u \simeq 72.5km\ sec^{-1}$. The recent work of Tifft and his associates on other samples of galaxies has suggested that the primary value of $c\Delta z_u$ may be $1/2$ or $1/3$ of the original number, i.e. about $36km\ sec^{-1}$ or $24km\ sec^{-1}$. Further analyses have been made by Guthrie & Napier (1988, 1992, 1994) of samples of nearby galaxies which have very accurate redshifts measured using the 21 cm line. Using 89 spiral galaxies with cz in the range $0 - 1000\ km\ sec^{-1}$ whose redshifts were accurately measured ($\sigma \le 4km\ sec^{-1}$) Guthrie & Napier (1992) have shown that when the redshifts are corrected for the optimum solar vector ($v_\odot = 227.9\ km\ sec^{-1}$, $l = 98.7, b = -2\overset{\circ}{.}8$) a periodicity is found at $37.22\ km\ sec^{-1}$, with a probability of finding this period by chance of 2.7×10^{-5}. Guthrie & Napier (1994) have extended this result to more galaxies within the supercluster out to $cz = 2600\ km\ sec^{-1}$ and have confirmed the result. Thus the Tifft effect is confirmed in a variety of nearby samples of normal galaxies.

It is very important to stress that the result of such high significance as that attained above is only obtained after the correction for our motion with respect to the Galactic Center i.e. the periodicity found with $c\Delta z_u = 37.2km\ sec^{-1}$ is associated with the difference in redshifts between the center of mass of our galaxy and the other systems.

4. Large Anomalous Redshifts in Normal Galaxies

There is evidence that individual galaxies can have redshifts very different from their companions, by amounts of the same order as the cosmological redshifts. Such galaxies are rare. Otherwise we would not find a good Hubble relation. The evidence is confined to values of $cz_u \stackrel{\sim}{<} 10000 \; km \; sec^{-1}$ or values of $z_u \stackrel{\sim}{<} 0.03$. For such galaxies $(1 + z_0) = (1 + z_c)(1 + z_r)(1 + z_u)$ If we suppose that $z_r \ll z_c$, and $z_c \sim z_u$, then $z_u = z_0 - z_c - z_u z_c$ The evidence for this effect comes from pairs, or compact groups of galaxies, which because of their proximity or obvious luminous connections, can be assumed to be at the same distance. Thus, for example, in the case of a physical pair, if the galaxies have grossly different redshifts, the redshift difference at least must have a non-cosmological origin. It is possible in such a case that the anomalous component is due to high kinetic velocity, i.e. it is possible to suppose that the galaxy is literally exploding away from the group, so that $z_u = z_r$, but we do not know whether or not this is the case.

The strongest evidence for this type of phenomenon comes from the compact groups of galaxies. By compact we mean isolated groups of 4-7 members with separations comparable to their diameters and low density surroundings. Three groups were the focus of discussion prior to 1985. They were Stefan's Quintet, Seyferts sextet, and VV116. Each of them contains one galaxy with a redshift very different from the means of the others. In the early 1980s a reasonably complete search for compact groups was carried out by HicksonHickson (1982). He found 100 such groups in the Palomar Sky Survey. Of these, 28 have one or more galaxies with a highly discrepant redshift with respect to the mean of the others. Most of the discussion of the compact groups has been centered about how these can exist, since based on the redshift dispersion among the members (excluding the galaxy with a discrepant redshift) the average time for mergers or expansion is only about $10^{-2} H_o^{-1} \sim 10^8$ years. However, the more important question is whether or not it is reasonable to argue that 28% of the groups should contain a discrepant redshift by accident. After the first three cases were discovered, strenuous attempts were made to argue that the discrepant galaxies were either foreground or background galaxies (cf. discussion of paper by Burbidge & Sargent 1970) but it was very clear from that discussion and in a badly flawed paper by Rose (1977), that from the early days there has been an extreme prejudice involved in the discussion in the sense that the "authorities" are always trying to find a way to explain the apparent associations as accidental. This even happened when analyses were done of the frequency of discrepant redshifts in the Hickson groups. Here the argument is clear-cut. Sulentic has done an extensive analysis. He has counted galaxies in circles with radii 0.5° and 1° about the compact groups (Sulentic 1987; see also Rood & Williams 1989; Kindl 1990) and used the local galaxy density to estimate the number of interlopers expected in each case. For the entire catalog of 100 groups he expects 6, and this is to be compared with the 28 found (Sulentic 1987). Thus he concluded and we concur that this is a highly significant result.

It was to be expected that attempts would be made to square the circle. This has been attempted by Hickson et al. (1988) who carried out a similar investigation to that of Sulentic but they concluded that the numbers expected by chance and actually seen were compatible. How did they manage this. By choosing a search radius which represented the *maximum* radius where an interloper would pass the selection criteria. Despite the fact that there are no compact groups in the catalog out at this larger radius this increases the area involved by a factor of about 250 over that used by Sulentic, and thus reduces the significance of the result. However even with this highly conservative approach the internally discordant redshifts cases can be used with the Sulentic search area, and there are 15 (out of 28) of these to be compared with the 6 expected.

All of the 28 cases are listed in Table 1. It is of some interest that of the 28, 20 have z_u positive and 8 have z_u negative. A remarkable feature seen of Table 1 is that most all of the galaxies with discrepant z are spirals, though the Hickson groups contain the normal mix of elliptical and spirals.

Apart from the compact groups of apparently normal galaxies there are also a small number of galaxies which have been discovered by Arp (1971, 1980) to have companions physically connected to them with very different values of z. The best example is NGC 7603 and its companion joined by a luminous bridge with values of cz of 8000 and $16000 km\ sec^{-1}$ respectively. The other cases are not as convincing, but we must accept that the NGC 7603 pair is direct evidence for a value of $cz_u \sim 10000\ km\ sec^{-1}$. This, together with the discrepant cases in the compact groups suggest that in rare cases apparently normal galaxies can have $cz_u \stackrel{\sim}{<} |10000|km\ sec^{-1}$.

TABLE 1. Compact Groups with at Least One Discrepant Redshift

Hickson Group No.	Type of Discrepant Galaxy	Group cz_c ($km\ sec^{-1}$)	z_c	$c(z_d - z_c)$ ($km\ sec^{-1}$)	$cz_u = \frac{c(z_d - z_c)}{1+z_c}$ ($km\ sec^{-1}$)
2	SBb	4320	0.0144	+17020	+16780
3	Sd	7650	0.0255	+3195	+3800
4	Sab	8400	0.0280	+10080	+9800
5	Sc	12300	0.0410	-4085	-3920
14	Sd	5490	0.0183	+2926	+2870
18	SOa	4175	0.0139	+5844	+5760
20	SOa	1420	0.0484	-3959	-3780
23	Sm	4830	0.0161	+5320	+5240
28	Sdm	11400	0.0380	+18805	+18120
29	CI	31410	0.1047	-18082	-16370
31	Sdm	4110	0.0137	+22790	+22480
38	SBa	8760	0.0292	+15522	+15080
43	Sc	9900	0.0330	+9605	+9300
52	Sdm	12900	0.0430	-6607	-6330
53	Sc	6180	0.0206	+2890	+2830
55	Sc	15780	0.0526	+21100	+20040
59	Scd	4056	0.0135	+15600	+15390
61	Im	3900	0.0130	-2773	-2740
63	SBbc	9330	0.0311	-4102	-3980
64	Sd	10800	0.0360	-4653	-4490
71	SO	9030	0.0301	+11560	+11220
72	Scd	12630	0.0421	+11420	+10960
78	SO	9380	0.0313	+8820	+8550
79	EO	4350	0.0145	+15459	+15240
84	EO	16680	0.0556	+15820	+14990
92	Sd	6450	0.0215	-5664	-5540
93	Sa	5040	0.0168	+3841	+3780
98	Sc	7980	0.0266	+6970	+6300

5. Periodicity in z in Faint Galaxy Surveys

Deep pencil beam surveys of galaxies show periodic redshift effects. The discoverers of this effect do not describe it in this way but they say that galaxies mapped in two or three dimensions are not distributed randomly, but show an excess correlation and apparent regularity in the galaxy distribution with a characteristic scale of $128h^{-1}Mpc$ for $z \geq 0.2$ (Broadhurst, et al. 1990; Broadhurst 1994). This corresponds to $cz_u = 12800km\ sec^{-1}$ or $z_u = 0.0426$ for $H_o = 50km\ sec^{-1}Mpc^{-1}$.

6. Summary of Results on Normal Stars and Galaxies

So far we have shown that there is evidence for the existence of an unexplained redshift component in ordinary stars and galaxies ranging from very small values in stars, to periodic differential effects in galaxies, larger components in discrepant systems and even larger periodicity with $z_u \simeq 0.043$.

None of these except possibly the last, is in conflict with the idea that the largest part of the redshift has a cosmological origin for most galaxies. It may have only a minor effect on the tightness of the Hubble relation. However understanding it may help us with the larger problem. One important effect may be that when we compute the random motions of galaxies in clusters and groups by taking each redshift away from the mean, part of the apparent random motion term may be due to a non-zero value of Δz_u. If this is true, the kinetic energy of the group may be over estimated. Thus the virial mass will be over-estimated. This in turn means that the dark matter present which is usually determined by using the virial theorem may be over-estimated.

7. Redshift of other Extragalactic Objects

We now turn to the evidence for an unexplained redshift term in the spectrum of extragalactic objects which are not simply composed of stars and hot gas. These are nearly all of the extragalactic objects which are called strong emission line radio galaxies, active galactic nuclei (AGN) and QSOs. It is in some of these objects that z_u may dominate and periodicities in z_u may also exist. Some may contain stars but in almost all cases the redshifts are measured from lines emitted by hot gas.

From the time of the original discovery of the QSOs it was clear that they did not follow a tight Hubble relation. As more and more objects were discovered it became clear that the Hubble diagram has largely the appearance of a scatter diagram (cf Hewitt & Burbidge 1993 Fig 1 for a recent demonstration of this). The conventional interpretation is that they show a very large scatter in their intrinsic luminosities.

8. Individual Redshifts of QSOs

The strongest evidence that a large part of z_0 for QSOs is of unknown origin is the clear association of many bright QSOs with comparatively nearby bright galaxies. Early work by Arp (1966, 1967 summarized by Arp 1987) and a detailed statistical study of the association of 3C QSOs with the Shapley Ames galaxies (Burbidge et al. 1971) showed that there are far more QSO-galaxy pairs with small separations than are expected by chance. For these systems z_g is very small, so that we must suppose that $z_Q \simeq z_u$. Many statistical studies have been made, and for bright galaxies and bright QSOs this result has been obtained many times. Apart from the pair NGC 4319 - Mk 205 where a clear luminous bridge joins the two objects (Sulentic & Arp 1987) the evidence is generally statistical, and is confined to the brightest QSOs ($\stackrel{\sim}{<} 18^m$) and galaxies brighter than $m = 15.5$. The surface density of QSOs on the sky is well established. Thus this evidence of physical association of QSOs with $z_0 \stackrel{\sim}{<} 2$ and galaxies with separations $\leq 3'$ is very strong at a level of at least 6σ (Burbidge 1979, 1980, Burbidge et al. 1990). There are now six of seven pairs where the morphology involving gas contours and other features as well as the statistics indicates that both components of the pair lie at the same distance (Burbidge 1994). Many still dispute the statistical evidence, but an unbiased analysis suggests that it is acceptable. The argument that the redshifts are due to gravitational lensing of background QSOs in the halos of the galaxies made originally by Canizares (1980) fails because there are not enough faint QSOs to be amplified by objects in the halos of galaxies (Ostriker 1990, Arp 1990).

These results directly imply that z_u can have large values at least up to $z \simeq 2$. At the same time there is evidence that some QSOs have associated faint galaxies with almost identical redshifts so that for some QSOs $z_u \approx 0$.

9. Periodicity in the redshifts of QSOs and Related Objects

Early in the studies of QSOs and related objects a sharp peak at $z = 1.955$ was reported (Burbidge & Burbidge 1967). Soon after this it was noticed that if we restrict ourselves to QSOs, and related objects distinguished from normal galaxies by their non-thermal continuum and emission line spectra which are similar to those of QSOs, the redshifts show a quantized appearance at values $z_u = n \times 0.061$ at least up to $n \simeq 10$. Since the redshifts are mostly very small compared with those of the QSOs most of the objects in the original survey (70 objects) were those in the second category. In this distribution a strong peak was seen at $z = 0.061$ and at multiples of it (Burbidge 1968).

As more QSO redshifts were obtained several additional peaks in the redshift distribution became apparent, particularly at $z = 0.30, 0.60, 0.96$ and 1.41 (Burbidge, 1978). Karlsson (1977) showed that these peaks are periodic with $\Delta \log(1 + z) = 0.089$ i.e. the ratio of successive peaks $(1 + z_0^{n+1})/(1 + z_u^n) = 1.227$. The first peak is at $z_u = 0.061$ and the last discernable peak at $z_u = 1.955$. This analysis referenced above was based upon about 600 QSO redshifts which are mostly comparatively bright radio QSOs. This result was confirmed using larger samples by Fang et al. (1982) and Depaquit, Pecker and Vigier (1985).

A new catalog of extragalactic emission line objects similar to QSOs was compiled recently by Hewitt and Burbidge (1991). It contains 935 objects. More than 700 have redshifts $z \leq 0.2$ and most are Seyfert galaxies, though many emission-line radio galaxies are included with $z \geq 0.2$. A histogram of these redshifts shows a large peak at $z = 0.06$ (Burbidge & Hewitt (1990). There are 89 objects out of about 500 with $z_u < 0.2$, in the very narrow redshift interval $\Delta z = 0.01$ between $z = 0.055$ and $z = 0.065$. Duari, DasGupta and Narlikar (1992) did a new analysis based on all the QSOs which have not been identified by any technique which determines to some extent the redshift range of the objects being discovered. This meant that they used 2146 objects out of the catalogs which contain more than 8000 objects (Hewitt and Burbidge 1991, 1993).

In a plot in their paper the peaks at $0.06, 0.18, 0.24, 0.30, 0.32, 0.36, 0.40$, 0.47, 0.55 and 0.62 can easily be seen. Duari et al. did a power spectrum analysis similar to that done originally on an earlier sample by Burbidge & O'Dell (1972) who confirmed the original peaks at 0.06 and 1.955. They also carried out the Kolmogoroff-Smirnoff test and the comb-tooth test and found strong evidence for the periodicity at 0.06 (the exact value is 0.0565, and its significance is increased when the redshifts are transformed to the Galactocentric frame). A second period of 0.0128 was also found at high significance. As far as the periodicity in large scale in $\Delta \log(1 + z)$ is concerned they were more cautious, and the reality of this periodicity has been recently questioned by Scott (1990).

To summarize these investigations, it appears that the peak at 0.06 and the periodicity up to about $n = 10$ which was first noted in 1968 has been shown to exist with something like 30 times as much data as existed then. The peak at 1.955 is also well established, and the larger scale periodicity may still need more study.

In discussing the reality of such effects it should always be borne in mind that if Δz_u is real and periodic quite a small range of values of z_c will be very effective in smearing out the peaks which would arise from the z_u term. Thus to find peaks at all in the observed data at multiples of 0.06 or at other values is remarkable. It strongly suggests that the cosmological components of objects in these redshift ranges must be very small ($z_c \ll 0.01$), or that

z_c and z_u are related.

10. Summary

We have shown that there is very good observational evidence for the existence in nature of a redshift component of unknown origin in stars, galaxies and QSOs. We summarize the evidence in Table 2.

TABLE 2.

TYPE OF	z_u	$cz_u(km\ sec^{-1})$	Comments
High Luminosity Stars	~ 0.00003	~ 10	K term observed
High Luminosity Stars	~ 0.0001	~ 30	K term corrected for stellar wind velocity
Normal Galaxies	$n_\times 0.000125$	$n_\times 37.6(\pm 0.2)$	Tifft effect - Guthrie and Napier
Normal Galaxies (rare)	$\sim \pm 0.03$	$\sim \pm 10000$	Discrepant redshifts in compact groups and pairs
Faint Galaxies $z \leq 0.2$	$n_\times 0.043$	$n_\times 12800$	Pencil beam studies of faint galaxies
QSOs	$\sim 0 - 2.5$		Non-cosmological redshifts based on QSO-galaxy associations
QSOs & AGN	$n_\times 0.06$	$\sim n_\times 18000$	In Seyfert galaxies, QSOs, and radio galaxies
QSOs and AGN	0.06, 0.30, 0.60 0.96, 1.41, 1.95		Peaks in redshift distribution

The repercussions on cosmology of this general result may be very considerable. When the values of $|u_r|$ are small it may be possible to treat them as minor perturbations on the cosmological redshifts expected in one's favorite cosmological model. But for the objects with the largest redshifts, the QSOs and the radio galaxies, it may very well be that for this class of objects at least only a small part of the observed redshift is cosmological in origin. Alternatively it may be that the large values of z_u found are rare among QSOs (it is only well established for the brighter QSOs) so that a large part of conventional cosmology will survive.

In any case we should take these phenomena seriously and try to understand them rather than ignoring them.

References

Arp, H.C. 1967, ApJ, 148, 321
Arp, H.C. 1971, Astr. Letters, 7, 221
Arp, H.C. 1980, ApJ, 239, 469
Arp, H.C. 1987, "Quasars, Redshifts & Controversies" (Interstellar Media, Berkeley)
Arp, H.C. 1990, A&A, 229, 93
Arp, H.C. 1992, MNRAS, 258, 800
Arp, H.C., Sulentic, J. 1985, ApJ, 29, 88
Broadhurst, T.J. 1994, Proc. of Cambridge Conf., July 1994
Broadhurst, T.J., Ellis, R.S., Koo, D. and Szalay, A.S. 1990, Nature, 343, 726
Burbidge, G. 1967, ApJ, 147, 851
Burbidge, G. 1968, ApJL, 154, L41
Burbidge, G. 1978, Physica Scripta, 17, 237
Burbidge, G. 1979, Nature, 282, 451
Burbidge, G. 1981, Ann. New York Acad. Sciences, 8, 123
Burbidge, G. 1994, to be published
Burbidge, G., Hewitt, A., Narlikar, J.V. and Das Gupta, P. 1990, ApJS, 74, 675
Burbidge, G. and O'Dell, S. 1972, ApJ, 178, 583
Burbidge, E.M., Burbidge, G., Solomon, P. and Strittmatter, P. 1971, ApJ, 170, 233
Burbidge, E.M. and Sargent, W.L.W. 1970, La Semaine & Etude sur Les Noyaux des Galaxies (Pontifical Academy) p. 351
Canizares, C.R. 1981, Nature, 291, 620
Depaquit, S., Pecker, J.C. and Vigier, J.P. 1985, Astron. Nach 306, 7
Duari, D., Das Gupta, P. and Narlikar, J.V. 1992, ApJ, 384, 35
Fang, L.Z., Chu, Y., Liu, Y. and Cao, C. 1982, A&A, 106, 287
Guthrie, B. and Napier, W.M. 1990, MNRAS, 243, 431
Guthrie, B. and Napier, W.M. 1991, MNRAS, 253, 533
Guthrie, B. and Napier, W.M. 1994, preprint
Hewitt, A. and Burbidge, G. 1990, ApJL, 359, L33
Hewitt, A. and Burbidge, G. 1991, ApJS, 75, 297
Hewitt, A. and Burbidge, G. 1993, ApJS, 87, 451
Hickson, P. 1982, ApJ, 255, 382
Hickson, P., Kindl, E. and Huchra, J., ApJL, 329, L65
Karlsson, K.G. 1977, A&A, 106, 287
Kindl, E. 1990, Ph.D. Thesis, U. British Columbia
Napier, W.M., Guthrie, B. and Napier, B. 1988 "New Ideas in Astronomy" (Cambridge Univ. Press) Ed. F. Bertola, J. Sulentic and B. Madore) p. 191
Ostriker, J.P. 1990, "BL Lac Objects" Proc. Workshop held in Como Sept 1988, (Springer-Verlag, Berlin) ed. L. Maraschi, T. Maccacara, M.-H. Ulrich, pp. 476, 477
Rood, H. and Williams, B. 1989, ApJL, 329, L65
Rose, J. 1977, ApJ, 211, 311
Russell, H.N., Dugan, R.S. and Steward, J.Q. 1938, Astronomy (Ginn & Co., Boston) p. 668
Scott, D. 1991, A&A, 242, 1
Sulentic, J.W. 1987, ApJ, 322, 605
Sulentic, J. and Arp, H.C. 1987, ApJ, 319, 687

Tifft, W. 1976, ApJ, 206, 38
Tifft, W. 1980, ApJ, 236, 70
Trumpler, R.J. 1956, Helvetica Physica Acta Suppl. IV, 106

FUTURE COSMIC MICROWAVE AND COSMIC INFRARED BACKGROUND MEASUREMENTS

JOHN C. MATHER

Code 685, Laboratory for Astronomy and Solar Physics
NASA Goddard Space Flight Center, Greenbelt, MD 20771
USA

Abstract.
Cosmic microwave and infrared background radiation (CMBR and CIBR) measurements in the near future have the potential to greatly advance our knowledge of the early universe. New instrument and space technology will soon enable much better measurements of both.

1. Cosmic Microwave Background Anisotropy

The CMBR anisotropy has been well measured on angular scales of $>7°$ by the COBE DMR instrument, and the data from the first two years of observation (Bennett et al. 1994) agree well with the maps from the first year (Smoot et al. 1992, Bennett et al., 1992; Wright et al. 1992). The maps of anisotropy can be most conveniently described in terms of their spherical harmonic representation. A wide variety of measurements at smaller ($\sim 1°$) angular scales suggest that there really is a detection of excess fluctuations at the "Doppler Peak". This is the name given to a bump in a plot of the c_ℓ^2 versus ℓ, where ℓ is the spherical harmonic order and c_ℓ^2 is the sum of the squares of the spherical harmonic coefficients of order ℓ. Doppler did not of course predict this peak, but our Russian colleagues tell us that it was discussed by Sakharov and should properly be called "Sakharov Oscillations".

In theory, these oscillations are the acoustic modes of the primordial material at the time of decoupling. Their scale size is the horizon length at that time, and their amplitude is driven by the density inhomogeneities frozen in from the Big Bang. Therefore, measurement of their detailed properties

419

M. Kafatos and Y. Kondo (eds.), Examining the Big Bang and Diffuse Background Radiations, 419–421.

would be extremely revealing, and could help determine the total density Ω, the cosmological constant Λ, and possibly the properties of the dark matter that govern damping of the acoustic waves.

Not only should there be large scale anisotropy and Sakharov oscillations, but the CMBR may be slightly polarized by them. This is the result of anisotropic radiation becoming polarized by its last Compton scattering.

New technology is now available to make these observations possible. Improved bolometer detectors and microwave amplifiers make better receivers with far better noise levels, bringing raw sensitivities within factors of 2 or 3 of the intrinsic photon noise of the CMBR. Observations from colder and dryer sites are now working well, and several groups are planning or proposing space missions. It now seems possible that the next decade will bring all–sky maps at angular resolutions of 0.5° and sensitivities limited by the cosmic fluctuations themselves. For smaller angular scales, the same sensitivity can be expected, but full sky coverage will be much more difficult and less important.

Measurements of this accuracy would enable us to test the tensor theories of gravity, which would modify the lowest order terms in the harmonic spectrum. On medium scales, the spectrum can be compared to predictions that $n = 1$ precisely, according to the simple versions of inflation. Searches for cosmic strings will also be possible. On smaller scales, the Sakharov oscillations could be detected, yielding the main unknown parameters of the Big Bang and dark matter. On even smaller scales, the Sunyaev-Zeldovitch effect and the peculiar velocities of clusters may be detectable, and with great persistence we may even find the transverse velocities of clusters from the polarization of their scattered light.

2. Cosmic Microwave Spectrum

The spectrum of the CMBR has been well measured by the COBE FIRAS instrument (Mather et al. 1994) but the accuracy was limited by systematic error problems. We believe that a factor of 3 improvement is possible using the existing data, but we do not know whether to expect a detection of a spectrum distortion or an upper limit. The energy required to keep the intergalactic medium ionized should also distort the spectrum, so a better measurement will restrict the epoch of reionization.

At wavelengths >5 mm, the FIRAS measurement is not restrictive, and direct measurements are needed. Two teams are preparing balloon experiments to measure the longer wavelength spectrum, and at least one satellite experiment has been proposed. It will be possible to measure the chemical potential distortion parameter μ much better than FIRAS could do, and to search for free-free emission from the reheating.

3. Cosmic Infrared Background Spectrum

The CIBR contains the accumulated emissions of generations of stars, galaxies, quasars, and other objects. It is even possible that some of the CIBR was produced before the decoupling, for instance by decay of unstable elementary particles. The present limits on the CIBR are weak, and it could contain an amount of energy comparable to the CMBR. A measurement of the CIBR would provide firm integral constraints on the evolution of most categories of luminous objects, especially galaxies.

The COBE DIRBE team is proceeding well in the analysis of the flight data, and limits on the CIBR are significantly below the foreground emission. The foreground from interplanetary and interstellar dust and starlight has many components and an accurate model requires much thought and analysis. The residuals from the fitting have a spectrum that resembles the foreground, so it is plausible that they represent errors in the foreground models. The calibrated sky maps have been delivered to the public and are available on the Internet.

Indirect measurements or limits have also been obtained by observations of very high energy gamma rays from a quasar. If the CIBR is sufficiently intense, then the gamma rays will be attenuated by pair production of electrons and positrons. Little attenuation is seen (Stecker and DeJager 1993, Slavin and Dwek 1994).

It is unlikely that a large signal-to-noise ratio detection of the CIBR will be possible from the DIRBE data, so more work will be needed. The main foreground source is interplanetary dust, so a space mission to observe outside the cloud would be very helpful. The solar flux falls as $1/r^2$ and the dust density falls roughly as $1/r$, so it helps a great deal to go even a little farther from the Sun. At least two such missions have been proposed to NASA and one to ESA, so it seems that they are technically feasible. The possibility to fly them will depend on on the level of scientific interest.

References

Bennett, C. L. et al. , 1992, ApJ, **396**, L7
Bennett, C. L., et al. , 1994, ApJ, **436**, 423-442
Dwek, E., and Slavin, J., 1994, ApJ, **436**, 696-704
Mather, J. C. et al. , 1994, ApJ, **420**, 439-444
Smoot, G. F. et al. , 1992, ApJ, **396**, L1
Stecker, F.W., & Dejager, O.C., 1993, ApJ, **415**, L71.
Wright, E. L. et al. , 1992, ApJ, **396**, L13

Philip Morrison

THREE COSMOLGICAL REMARKS

In traditional societies the remarks of seniors were prized, for their memories often spanned events rare enough to have remained unseen by most yet frequent enough to offer eventual challenges to all. Even here among friends who are self-styled cosmological "old radicals", I think I am the senior.

But I am not persuaded that astronomy, for all its antiquity, ought to share the values of traditional societies. Orbital radiometers and CCDs are not very old; it is the news that is our meat, if not our bread and butter. I will therefore tell three small stories old and new.

1. One true old tale: (I suppose its effect is to praise famous men.) In the spring of 1937 I was an eager and naive graduate student under Robert Oppenheimer at Berkeley. It was Niels Bohr himself who held forth there for several evenings in a lecture series of considerable depth, though aimed at a wide university audience. His topic was fundamental physics, in particular the quantum theory and its applications, One question from the audience remains in my mind for the prescience of Bohr's answer, a guide to my own views ever since.

The questioner asked for Bohr's opinion of the cosmological theories of the day, the universes of LeMaitre, de Sitter, Milne and others, none of them much beyond a powerful but purely geometrical stage. Bohr's answer came firmly though not quickly. He replied that he felt it premature to form cosmological judgments. Two great relevant domains of physics would allow real progress towards the grand questions of origins and endings only if they were taken together.

On the one hand they were the relativistic ideas of universal space-time-gravity; on the other, the nature of the fundamental particles of physics. Until these two domains should interact in some observable physical context, no sound advance could be made. In 1937 no such context had yet been recognized. The question was still premature.

That context was first dimly seen at the end of WWII by George Gamow, and by now this symposium, and every one like it is largely concerned with physical cosmology, some evolving space-time chockfull of particles known and surmised, and what they might or might not have done. We have the right context and within it much admirable data, especially from COBE, though not yet many firm answers.

M. Kafatos and Y. Kondo (eds.), Examining the Big Bang and Diffuse Background Radiations, 423–425.
© 1996 *IAU.*

2. That context is of course the early and not-so-early universe, mainly before plasma recombination, but with a long tail into the present, for instance the form of the Hubble flow and the amount and kind of dark matter. Our Symposium 168 opened with redshifts and went soon to COBE and its current ground-based augmentation.

For me it is obligatory to weigh COBE very highly, since its astonishing precision is unmatched anywhere else in cosmology. That the large-angle microwave isotropy—a neat dipole convincingly removed—is as good as 5 or 10 ppm, that the spectrum is so astonishingly Planckian, its polarization so tiny, its energy density so high—these are rock-solid results. They predispose me to any theory that makes them the natural outcome of fundamental processes, and so avoids the ubiquitous astrophysical realities, all the plausible imperfections, clustering, clumping, motions, alignments, relaxation times, local gravitational potential wells, and all the rest. The inflation idea sets them all aside by a simple process, over many orders of magnitude. That seems to me the source of its appeal. It gets first things right first. Evidently that is not yet a proof, but only the promise of an understanding of simplicity.

3). I want to add another name beside Bohr's, that of a colleague who might have been here, Sir Fred Hoyle. His direct contributions are not exceeded by those of any theorist in cosmology since Einstein. Why, it was he who even gave the name to our Symposium: "Examining the Big Bang..." Fred formed that phrase on the air in 1950 in a rather derisory mood. Yet it has stuck firmly not only among us, the happy few astronomers, but far, far beyond our IAU, to the comic strips, even into everyday slang.

I believe The Big Bang has become a term of dangerous ambiguity. Here in our Symposium 168 it is used over and over again to name all the properties we have strong evidence for: the universal spatial expansion with its dilution of matter and radiation density, long evolution, and an outrush of much uniformity at high speed. But that was not the original thrust of the term, and by far it is *not* what even well-informed science journalists, not to mention their viewers and readers, now understand by it.

Of course The Big Bang does in fact entail all those fine observable processes. We like that part. But most of those who watch us and hear us use the term freely don't really care so much about observable physical processes, grand as they are. What the term means to them—and it meant this to Hoyle too—was something singular. To Hoyle it was a true singularity of the equations of the field theory, to most others a truly singular event, a metaphysical event without a physical cause. Most outsiders would still agree with that reading.

But now we insiders usually attach the same name to quite reasonable if extraordinary *consequences* of an initial expansion of a small parcel of matter and field, not to the singular point itself. I submit that makes for real trouble.

For I do not think we have any evidence one way or another of any singularity. The popular inflation scenario imagines a region before inflation that was non-singular and fully covered by the Einstein equation. The quasi-steady state alternative, for one, postulates a C-field present, to give some sort of causal action-like account of what came before. Certainly the true singularity, The Big Bang, is not at all excluded. It might be there, just before the delta-t moment of the inflation. Who knows?

But no longer is a Big Bang out of nothing the direct extrapolation of what we see and reckon from physics. It is now only one postulate open to theorists, and not one to be expected in every theoretical account.

Symposium 168 quite properly did not much discuss *The Bang*, and I much wish we hadn't labelled all the rest of what we do in cosmology by the powerful old phrase. If we do not find a more careful way of talking—and I concede it isn't easy—I fear that Sir Fred will have had the last word with his witticism. That is one victory I doubt that he wants!

The public will continue to think that we see the First Uncaused, right there in the black-body intensity variations. Surely that inference cannot content us, even those who hold what is certainly possible, but equally certainly not proved, that the Big Bang came only just before the time horizon for the first inflation. We owe to the public appreciation of modern cosmology a clearing up of our ambiguous use of Hoyle's infectious metaphor.

An expository invention is badly needed, or at least repeated brief clarifications by the many who write on these matters—for instance, in the coming preface to Symposium 168!

PANEL CONTRIBUTION–IAU SYMPOSIUM 168

R.B. PARTRIDGE

Haverford College

These brief comments will reflect the point of view of an observer, rather than a theorist. They will also be quite informal, as were the remarks I made while at the symposium in The Hague. Finally, I will restrict myself to relatively low energy proton backgrounds–roughly speaking from $10^{-5} - 10^1$eV. I fear that will slight the important work being done by our colleagues on the X-ray and gamma ray backgrounds, but those topics were nicely covered in the symposium itself.

The last IAU symposium to deal with questions of cosmic (and Galactic) backgrounds was held 5 years ago in Heidelberg. If I may use that earlier meeting as a benchmark, I am struck by how much progress we have made in the past 5 years, and, frankly, by how uneven that progress has been.

At the risk of seeming parochial by placing an area of my own interest at the top of the list, I would begin by pointing out one spectacular success: the determination of the spectrum of the cosmic microwave background radiation (CBR); see Mather *et al.* (1990), Gush *et al.* (1990) and Mather *et al.* (1994). As another contributor to this panel has noted, in many ways this precision measurement of the CBR spectrum is an even more crucial result than the long-awaited discovery of fluctuations in the intensity of the CBR. I do not wish to minimize the latter, however. Observers like myself have been seeking measurable fluctuations in the angular distribution of the CBR for more than a quarter century. With the exception of a few false alarms, and the robust detection of a dipole moment ascribed to the velocity of the earth, no variations in the angular distribution were detected until 1992 (Smoot *et al.*). That paper has unleashed a flood of additional reports of positive detections, many nicely reviewed by Lubin in this volume. I would say that the observational situation is at the moment a little uncertain on angular scales smaller than the 7° beam of the COBE-DMR instruments. As the dust settles (and workers in the field will realize this is a pun with some point), I suspect we will have found that we do have robust detections of CBR fluctuations on degree scales as well as the larger angular scale

M. Kafatos and Y. Kondo (eds.), Examining the Big Bang and Diffuse Background Radiations, 427–430.
© 1996 *IAU.*

variations reported by the COBE team. These results, combined with the very tight limits on spectral distortions, I believe, will greatly enhance the astrophysical and cosmological utility of the CBR. In fact, the best studied of all cosmic backgrounds has become even better characterized, and has much more to contribute to astrophysics and cosmology.

Moving up a step in frequency, we seem to be tantalizingly close to reaping the great riches of far infrared background astronomy. That, at least, is the optimistic conclusion I draw from the fine review by Hauser in this volume. It is interesting to contrast the spectacular success of Mather and his team with the much tougher task faced by Hauser and his. Do remember that the graph of the CBR spectrum so many of us have seen so often was determined from *11 minutes'* worth of COBE data taken in late 1989. By contrast, Mike Hauser and his colleagues are still trying to sort out local foreground contributions in the far infrared, years after the COBE instrument that took the data has shut down. I'll come back to this point in a moment.

In the meantime, however, let me move to the optical. It is intriguing that the cosmological background in the optical, which was the subject of considerable discussion in Heidelberg 5 years ago, received essentially no attention here. Instead, interest focused on counts of galaxies made in the optical (or near infrared K band) and on questions of galaxy evolution. My non-expert reading of the field is that the counts of galaxies are now in good shape, and that the apparent disagreement between counts made in the B and K bands is no longer a cause for concern. On the other hand, as Koo among others noted here, the question of galaxy evolution is still far from settled. Simon White's talk made that clear; so, too, did Dave Koo's conservative, reductionist suggestion. It is intriguing that some 30 years after the paper by Eggen, Lynden-Bell and Sandage (1962) and the various models of galaxy formation that followed from it (eg. Partridge and Peebles, 1967), we still can't say whether we have detected bona fide primeval galaxies or not. The poster by Pritchett *et al.* here says no; the work by Miley and Chambers here and elsewhere says yes. I should go on to say that the problem with galaxy evolution is not just a problem with high redshift objects; there are plenty of open questions about galaxy evolution even at modest redshifts of order $0.5 - 1$.

Here it is appropriate to insert a word of praise for those doing redshift and other large-scale surveys in the optical. There has been spectacular progress on this front in the past 5 years and spectacular promise for the years to come. That some of us are talking about the possibility of a new IAU Commission on Large-Scale Structure is one reflection of the fine work by observers on these teams as well as those making more and more sophisticated computer models of large-scale structure and its evolution.

In my view, even more dramatic results have been derived from the use of gravitational lensing to allow us to see high redshift or faint background objects as well as to trace out the mass distribution in foreground sources. While most of that work has been done in the optical, radio astronomers are now beginning to make their contribution to the astrophysical and cosmological results from such studies, as well as to the location of lens candidates. Since gravitational lensing was beautifully reviewed in this symposium by Peter Schneider, I won't pursue the topic in detail, but I'd put substantial money down that gravitational lensing will be a more and more useful tool in cosmology over the next 5 years.

Before ascending further up the frequency ladder, I do want to point to the contributions made by radio astronomers in the characterization of moderate redshift galaxies as well as the discovery and characterization of gravitational lens sources. That work is nicely reviewed in a brief paper by Wall here. Radio astronomers have now pushed the counts down to nearly microJansky levels (Windhorst et al., 1994), and are beginning to discover the sky is "paved" with radio sources in the same way it is with faint, 26^m, optical Galaxies.

Finally, the situation in the UV strikes me, as a low-energy photon person, as still quite complex and even disputatious. Is there an overall cosmological background in the UV; and, if so, is it relatively bright or relatively faint? It seems to me that we do not have convincing and widely accepted answers to those questions. It is equally clear that we have very able and innovative observers working on the questions, as the reviews by Jakobsen and Bechtold indicate. My hunch is that we'll have a much clearer picture of the ultraviolet background by the next such IAU symposium .

Now for a few generic conclusions. The first of these is that many of the remaining problems in all of the fields I've touched on above involve rather ordinary, messy, astrophysical issues. I hasten to add that I am *not* saying that all the basic problems are solved and that we now find ourselves tidying up the loose ends. Quite the contrary; the large problems have *not* been solved, but I believe the solution may well involve grubbing about in the messy details.

The most salient example is sorting out the foreground contributions to the infrared sky brightness. In different wavelength bands of the infrared, one has scattered sunlight, reemission from interplanetary dust, Galactic emission, possible band and line emission from PAH's, and much else. As Hauser's talk suggested, the goal in sorting out these foregrounds is partly to understand them in their own right, and partly to pare them away so that we can get at the truly cosmological background (which itself may be complex). We will need much better characterizations of the interplanetary dust and of the dusty emission from our own Galaxy before we can get to the kernel of the cosmological issues.

Many other unsolved problems, I suspect, will involve the same kind of careful work. I've already alluded to some of the issues in the ultraviolet background, and there are plenty of other examples scattered throughout the talks at this symposium. Let me provide a quick and certainly incomplete list: indirect upper limits on the ultraviolet flux, the question of wide-spread dust in the Universe (as revealed, for instance, by the newly discovered "red quasars"), the evolution of galaxies at both high and moderate redshifts, the galaxy luminosity function at the faint end, and the contamination of the CBR by foreground sources.

The fact that there are so many problems remaining I would regard as both good news and bad news, if you'll allow me a rather sociological comment. The bad news, of course, is that in some fields the glory days are over. The next steps may be hard and unglamorous. Again, I wish to repeat that I am not saying that we've been reduced to straining for the last decimal place. Rather, I'm saying that in order to answer the big questions, we may have to do some rather conventional astronomical work.

The good news is essentially the same–that there *is* a lot of work yet to be done. Much of it involves relatively straightforward, if painstaking, astronomical observations, of the kind that all astronomers, not just those with multimillion dollar satellites or state-of-the-art telescopes, can engage in. We stand to learn a great deal, for instance, from further study of gravitational lens sources, from confirmation of reported detections of CBR fluctuations, from a deeper understanding of the role of dust at moderate redshifts, and from a more careful characterization of Galactic emissions at ultraviolet, radio, submillimeter and infrared wavelengths. Here lies the future of the field, in my view, and here lies my optimism.

References

Eggen, O. J., Lynden-Bell, D., and Sandage, A. R. 1962, *Ap. J.*, **136**, 748.
Gush, H. P., Halpern, M., and Wishnow, E. H. 1990, *Phys. Rev. Lett.*, **65**, 537.
Mather, J. C., *et al.* 1990, *Ap. J. (Letters)*, **354**, L37.
Mather, J.C., *et al.* 1994, *Ap.J.*, **420**, 439.
Partridge, R. B., and Peebles, P. J. E., 1967, *Ap. J.*, **147**, 868.
Smoot, G.F., *et al.* 1992, *Ap.J.(Letters)*, **396**, L1.

Other papers referred to are in this volume.

KNOWLEDGE LIMITS IN COSMOLOGY

MENAS KAFATOS
Earth Observing and Space Research Program
Institute for Computational Sciences and Informatics
and Department of Physics
George Mason University
Fairfax, VA 22030

ABSTRACT.   In cosmology one faces the observational challenge that knowledge about distant regions of the universe is dependent on assumptions one makes about these regions which are themselves coupled to the observations. Within the framework of the Friedmann-Robertson-Walker big bang models the universe becomes opaque to its own radiation at $z \approx 1,000$ and the earlier, and more distant, regions of the universe are not directly accessible through observations. Other challenges exist such as possible merging of extended distant sources and confusion of spectra from distant galaxies. One, therefore, encounters horizons in our understanding of the universe. Such horizons exist in any mode of description. To use the quantum analogy, the observer is always part of the system under study, the universe, and a description of the universe entails including the observer and observing apparatus. Since the early universe should be described in quantum terms, it follows that non-locality in the universe is not an a-priori requirement but the outcome of the observing process itself. As such, the flatness and horizon problems may not be preconditions on theoretical models.

1. Introduction

It was Einstein's general theory of relativity that allowed the possibility of an evolving, non-static, universe. The cosmological revolution would prove of equally great significance as the quantum revolution -- both broke away from traditional thinking. With the theoretical framework of general relativity already in place, the Belgian cosmologist Abbe Lemaitre and the Russian mathematician Alexander Friedmann postulated in the early 20's a dynamic, expending and evolving universe. The Friedmann models obeyed the *cosmological principle*, which states that the universe is isotropic -- the same in all directions -- and homogeneous -- of equal density, on the average, everywhere. It then followed that as the universe expands, the average density of matter would decrease. To accommodate the obvious observational picture of Hubble's expanding universe with a framework of an eternal universe, Herman Bondi, Thomas Gold and Fred Hoyle

M. Kafatos and Y. Kondo (eds.), Examining the Big Bang and Diffuse Background Radiations, 431–438.
© 1996 *IAU.*

proposed in the late 40's the steady state theory, describing a universe which although expanding would obey the *perfect cosmological principle*: The universe would appear the same to all observers at all times. What Bondi, Gold and Hoyle were attempting to avoid was the question of the origins of the universe: If the universe is expanding, Lemaitre and the other cosmologists of the early 20's reasoned, it must have been much denser and hotter in the distant past, it must have started from a primordial singularity. Bondi, Gold and Hoyle avoided the problem of a primordial dense universe at the expense of a cherished principle of physics, the principle of conservation of mass and energy. The steady state model demanded the creation of matter from nothing, new matter had to be created to fill the voids of expanding space.

In the early 50's, cosmologist George Gamow extended Lemaitre's and Friedmann's original ideas by incorporating nuclear physics. Cosmologists could now use nuclear physics to speculate what might have happened in the early lifetime of the universe when it was very hot and prone to nuclear interactions.

Soon observational astronomy armed with the new branch of millimeter and radio astronomy and sophisticated optical spectroscopy provided strong evidence in favor of the big band model, through existence of the 3°K black body radiation and the existence of quasars. In the big bang picture, the microwave background is the relic radiation from the initial big bang. The steady state theory could not as easily account for the background. The microwave radiation can be easily accounted for by the big bang theory and, it appeared no extra assumptions were needed.

The second cosmological observation that challenges steady state is the existence of the quasars (Berry 1976). These objects appear to be very distant, some of them receding away from us at speeds exceeding 80% of c. In the general "standard" scenario, at the distance of a quasar, only the brilliant star-like nucleus can be seen. These bright nuclei of galaxies were very brilliant in the past compared to the present era, indicating that sources evolved as time went on. If the interpretation is correct, quasars would violate the *perfect cosmological principle* because the universe would not look the same at all times. Recently, it has become obvious that the big bang theory itself faces theoretical challenges not appreciated before. Yet, the vast majority of astronomers, cosmologists and particle physicists still adhere to the big bang theory, which has become the gospel of cosmological theory.

One should, however, be cautious in embracing a single vision of the universe. Even though the standard theory has had substantial successes, the mystery of quasars has still not been completely unravelled. Moreover, modern challenges have emerged, such as the remarkable smoothness of the microwave background, the increasing complexity of assumptions tied to cold dark matter (CDM) theory and as recent HST observations indicate, the perplexing emergence of a relatively young universe at odds with both stellar evolution and the inflationary predictions. As such, perhaps a re-examination of cosmological theories and a possible convergence and acceptance of, seemingly, opposing views may be warranted. It is my purpose here to examine the limits of both cosmological theories and observations and to provide some new approaches to cosmology.

2. The Early Universe and Inflation

Recent advances in particle theory have afforded us unique opportunities to describe the conditions of the early universe. The early universe was in a quantum state and moreover, the early universe can be used as a cosmic laboratory to push physics to the ultimate frontier of quantum gravity.

Very near the original singularity, the space-time description breaks down entirely. This is followed by the so-called "inflationary era" at about 10^{-35} sec (Guth and Steinhardt 1984). During the inflationary era, the universe underwent an extremely rapid expansion doubling in size every 10^{-35} of a second. By the time the universe had finished through this phase, it had expanded in size by a staggering factor of 10^{50} of more. Prior to inflation, the universe was in a phase of symmetry with respect to the so-called Higgs fields -- and the strong, weak and electromagnetic interactions were unified. In that situation the Higgs fields, members of a special set of quantum fields postulated in Grand Unified Theories (GUT) to account for spontaneous symmetry breaking, were all zero. At a temperature of about 10^{27} K, the universe underwent a phase transition from the false vacuum where all the Higgs fields were zero to a less energetic phase, the true vacuum of quantum theory. In the true vacuum state (Barrow 1988), the Higgs fields acquired non-zero values and the GUT symmetry broke down. During the inflationary era, the non-zero values of the Higgs fields broke down the GUT symmetry: The strong force separated from the electroweak force.

The false vacuum of GUT has many peculiar properties, the most peculiar being perhaps the form of its negative pressure. In those conditions, general relativity predicts that gravity rather than pulling together would be pushing away. The vast energies locked up in the negative pressure of the false vacuum were released and all matter formed.

The inflationary model was originally proposed not because of a compelling theoretical reason but in order to address some observational problems faced by standard big bang. For example, the horizon of the universe, within which parts of the expanding primordial matter were in contact among themselves, expanded becoming much larger than the radius of the observable universe. In the big bang cosmology without inflation the horizon is always less than the radius of the observable universe. The "true" universe would then be expected to be much larger than the universe we can see observationally. This fact is critical to resolve one of the observational problems of the standard big bang without inflation namely, the *horizon problem*.

After inflation released the vast amounts of energy that would later on coalesce to form all the observable objects in the universe, the expansion proceeded according to the original version of the big bang theory. Between 10^{-30} sec and 10^{-6} sec the universe was filled with a primordial soup of quarks and leptons. After about 10^{-6} sec, the quarks, combined together to form the nucleons. In the next phase, between 1 sec and 3 minutes, protons and neutrons underwent nuclear reactions forming nuclei of helium and its isotopes.

Until about ~ 100,000 years after the big bang, photons and matter were coupled together. The 2.735°K black body photons were also emitted at this time but provided no opportunity following this era to probe the earlier periods where fundamentally important physical processes were taking place. It follows that events in these earlier periods cannot be verified by means of the main tool of observational astronomy -- quanta of light.

3. Observational Limits in Cosmology

The big bang theory of the universe could not account for a number of features revealed by observational cosmology such as the so-called *flatness problem*. Quantitatively this is expressed as $\Omega = \rho/\rho_{crit}$ and the critical density can be expressed as (Barrow 1988)

$$\rho_{crit} = 5 \times 10^{-30} \ (H_o/50 \ \text{km/sec/Mpc})^2 \ \text{gr cm}^{-3}$$

where H_o is the Hubble constant. It is obvious that the rate of expansion cannot increase and infinitum. There would be a point when the rate of expansion as seen from the earth would reach the speed of light. This would constitute the horizon of the universe. For $H_o = 50$ km/sec/Mpc that horizon lies ~ 20 billion light years away from the earth. The precise value of H_o remains the most fundamental challenge of observational cosmology (see present volume). Were Ω turn out to be precisely equal to unity the geometry of the universe would be exactly flat. Inflation nicely accounts for the apparent flatness since no matter what the original curvature was, the inflationary era washed it out into perfect flatness.

Current observations cannot unequivocally distinguish the type of the universe we line in. Values of Ω for luminous matter are in the range ≤ 0.1 to even as large as 2 although most observers favor values ≤ 0.1. If the only type of matter existing in the universe is luminous matter, this result would favor an open universe. Even though present observations only indicated an approximate range of the mean density of the universe, this range is so close to the value of the critical density required for a flat geometry that many astronomers assume as a working hypothesis that the universe is exactly flat, hence the need for unseen forms of matter such as cold dark matter (CDM).

The second problem facing big bang cosmology has to do with the uniformity of the 3°K (2.735°K to be more exact) black body radiation (see discussion in present volume). The microwave radiation has the same temperature to within one part in $\geq 10,000$ in every direction of the sky. In the hot big bang, though, opposite parts in the sky at the time that the microwave background formed 10^5 years from the beginning, were separated by distances of 10^7 light years (Schramm 1983). Given the near identity of temperatures from all parts of the sky and presuming that classical causality holds, one would conclude that opposite parts of the sky had to be in casual contact. Relativity, though, states that no signal can travel faster than light, and opposite parts of the sky were spacelike in their separation. This is known as the *horizon problem*, and as with the *flatness problem*, it represents incredibly fine tuning in the conditions prevailing in the early universe.

Big bang models of the universe assume that the universe is isotropic and homogeneous. This may be difficult to achieve given the large number of possibilities in the initial conditions. In fact, quantitative calculations show that slight anisotropies would not die away but on the contrary would get amplified. This is known as the *isotropy problem.* The problem becomes even more severe in the steady state scenario which requires that whatever fluctuations in isotropy exist should be present at all levels. Whether the universe is isotropic and homogeneous is an observational question. The microwave radiation is highly isotropic. The universe is also presumed to be expanding the same way in all directions, i.e. to be isotropic in matter as it is in background radiation. And the distribution of matter is presumed to reach homogeneity beyond the largest structures seen, the superclusters, i.e. beyond hundreds of Mpc.

Recent observations challenge both these pillars of traditional cosmological thinking: The universe may not be homogeneous, larger and larger structures have been found. Galaxies seemingly cluster themselves to increasing hierarchies of clusters, superclusters and maybe even super-superclusters. Often these structures assume the form of filaments on the surface of very large bubbles with large "voids" in between (Geller and Huchra 1989). The universe may also not be isotropic, galaxies have been found which do not follow the Hubble flow (Dressler et al. 1987).

One would, therefore, conclude that the universe requires incredibly fine tuning at the beginning, i.e. the universe represents a very unlikely "accident". Paul A.M. Dirac (1937) first noticed in the 30's that certain ratios involving fundamental constants of nature and physical parameters obey simple numerical relations. Some of these coincidences yield very large numbers which are not random as one might have expected. Dirac believed these could not be coincidences and formulated his *large number hypothesis.* Simply put, he reasoned that since as the universe expands its radius changes in value; in order for various ratios to be equal today, a physical quantity such as Newton's gravitational constant must change in time. Attempts to verify Dirac's hypothesis have failed.

Today some physicists, primarily Dicke, Carter, Barrow and Tipler favor another approach. They have postulated that our existence as observers requires a fine tuning and that the seemingly unrelated radios of different quantities point to our existence as necessitating the kind of universe we live in. Put differently, the universe is unique because it contains conscious observers. This is known as the *anthropic principle* (Barrow & Tipler 1986).

4. Horizons of Knowledge in Cosmological Theory

As we study the observations pertaining to the early universe, we encounter a number of *observational horizons of knowledge* (Kafatos, 1989). These observational horizons are tied to the quantum nature of photons. For example, to obtain the distance of a faint galaxy requires that we obtain its spectrum, which in turn requires that we disperse the light. This requires isolating the light from the galaxy by means of a narrow slit. At larger distances, when few photons are involved, one cannot disperse the light without limit. Attempting to obtain more photons by decreasing the dispersion would, on the other hand, cause an observational confusion as light from neighboring galaxies in that part of the sky would also fall onto the spectrograph. There is then a complementary inverse relationship between dispersion and brightness which does not permit accurate spectra of faint, very distant galaxies to be obtained (practically, this limit does not apply to the current telescopes, including HST).

When we study the predictions of various competing models of the universe we find that the observational horizons complicate the theoretical picture. Moreover (Kafatos 1989), theoretical models present us with their own limitations, what one may call *theoretical horizons of knowledge.*

For example, big bang cosmology itself imposes a fundamental limit on the observability of the early universe. Direct observations of the early universe based on photons provide information only after a timescale of $\sim 100,000$ years from the beginning. One hundred thousand years after the beginning corresponds to z of ~ 1000, i.e. when the universe was only 0.1% of 1% of its present age. On the other hand, the most distant quasars are seen at a redshift of ≥ 4 and emitted their light received by us today when the universe was about 10% of its present age. Radiation can in principle tell us much more about the early universe than matter. However, the opaqueness of the universe prior to $z \sim 1000$ simple does not allow us to confirm or deny big-bang cosmology based solely on photons. At $z \sim 1000$ we encounter the first theoretical horizon of knowledge about our universe and as long as we are constrained to observe photons, that horizon is impregnable (Kafatos 1989).

It is unlikely that any other means will provide as clear-cut evidence as light about the universe we live in. Even though neutrino astronomy is an exciting new branch of observational astronomy, it cannot replace traditional astronomy and its main tool, light. In principle, primordial neutrinos emitted a few seconds after the beginning of the universe or at a redshift $z \sim 10^9$ present, therefore, the ultimate horizon from which we can access direct information about he universe (Kafatos 1989).

Relying on ordinary matter can yield a lot of detailed information about the hypothesis of element formation in the early universe (Schramm 1983). The actual situation is in reality difficult, since uncertainties in the abundances of the primordial elements deuterium, helium and ^7Li are large and the details of big bang models are least sensitive to the abundance of helium, by far the most abundant of primordial elements formed from hydrogen. Present results, if no CDM exists, imply a mean density of baryons $\approx 3 \times 10^{-31}$ gr/cm^3, 2 orders of magnitude less than the critical density, implying an open universe. The corresponding horizon of knowledge associated with nucleosynthesis is $z \sim 10^8$.

Of all the horizons of knowledge at $z \sim 10^9$, 10^8, and 1000, it is the last that is likely to remain the only one we can explore for the foreseeable future. The neutrino horizon at 10^9 is not going to be qualitatively different from the photon horizon because both primordial neutrinos and photons follow a black body radiation law. Whatever problems of interpretation we are facing today with regards to the background photons will not go away with neutrinos, except that the problem of detection and interpretation will only be much worse. And the element horizon going back to the first three minutes will not yield much better information either due to the complexity of nuclear reactions applicable at that time or the uncertainties of whether these elements represent truly primordial matter.

Considerations from classical cosmological theory can shed more light as to the type of the universe we live in, by studying the Hubble Diagram for distant galaxies.

Redshifts in the range $5 \sim 10$ would be particularly important to study since astronomers suspect that at these redshift galaxies began to form. It is conceivable though that distant galaxy images would merge at large, ≥ 30, redshifts, and it would be virtually impossible to obtain accurate spectra to study the geometry of the universe. The curvature of the universe causes the image to decrease for redshifts greater than the minimum value which occurs for $z \sim 1$ (Narlikar 1983). Although presently inaccessible, this "galaxy image" theoretical limit may one day be reachable, if indeed there were any galaxies out to $z \geq 30$.

These horizons which occur inherently in the particular theoretical picture used become worst when the quantum nature of light is considered: Spectra of different sources in the sky would themselves be blended together as one looks at fainter and fainter sources. Eventually, the background from different faint galaxies would dominate the spectrum from a single distant galaxy and reliable spectra could not be obtained. It is for these reasons, that we encounter cosmological horizons of knowledge which prevent us from ultimately deciding unequivocally how these tests confirm or reject particular theoretical models (Kafatos 1989). This is precisely the case, where Bohr insisted, complementarity acquires great importance. One would then view the various cosmological models not as rival theories of which one day only one will emerge as *the theory of* the universe, but as competing complementary constructs. Coupled with the fact that the early universe should be described in quantum terms, one would conclude that these emergent complementary models and the implied underlying wholeness are not an a-priori philosophical preference but the very outcome of the observing process for the early universe.

As such, the flatness, horizon and isotropy problems may be tied to the observing process itself rather than as preconditions for theory. As we look at more and more distant galaxies, the universe may be appearing as flat not because of inflation but rather, because such a universe would naturally emerge as the boundary between complementary constructs: The open versus closed universe constructs. One should perhaps view big bang and steady state models as complementary constructs as well. Although clearly favored at present, big bang theory may reveal further weaknesses as the observational limits are extended.

One final point that I would like to raise is the intriguing possibility that quantum-like non-locality prevailed even after the early quantum gravity and quantum inflationary any era. This would provide an alternative path to the correlations implied by the COBE results. As such, Bell-type quantum correlations may be frozen in (Kafatos 1989). We may indeed have to take our own views of the quantum nature of the early universe much more seriously. In that case, the implications not just for cosmology but for the general epistemology of science would be staggering (Kafatos and Nadeau 1990).

438

5. References

Barrow, J.D. (1988), *Q.Jl. R. Astr. Soc.* **29**, 101-117.

Barrow, J. D. and Tipler, F.J. 1986, *The Anthropic Cosmological Principle*, Oxford Univ. Press, Oxford.

Berry, M. (1976), *Principles of Cosmology and Gravitation*, Cambridge Univ. Press, Cambridge.

Dirac, P.A.M. (1937), *Nature* **139**, 323.

Dressler, A. et al. (1987), *Ap. J.* **236**, 531.

Geller, M. J. and Huchra, J. (1989), *Science* **246**, 897.

Guth, A. and Steinhardt, P.J. (1984), *Scientific American*, June 1984, 116-128.

Kafatos, M. (1989) in *Bell's Theorem Quantum Theory and Conceptions of the Universe*, ed. M-Kafatos, 195, Kluwer Academic.

Kafatos, M. and Nadeau R. (1990), *The Conscious Universe: Part and Whole in Modern Physical Theory*.

Narlikar, J. (1983), *Introduction to Cosmology*, Jones and Bartlett, New York.

Rowan-Robinson, M. (1985), *The Cosmological Distance Ladder*, W.H. Freeman and Co., New York.

Schramm, D.N. (1983), *Physics Today*, April 1983, 27-33.

PART II

Contributed Papers

TEMPERATURE FLUCTUATIONS OF THE
MICROWAVE BACKGROUND IN
PRIMEVAL ISOCURVATURE BARYON MODELS

S. N. DUTTA AND G. EFSTATHIOU
Department of Physics
University of Oxford
OX1 3NP, United Kingdom

Abstract. We calculate the temperature fluctuations in the microwave background in open primeval isocurvature baryon models (Peebles, 1987) with cosmological densities in the range $0.05 \leq \Omega \leq 0.2$ We assume that the power spectrum of fluctuations is a power law with the index varying between $-1 \leq n \leq 0$, as indicated by observations of large scale structure in the Universe. The Universe is assumed to be always fully ionized. The South Pole 13 field point experiment (Schuster *et al.*, 1993) is compared to our theoretical predictions, and we find that the models predict larger temperature fluctuations than are observed. The observed temperature fluctuations on intermediate scales of $\lesssim 1°$ thus seem difficult to reconcile with the isocurvature baryon model.

1. Introduction

The Cosmic Background Radiation (CBR) remains, one of the most powerful diagnostics of the Universe on large scales. The discovery of the temperature anisotropies in the CBR at various angular scales allows us to place new constraints on the formation of structure in the Universe. One of the models of structure formation that has been around for a long time is the Primeval Isocurvature Baryon (PIB) model (Peebles, 1987). In this model, the universe is open ($\Omega \lesssim 0.2$) and is uniform at early times except for small fluctuations in the entropy per baryon. The main attraction of the picture is, of course, that there is no need for exotic dark matter (Peebles, 1987).

M. Kafatos and Y. Kondo (eds.), Examining the Big Bang and Diffuse Background Radiations, 441–444.
© 1996 *IAU.*

Most dynamical measures of Ω indicate a value around 0.2, suggesting the Universe is open. The limit on Ω_b from standard Big Bang nucleosynthesis is significantly less at $0.0125h^{-2}$ (Walker *et al.*, 1991), where h is the Hubble's constant in units of $100\,\mathrm{km\,s^{-1}\,Mpc^{-1}}$. With a low h it is may be possible to explain the light element abundance via inhomogeneous nucleosynthesis with a value Ω_b as high as 0.1 (Gnedin & Ostriker, 1992), overlapping with the lower range of Ω determined from dynamical studies (eg. Davies & Peebles, 1982). Further, new observations of microlensing of the stars in the Large Magellanic Cloud (Udalski *et al.*, 1994) indicate the possibility of the existence of a large population of "dark" compact objects. For these reasons, there has been some renewed interest in the PIB model (Cen *et al.*, 1993).

One inelegant aspect of the PIB model is that, as yet, there is no compelling theory for the origin of the entropy fluctuations. The power spectrum of the initial entropy fluctuations is therefore not fixed, and can be varied "by hand" to fit observations. This arbitrariness makes testing of the model difficult. One way around the problem is to use the observational constraints on the power spectrum on scales at which the temperature fluctuations are measured. This is the approach we adopt in the present work.

The PIB models are distinguished by the early formation of structure, which can produce enough ionizing radiation to reionize the universe at early times. Previous work (eg. Efstathiou & Bond, 1987) has shown that reionization is required in PIB models to satisfy the constraints on temperature anisotropies on scales of a few arcminutes. We therefore assume that the Universe is always ionized and compute the temperature anisotropies on intermediate angular scales.

2. The Temperature Fluctuations

2.1. THE POWER SPECTRUM

The temperature fluctuation of the CMBR in a PIB model is a combination of the initial entropy fluctuations, a small Sachs-Wolfe effect and a large contribution from Thomson scattering off moving electrons. We will consider angular scales where the Sunyaev-Zeldovich (Zeldovich & Sunyaev, 1970) and second order effects (Vishniac, 1987, Efstathiou, 1988) are utterly negligible. The standard Boltzmann equations (Bond & Efstathiou, 1987) for the perturbed distribution functions for the radiation are solved by expanding the relative temperature fluctuations $(\Delta T/T)$ in terms of Legendre polynomials. We assume the universe to be spatially flat in calculating the connection between the angular scale and the different orders of the polynomial expansion. This should be accurate at the angular scales we are interested in.

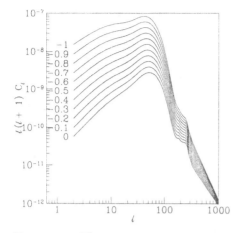

Figure 1. The power spectrum of the temperature fluctuations of the CMBR. $\Omega_b = 0.1$ and $h = 0.5$. The entropy fluctuations have been normalised to $\sigma_8 = 1$. The numbers next to each line give the spectral index of the initial entropy perturbations.

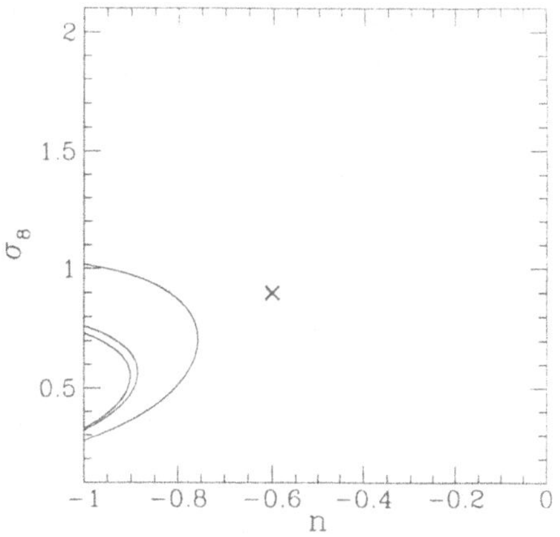

Figure 2. The contours of constant likelihood in the $n - \sigma_8$ plane in the Universe with $\Omega_b = 0.1$ and $h = 0.5$. The contours are the confidence levels with 99%, 98% and 68% probability. The cross indicates the best values from observations of large scale structure.

The power spectrum of the CBR temperature fluctuations is given in terms of the coefficients of the expansion as,

$$C_l = \frac{V_x}{8\pi} \int_0^\infty k^2 dk |\Delta_{Tl}(k, \tau_0)|^2.$$

As mentioned before this needs to be normalised to some initial value of the entropy fluctuations. We normalise the spectrum such that the r. m. s. fluctuations in the mass distribution in spheres of radius $8h^{-1}$ Mpc, σ_8, is unity. The power spectrum of the temperature fluctuations for $\Omega_b = 0.1$ and $h = 0.5$ is shown in Fig. 1, for various values of the primordial spectral index, n.

2.2. COMPARISON WITH OBSERVATIONS

We have compared the models with the temperature anisotropies observed in the South Pole experiment of Schuster *et al.*, (1993). This was a balloon borne and ground based experiment that used a cryogenic HEMT amplifier on a 1 m telescope. Measurements were made on four bands of equal (2.5 GHz) width. Several scans were made, each of which consisted of stepping between 15 target points on the same elevation, seperated by $2°.1$. The full width at half-maximum was $1°.5$.

We perform a likelihood analysis, assuming the fluctuations follow a Gaussian distribution. For each model we maximise the likelihood function,

$$L \propto \frac{1}{\sigma_8^N |C_{ij}|^{1/2}} \exp\left\{ -\frac{\sum x_i x_j C_{ij}^{-1}}{2\sigma_8^2} \right\}$$

with respect to the parameter σ_8. For the SP13 experiment, the covariance matrix C_{ij} is given by ,

$$C_{ij} = 2 \sum_l C_l F_l^2 \sum_{m>0} Y_l^{|m|}(\cos\theta_i, 0) Y_l^{|m|}(\cos\theta_j, 0) \cos[m(\phi_i - \phi_j)] H_0^2(m\phi^*)$$

where $F_l = \exp[-l^2\theta_s^2/2]$ and H_0 are the Struve functions and θ_s is the beam resolution.

Fig. 2 shows the likelihood contours in the $\sigma_8 - n$ plane for the PIB model with $\Omega_b = 0.1$ and $h = 0.5$. The temperature fluctuation of the microwave background favours low σ_8 around 0.6 and n around -1. However, observations of large scale structure (Cen et al., 1993) favour $\sigma_8 \simeq 0.9$ and $n \simeq -0.6$, (indicated by a cross in Fig. 2), so it appears that the model fails to match the observations. We find similar results for models with $\Omega_b = 0.2$ and 0.05 and $h = 0.5$.

The SP13 experiment thus sets strong constraints on the PIB model. We are currently investigating a wider range of model parameters and comparisons with other experiments on intermidiat angular scales of $1°$. However, this analysis does indicate that anisotropy experiments on intermidate angular scales can provide strong constraints on this class of models.

References

Bond, J.R. & Efstathiou. G. (1987) *M.N.R.A.S.*, **226**, pp. 655-687

Cen, R., Ostriker, J.P. & Peebles, P.J.E (1993) *Ap. J.*, **415**, pp. 423-444

Davis, M. & Peebles, P.J.E. (1983) *Ap. J.*, **267**. pp. 465-482

Efstathiou. G. (1988) in Large Scale Motions in the Universe: A Vatican Study Week, ed. Rubin, V.C. & Coyne. G.V, S.J., Princeton University Press, USA.

Efstathiou. G. & Bond. J.R. (1987) *M.N.R.A.S.*. **227**, Short Communication, pp. 33P-38P

Gnedin, N.Y. & Ostriker. J.P. (1992) *Ap.J.*, **400**. pp. 1-20

Peebles, P.J.E. (1987) *Nature*, **327**, pp. 210-211.

Schuster, J., Gaier, T.. Gundersen, J., Meinhold. P., Koch, T., Seiffert, M., Wuensche, C.A. & Lubin, P. (1993) *Ap. J.*, **Vol no 412**. pp. L47-L50

Udalski, A.. Szymański. M., Kaluzny, J., Kubiak, M., Mateo. M. & Krzmiński, W. (1994) *Ap.J.*, **Vol no 426**. pp. L69-L72

Vishniac, E.T. (1987) *Ap. J.*, **322**. pp. 597-604

Walker, T.P., Steigman. G., Schramm, D.N., Olive. K.A. & Kang, H. (1991) *Ap. J.*, **376**. pp. 51-69

Zeldovich, Ya. B. & Sunyaev, R.B. (1970) *Ap. Sp. Sc.*, **7**, pp. 3-19

CMB: ANISOTROPIES DUE TO NON-LINEAR CLUSTERING

E. MARTÍNEZ–GONZÁLEZ AND J. L. SANZ

*Dpto. de Física Moderna, Universidad de Cantabria and
Instituto de Estudios Avanzados,
CSIC–Universidad de Cantabria
Facultad de Ciencias, Avda. de Los Castros s/n,
39005 Santander, Spain*

Most of the studies on the anisotropy expected in the temperature of the cosmic microwave background (CMB) have been based on linear density perturbations. The anisotropies at angular scales $\geq 1°$ (horizon at recombination) are preserved during the evolution of the universe, whereas for smaller scales new effects can appear, generated during the non-linear phase of matter clustering evolution: i) the Sunyaev-Zeldovich effect due to hot gas in clusters (Scaramella et al. 1993), ii) the Vishniac effect (Vishniac 1987) due to the coupling between density fluctuations and bulk motions of gas and iii) the integrated gravitational effect (Martínez–González et al. 1994) due to time-varyng gravitational potentials. A single potential $\phi(t, \mathbf{x})$, satisfying the Poisson equation, is enouph to describe weak gravitational fields associated to non-linear density fluctuations when one considers scales smaller than the horizon and non-relativistic peculiar velocities. The temperature anisotropies, in a flat universe, are given by the expression (Martínez–González et al. 1990)

$$\frac{\Delta T}{T} = \tfrac{1}{3}\phi_r + \mathbf{n}{\cdot}\mathbf{v_r} + 2 \int_r^o dt \, \frac{\partial \phi}{\partial t}(t, \mathbf{x}),$$

where $\mathbf{n}$ is the unit vector in the direction of observation, the subscript r denotes the recombination time and we have chosen units such that $8\pi G = c = 1$. The temperature correlation function $C(\alpha)$ can be obtained from the previous equation by averaging over all direction pairs separated by an angle α and it is given by the following integrated effect

$$C(\alpha) = 4 \int_0^1 d\lambda_1 \int_0^1 d\lambda_2 \frac{\partial^2}{\partial \lambda_1 \partial \lambda_2} C_\phi(\lambda_1, \lambda_2; x),$$

445

M. Kafatos and Y. Kondo (eds.), Examining the Big Bang and Diffuse Background Radiations, 445–446.
© 1996 IAU.

where $x^2 = \lambda_1^2 + \lambda_2^2 - 2\lambda_1 \lambda_2 \cos\alpha$, λ_1 and λ_2 represent the distances to the two photons from the observer and C_ϕ is the correlation of the gravitational potential. As a model for the evolution of the correlation of matter $\xi(z_1, z_2; r)$, we will assume

$$\xi(z_1, z_2; r) = [(1 + z_1)(1 + z_2)]^{-\frac{3+\epsilon}{2}} \left(\frac{r_o}{r}\right)^\gamma, \quad r \leq r_m,$$

and $\xi = 0$ otherwise. The values we will consider are $\gamma \simeq 1.8$ and $r_o \simeq 5\, h^{-1} Mpc$ based on observations of the correlation function of galaxies at the present epoch. The evolution in redshift is parametrized in terms of ϵ, the value $\epsilon = 0, -1.2$ represent stable and comoving clustering, respectively.

The main results we have obtained can be summarized as follows: i) the temperature correlation function $10^{-6} \leq C(\alpha)^{1/2} \leq 10^{-5}$ for values of the evolution parameter $-1.2 \leq \epsilon \leq 3$, initial redshift $z_m \geq 3$ and cut-off $r_m \geq 10\, h^{-1} Mpc$ at present. ii) $\Delta T/T$ is maximum for comoving clustering ($\epsilon = -1.2$), decreases for stable clustering ($\epsilon = 0$) and slowly increases for $\epsilon \geq 0$. iii) More than 80% of the effect is produced in the interval $z \leq 3$ for $\epsilon \geq 0$, whereas for comoving clustering the effect is created near the initial redshift. iv) Considering present experimental limits on degree scales ($\Delta T/T \leq 10^{-5}$), the non-linear evolution of the correlation function for comoving clustering must have started at $z \leq 10$ if $r_m = 10\, h^{-1} Mpc$. Extrapolating to $r_m = 19\, h^{-1} Mpc$, those limits imply that clusters must have formed by $z \leq 4$ for the same type of evolution. v) The stable clustering model or models with a faster evolution ($\epsilon > 0$) give $\Delta T/T$ of the order of 10^{-6}, so they do not violate present upper-limits.

References

Martínez-González, E., Sanz, J.L. and Silk, J.(1990), *The Astrophysical Journal Letters*, **Vol. no. 355**, L5.

Martínez-González, E., Sanz, J.L. and Silk, J.(1994), *The Astrophysical Journal*, in press.

Scaramella, R., Cen, R. and Ostriker, J. P. (1993), *The Astrophysical Journal*, **Vol. no. 416**, 399.

Vishniac, E. T. (1987), *The Astrophysical Journal*, **Vol. no. 322** ,597.

PLANNING FUTURE SPACE MEASUREMENTS OF THE CMB

J.L. PUGET, N. AGHANIM AND R. GISPERT
Institut d'Astrophysique Spatiale, Orsay, France

AND

F.R. BOUCHET AND E. HIVON
Institut d'Astrophysique, Paris, France

1. Small scale anisotropies of the Cosmic Microwave Background (CMB)

A central problem in cosmology is the building and testing of a full and detailed theory for the formation of (large-scale) structures in the Universe. It is widely believed that the observed structures today grew by gravitational instability out of very small density perturbations. Such perturbations should have left imprints as small temperature anisotropies in the cosmic microwave background (CMB) radiation.

The COBE satellite has opened up a new era by reporting the first detection of such temperature anisotropies. However, COBE was limited by poor sensitivity, and restricted to angular scales greater than 7° , much more extended than the precursors of any of the structures observed in the Universe today. As a result, these measurements are compatible with a wide range of cosmological theories. On the other hand, high precision measurements at the degree scale will be extremely discriminating, allowing strong constraints on large scale structure formation theories, the ionization history, the cosmological parameters (such as the Hubble and cosmological constants, or the type and amount of dark matter), and very early Universe particle theories. These goals could be met by producing nearly full-sky maps of the background anisotropies with a sensitivity of $\frac{\Delta T}{T} \sim 10^{-6}$ on scales from 10 arc minutes to tens of degrees.

More precisely, the **statistics of the background anisotropies**, if Gaussian, would favor inflationary models, whereas non-Gaussian fluctuations would rather favor models in which irregularities are generated by topological defects such as strings, monopoles and textures. The **shape of**

447

M. Kafatos and Y. Kondo (eds.), Examining the Big Bang and Diffuse Background Radiations, 447–452.
© 1996 *IAU.*

the primordial fluctuation spectrum, according to most theories of the origin of the fluctuations in the Universe corresponds to potential fluctuations $\delta\phi$ which should be independent of the scale of the irregularities (λ). Even a few percent deviation from this prediction would have extremely important consequences for the inflationary models. These models also predict a specific **ratio of tensor*vs. scalar mode anisotropies** depending notably on the scale of the fluctuations. The best way to learn about the **reionization history of the universe** would be through the small-scale anisotropies which may be partly erased if the inter-galactic medium was re-ionized at high redshift. These small scale anisotropies depend also **on the baryon density, the nature of the dark matter and the geometry of the Universe** (and of course on the initial spectrum of irregularities). Observations of the structure of the CMB on scales down to $10'$ will resolve structures comparable in scale to clusters of galaxies ($\sim 10h^{-1}\mathrm{Mpc}^\dagger$) and so allow a much more direct link between observations of **galaxy clustering, galaxy peculiar velocities and temperature anisotropies**. Additionally, the detection of at least 1000 rich clusters[‡] by the **Sunyaev-Zeldovich effect**[§], in combination with X-ray observations would constrain the evolution of rich clusters of galaxies and provide another handle on H_0. Finally, the Doppler effect on the same clusters might provide a unique measurement of **their peculiar velocity dispersion**.

Measuring the cosmological background anisotropies with an accuracy better than a part in a million on scales down to ten arc minutes is a very ambitious goal. Although difficult, it is achievable in the coming decade with a dedicated satellite experiment using the best detectors available today. Several concepts of missions are currently studied and two of them are presented below. The accuracy of the measurements of the cosmological anisotropies will rely on the ability to separate these fluctuations from those coming from the galactic background and from unresolved extra galactic sources. The different contributions to the microwave anisotropies can be non-Gaussian, and the separation of the components is a non-linear process. This calls for numerical simulations of the sub-millimeter and microwave sky.

---

[*]generated by gravitational waves.

[†]h is Hubble's constant in units of 100 km s^{-1}Mpc^{-1}.

[‡]assuming a photon noise limited space experiment.

[§]caused by the frequency change of microwave background photons scattered by hot electrons in the gaseous atmospheres of rich clusters of galaxies

2. Sky simulations

By using both theoretical modeling and data extrapolations, the RUMBA group[¶] aims at creating maps as realistic as possible of the relevant physical phenomena which may contribute to the detected signals The useful wavelengths for such studies are in the range from 200 μm to 2 cm where sources of flux anisotropies which may contribute in addition to the Primary $\frac{\Delta T}{T}$ of the 2.726 K background at comparable levels are :

- Effects of clusters, through the Sunyaev-Zeldovich effect, and the Doppler shift due to their peculiar velocities.
- Secondary effects like topological defects created during an early Universe Symmetry–breaking phase transition (*e.g.* Cosmic Strings, Textures, Global monopoles, etc...), or Rees-Sciama effect, Vishniac effect, etc...
- Sub-millimeter emission from the galaxies (including starburst galaxies, radio galaxies and AGN's)
- Emission from the Milky Way due mainly to three components, the interstellar dust (even at high galactic latitude : galactic cirrus), the bremsstrahlung and synchrotron emissions.

At this stage, our modeling of the various physical phenomena has been done as follows:

- For the primary $\frac{\Delta T}{T}$ and the effects from clusters, we have created realizations corresponding to specific theoretical models (*e.g.* standard CDM). This was also done for the possible secondary fluctuations from cosmic strings.
- For the galactic emission we use extrapolations of far-infrared data (e.g. IRAS maps and catalogs) by a single temperature model and of radio data at 408 MHz.
- For the unresolved component of the emission of galaxies, we are currently modeling the fluctuations by Monte-Carlo simulations.

All these processes are stored as elementary maps (presently 12.5° × 12.5°, with 500^2 pixels, each of size 1.5 arc minutes). The performances and properties of the instrument itself are included in order to model the expected signals and noises. The structure of our tool is designed to deal easily with further modifications of the experimental configuration. At this stage we assume simplified configurations with top-hat spectral filters, Gaussian lobes and a $1/f$ detector noise with a low frequency cutoff. Thus the final outputs of our simulations are a set of simulated signals as expected to be received

---

[¶]The RUMBA group currently includes N. Aghanim, R. Bond, F. R. Bouchet, A. De Luca, F.-X. Désert, M. Giard, R. Gispert, B. Guiderdoni, E. Hivon, F. Pajot and J.-L. Puget.

from the satellite. They can then be used as test beds for the assessment of the feasibility of various signal-separation techniques. Furthermore, we use this tool in an iterative fashion to optimize the instrumental characteristics of the planned experiments.

Of course, our current modeling is far from perfect. For the high–z sources of fluctuations, we have to rely on theoretical models. In some cases, years of effort have led to the development of reliable tools, such as for the primary $\Delta T/T$ fluctuations in Gaussian models. In other case, it is technically hard to make definite predictions, even though the model is perfectly well specified (e.g. for strings, or the SZ effect which depends on the gas behavior in very dense environments). We also have only weak constraints on the high–z distribution of galaxies. For lower–z sources of fluctuations, like those coming from the dust in galaxies, we cannot do much more at this time than to extrapolate the dust emission from the IRAS and COBE measurements. But there might very well be another cold dust population, spatially uncorrelated with the hot one... In any case, the database structure adopted should permit us to incorporate all the latest developments in relevant theoretical modeling and/or new observations, as they appear.

3. Mission Concepts

The European Space Agency is studying a mission (COBRAS-SAMBA) which is aimed at the measurements described above. This mission will be up for selection in the spring of 1996 to become the third medium-size mission of the Horizon 2000 plan. This concept results from the work of the ESA Science Team[‖].

The payload will be passively cooled; preliminary thermal models indicate that the focal plane assembly will reach an average temperature of $\sim$100 K, while the telescope optical surfaces will stabilize at $\sim$120 K. These temperatures, in addition to the absence of the atmospheric noise and heat load due to the chosen orbit, insure that the conditions for a low and stable background are met.

An off-axis 1.5 meter gregorian telescope provides the necessary angular resolution and collecting area to meet the objectives of the mission. The separation of the different components of the foreground emissions requires a broad frequency coverage. The proposed mission will incorporate eight frequency bands provided by four arrays of bolometers at high frequency

[‖]The COBRAS/SAMBA Science Team which carried the study was : M. Bersanelli, C. Cesarsky, L. Danese, G. Efstathiou, M. Griffin, J.-M. Lamarre, M. Mandolesi, H.U. Norgaard-Nielsen, O. Pace, E. Pagana, J.-L. Puget, J. Tauber and S. Volonté. Requests for copies of the relevant report can be obtained from S. Volonté, ESA HQ, 8-10 rue Mario Nikis, PARIS CEDEX 15, FRANCE.

and four arrays of passively cooled tuned radio detectors at low frequency. The 50 bolometers require cooling to ∼0.1 K, which will be achieved with an active system similar to the one foreseen for the ESA cornerstone mission FIRST. It combines active coolers reaching 4 K with a dilution refrigeration system working at zero gravity, both being developed and space-qualified in Europe. The refrigeration system will include two pressurized tanks of ^{3}He and ^{4}He, giving an operational lifetime of at least two years.

On the low frequency side, HEMT (High Electron Mobility Transistors) amplifiers give the required sensitivity to detect temperature anisotropies of the cosmological background at all angular scales larger than 0°.5. The frequency and angular coverage will provide good overlap with the COBE-DMR maps.

The frequency band at which the foreground emission is the lowest (∼130 GHz) will be covered by both the tuned radio receivers and the bolometers. The fact that the tuned radio detectors are passively cooled insures that the low frequency channels can be operated for a duration limited only by spacecraft consumables.

COBRAS-SAMBA will be in a small size orbit around the L5 Lagrangian point of the Earth-Moon system, at a distance of about 400 000 km from both the Earth and the Moon. The choice of a far-Earth orbit is a distinctive feature of this mission and allows to reduce below significance level the potential contamination from Earth radiation and straylight, which is critical at the goal sensitivities. Compared to a low-Earth orbit, the requirement on side-lobe rejection drops from 10^{-13} to 10^{-9}. The L5 orbit is also very favorable from the point of view of passive cooling and thermal stability as the satellite will always point within ±40° of the anti-solar direction.

The requirement of coverage of a large fraction of the sky is obtained by offsetting the optical axis of the telescope by 30° with respect to the spin axis of the satellite. In the basic configuration, the spin axis is the major axis of symmetry of the spacecraft, and is directed along the line joining the satellite and the Sun. In this nominal position, the optical axis scans a circle of diameter 60° , centered on a point in the ecliptic plane in the anti-Sun direction. In one year, the observed circle sweeps the whole ecliptic. To increase the sky coverage, the spacecraft rotation axis will be moved away from the ecliptic plane by up to ±40° .

Center Frequency	31.5	53	90	125	140	222	400	714
Detector Technology	HEMT arrays				Bolometer arrays			
Detector Temperature	~100 K				0.1-0.15 K			
Number of Detectors	13	13	13	13	8	11	16	16
Angular Resolution	30	30	30	30	10.5	7.5	4.5	3
Optical Transmission	1	1	1	1	0.3	0.3	0.3	0.3
Bandwidth ($\frac{\Delta \nu}{\nu}$)	0.15	0.15	0.15	0.15	0.4	0.5	0.7	0.6
$\frac{\Delta T}{T}$ Sensitivity (1σ, 10^{-6} units, 2 years)								
90% sky coverage	1.7	2.7	4.1	7.2	0.9	1.0	8.2	10^4
2% sky coverage	0.6	0.9	1.4	2.4	0.3	0.3	2.7	5000

Table I: Payload Characteristics. Frequencies are in GHz, Angular Resolutions in arc minute.

The coverage of the sky is not uniform in terms of integration time per pixel. However, the motion of the spin axis can be chosen in such a way as to give deeper integrations in chosen parts of the sky. The regions of lowest galactic emission are of particular interest for the anisotropy measurement, and many of them will be observable during the mission. Table I shows the estimated average sensitivities per pixel at the end of the two year baseline mission, for each frequency channel, assuming an observing strategy as sketched above. The pixel size for each channel is also given in Table I.

A less ambitious version is studied by the French Space Agency (CNES) in collaboration with laboratories in UK, Italy and USA. It relies only on bolometers cooled to 100 mK with the same technique as in COBRAS-SAMBA mission and plans to include only 5 wavelengths channels. It is based on a 80-cm telescope placed in a polar sun-synchronous orbit.

NEW RESULTS ON CMB STRUCTURE
FROM THE TENERIFE EXPERIMENTS

R.D. DAVIES AND C.M. GUTIÉRREZ
University of Manchester, Nuffield Radio Astronomy Laboratories,
Jodrell Bank,
Macclesfield SK11 9DL, UK

R.A. WATSON AND R. REBOLO
Instituto de Astrofísica de Canarias,
38200 La Laguna, Tenerife, Spain

A.N. LASENBY AND S. HANCOCK
Mullard Radio Astronomy Observatory, Cavendish Laboratory,
Madingley Road, Cambridge CB3 OHE, UK

Abstract. Temperature fluctuations in the CMB (Cosmic Microwave Background) are a key prediction of cosmological models of structure formation in the early Universe. Observations at the Teide Observatory, Tenerife using radiometers operating at 10, 15 and 33 GHz have revealed individual hot and cold features in the microwave sky at high Galactic latitudes. These well-defined features are not atmospheric or Galactic in origin; they represent the first detection of individual primordial fluctuations in the CMB. Their intensity is defined by an intrinsic *rms* amplitude of 54^{+14}_{-10} μK for a model with a coherence angle of 4°. The expected quadrupole term for a Harrison-Zel'dovich spectrum is $Q_{RMS-PS} = 26 \pm 6$ μK. When our data at Dec=+40° are compared with the COBE DMR two-year data, the presence of individual features is confirmed. New experiments to detect structure on smaller scales are described.

1. Introduction

The study of the CMB radiation gives a unique insight into physical conditions in the early Universe, providing direct access to the epoch $\sim 300,000$ years after the initial singularity. By observing different angular scales, we investigate different linear and mass scales; for example on the 5° scale of

M. Kafatos and Y. Kondo (eds.), Examining the Big Bang and Diffuse Background Radiations, 453–460.

our Tenerife beamswitching experiments, we are probing the equivalent of 300−500 Mpc structure in our contemporary Universe, which corresponds to the largest scales identified in the distribution of galaxies.

Much attention is being directed both observationally and theoretically, towards a comparison of the amplitude of CMB structure as a function of angular scale. The standard picture envisages two components of the structure on scales $\gtrsim 2°$. The first is scalar fluctuations arising from the Sachs-Wolfe effect, while the second is a tensor component due to gravitational wave radiation coming from the inflationary era. A detailed measurement of the spectrum of fluctuation amplitude on such scales will then establish the tensor contribution. A second feature of the angular spectrum of fluctuation amplitudes is the "Doppler peak" centred on scales of $1° − 2°$. These fluctuations are due to the motion of massive structures in the epoch of recombination, and are expected to have an amplitude of $2 − 3$ times that on large angular scales.

Now that the *rms* amplitude of CMB fluctuations is established to be at least 30 μK over a large angular range (Smoot *et al.* 1992, Strukov *et al.* 1993, Hancock *et al.* 1994, Ganga *et al.* 1993, Clapp *et al.* 1994, Schuster *et al.* 1993, Wollack *et al.* 1993, Cheng *et al.* 1994, De Bernardis *et al.* 1994), it is possible to undertake observations with adequate sensitivity to obtain detailed maps on a reasonable timescale. We describe here a programme of actual and planned observations covering the angular range from a few arcminutes to $\sim 15°$. Our beamswitching observations on an angular scale of 5° have now detected structure, with the unambigous identification of hot and cold features on the Dec=+40° scan (Hancock *et al.* 1994), while scanning observations at adjacent declinations are now reaching the sensitivity required to make maps.

2. The experimental arrangement

The choice of radio frequency for observation is a critical parameter in the design of a system capable of detecting intrinsic CMB structure. A ground-based system is limited by the atmosphere to operate at frequencies below 40 GHz, and the presence of Galactic emission requires that observations should be made over a substantial range of frequencies to correct for that contribution. In our experiments the lowest frequency was chosen to be 10.45 GHz where the dominant Galactic emission is likely to be synchrotron or free-free with spectral indices $\sim −3.0$ (Lawson *et al.* 1987, Watson *et al.* 1992) and −2.1 respectively. An intermediate frequency was chosen at 15 GHz, and the highest at 33 GHz where the synchrotron and free-free contributions are respectively 10 and 30 times less than at 10.45 GHz. At 33 GHz the variability of atmospheric water vapour limits the amount of time for which good data can be obtained. Our experience at the Teide Observatory at 2400 m altitude showed that the atmospheric conditions allow us to operate for a significant amount of time at this frequency (Davies *et al.* in preparation).

Central to the design of our radiometers is the triple beam switching technique (Davies *et al.* 1992). This is achieved by fast switching (61 Hz) between two adjacent beam horns combined with a slow ($\sim$ 8 sec) wagging of the double-beam response. The resulting beam response has the form (-0.5, $+1.0$, -0.5) with the central beam (FWHM$\sim$ 5°) lying in the meridian and the negative beams displaced $\sim \pm 8°$ in azimuth. The fast switching removes receiver instabilities on timescales $\lesssim 0.1$ sec, while the beam wagging greatly reduces the long-term atmospheric and radiometer drifts. Systematic effects arising from the environment of the equipment are reduced to a minimum by keeping the system fixed and scanning the sky using Earth-rotation.

The sensitive amplifier element in each receiver is a high electron mobility transistor (HEMT) amplifier. Each system has a bandwidth of $\sim$ 10% and operates at an equivalent system brightness temperature in the range 70 – 100 K; this latter figure includes the effects of circulator and feed losses as well as atmospheric emission. Each instrument consists of two independent channels giving further confidence in identifying atmospheric and systematic effects. The resulting theoretical sensitivities, including both channels, are 5.6, 3.4 and 2.2 mK$\times$Hz$^{-1/2}$ at 10, 15 and 33 GHz respectively.

3. Observations and analysis of the results

Our objective is to construct a map of the sky covering declinations 30° to 45° by observations in strips separated 2°.5 in declination. By observing N times at each declination we obtain a very sensitive scan with a reduction in noise by a factor $\sqrt{N}$ compared with an individual scan. Figure 1 shows these stacked scans at 15 GHz covering the declination range 30° to 45°. The strong emission at RA$\sim$ 300° is the most intense crossing of the Galactic plane; the weak crossing which contains several components is also evident in the RA range 60° to 80°. A clear correlation exists between the Galactic signals on adjacent declinations at 2°.5 separation. The most intense signal comes from the Cygnus X H II region at Dec$\sim$41°, RA=305°. No structures other than the Galactic crossings are seen on this intensity scale. The search for CMB fluctuations must be made at high Galactic latitudes. Table 1 presents the sensitivity per beam area reached at each declination and frequency in the section of our data in the range RA=161°$-$250° corresponding to high Galactic latitude. On the basis of the *rms* of the CMB

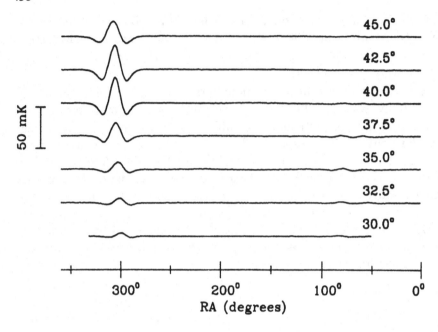

Figure 1. Stacked scans for each declination at 15 GHz, showing the strong (RA~ 300°) and the weak (RA ~ 60° − 90°) crossings of the Galactic plane.

fluctuations claimed by several experiments (see Section 1), we can see that the sensitivity reached at some declinations and frequencies is enough to detect and map features in the CMB. In particular, at Dec=+40° we have reached a good sensitivity in our scans at 15 and 33 GHz; these are shown in an expanded scale in Figures 2a and 2b respectively. Common features are clearly seen in both scans with the most intense centred at RA~ 185° having a similar amplitude at both frequencies. Our scan at 10 GHz has not reached enough sensitivity to clearly delineate the CMB features; at this low frequency some degree of Galactic contamination is expected. Nevertheless the results at 10 GHz are important in that they can be used to put limits on the possible Galactic contribution at the higher frequencies. We have computed a maximum Galactic *rms* contribution at 33 GHz of 4 μK which is a small fraction of the detected signals at this frequency. Our conclusion is that most of the signal present in these scans at 15 and 33 GHz is cosmological in origin. Assuming that all signals present at both frequencies are real CMB fluctuations, we can add them to obtain the very sensitive scan shown in Figure 2c. Using a likelihood analysis we have estimated the amplitude of the signals present in our data. For a model of Gaussian fluctuations with a coherence angle of 4°, we obtain an

TABLE 1. Standard error (in μK) per beam-sized area over the RA range 161°−250°.

Experiment	Declination						
	30°0	32°5	35°0	37°5	40°0	42°5	45°0
10 GHz	−	69	97	62	57	75	77
15 GHz	22	27	30	19	30	25	24
33 GHz	−	38	54	47	21	−	−

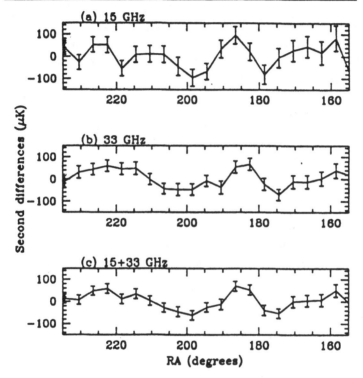

Figure 2. The stacked data scans at Dec=+40° and their one-sigma error-bars in a 4° bin in RA at (a) 15 GHz, (b) 33 GHz and (c) (15+33) GHz.

intrinsic *rms* signal of 54^{+14}_{-10} μK. Assuming a Harrison-Zel'dovich spectrum for the primordial fluctuations we infer an expected quadrupole amplitude $Q_{RMS-PS} = 26 \pm 6$ μK (Hancock *et al.* 1994).

4. Comparison with COBE

A most important outcome of a comparison between the CMB amplitudes on the angular scales represented by the Tenerife and COBE measurements is the determination of the spectrum of primordial fluctuations. The statistic most suitable for this comparison is the equivalent quadrupole component of the power spectrum Q_{RMS-PS}. The COBE first-year data gave the value of 16.3 ± 4.6 μK for a model with an spectral index $n = 1.15$ (Smoot et al. 1992), which may be compared with our value of 26 ± 6 μK at $n = 1$. Both these results are affected by random error, as well as cosmic variance and sample errors. All these factors can be taken into account using Monte Carlo techniques to produce a large number of sky realizations described by a power law spectrum with a range of power law indices n. By combining the Tenerife and COBE detections it is possible to restrict the range of n to $0.87 \leq n \leq 1.6$ as compared with either experiment alone. This lower limit is particularly significant in the context of inflationary theories of galaxy formation, since $n \approx 0.8$ is the value where the tensor mode contribution to the CMB fluctuations is equal to the scalar contribution. For $n \gtrsim 0.8$ the tensor contribution declines rapidly. Clearly further observations with COBE using the full 4 years of data along with Tenerife observations at adjacent declinations, will restrict the range of n even further and provide one useful test of inflationary models of structure formation in the early Universe.

Our Tenerife experiments have unequivocally detected hot and cold features in the CMB field, whereas COBE DMR first year of data did not have enough sensitivity to identify particular spots and consequently gave only a statistical result representative of the whole sky. Certain areas of the DMR sky survey have more sensitivity than the average point; it is fortunate that such an area near the North ecliptic pole coincides with the high Galactic latitude section of the Tenerife Dec=+40° scan. That makes possible a direct comparison of features in the COBE DMR two-year data (Bennett et al. 1994) and in the Tenerife data.

The RA=160° to 230° part of the Dec=+40° scan was used to investigate the presence of structures in both data sets which had the properties of CMB fluctuations. The most prominent feature in the Tenerife data (at RA~ 185°) is evident in the 53 and 90 GHz maps (the 31 GHz has lower sensitivity) and has the Planckian spectrum expected for CMB anisotropy. Futhermore, the correlation function between the Tenerife and COBE DMR scans is also indicative of common structure across the range RA=160°−230°. The combination of the spatial and spectral information from the two data sets is consistent with the statistical level of fluctuations claimed by each experiment and strongly supports the cosmological origin of this structure. The detection covers a range of 10 to 90 GHz. This is the first confirmation of actual CMB structural features detected in two independent experiments.

5. The future programme

The present situation in the beamswitching radiometer programme operated by NRAL Jodrell Bank and IAC at Teide Observatory, Tenerife, is summarized in Table 1. Data collection will continue in order to give a coverage of Dec=+30° to 45° at the full sampling separation of 2°5 (half the FWHM). We plan to reach *rms* sensitivities in a 5° beam of 20 μK at 15 and 33 GHz, and 50 μK at 10 GHz. This combination of sensitivities will enable us to detect CMB fluctuations, and at the same time to determine the Galactic contribution to better than 5 μK at the highest frequency.

A 33 GHz two-element interferometer is being constructed at Jodrell Bank in preparation for installation at Teide Observatory in collaboration with IAC. This interferometer will have a resolution of 2°5 with full sine and cosine correlation in a 3 GHz bandwidth. The low noise amplifiers used are cryogenically cooled HEMTs, and the anticipated sensitivity is 0.7 $mK \times Hz^{-1/2}$. A 5 GHz interferometer of similar correlator design but using a beamwidth of 8° has been operating at Jodrell Bank for several years. Observations have been taken over the declination range 30° to 50° at separations of 11λ and 33λ. These data are being used to estimate the contribution of Galactic emission and extragalactic sources at 5 GHz.

The Cosmic Anisotropy Telescope (CAT) has been developed by the MRAO group. It is operating at Cambridge making synthesis maps of CMB structure over the angular range 10′ to 2°5. The frequency range is 12 to 18 GHz, and the sensitivity achieved per resolution element is approaching 20 μK. Point source contribution is removed by making observations with the Ryle telescope array at the same frequency. CAT is promising to have a significant contribution to CMB structure studies in the next few years.

The collaboration between MRAO, NRAL and IAC continues with a proposal to build the Very Small Array (VSA) for CMB structure studies. Operating at 33 GHz on the Tenerife site, the VSA will have the capability of imaging primordial CMB structure to a sensitivity of 5 μK over the angular range 10′ to 2°5. Funding is being sought for this project.

Acknowledgements

The Tenerife experiments are supported by the UK PPARC, the European Community "Science" programme SC1-CT92-830, and the Spanish DGI-CYT science programmes.

References

Bennett, C.L. *et al.* 1994, *Astrophys. J.*, (*in press*).
Cheng, E.S. *et al.* 1994, *Astrophys. J.*, **422**, L37.
Clapp, A. *et al.* 1994, *Astrophys. J.*, **433**, L57.

460

Davies, R.D., Watson, R.A., Daintree, E.J., Hopkins, J., Lasenby, A.N., Sanchez-Almeida, J., Beckman, J.E., & Rebolo, R. 1992, *Mon. Not. Roy. Soc.*, **258**, 605.

De Bernardis, P. *et al.* 1994, *Astrophys. J.*, **422**, L33.

Ganga, K., Cheng, E., Meyer, S., & Page, L. 1993, *Astrophys. J.*, **410**, L57.

Lawson, K.D., Mayer, C.J., Osborne, J.L., & Parkinson, M.L. 1987, *Mon. Not. Roy. Soc.*, **225**, 307.

Hancock, S., Davies, R.D., Lasenby, A.N., Gutierrez de la Cruz, C.M., Watson, R.A., Rebolo, R., & Beckman, J.E. 1994, *Nature*, **367**, 333.

Schuster, J. *et al.* 1993, *Astrophys. J.*, **412**, L47.

Smoot, G.F. *et al.* 1992, *Astrophys. J.*, **396**, L1.

Strukov, I.A., Brukhanov, A.A., Skulachev, D.P., & Sazhin, M.V. 1993, *Phys. Lett.*, **B**, **315**, 198.

Watson, R.A., Gutierrez de la Cruz, C.M., Davies, R.D., Lasenby, A.N., Rebolo, R., Beckman, J.E., & Hancock, S. 1992, *Nature*, **357**, 660.

Wollack, E.J., Jarosik, N.C., Netterfield, C.B., Page, L.A., & Wilkinson, D., 1993, *Astrophys. J.*, **419**, L49.

On the Use of COBE Results

M.Melek

Astronomy and Meteorology Department, Faculty of Science,Cairo University, , Giza, Egypt.

Abstract

The aim of this brief report is to find a lower limit of the Hubble parameter using the COBE's detected fluctuations in the temperature of the CMBR.

1.Mathematical Considerations

If T is the temperature of the CMBR and $T_\mu = \frac{\partial T}{\partial x^\mu}$ is a time-like covariant gradient vector, then it is possible to define the temporal variation of the magnitude of the gradient of T as follows [1]:

$$F = \frac{d}{dt}(g^{\mu\nu}T_\mu T_\nu)^{1/2} = \frac{dG}{dt},$$

(1)

where t is the cosmic time, $g^{\mu\nu}$ is the contravariant metric tensor which represents the background gravitational field in which the microwave background radiation exists and $\mu, \nu = 0,1,2,3$.It is possible to prove that F can be written as follows (Melek (1992)):

$$F = \frac{1}{G}g^{\mu\nu}T_{\mu;\sigma}T_\nu U^\sigma,$$

(2)

where $T_{\mu;\sigma}$ is the usual covariant derivative with respect to x^σ and $U^\sigma = \frac{dx^\sigma}{dt}$.

2.Physical and Cosmological Considerations

Since the COBE results confirmed that the universe's anisotropy, regarding the CMBR, is of the same order as the anisotropy which can be produced by a small spatial perturbations of the FRW metric ([2], [3],

M. Kafatos and Y. Kondo (eds.), Examining the Big Bang and Diffuse Background Radiations, 461–463.

[4], [5] and references therein) therefore it is siutable to represent the universe's gravitational field by the spatially perturbed FRW metric, which is given as follows in the spherical polar coordinates:

$$d\tau^2 = R^2(t)dt^2 - \frac{R^2(t)(1+h_{11})}{1-kr^2}dr^2 - R^2(t)h_{12}rdrd\Omega - R^2(t)r^2(1+h_{22})d\Omega^2,$$

(3)

where Ω is a solid angle defined in terms of θ,ϕ and $R(t)$ is the scale (expansion) factor. The parameter k is a constant which takes the values $+1$ or 0 or -1, if the universe is closed or flat or open respectively. Finally h_{11}, h_{12} and h_{22} are the spatial perturbat- ions which are assumed to be small quntities, i.e. their qudratic terms are neg- ligable.

Since the FRW universe is an isotropic and homogeneous,therefore

$$F_{(FRW)} = \frac{d}{dt}(g^{\mu\nu}{}_{(FRW)}T_\mu T_\nu)^{1/2} = 0,$$

(4)

where $g^{\mu\nu}{}_{(FRW)}$ is the metric tensor of the FRW world line.

Since the general motion in the universe is only due to its general expansion, then:

$$\frac{dr}{dt} = \frac{d\Omega}{dt} = 0,$$

(5)

Besides that, at any fixed cosmic time since the CMBR was separated from the matter, its temprature is independent of the radial coordinate r, i.e.

$$T_1 = \frac{\partial T}{\partial r} = 0,$$

(6)

Using (5), (6) and the fact that the temprature of the CMBR is decreasing only due to the expansion of the universe, one can conclude that:

$$\frac{dT_1}{dt} = 0$$

(7)

3.Expression of F in the Spatially Perturbed FRW Universe and its Consequences

Using the metric (3) and inserting the physical and the cosmological constraints (4), (5), (6) and (7) in (2), therefore the expression of F is given as follows:

$$F = \frac{h_{22}}{R^3(t)Gr^2} T_2 [\frac{dT_2}{dt} - HT_2], \qquad (8)$$

This expression gives the temporal variation of the magnitude of the angular gradient of the temprature of the CMBR. This magnitude should decrease with respect to the cosmic time due to the universe's expansion. Therefore the expression (8) of F should be negative. Since T_2 is positive, then F will be negative iff the following inequality is satisfied:

$$\frac{\frac{dT_2}{dt}}{T_2} < H, \qquad (9)$$

where H is the Hubble parameter.

4.Conclusion

The iequality (9) gives a lower limit of the Hubble parameter in terms of the ratio between the cosmic time variation of the angular gradient and the angular gradient itself of the temperature of the CMBR. The author hopes that, using the COBE data, it might be possible to calculate the left hand side of the iequality (9).

References

[1] Anile A.M. and Motta S. (1976), Ap.J. 207, 685.

[2] Melek M. (1992), ICTP preprint NO. IC/92/95.

[3] Peebles P.J.E. and Yu J.T. (1970), Ap.J. 162, 815.

[4] Sachs R.K. and Wolfe A.M. (1967), Ap.J. 147, 73.

[5] Wright E.L., etal. (1992), Ap.J. 396, L13.

Is the Early Universe Fractal?

E. M. de Gouveia Dal Pino[1], A. Hetem[1], J. E. Horvath[1], & T. Villela[2]

1. Universidade de São Paulo, IAG, São Paulo, SP 9638, 01065-970, Brazil
2. INPE, São José dos Campos, SP, 12201-970, Brazil

1. Introduction

It is generally believed that on *very* large scales the distribution of matter in the universe is homogeneous and effectively all the existing theoretical approaches are based on this assumption. However, recent re-analysis of galaxy-galaxy and cluster-cluster correlations by Coleman and Pietronero (1992, hereafter CP) and Luo and Schramm (1992, hereafter LS) indicates that the distribution of the visible matter in the universe is fractal or multifractal up to the present observed limits (~100 h^{-1} Mpc for H_0=100 h km s^{-1} Mpc^{-1} and 0.5≤h≤1) without any evidence for homogenization on those scales. The fractal dimension obtained from these analyses is D ~ 1.2 - 1.3 (CP; LS).

The temperature fluctuations of about 13 μK recently detected by the COBE team in the 2.7 K cosmic microwave background radiation (CMBR) over angular scales larger than 7° (e.g., Smoot et al. 1992) correspond to linear sizes of ~1000 h^{-1} Mpc. According to the Sachs-Wolfe effect, they were imprinted on the CMBR by primordial density fluctuations of the matter at the recombination layer and thus, provide a signature of the seeds of the structures observed in the present universe. In this work, we analyse these temperature fluctuations and show that they are also consistent with a fractal structure with fractal dimension D=1.43±0.07.

2. The Fractal Analysis

A fractal consists of a system in which more and more structure appears at smaller and smaller scales and the structure at small scales is similar to the one at large scales. The relation between the perimeter p and the area a of a two-dimensional fractal distribution is (e.g., Gouveia Dal Pino et al. 1994): $a^{1/2} = Fp^{1/D}$, where D is the fractal dimension and F is the shape factor which is related to the form of the distribution. For fractals, D is in the range $1 < D < 2$.

To obtain the fractal dimension from the CMBR image (Smoot et al. 1992), we measured the perimeter and the area of regions situated within iso-temperature contours (Gouveia Dal Pino et al. 1994). The points in the logarithmic plot of perimeter versus area, obtained from the various temperature contours of the distribution, are fitted by a straight line where D/2 is the slope, in agreement with the equation above. The result is shown in Fig. 1. The corresponding fractal dimension is D = 1.43 ± 0.07 and the shape factor is F = 0.42 ± 0.04.

The fractal dimension evaluated above is in reasonable agreement with the fractal dimension evaluated by CP and LS from galaxy-galaxy and cluster-cluster correlations up to ~100 h^{-1} Mpc. The fact that the observed temperature fluctuations correspond to scales >> 100 h^{-1} Mpc and are signatures of primordial density fluctuations suggests that the structure of the matter at the early universe was already fractal and thus non-homogeneous on those scales.

If we accept the results above, we can inquire if the fractal structure imprinted on the CMBR by the density fluctuations at t_{rec}, on scales >>100 h^{-1} Mpc, still persists at the present large scale universe as it seems to occur on smaller scales. If so, how far does the fractal correlation extend or *where does the universe become effectively homogeneous?* We can also argue if the observed fractal structure on large scales at t_{rec} also occurs on small scales at that epoch. We certainly do not have the answers to all these questions but we can try to make some tentative predictions.

465

M. Kafatos and Y. Kondo (eds.), Examining the Big Bang and Diffuse Background Radiations, 465–466.
© 1996 IAU.

466

Let us borrow from LS, the assumption that some kind of growth process provides the fractal correlation while gravity enhances the correlation amplitude on small scales. Considering some initial primordial seeds (or density fluctuations), the growth process will provide the

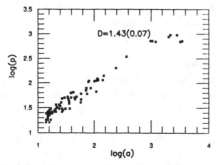

Figure 1. Perimeter versus area relation obtained from the temperature contours of the observed distribution of the CMBR fluctuations.

aggregation of matter to the seeds, and the growth rate is limited by the diffusing flux of matter onto them. Ball and Witten (1984) find that the fractal grown from a diffusion-limited process implies that $D \geq d-1$, where d is the dimension of the growth space of the aggregate. In the present work, the derived fractal dimension $D \approx 1.4$ results d < 2.4. In other words, the growth space should involve a 2-dimensional sheetlike object and favor, for example, light domain wall, pancake, or cosmic string seed models (see, e.g., LS). Since the growth process is limited by the diffusion of particles onto the aggregate, its rate can become smaller than the expansion rate of the universe at some extension. At this point we should expect the end of the fractal structure in the universe and the beginning of homogeneity. To evaluate the breakdown scale of the fractal correlation (L), LS assume that the aggregation of matter onto the seeds, during the growth of a fractal up to L, will perturb the CMBR up to the observed maximum amplitude $\delta T/T \sim 10^{-5}$. Then, they find a lower limit for the fractal correlation length $L \geq 100h^{-1}$ Mpc. If correct, this result indicates that the fractal structure obtained in the present work for primordial seeds at large scales (>1000 h^{-1} Mpc) has probably been diluted as a consequence of the diffusion-limited growth process in the expanding universe. However, further observations on different angular scales are still required to tell us the real value of L and thus, where the fractal scale ends up and homogeneity begins.

Finally, the cosmological principle implies the condition of isotropy and homogeneity. But, according to the results of the present work (and also of CP and LS), this assumption is not supported since the distribution of matter in the present visible universe on scales up to ~100 h^{-1} Mpc and also the distribution in the early universe on scales >1000 h^{-1} Mpc, appear to be fractal. As pointed out by CP, the fractal structure implies an asymmetry between space points occupied by the structure and empty points and suggests a modification of the cosmological principle: *in a simple fractal structure, from every point of the system one observes statistically the same kind of local overdensity characterized by a single exponent D.* Clearly, this modified cosmological principle is still satisfactory and does not imply a preferred point in the universe.

References

Ball, R.C. and Witten, T.A. 1984, Phys. Rev. A29, 2966
Coleman, P.H. & Pietronero, L. 1992, Phys. Reps., 213, 311 (CP).
Gouveia Dal Pino, E.M., et al. 1994, Ap. J. (submitted)
Luo, X. & Schramm, D.N. 1992, Science, 256, 513.
Smoot, G.F. et al. 1992, Ap.J., 396, L1.

A FRACTAL MODEL OF THE UNIVERSE

Do Large Galaxies Interact by Quasar Exchange ?

P. DRIESSEN
Av. Montjoie 24, Bte 6, 1050 Brussels, Belgium.

A fractal universe model has been proposed elsewhere (Driessen, 1991, 1994), based on the assumption of perfect self-similarity between some definite scales, and in particular between particles and galaxies. Its motivation rested essentially on two ideas: first, if striving for unification is justified, then the fractal idea makes sense, for in this case one single theory of particles could encompass all self-similar structures at once. Secondly, it could be demonstrated that, provided some reasonable hypothesis were made about the structure of spiral galaxy cores, they came back to the same topological state after two complete revolutions (Driessen, 1991). These ideas combined suggested that spiral galaxies might correspond to electrons on the cosmic scale. The main test of the coherence of the fractal picture was then to discover if other galaxy types and quasars do find a place in this model. Now, a small group of dissident has sustained for years that QSOs (or at least a certain class of QSOs) are local objects emitted by galaxies. This class would contain very small (d ≈ 0.3 kpc), faint ($M_V \approx$ -12), and most often radio-quiet objects with large (non cosmological) excess redshift. The analogy between the two statements "electrons emit photons" and "galaxies emit QSOs" hinted that QSOs embody cosmic scale photons, and entails that galaxies experience quasar exchange interactions. It was claimed further that galaxy clusters reproduce atomic and molecular-like systems with one or a few central giant ellipticals (the cosmic scale nuclei) interacting via quasar exchange with their satellite spirals (the cosmic scale electrons). The flock of dwarf galaxies around large ellipticals and spirals would result from condensation of dusty gas clouds entrained by ejected QSOs.

The model is falsifiable in the sense that it yields several strong predictions: large galaxies should always be surrounded by QSOs and attached to them by filaments as those emerging from NGC4319 or NGC1097. We call these "virtual" QSOs because they are still tied to their progenitor and hence do not constitute fully separate objects. The situation presumably mimics the cloud of virtual photons surrounding each electron. If the filament exerts a strong restoring force, the fate of most "virtual" QSOs is to fall back onto the galactic core, a process that could be depicted by a self-energy Feynman diagram. Due to the successive recoils created by these quasi periodic ejections and recaptures, galaxies should be submitted to a kind of *Zitterbewegung*. Sometimes a virtual quasar should be captured by a colliding galaxy and this would represent an "electrodynamical" type of galactic interaction, as in the photon exchange diagrams. Furthermore, this quasar exchange interaction should exert a repulsive force between spirals, and an attractive one between spirals and giant ellipticals. This clearly would have important consequences for the formation of clusters and larger structures, although they have not yet been worked out in detail. Finally, galaxies should display a quantum behaviour.

The supporting observational evidence is scant but it does exist and will be discussed in another paper. Let us simply note here that there is strong evidence for ejection (see Arp, 1987 and references therein). In Arp's famous system for example, an undulating filament emerging from the core of NGC4319 is directed

M. Kafatos and Y. Kondo (eds.), Examining the Big Bang and Diffuse Background Radiations, 467–468.
© 1996 *IAU.*

toward MK205 (Sulentic, 1983). In fact I pointed out that the situation may be more complex than suspected hitherto. Because the halo of MK205 is about 10 times larger than those of other QSOs found very close to galaxies, I hypothesized that MK205 is not a quasar, but a Seyfert galaxy exchanging quasars with NGC4319 (Driessen, 1994). In that respect it would be interesting to investigate if the red companion 3" NNE of MK205 (Stockton et al., 1979) could be such an object. Interestingly, recent ROSAT X-ray observations (Arp, these proceedings) revealed compact X-ray sources connected to MK205 by lower surface brightness X-ray filaments. Three of the compact sources turned out to be QSOs of much higher redshift than MK205. This implies QSO ejection from MK205 and would, in my model, represent the virtual QSOs surrounding MK205. As for QSO absorption by a colliding galaxy, there is as yet no direct evidence but close inspection of the plates 103-4-5 of Arp's atlas (Arp, 1966) show interacting galaxies (spirals and ellipticals) whose *cores* are connected by extremely long, thin and straight filaments of unvarying width. These characteristics do not fit well with those expected from interaction plumes. In our model the filament connects the black-hole core of the parent galaxy to the core of the QSO which is currently swallowed by the receiving galaxy. Thus the filament appears as a direct connection between the two galaxies, and this illustrates a one-quasar exchange diagram. Allowing to the model, giant ellipticals also interact, and in this respect, it should be remembered that Arp has discovered that the jet in M87 points toward M84, another giant in the Virgo cluster (Arp, 1968).

Once the model is accepted as a working hypothesis, scaling factors can be calculated. Four determinations of the length scaling factor do not diverge for more than a factor 10 from the median which is $\Lambda_L \approx 2.4 \ 10^{34}$. Unfortunately, only one determination of the time scaling factor was possible and it yielded $\Lambda_T \approx 3.1 \ 10^{36}$ yr. The ratio gives the velocity scaling factor $\Lambda_V = \Lambda_L / \Lambda_T \approx 1/127$ which incidentally falls closes to $\alpha \approx 1/137$ (this may not be ε simple coincidence). It implies that free quasars should move at the constant speed $C = c \ \Lambda_V \approx 2200$ km s^{-1} with respect to the cosmic background radiation. This would represent the strongest possible deviation from the Hubble flow for any galactic scale object. The quantum of angular momentum is found to be $H \approx 2.2 \ 10^{77}$ gcm^{2}s^{-1} which is much larger than the values computed for spirals with truncated rotation curves, so that the missing angular momentum is quite large as could be expected. A rough calculation shows that a value as high as $J = H/2$ could result from a dark matter halo rotating at the asymptotic constant velocity. The strength of the quasar exchange force between two spiral galaxies is measured by the quantity $E^2 = 3.79 \ 10^{83}$ g cm^3 s^{-2} which can be interpreted as the square of the spiral galaxy "charge". Now, the gravitational constant scales with dimensions so that distinct "constants" must be defined for each self-similar scale. We find new values of order $G \approx 7 \ 10^{-12}$ cm^{3}g^{-1}s^{-2} on the galactic scale and $g \approx 5 \ 10^{30}$ cm^{3}g^{-1}s^{-2} on the particle scale. Using G and the mass of the typical spiral, it can be shown that the quasar exchange force is greater than the gravitational interaction. The value of g is of the same order as that proposed by Oldershaw (1987) in a somewhat similar fractal model. Using g together with $\hbar$ and c in Planck's units furnishes a Planck mass very close to the proton mass, a Planck length close to the proton Compton wavelength and a Planck time close to the proton *Zitterbewegung* characteristic time. Similarly, when G, H and C are inserted into Planck's units, the latter fall close to characteristic quantities associated with giant ellipticals (our cosmic scale nuclei). It is clearly a very good point for the model that Planck's unruly values here reduce to those familiar quantities.

REFERENCES

Arp, H. (1966) Atlas of Peculiar Galaxies, *Astrophysical Journal Suppl. Series* **14**, 1.
Arp, H. (1968) *Publication of the Astronomical Society of the Pacific* **80**, 129.
Arp, H. (1987) *Quasars, Redshift and Controversies*, Interstellar Media, Berkeley.
Driessen, P. (1991) A Fractal Space-Time Model for Spinning Structures, unpublished.
Driessen, P. (1994) A Fractal Universe Model, submitted to *Astrophysics and Space Science*.
Oldershaw, R.L. (1987) *Astrophysical Journal* **322**, 34.
Stockton, A.N., Wyckoff S., and Wehinger P.A. (1979) *Astrophysical Journal* **231**, 673.
Sulentic, J.W. (1983) *Astrophysical Journal* **265**, L49.

POWERFUL EXTENDED RADIO SOURCES:
A GOLDMINE FOR COSMOLOGY

R. A. DALY
Princeton University
Department of Physics
Princeton, NJ 08544
U. S. A.

Abstract.
The radio properties of powerful extended radio sources may be used to study the environments of the sources, the source energetics, a characteristic length-scale that can be used as a cosmological tool, and the relation between radio galaxies and radio loud quasars. Thus, these sources offer a rich variety of diagnostics, both direct and indirect, of the cosmological model that describes our universe. They, indeed, are a goldmine for cosmology.

Perhaps the most significant result of the investigations mentioned here is the use of the radio properties of the sources to estimate the ambient gas density in the vicinity of the radio lobe. As discussed below, the ambient gas densities estimated using the strong shock jump conditions across the radio lobe indicate that these sources lie in relatively dense gaseous environments, similar to the intracluster medium found in clusters of galaxies at low redshift. Thus, the observations suggest that at least some clusters with their intracluster medium in place exist at high redshift.

1. Introduction

Several key cosmological questions remain unanswered at the present time. One piece of the cosmological puzzle is the history of the formation and evolution of structure in the universe. Radio sources can be used to study gas-rich environments and the evolution of these environments with redshift. This has implications for the epoch of cluster formation, which in

M. Kafatos and Y. Kondo (eds.), Examining the Big Bang and Diffuse Background Radiations, 469–472.
© 1996 IAU.

turn has implications for models of structure formation and evolution. For example, structure formation is expected to occur rather late in models with $\Omega = 1$, and relatively early in open models with low values of Ω.

To study the distant universe, it is necessary to understand the sources that we use as probes of the state and structure of the distant universe. Radio observations may be used to estimate the rate at which energy is deposited to the radio hotspot, and to study whether the source energetics vary systematically with redshift, as described below.

At the present time it is unknown whether the universe is open and will continue to expand forever, $\Omega < 1$, the universe is of critical density, $\Omega = 1$, or the expansion will halt and the universe recollapse, $\Omega > 1$; here Ω refers to the ratio of the mean mass density of the universe to the critical density. The remarkable properties of powerful extended radio sources suggest that these sources may be used to define a calibrated yardstick for each source (Daly 1994a). The comparison of the physical size of the radio source, estimated from the angular size of the source and the coordinate distance to the source, with the calibrated yardstick indicates whether the universe is open or closed, since the calibrated yardstick is independent of the angular source size and is either independent of or inversely proportional to the coordinate distance to the source (Daly 1994a,b).

Finally, the detailed comparison of radio galaxies and radio loud quasars sheds light on the similarities and differences between these types of active galactic nuclei (AGN). It is of interest to know whether both types of source are drawn from a single parent population. If the sources emit radiation anisotropically, then the observer might identify the source as a radio galaxy or a radio loud quasar depending on their orientation relative to the source. Alternatively, it is possible that some radio galaxies are intrinsically different from radio loud quasars, and that orientation effects plus other parameters are needed to describe the differences between powerful extended radio sources that are galaxies and quasars.

Many of the results discussed here are presented and discussed in more detail by Daly (1990, 1994a,b).

2. The Basic Model and Summary of Results

The basic model for double-lobe radio sources has been discussed by many authors (see Begelman, Blandford, & Rees 1984 and references therein). The AGN emits a highly collimated outflow in two opposite directions; this causes a strong shock wave to propagate into the ambient medium and the source size grows with time. This also provides the energy needed to produce relativistic electrons and magnetic fields that cause the source to be a powerful radio emitter. Radio sources with 178 MHz radio powers greater than about $(10^{26}$ to $10^{27})$ h^{-2} W Hz^{-1} sr^{-1} have morphologically simple and regular radio bridges indicating little backflow and little lateral expansion (Leahy & Williams 1984), in which case the rate of growth of the source estimated from the synchrotron and inverse Compton aging of the relativistic electrons across the radio bridge is a measure of the rate of growth of the radio bridge, which is referred to as the lobe propagation velocity. The lobe propagation velocities of these powerful sources indicate that they propagate supersonically relative to the ambient medium (Alexander & Leahy 1987). Thus, the strong shock jump conditions can be applied to these types of sources.

The strong shock jump conditions indicate that: $n_a v_L^2 \propto P_L \propto B^2$ where n_a is the ambient gas density just beyond the radio lobe, v_L is the lobe propagation velocity, P_L is the pressure in the lobe, and B^2 is a measure of the lobe pressure when the magnetic field B in the radio lobe is in rough equipartition with the relativistic electrons. Thus, n_a can be estimated from

$$n_a \propto (B/v_L)^2 \ . \tag{1}$$

Daly (1994a,b) uses this method to estimate ambient gas densities for the radio sources in the Leahy, Muxlow, & Stephens (1989) data set and finds that: many powerful extended radio sources, including radio galaxies and radio loud quasars, are similar to Cygnus A in that they sit near the base of the gravitational potential well of a cluster of galaxies that has a high-density gaseous intracluster medium in place. Wellman & Daly (1994) have applied eq. (1) to the Liu, Pooley, & Riley (1992) data and find very similar results.

The observations suggest that radio galaxies and radio loud quasars inhabit very similar environments, and these environments exhibit little evolution with redshift over the redshift range from about zero to 2 (Daly 1994a,b). This may be related to the conditions required for the existence of a powerful extended radio source; it could well be that if a source with a given highly collimated outflow is moved to a lower density, lower pressure environment, it will not produce a powerful radio source. However, irrespective of selection effects, the observations suggests that at least some clusters with their intracluster medium in place exist at fairly large redshift. The existence of cluster-like, extensive, gaseous environments at large redshift suggests that structure formation is well under way at early epochs. This is expected in models of structure formation in an open universe, that is, in a universe with a relatively low value of the density parameter Ω.

The physics of strong shocks also indicates that the rate at which energy is deposited to the hotspot L_j and drives the shock wave is very simply related to observable quantities: $L_j \propto v_L^3 \, n_a \, a_L^2$ or

$$L_j \propto v_L \, B^2 \, a_L^2 \ , \tag{2}$$

where a_L is the lobe radius (Daly 1990, 1994a,b). Note that this equation is completely consist with the familiar expression $L_j \propto v_j \, v_h^2 \, n_a \, a_h^2$, which involves parameters that are not accessible observationally: the jet velocity v_j, the velocity of the head or hotspot v_h and the area of the head a_h^2, where a_h is the radius of the hotspot or head of the jet. The consistency of the two equations is discussed in detail by Daly (1994b), and can be seen by noting that $v_h \propto v_j$; Daly (1994b) shows that several models are simultaneously consistent with both equations.

Equation (2) has been applied to the data of Leahy, Muxlow, & Stephens (1989) to study source energetics (Daly 1994b). The range and upper envelope of L_j for the 10 radio galaxies in the sample, that spans redshifts from zero to 2, does not evolve with redshift, and the range of L_j spans about an order of magnitude. This suggests that the sources are remarkably homogeneous, and exhibit little evolution with redshift. Further, it is interesting to note that L_j for the radio galaxies is independent of lobe-lobe separation, suggesting that L_j is relatively constant over the lifetime of an individual radio source.

Powerful extended radio sources exhibit remarkable properties that suggest a time-independent characteristic source size can be estimated for each source, and this source size is independent of the angular source size. The comparison of this calibrated yardstick with the physical sizes of sources estimated from their angular sizes may provide a very useful cosmological tool (Daly 1994a,b). This method is still undergoing extensive testing; preliminary results indicate that the model is internally consistent, and applications to various data sets indicate that an open universe with a relatively low value of Ω is favored by the data analyzed to date (Daly 1994a,b; Guerra & Daly 1994).

It is a pleasure to thank Roger Blandford for helpful discussions. This work is supported in part by the U. S. National Science Foundation, the Independent College Fund of New Jersey, and Princeton University.

References

Alexander, P, & Leahy, J. P. 1987, MNRAS, 225, 1
Begelman, M. C., Blandford, R. D., & Rees, M. J. 1984, Rev. Mod. Phys., 56, 255
Daly, R. A. 1990, ApJ, 355, 416
Daly, R. A. 1994a, ApJ, 426, 38
Daly, R. A. 1994b, ApJ, submitted
Guerra, E. J., & Daly, R. A. 1994, in preparation
Leahy, J. P., Muxlow, T. W. B., & Stephens, P. W. 1989, MNRAS, 239, 401
Leahy, J. P., & Williams, A. G. 1984, MNRAS, 210, 929
Wellman, G. & Daly, R. A. 1994, in preparation

ON THE USE OF FRACTAL CONCEPTS IN ANALYSIS OF DISTRIBUTIONS OF GALAXIES

I.B.VAVILOVA
Astronomical Observatory of Kiev University
Observatornaya str. 3
Kiev 254053 Ukraine
e-mail: vil%rosa.kiev.ua@relay.ua.net

The well- grounded polemics about the fractal structure of the Universe and a new cosmological picture which appears in connection with this , in first instance the absence of any evidence for homogenization up to present observational limits 200h^{-1} Mpc, have been detailed at the work by Coleman, Pietronero (1992). Two versions on nature of fractal pattern of the galaxy distribution in the observed universe also are now: it behaves like a simple homogeneous fractal (Pietronero 1987; Coleman et al. 1988) and as a multifractal - fractal having more than one scaling index (Jones et al. 1988; Martinez, Jones 1990; Martinez et al. 1990; Borgani et al. 1993 (with the good review for matter of above)).

This work does not play decisively into hands of these versions so the fractal concepts, exactly a selfsimilarity and multifractal, were applied here for the analysis of *two - dimensional* distribution of the *bright* galaxies and dwarf galaxies of the low surface brightness (*LSBD*) belonging to the Local Supercluster (LS). But if the observed universe holds the fractal structure, so it is useful to trace over the lower fractal pattern on the small scales of clustering of galaxies within the framework of the known superclusters and, in the first instance, within the local overdensity of galaxies. This work is a preliminary before preparing one with the same analysis of three- dimensional distribution of galaxies of LS.

We included to consideration the bright and LSBD galaxies compiled from the Catalogue of LSBD galaxies (Karachentseva, Sharina, 1988), VCC (Binggeli et al. 1987) and FCC (Ferguson, Sandage 1990,1991) with the $V_h \leq$ 2000 km/s: bright and LSBD's of the Virgo, N = 974; bright and LSBD's of the Fornax, N = 371, LSBD's of all the Local Supercluster, N=1714. Concerning principles of selection and results on the clustering for these samples we refer to works by Karachentseva, Vavilova (1994a,b). We also generated a point distribution of galaxies according to the law of "Koch-curve" (pure fractal) to have a handle for the comparison with our results. The spherical coordinates of galaxies of all samples were transformed by using an equigraphic Mol'veide projection that is topologically invariant for this task.

For calculation of measure of the homogeneous fractal we chose a box-counting method providing a good accuracy for a finite point set. It is known that the fractal holds a straight-line behaviour at the plot of log N(ε) vs. (log 1/ε) for the certain meanings of size ε of box, here N(ε) is the number of occupied boxes which are received by partitioning of sample. The slope for this part of curve is the fractal dimension D. Samples of galaxies of the Virgo and the Fornax clusters demonstrate scaling - dependent behaviour from the size ε but have a little plateau (Fig.1) which is a characteristic for the homogeneous fractal (D ~ 1.4 and 1.15 accordingly). The distribution of LSBD's of all the LS appears a scale - independence at the scales ε ~ 0.8° ÷10 ° when a statistical weight of group or cluster of galaxies coincides to the weight of field galaxy.

M. Kafatos and Y. Kondo (eds.), Examining the Big Bang and Diffuse Background Radiations, 473–475.
© 1996 *IAU.*

Second fractal concept has been considered is the multifractal (we applied the work by Feder 1988). Analogously dividing in the box-counting method we calculated the number of galaxies N_i (ϵ) within each cell and assigned a probability measure $P_i(\epsilon)$ to occupy the i- cell. Neglecting the known formulas for breviary, note that moments of this measure define the function $\tau(q)$ which is related to the generalized dimension $D_q=\tau(q)/(q-1)$, where D_0 is the Hausdorff dimension, D_1 - information dimension and D_2 - correlation dimension. D_q is always decreasing function and large values of q coincide largest values of $P_i(\epsilon)$, by other words the denser regions of the galaxy distribution. We also calculated the $f(\alpha)$- spectrum which is connected with the variables $\tau(q)$ and q through the Legendre transformation likewise to Gibbs formalism of thermodynamics. The curve $f(\alpha)$ has an unique maximum and value of the $f(\alpha)$ - spectrum in that point is the fractal dimension $f(\alpha(q=0))=D_0$. Also α_{min} is the scaling exponent for the region of max concentration of galaxies and , opposite, α_{max} is the scaling index of the poorer region. In the figure 2 we present the dependencies D(q) vs. q and $f(\alpha)$ vs. α for considered samples of galaxies on some length scale. All samples of galaxies demonstrate a weak dependence D(q) from q. The similarity of the curves indicates on similarity of clustering properties of galaxies of the LS and maximum of the $f(\alpha)$ curves corresponds to $D_0 \approx 1.2$ for the law of similarity with the $\gamma \approx 1.8$ (power law exponent of N-order correlation functions). It would be an inappropriate to suggest an existence of any fractality for these samples at this stage without decision of question for completeness of these samples and working of three-dimensional analysis

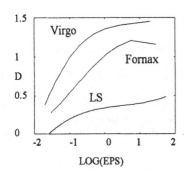

Fig.1.The scaling behaviour of fractal dimension D with the size ϵ for samples of galaxies of the Virgo and Fornax clusters and LSBD's of Local Supercluster. These graphes are received after the numerical differentiation and smoothing of curves log $N(\epsilon)$ vs. log $(1/\epsilon)$.

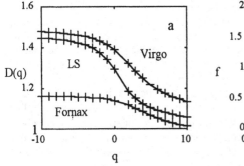

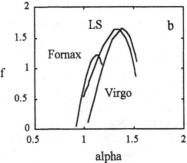

Fig.2. a) Generalized dimensions D(q) and b) the $f(\alpha)$- spectrum for the same samples.

Acknowledgment

The author thanks Dr.V.E.Karachentseva, our joint working stimulated this work.

References

Binggeli, B., Tammann, G.A. & Sandage, A., 1987. *Astron.J.*, **94**, 251.
Borgani, S., Plionis, M., & Valdarnini, R., 1993. *Astrophys. J.*, **404**, 21
Coleman, P.H., Pietronero, L., 1992. *Physics Rep.*, **213**, 313.
Coleman, P.H., Pietronero, L. & Sanders, R.H., 1988. *Astron. & Astrophys.*, **200**, L32.
Feder J., 1988. Fractals (Plenum Press, New York).
Ferguson, H.C., Sandage, A., 1990. *Astron.J.*, **100**, 1.
Ferguson, H.C., Sandage, A., 1991. *Astron.J*, **101**, 765.
Jones, B.J.T., Martinez, V.J., Saar, E., Einasto, J., 1988. *Astrophys. J.*, **332**, L1.
Karachentseva, V.E., Sharina, M.E., 1988. *Commun. of SAO*, **88**, 1.
Karachentseva, V.E., Vavilova, I.B., 1994a. *Bulletin of SAO*, **37**, 98.
Karachentseva, V.E., Vavilova, I.B., 1994b. In "Dwarf galaxies", ESO/OHP Workshop, France, 6-9 Sept., 1993, eds. J.Meylan and P.Prujniel, ESO Conf. & Workshop Proc., No. 49, p.91.
Martinez, V.J. & Jones, B.J.T., 1990. *Mon. Not. R. astron. Soc.*, **242**, 517.
Martinez, V.J., Jones, B.J.T., Domingues-Tenreiro, R., van de Weygaert, R., 1990. *Astrophys. J.*, **357**, 50.
Pietronero, L., 1987. *Physica A*, **144**, 257.

MILLIARCSECOND RADIO STRUCTURE OF AGN
AS A COSMOLOGICAL PROBE

L.I. GURVITS
Joint Institute for VLBI in Europe and
Netherlands Foundation for Research in Astronomy
Postbus 2, 7990 AA, Dwingeloo
The Netherlands
and
Astro Space Center of P.N.Lebedev Physical Institute
Moscow, Russia

Abstract.
Recent achievements in Very Long Baseline Interferometry (VLBI) allow the reconsideration of milliarcsecond radio structure of Active Galactic Nuclei (AGN) as a "standard object". The basic concept of this approach is described and illustrated.

1. Introduction

An idea to use extended radio sources as cosmological standard rods in order to measure parameters of cosmological models had been first suggested by Hoyle (1959). Since then a number of studies of the behaviour of radio source linear size on kiloparsec scales as a function of redshift resulted in a conclusion on the dominance of intrinsic evolutionary effects in source structures rather than universal cosmological effects (*eg.* Kapahi 1989; Barthel and Miley 1988; Singal 1993 and references therein). An application of kiloparsec radio structures in cosmology remains a challenge, though not a hopeless one (Daly, these Proceedings).

Recently a few attempts have been made to apply Hoyle's idea to milliarcsecond (*i.e.* parsec scales) radio structures of AGN (Kellermann 1993; Gurvits 1993, 1994; Pearson *et al.* 1994). The approach is based on measurements of sizes of AGN obtained with VLBI. A number of follow up

M. Kafatos and Y. Kondo (eds.), Examining the Big Bang and Diffuse Background Radiations, 477–480.

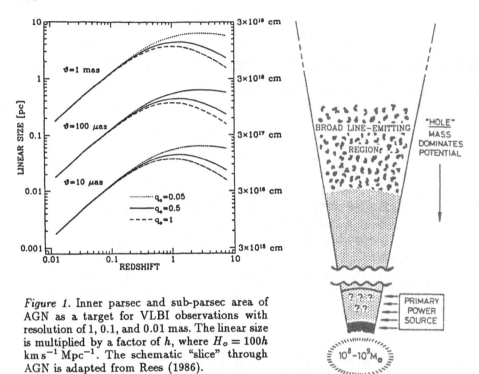

Figure 1. Inner parsec and sub-parsec area of AGN as a target for VLBI observations with resolution of 1, 0.1, and 0.01 mas. The linear size is multiplied by a factor of h, where $H_o = 100h$ km s^{-1} Mpc^{-1}. The schematic "slice" through AGN is adapted from Rees (1986).

publications indicate a reasonable interest in the approach (Krauss and Schramm 1993; Kayser 1994; Stelmach 1994; Jackson and Dodgson in these Proceedings). Leaving aside various technical details of the use of milliarc-second radio structures of AGN as a standard rod, I concentrate here on the basic concept of the approach.

2. Inner Parsecs of AGN as a Standard Object

At present a typical resolution of about 1 mas in VLBI at GHz frequencies allows to "see" the inner part of AGN in the range of linear sizes approximately 0.1 – 10 pc for redshifts $0.01 < z < 10$ (Fig. 1).

Various features of continuum radio emission from AGN are well described in a "zero" approximation by rather simple theoretical models (*eg.* Slysh 1963; Pacholczyk 1970; Kellermann and Owen 1988). These models are based on three assumptions: *(i)* compact radio components emit synchrotron radiation; *(ii)* synchrotron self absorption is a typical feature of their radio continuum spectra; *(iii)* Doppler boosting may play a significant role in sources structural and spectral appearance. Using these assumptions the source linear size, l, can be expressed as

$$l \sim S_m^{\frac{1}{2}} \cdot B^{\frac{1}{4}} \cdot (1+z)^{\frac{1}{4}} \cdot \nu_c^{-\frac{5}{4}} \cdot f(p)^{\frac{5}{4}} \cdot \delta^{-\frac{1}{4}}$$

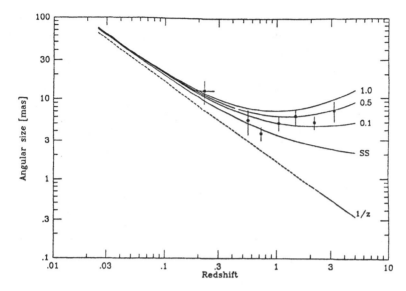

Figure 2. An example of the dependence "apparent angular size – redshift" for a sample of radio cores of 71 quasars (Gurvits *et al.* 1994). Solid curves present cosmological models with $\Lambda = 0$ and various values of q_o.

where S_m is the maximum flux density of the component, B is the magnetic field strength, ν_c is the self-absorption cutoff frequency, f is a weak function of the index p of power low distribution of synchrotron electrons, and $\delta = \gamma^{-1}(1 - \beta \cos \alpha)^{-1}$ is the correction for the relativistic Doppler factor with the Lorentz factor γ, and the angle between the plasma motion direction and the line of sight α (see Kellermann and Owen 1988 for more discussion).

Just a glance at this formula can provide a meaningful conclusion, *i.e.* source size is a weak function of key physical parameters. Indeed, the least known parameters, the magnetic field strength B and the Doppler correction factor δ are to the power of $\pm 1/4$. The function $f(p)$, as noted above, is a weak function of its only single parameter p. Other parameters are better constrained to some extent, as the flux density S_m and the cutoff frequency ν_c are direct observables. Thus, a careful selection of sources can minimize a bias due to a spread in these two parameters. Therefore, one can expect a not too wide range of source linear size, that are emitting in accordance with the formula above.

This conclusion points to the possible usefulness of AGN's parsec scale radio structure in cosmological studies. Inevitable complications of this suggestion, due to deviations from the simple model for AGN radio emission, will be conclusive for better understanding of AGN radio emission.

A number of recent observational programmes will provide a reasonable number of VLBI images (more than 200) of AGN in the redshift range of $0.01 < z < 4.0$ (Taylor *et al.* 1994, Thakkar *et al.* 1994, Gurvits *et al.* 1994 and references therein). Fig. 2 presents an example of observational data of the "$\theta - z$" dependence, adapted from Gurvits *et al.* (1994). The sample presented is composed mainly from the Kellermann (1993) data with an addition of three quasars at $z > 3$. A simple "demonstrational" regression analysis applied to these data implies the best fit for the deceleration parameter value $q_o = 0.5 \pm 0.4$ with the postulating value $\Lambda = 0$. This example as an indication on the fruitfulness of the approach rather than a cosmologically meaningful conclusion. The latter one is a subject for the analysis of future results of VLBI surveys.

Acknowledgements. I would like to thank Ken Kellermann for stimulating discussion of the subject and Hardip Sanghera for useful remarks on this contribution. I acknowledge a grant from the Netherlands Foundation for Space Research. This work is a part of the Early Universe programme of the NFRA. NFRA is supported by the Netherlands Foundation for Scientific Research (NWO).

References

Barthel P.D. and Miley G.K. 1988, Nature 333, 319

Hoyle F. 1959, in Paris Symposium on Radio Astronomy, IAU Symp. No. 9 and URSI Symposium No. 1, ed. R. N. Bracewell, Stanford University Press, 529

Daly R., these Proceedings

Gurvits L.I. 1993, in Sub–Arcsecond Radio Astronomy, eds. R.J.Davis & R.S.Booth, Cambridge Univ. Press, 380

Gurvits L.I. 1994, ApJ 425, 442

Gurvits L.I., Schilizzi R.T., Barthel P.D., Kardashev N.S., Kellermann K.I., Lobanov A.P., Pauliny-Toth I.I.K. and Popov M.V. 1994, AA, in press

Jackson J.C. and Dodgson M. 1994, these Proceedings

Kapahi V.K. 1989, AJ 97, 1

Kayser R. 1994, AA, in press

Kellermann K.I. and Owen F.N. 1988, in Galactic and Extra-Galactic Radio Astronomy, eds. G.L.Verschuur and K.I.Kellermann, Springer-Verlag (New York), 320

Kellermann K.I. 1993, Nature 361, 134

Krauss L.M. and Schramm D.N. 1993, ApJ 405, L43

Pacholczyk A.G. 1970, Radio Astrophysics, Freeman and Company (San Francisco)

Pearson T.J., Xu W., Thakkar D.D., Readhead A.C.S., Polatidis A.G. and Wilkinson P.N. 1994, in Compact Extragalactic Radio Sources, eds. J.A.Zensus and K.I.Kellermann, NRAO, 1

Rees M.J. 1986, in Quasars, Proceedings of the IAU Symposium No. 119, eds. G.Swarup and V.K.Kapahi, Reidel (Dordrecht) 1

Singal A.K. 1993, MNRAS 263, 139

Singal A.K., these Proceedings

Slysh V.I. 1963, Nature 199, 682

Stelmach J. 1994, ApJ 428, 61

Taylor G.B., Vermeulen R.C., Pearson T.J., Readhead A.C.S., Henstock D.R., Browne I.W.A. and Wilkinson P.N. 1994, ApJS, in press

Thakkar D.D., Xu W., Readhead A.C.S., Pearson T.J., Polatidis A.G. and Wilkinson P.N. 1994, ApJS, in press

MAPPING LARGE-SCALE STRUCTURE WITH RADIO SOURCES

J.V. WALL
Royal Greenwich Observatory
Madingley Road, Cambridge, U.K. CB3 0EZ

C.R. BENN
Isaac Newton Group, Royal Greenwich Observatory
Apartado 321, 38780 Santa Cruz de La Palma, Spain

AND

A.J. LOAN
Institute of Astronomy
Madingley Road, Cambridge, U.K. CB3 0HA

1. Introduction

In regions away from the Galactic plane, formal tests indicate an isotropic, random and independent distribution of radio sources on the sky (*e.g.* [1], although there are strong indications of large-scale anomalies (*e.g.* [2], [3]). An accommodation of these results was suggested by Shaver and Pierre [4] and by Shaver [5] who showed that the large-scale deviations could be due to the supergalaxy, a possibility which had been noted by Pauliny-Toth *et al.* in 1978 [6]. As to the influence of *other* superclusters, or indeed the cellular structure of the universe in which galaxies cluster on scales up to at least 100 h^{-1} Mpc (*e.g.* [7]), at what flux-density level does this large-scale structure become apparent? Conversely, what can be learnt about structure on the largest scales through the sky distribution of radio sources? Here we describe three investigations in various stages of completion which consider these issues.

2. A Toy Voronoi Universe

To determine what kind of radio survey (area, flux-density level) we need in order to see the imprint of the skeleton of the universe, we modelled the large-scale structure as a three-dimensional Voronoi tessellation [8]. Poisson

481

M. Kafatos and Y. Kondo (eds.), Examining the Big Bang and Diffuse Background Radiations, 481–485.

Voronoi tessellations are constructed by dividing space into cells around randomly-distributed seeds, such that every point in a given cell is nearer to its seed than to any other seed. The tessellations are thus characterised by one parameter: the mean space density of seeds ρ. We allowed this parameter the range $50\ h^{-1}$ Mpc $< \rho^{-1/3} < 800\ h^{-1}$ Mpc. Pencil-beam surveys slicing through typical Voronoi tessellations are shown in Fig. 1.

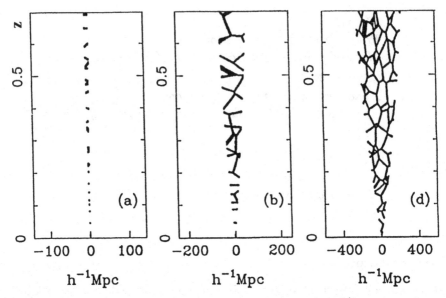

Fig. 1   Slices through Voronoi-foam simulations with $\rho^{-1/3} = 100\ h^{-1}$ Mpc, for pencil-beam surveys (a) $0.5°$ in diameter, (b) $5°$ in diameter, (c) $20°$ in diameter. The horizontal axis is comoving h^{-1} Mpc.

We found the predictions of the model to depend only weakly on the other main parameters: the redshift-dependent luminosity function of the radio sources to be sprinkled on the cell walls; the density parameter Ω; and the distribution of radio sources within the cells.

We tested the isotropy of the sky distribution for each value of $\rho^{-1/3}$, at each of a range of flux densities between $10\ \mu$Jy and 10 Jy, comparing the area-to-area fluctuations in counts with those expected from statistical noise for a range of survey opening angles. The results, reported in detail elsewhere [9], are as follows:

(1) The observed isotropy of the radio-source distribution at Jy, mJy and μJy levels implies $\rho^{-1/3} < 150\ h^{-1}$ Mpc, with the strongest constraints currently being provided by mJy (e.g. 5C) and μJy surveys. This limit in the context of other constraints on large-scale structure occupies a critical range between those provided by galaxy surveys and by the COBE results.

(2) If $\rho > (100\ h^{-1}$ Mpc$)^{-3}$, area-to-area fluctuations are expected to be

significant ($>$ statistical noise) for *e.g.* $5°$-diameter surveys at flux densities $S_{408MHz} < 10$ mJy, or $0.5°$-diameter surveys at flux densities $S_{408MHz} < 0.3$ mJy. These are the flux-density limits of the deepest pencil-beam radio surveys yet carried out. If the largest structures are indeed $\sim 100\ h^{-1}$ Mpc as optical surveys suggest, we must now be on the threshold of detecting the imprint of large-scale structure in the low-frequency radio sky.

3. Direct sensing of cellular structure from redshift surveys of radio sources

Radio surveys select sources over a wide range of redshifts and can thus sample the superclustering structure over very large volumes. For bright surveys, the sampling is very sparse ($<< 1$ source per supercluster), but surveys which detect sources with space density $>\sim (100\ h^{-1}$ Mpc$)^{-3}$ (corresponding to $P_{408MHz} < 10^{25.5}$ WHz^{-1}) can be used to trace individual structures. What are the ideal parameters of such a radio/optical survey?

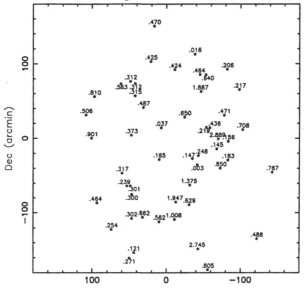

Fig. 2   Distribution on the sky of 5C12 sources with measured redshifts, overlapping points slightly displaced for clarity.

It is relatively simple to obtain redshifts of radio galaxies out to $z \sim 0.5$. The radio survey should be $>\sim$ one cell in diameter at this redshift, in order that large fractions of individual cells can be traced. This suggests a survey opening angle θ of a few degrees (given that 100 comoving h^{-1} Mpc at $z = 0.5$ corresponds to $\theta = 5°$). The radio survey should also be deep enough to detect sources at this redshift with $P_{408MHz} << 10^{25.5}$ WHz^{-1},

$i.e. S_{408MHz} \ll 100$ mJy.

The 5C12 radio survey of a $4°$-diameter patch of sky reaches S_{408MHz} $= 10$ mJy; the catalogue contains ≈ 600 sources. Redshifts have now been measured (mainly with the WHT faint-object spectrograph) for about 60 of the sources, and Fig. 2 shows their distribution on the sky. Several groups and pairs are visible. The group at $z \approx 0.315$ has projected comoving diameter 25 h^{-1} Mpc, while others have projected diameters ranging 10 - 70 h^{-1} Mpc.

4. Large-area surveys: two-point correlation

The third investigation is two-point correlation-function analysis, particularly well-suited to surveys of irregular areas, such as all-sky surveys after excision of regions of low Galactic latitude and confused regions near bright/extended sources. Two surveys at 5 GHz, each covering $\approx$ one hemisphere (*e.g.* [10],[11]), now catalogue $\approx 10^5$ sources. New correlation results (Fig. 3) confirm the apparent signal previously noted in this type of analysis from the northern region alone [12][13].

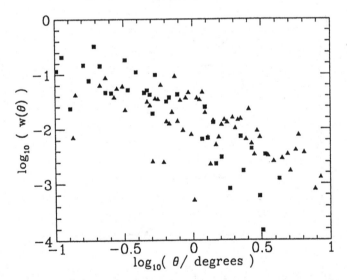

Fig. 3   Preliminary two-point correlation functions for sources of flux density > 50 mJy in different areas of the Greenbank - PMN 5-GHz survey.

Further analysis is needed to distinguish intrinsic non-randomness from instrumental and geometrical effects [14]. If intrinsic non-randomness is confirmed, Limber's equation (*e.g.* [15]) can be applied to derive the power spectrum of the inherent clustering, because the nature and redshift distribution of the 5-GHz sources are reasonably well established [12][16].

References

1. Webster, A.S., 1977, in *Radio Astronomy and Cosmology*, Proc IAU Symp. 74, ed. Jauncey, D.L., Reidel, 75.
2. Wall, J.V., 1977, in *Radio Astronomy and Cosmology*, Proc IAU Symp. 74, ed. Jauncey, D.L., Reidel, 55.
3. Pauliny-Toth, I.I.K., 1977, in *Radio Astronomy and Cosmology*, Proc IAU Symp. 74, ed. Jauncey, D.L., Reidel, 63.
4. Shaver, P.A. and Pierre, M., 1989, Astron. Astrophys., **220**, 35.
5. Shaver, P.A., 1991, Austr. J. Phys., **44**, 759.
6. Pauliny-Toth, I.I.K., Witzel, A., Preuss, E., Kühr, H., Kellermann, K.I., Fomalont, E.B. and Davis, M.M., 1978, Astron. J., **83**, 451.
7. Giovanelli R. and Haynes M.P., 1991, Ann. Rev. Astron. Astrophys., **29**, 499.
8. Icke V. and van de Weygaert R., 1991, Q. J. R. astr. Soc., **32**, 85.
9. Benn C.R. and Wall J.V., 1994, Mon. Not. R. astr. Soc., in press.
10. Gregory, P.C. and Condon, J.J., 1991, Astrophys. J. Suppl. Ser., **75**, 1011.
11. Griffith, M.R. and Wright, A.E., 1993, Astron. J., **105**, 1666.
12. Wall, J.V., Rixon, G.T. and Benn, C.R., 1993, in *Observational Cosmology*, eds Chincarini, G. *et al.*, ASP Conf. Ser. **51**, 576.
13. Kooiman, B.L., Burns, J.O. and Klypin, A.A., 1994, Astrophys. J., in press.
14. Loan, A.J., Wall, J.V. and Lahav, O., in preparation.
15. Fall, S.M., 1979, Rev. Mod. Phys., **51**, 21.
16. Dunlop, J.R. and Peacock, J.A., 1990, Mon. Not. R. astr. Soc., **247**, 19.

THE RHODES/HARTRAO 2300 MHZ HORN TELESCOPE

J.L. JONAS

Rhodes University, Grahamstown 6140, South Africa

phjj@hippo.ru.ac.za

Abstract. The recently completed Rhodes/HartRAO 2300 MHz radio continuum survey is the highest resolution and highest frequency survey of the entire southern sky. This makes it useful for those wishing to model the Galactic foreground in order to study cosmic background emission. Unfortunately there are low-amplitude, large-scale gradients in the survey data which make it difficult to interpret the very high latitude Galactic emission. We describe a small, low-cost horn telescope which we hope to use to correct the gradients in the survey data.

1. Introduction

The estimation of the Galactic foreground contribution to measurements of the CMB is becoming very important now that positive detections of multipole anisotropies are being reported. Large-scale centimetre-wavelength radio continuum surveys at a number of frequencies are necessary to obtain accurate information about the spatial distribution of the brightness and spectral index of the Galactic emission. Results from these surveys may be extrapolated to the millimetre-wavelength regime of CMB measurements.

The Rhodes/HartRAO 2300 MHz (13 cm) survey (Jonas, Baart & Nicolson, in preparation) provides unique centimetre-wavelength coverage of the southern hemisphere. Being the highest-frequency (and resolution) radio continuum survey made of the southern sky using a terrestrial telescope it is well suited to the extrapolation necessary to estimate Galactic emission at higher frequencies. The large range of extrapolation places stringent requirements on the calibration and baseline accuracy of the low frequency data. Large-scale gradients in the 2300 MHz data introduce relative baseline errors of up to 50 mK, and the brightness temperature calibration is of

M. Kafatos and Y. Kondo (eds.), Examining the Big Bang and Diffuse Background Radiations, 487–488.
© 1996 IAU.

order 5%. In order to improve these parameters of the survey it was decided that unblocked aperture measurements had to be made.

2. Telescope Structure

The telescope is constructed almost entirely from wood to ensure low cost. The horn antenna and all of the electronic instrumentation are contained within a rectangular "tube", which is internally clad with polystyrene sheets for thermal insulation. The tube may be fixed at any elevation angle, and the alidade has a motorized azimuth drive which allows computer control. The telescope baseplate is mounted on a road trailer, allowing the telescope to be transported between different sites.

3. Instrumentation

Currently the telescope uses a simple rectangular horn antenna, which has a HPBW of 15° in both the E- and H-planes. The sidelobe performance is not well suited to this experiment, but funds are not available for a purpose-built horn. Ambient temperature GaAsFET amplifiers provide about 60 dB front-end gain to the receiver. The radiometer is a noise adding, gain-stabilized design which is well suited to provide the stability necessary for this experiment. All of the receiver components are housed in the "tube", where the air temperature is controlled to within 0.1° C.

A PC-type computer controls all aspects of the telescope operation, allowing unattended all-night observations. Modular design of the instrumentation ensures that the telescope can be set up quickly at the observing site.

4. Technique

The experiment has been designed to make differential sky temperature measurements, using the south celestial pole (SCP) as a reference point. The antenna is fixed at the elevation of the SCP and made to scan alternately clockwise and anti-clockwise by 360° in azimuth. Each scan consists of slewing steps between positions on a 5° declination grid, and each scan begins and ends at the SCP reference position. The sky temperature is integrated for 10 seconds at each position, resulting in one complete 360° scan taking about 12 minutes. The short scan time allows baseline drifts to be removed by linear interpolation of the SCP temperatures.

By stacking the data from observations made throughout the year and from different sites we will obtain a low resolution map of the sky brightness relative to the SCP. The main survey data will be convolved with an appropriate beam so that it may be compared with the horn data.

Vacuum-energy and the angular-size/redshift diagram for milliarcsecond radio sources

J.C. Jackson and Marina Dodgson

Department of Mathematics and Statistics, University of Northumbria at Newcastle, Ellison Building, Newcastle upon Tyne, NE1 8ST, UK

We have considered two compilations of angular-size/redshift data for ultra-compact radio sources, due to Kellerman (1993) and Gurvits (1994) respectively, and a family of homogeneous isotropic universes with two degrees of freedom, represented by pressure-free matter and vacuum-energy. Kellerman's results seem to support the canonical model $q_0 = 1/2$, $\Omega_0 = 1$; the question posed here is: can we produce significant deceleration without dark matter?

The usual cosmological parameters are related by the dimensionless versions of Friedmann's equations

$$q_0 + \Lambda_0 = \Omega_0/2 \tag{1}$$

and

$$\Omega_0 + \Lambda_0 = 1 + K_0. \tag{2}$$

If we believe that q_0 is large, equations (1) and (2) allow a clear alternative to large Ω_0, namely $\Omega_0 \sim 0$, $\Lambda_0 \sim -q_0$, and $K_0 \sim -1 - q_0 < 0$, that is a violently decelerating open Universe, with dynamics dominated by a large negative cosmological constant. It turns out that the angular-size/redshift relationship is particularly sensitive to these parameters. In the limit $\Omega_0 = 0$ exactly there is an analytical expression giving the corresponding angular-diameter distance $d_A(q_0, z)$ (Jackson 1992), which is

$$d_A = \frac{1}{q_0 H_0}\{[1 + q_0 - q_0(1 + z)^{-2}]^{1/2} - 1\}. \tag{3}$$

Figure 1 compares the corresponding angular-size/redshift curves (labelled by values of q_0) with Kellerman's data, comprising 79 high-luminosity sources in 7 redshift bins, assuming a fixed linear size of $20.5h^{-1}$ parsecs ($h = 1 \Rightarrow H_0 = 100$ km s^{-1} Mpc^{-1}). Clearly there are curves which are compatible with the data, if we are prepared to contemplate values of $q_0 \geq 3$. At this stage we make no claim that these vacuum-dominated universes are uniquely favoured by the data; the point is that there is in this context an acceptable model which invokes neither dark matter nor evolutionary effects. The crucial difference between these models and the canonical matter-dominated one is qualitative; in the latter case a pronounced increase in angular size is expected, beyond $z = 1.25$, whereas in our models such behaviour is absent, corresponding to an asymptotic value $d_A \to (q_0 H_0)^{-1}[(1 + q_0)^{1/2} - 1]$ as $z \to \infty$.

Tentatively, we can make a more positive statement than this. Very recently a large data compilation has appeared (Gurvits 1994), comprising 270 ultra-compact high-luminosity sources in 12 redshift bins in the range $0.5 \leq z \leq 3.8$. It is difficult to combine the two samples, as they use different definitions of size and correspond to different frequencies (5 Ghz and 2.3 Ghz respectively); nevertheless, we have attempted to do just this. Gurvits notes a reasonably systematic difference between his sizes and Kellerman's, the latter being

489

M. Kafatos and Y. Kondo (eds.), Examining the Big Bang and Diffuse Background Radiations, 489–490.
© 1996 IAU.

490

larger than the former by a factor of about 3 over the high-redshift flat part of each diagram. Figure 2 shows a composite sample, comprising Kellerman's low-redshift points (at $z = 0.047$ and $z = 0.228$), and Gurvits' high-redshift points, scaled up in size by an appropriate fixed factor. At face value this diagram strongly favours the models proposed here, as there is no sign of the turn-up at high redshift expected in matter-dominated universes. A naive least-squares fit, giving free rein to q_0 and Ω_0, and to the linear dimension d characterising the sources, gives $q_0 = 8.34$, $\Omega_0 = 0.17$, and $d = 15.5h^{-1}$ parsecs. The theoretical curve (which is exact, i.e. does not use approximation (3)) shows a very gentle turn-up beyond $z \sim 4$, as the corresponding universe becomes matter-dominated.

One might expect the age problem associated with such high values of q_0 to be beyond salvation, but this class of models turns out to be remarkably forgiving in this context. This is illustrated by Figure 3, which plots age in Hubble units against q_0 for the limiting case $\Omega_0 = 0$ (solid line), to be compared with the standard matter-dominated case (dashed line); at a given value of q_0, ages are some 40% longer in the former case. The range $3 \leq q_0 \leq 8$ requires h to be in the range 0.3 to 0.4, to give an age of 15×10^9 years, compared with $h = 0.44$ in the canonical case.

REFERENCES

Gurvits L.I., 1994, ApJ, 425, 442
Kellerman K.I., 1993, Nature, 361, 134
Jackson J.C., 1992, QJRAS, 33, 17

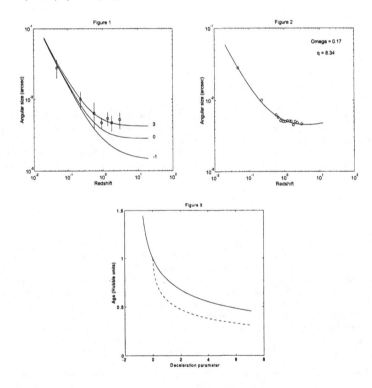

A Hint to Possible Anisotropy in Radio Universe

Rendong Nan and Zhengdong Cai *
Beijing Astronomical Observatory, Beijing 100080, P.R. China

Abstract. This paper presents our investigations of statistical behavior of the small-scale jet orientations and the intrinsic polarization position angles. We found that the distribution of the milliarcsecond-scale jet orientations and intrinsic polarization position angles of extragalatic radio sources appears anisotropic with high significance levels in most cases.

Key words: Radio sources: Isotropy – Jet – Polarization position angles

1 Projected Orientation of Radio Jets in the sky

We started with an analysis of the data which came from the results of a long-term study by Aller *et al.* (1992) on the sources in the Pearson-Readhead VLBI survey (1981). The distribution of jet position angles on small scale determined by VLBI is summaried in figure 1. in which, two-thirds of the sample sources (26 of 39 sources) have the righthanded structures, and only 3 out of 39 sources are within the second quadrant. We apply binomial and χ^2 tests to the righthanded event and the unusual sparseness of jet orientations in the second quadrant, which yield significance levels $\alpha = 1.3 \times 10^{-2}$ and 0.1 respectively. The similar results are also found in the Polatidies' 1 Jy sample (1993) for which measurements of the small-scale position angles are available. observations. Probably a description of the anisotropic orientation of jets, in more precise terms, should be revised as that the radio jets in the survey seem to avoid the second quadrant.

The relation between bright cores and relativistic beaming prompts us to check the jet orientation in those superluminals. There are 13 superluminal sources in the samples, among which, 9 out of 13 are righthanded. None is found in the second quadrant. We have also made statistics of all superluminal sources whose orientation information is available to us. Figure 2 presents the orientations of radio jets of the superluminal sources. 22 out of 36 sources are righthanded and only 2 are located in second quadrant, suggesting rather identical feature compared to figure 1.

2 Polarization Position Angles

The relations between polarization direction and structures of different types of sources on mas-scale also turn to exist (Gabuzda *et al.* 1989). For example, the magnetic fields in the mas-scale jets are perpendicular to the jet orientations in BL Lacs and parallel in radio quasars. Considering the geometrical relation between radio jets and magnetic fields, we examined the distribution

---

* The project is supported by the National Natural Science Foundation of China.

M. Kafatos and Y. Kondo (eds.), Examining the Big Bang and Diffuse Background Radiations, 491–492.

492

of polarization position angles to take the possible chance in verifying the isotropy of jet orientation. The data from the Aller's paper are graphically presented in figure 3. The distribution of the polarization position angles corrected for Faraday rotation is dramatically peaked at around 90° and clusters the angles of about 80 percent of sources within 60 degrees from 60° to 120°. The χ^2 test of figure 3 for eight degrees of freedom yielding a significance level of 5.0×10^{-3}. We extended the studies to a larger sample of 555 radio sources (Simard-Normandin *et al.* 1981). In figure 4, the intrinsic polarization position angles obviously tend to prefered the horizontal and vertical directions with the significance levels of $\sim$ 0.045 and 0.38 respectively.

The results are primitive. Elaborative inspection of large sample is definitely required for a more convincing conclusion. The main purpose of this work is to draw attention to those careless omissions in observations, and to invite the verification of the imperfections in method and technique which may cause the irregularity in spite of the implausibility. Although the reality of the phenomena is hardly thinkable, "the universe is not only stranger than we know, it is stranger than what we can know".

References

Aller, M.F., Aller, H.D., and Hughes, P.A., 1992, *Ap.J.*, **399**, 16
Esko Valtaoja and Mauri Valtonen, 1992, *Variability of Blazars,* Combridge Univ. Press.
Gabuzda, D.C., Cawthorne, T.V., Roberts, D.H., and Wardle, J.F.C., 1992, *Ap.J.*, **388**, 40
Polatidis, A.G., 1993, Ph.D thesis, *Univ. of Manchester*
Simard-Normandin,M., Kronberg, P.P., and Button, S., 1981, *Ap.J.Supple.*, **45**, 97
Zensus J.A. and Pearson T.J., 1987, **Superluminal Radio Sources**, Cambridge Univ. Press.

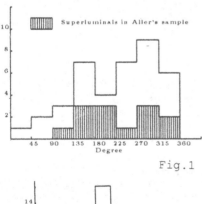

Fig.1

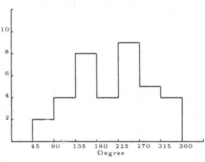

Fig.2

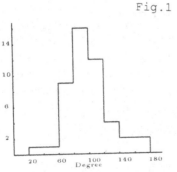

Fig.3

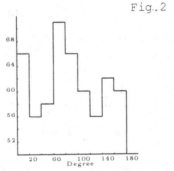

Fig.4

PROJECTED CLUSTERING AROUND $1 < Z < 2$ RADIOGALAXIES

E. MARTÍNEZ-GONZÁLEZ, N. BENÍTEZ AND J. I. GONZÁLEZ-SERRANO
Dpto. de Física Moderna, Universidad de Cantabria and
Instituto de Física de Cantabria, CSIC–Universidad de Cantabria,
Facultad de Ciencias, Avda. de Los Castros s/n, 39005 Santander,
Spain

AND

L. CAYÓN
Lawrence Berkeley Laboratory and Center for Particle Astrophysics,
Berkeley, CA 94720, USA

Abstract. We have taken deep R-band images of fields around five radio-galaxies: 0956+47, 1217+36, 3C256, 3C324 and 3C294 with $1 < z < 2$. We found a statistically significant excess of bright ($19.5 < R < 22$) galaxies on scales of 2 arcmin around the radiogalaxies. The excess has been determined empirically to be at $\gtrsim 99.5\%$ level. It is remarkable that this excess is not present for $22 < R < 23.75$ galaxies within the same area, suggesting that the excess is not physically associated to the galaxies but due to intervening groups and then related to gravitational lensing.

1. Introduction

Observations of the environment of AGNs indicate that these objects tend to lie in regions with galaxy density richer than average. The amplitude of this enhancement increases with the radio luminosity of the AGN, as has been found for FR I and II radiogalaxies (Lilly and Prestage 1987, Hill and Lilly 1991) and for radio-loud and radio-quiet QSOs (Yee and Green 1987, Ellingson et al. 1991) for $z \lesssim 0.5$. Recent observations of both types of QSOs (Boyle and Couch 1993, Hintzen et al. 1991) seem to indicate that this trend is also present at $0.9 < z < 1.5$. The environments of radio-selected BL Lac objects have been studied by Fried et al. (1993). They found clustering around objects at redshifts $z \lesssim 0.7$ but not for the $z \approx 0.9$ ones.

493

M. Kafatos and Y. Kondo (eds.), Examining the Big Bang and Diffuse Background Radiations, 493–496.
© 1996 IAU.

There have been several reports of statistically significant associations of high redshift QSOs and radiogalaxies with foreground objects: galaxies (Hammer and Le Fevre 1990), Zwicky clusters (Seitz and Schneider 1994), IRAS galaxies (Bartelmann and Schneider 1993), X-ray photons taken from the ROSAT All Sky Survey (Bartelmann et al. 1994). These associations are interpreted as a lensing effect with the lens being either a single galaxy or clusters of galaxies. Powerful radio sources are particularly well suited to detect associations due to gravitational lensing, because of the steep slope of their number counts (Wu and Hammer 1994).

Here we search for excesses of objects around $1 < z < 2$ radiogalaxies with apparent magnitudes $R \approx 21.4$ and spanning a range of one order of magnitude in radio luminosity.

2. Observations and results

Observations were carried out during the night of 1992 April 1 at the prime focus of the 2.5m Isaac Newton Telescope (INT) on the island of La Palma (Canary Islands, Spain). The detector used was an EEV 1280×1180 CCD camera with a scale of 0.57 arcsec/pixel and readout noise of 6 electrons. We took 7 exposures of 500 seconds each in the direction of the radiogalaxies 0956+47, 1217+36, 3C256, 3C294 and 3C324 using the Kitt Peak R-band filter. All the fields have $b \gtrsim 50$ deg. Photometric standard stars in NGC4147 were also observed in order to calibrate the frames. The night was not photometric and the seeing was approximately 1.2 arcsec FWHM. There is no vignetting in the field. The final images, obtained by coadding the individual frames for each object, were flat-fielded using sky and dome exposures. Their final sizes, after removing the borders, are 1200×1090 pixels ($11.4 \times 10.3 \mathrm{arcmin}^2$). We searched for objects in the frames using the standard package PISA (Position Intensity and Shape Analysis, Draper and Eaton 1992). Our detection threshold was of 9 connected pixels above 1σ level per pixel. In order to test our detection algorithms we generated frames with Poisson noise. From these simulations we estimate the number of spurious detections as $\lesssim 7\%$. This estimation was also confirmed by the results of the search for 'negative' objects in the real images.

We have studied the clustering of objects in 2 arcmin $\times$ 2 arcmin boxes around each radiogalaxy which corresponds to a scale of ≈ 1 Mpc ($\Omega_o = 1, h = 0.5$), at redshift $1 < z < 2$. We have divided the objects into two groups, one with magnitudes similar or brighter than the radiogalaxies: $19.5 < R < 22$, and another with fainter magnitudes: $22 < R < 23.75$. The main results for the 2 arcmin $\times$ 2 arcmin boxes are listed in the table below. In order to find the expected background value (ne in the table) we multiply the 'unspoilt' surface within the considered central region by the average density of galaxies within the given range of magnitudes on that field. We determined the rms empirically, although many of the quoted authors find the statistical significance taking the variance as if the distribution of galaxies were poissonian, what certainly it is not the case. We find that the empirical rms number of objects in the boxes are $\approx 1.5 \times \sqrt{N}$.

The sum of the number of objects found within the central regions of the five fields give us 91 objects against 59.7 ± 12.5 with $19.5 < R < 22$ (i.e. an empirical excess of 2.5σ) and 215 objects against 220.4 ± 22.3 with $22 < R < 23.75$. Therefore, no significant excess is found at fainter magnitudes. In order to determine the statistical significance of the excess found at brighter magnitudes, we have constructed an empirical probability distribution law. This distribution is formed by the sum of the objects contained in all the possible combinations of five boxes, each of them from a different field. We have normalized the results of each box by the useful surface of the radiogalaxy box of that field. Then we have excluded all the combinations of boxes that have an added useful surface smaller than a given threshold which is less or equal than the sum of the useful surfaces of the radiogalaxy boxes (94% of 5 2 arcmin $\times$ 2 arcmin boxes). The shape of the distribution so obtained is very close to a Gaussian. We obtain that the excess is significant at a $\gtrsim 99.5\%$ level for the 2 arcmin scale.

Table: Projected number of objects in 2×2 arcmin^2 boxes

Field	N_b	N_f	s(%)	n_b	ne_b	n_f	ne_f
1217+36	11.7± 4.5	50.5± 8.8	100	25	11.7± 4.5	55	50.5± 8.8
0956+47	13.2± 5.7	47.4± 9.5	100	21	13.2± 5.7	54	47.4± 9.5
3C256	13.8± 7.8	54.2± 12.1	94	21	12.8± 7.6	50	50.8± 11.7
3C324	14.2± 4.9	44.6± 9.3	87	14	12.3± 4.6	32	38.8± 8.6
3C294	10.9± 4.9	36.8± 11.5	89	10	9.7± 4.6	24	32.9± 10.9

Notes: N_b and N_f are the average of bright ($19.5 < R < 22$) and faint ($22 < R < 23.75$) objects within a box (the errors are the empirically found rms). s(%) is the percentage of useful surface of the box centered on the radiogalaxy. n_b, n_f are the number of bright and faint objects found and ne_b, ne_f the expectation within that box.

3. Discussion and conclusions

It is remarkable that no excess is detected at fainter magnitudes: for $22 < R < 23.75$ we only found 215 objects against 220.4 within a 2 arcmin box. Assuming that the spectral distribution of cluster galaxies is consistent with

496

a model with no evolution then the magnitude expected for these galaxies at $z \gtrsim 1$ would be $R > 24$ (as it is the case for the galaxies in the cluster found around 3C324, Dickinson 1994). Thus we do not expect to see in our images any early type galaxy at those high redshifts and therefore the only possibilities are either that the objects forming the excess are at $z > 1$ and then blue or they are at an intermediate redshift $z \sim 0.5$. Considering the sizes of the objects the most likely possibility is the later one.

Finally, the previous discussion strongly suggests that the excess is not physically associated to the radiogalaxies but due to intervening groups and then related to gravitational lensing.

References

Bartelmann, M. and Schneider, P. 1993, preprint.
Bartelmann, M., Schneider, P. and Hasinger, G., 1994, *Astron. Astrophys.*, in press.
Boyle, B.J. and Couch, W.J. 1993, *Mon. Not. R. astr. Soc.*, 264, 604
Draper, P.W. and Eaton, N. 1992, Starlink Project User Note 109.5, Rutherford Appleton Laboratory
Dickinson, M. 1994, private communication
Ellingson, E., Yee, H.K.C. and Green, R.F. 1991, *Ap. J.*, 371, 49
Fried, J.W., Stickel, M. and Kühr, H. 1993, *Astron. Astrophys.*, 268, 53
Hammer, F., and Le Fevre, O. 1990, *Ap. J.*, 357, 38
Hill, G.J. and Lilly, S.J. 1991, *Ap. J.*, 367, 1
Hintzen, P., Romanishin, W. and Valdés, F. 1991, *Ap. J.*, 366, 7
Lilly, S.J. and Prestage, R.M. 1987, *Mon. Not. R. astr. Soc.*, 225, 531
Seitz, S. and Schneider, P. 1994, preprint
Wu, X.P, Hammer, F. 1994, to appear in *Astron. Astrophys.*,
Yee, H.C. and Green, R.F. 1987, *Ap. J.*, 319, 28

UPPER LIMITS ON THE LYα EMISSION AT Z = 3.4

E. MARTINEZ-GONZALEZ, J.I. GONZALEZ-SERRANO, J.L. SANZ AND
J.M. MARTIN-MIRONES
Dpto. Física Moderna, Univ. Cantabria and
Instituto de Estudios Avanzados, CSIC-Univ. Cantabria,
Facultad de Ciencias, Avda. Los Castros s/n, 39005 Santander,
Spain

AND

L. CAYON
Lawrence Berkeley Laboratory,
1 Cyclotron Road, Berkeley, CA 94720, USA

Many searches have been carried out to detect emission from massive primeval hydrogen clouds at high redshift. By observing the 21 cm line it has been possible to impose strong upper limits on the mass and number of protoclusters at high redshift (Wieringa, et al. 1992). Since strong Lyα emission is expected from primeval galaxies undergoing their first burst of star formation, many attempts to detect this emission have been made, but no positive detection has been reported, imposing strong constraints on models of galaxy formation (see Djorgovski et al. 1993 for a review).

We have carried out deep CCD imaging of a particular region using a narrow-band filter isolating the Lyα line at a redshift of 3.4 at the 2.m Isaac Newton Telescope (Canary Islands, Spain). The seeing was poor $\approx$ 3".4. We have used PISA to search for objects in the frames.

We do not find any Lyα emitting object in an area of $\approx$ 120 arcmin^2 within the redshift interval $3.37 - 3.44$. The surface brightness limit obtained is 4.3×10^{-18} ergs cm^{-2} s^{-1} arcsec^{-2} (1σ confidence level). We conclude that no source of size in the range 5"-10" (75-150 h^{-1}Mpc at z=3.4 for $\Omega = 1$) with line flux greater than 2.1×10^{-17} ergs cm^{-2} s^{-1} was present in the observed field at the 5σ confidence level. In other words, this limit implies that only unobscured Lyα emission objects 5 magnitudes fainter than the nearby radio galaxy 0902 + 34 can exist within the searched area. We have arrived to a flux limit a factor 2 below the one given by Rhee et al. (1993) with no restriction on the size of the objects to be less than 5" and

M. Kafatos and Y. Kondo (eds.), Examining the Big Bang and Diffuse Background Radiations, 497–498.
© *1996 IAU.*

within an area $\approx$ 5 times bigger, being therefore the most stringent limit on Lyα emission at redshifts around ≈ 3.4. Moreover, considering limits on emission at other redshifts our flux limit slightly improves upon publised works with number density $\lesssim$ 50Lyα emitters per square degree (see figure 1 of the recent review by Djorgovski, Thompson and Smith 1993 for 1σ flux limits).

The upper limit found on the Lyα flux can be used to constrain the models for the appearance of primeval galaxies proposed by Baron and White (1987).

A plausible explanation of the absence of Lyα emission from these observations and other sky surveys and from damped high redshift Lyα systems could be attenuation by dust. The Lyα photons can suffer a large number of resonant scattering in the ambient neutral atomic hydrogen, so this phenomenon can increase the chance for these photons to be absorbed by dust grains in contrast to the UV continuum ones. In such conditions, a 90 Å width filter as we have used in our observations may not be narrow enough to single out any Lyα feature.

Finally, we can obtain the expected Lyα emission from an HI cloud placed at $z = 3.4$ and with a 21 cm emission consistent with the upper limit given by the most recent radio observations of ≈ 2.3 mJy. Assuming a characteristic size for the cloud of 5$'$ and a gaussian line-width of ~ 100 km s^{-1} the column density should be $N_{HI} \lesssim 10^{20}$ cm^{-2}. We have calculated the Lyα emission expected from a hydrogen cloud with the previous neutral column density, total hydrogen number density n_H ranging from $0.1 - 1$ cm^{-3} and a primordial abundance for the cloud of 24.5% of He by mass. For a quasar-dominated ionizing energy distribution ($I_\nu = J^0_{-21} \times 10^{-21}(\nu_c/\nu)$, with $\lambda_c = 912$ Å) the Lyα emissivity expected is $I_\alpha/J^0_{-21} \lesssim 3 \times 10^{-19}$ erg s^{-1} cm^{-2} arcsec^{-2}. This last upper limit is an order of magnitude below our observational upper limit. Therefore, the existence of HI clouds with the physical properties indicated above and at $z \approx 3.4$ is still possible considering present radio and optical observational constraints.

References

Baron J., White S., 1987, ApJ, 322, 585

Davies R. D., Pedlard A., Mirabel I. F., 1978, MNRAS, 182, 727

Djorgovski S., Thompson D., Smith J. D., 1993, 16th Texas Symposium on Relativistic Astrophysics, eds. C. W. Akerlof & M. A. Srednicki, Ann. N. Y. Acad. Sci., 688, p. 515

Rhee G., Loken C, Le Fevre O., 1993, ApJ, 406, 26

Wieringa, M. H., de Bruyn, A. G., Katgert, P., 1992, A&A, 256, 331

INTERMEDIATE RESOLUTION SPECTROSCOPY OF THE RADIO GALAXY B2 0902+34 AT $Z \approx 3.4$

J. L. SANZ, J. M. MARTíN–MIRONES, E. MARTíNEZ–GONZÁLEZ AND
J. I. GONZÁLEZ–SERRANO
Dpto. de Física Moderna, Universidad de Cantabria and
Instituto de Estudios Avanzados,
CSIC–Universidad de Cantabria
Facultad de Ciencias, Avda. de Los Castros s/n,
39005 Santander, Spain

We have carried out optical spectroscopic observations at intermediate spectral resolution of the massive high redshift radio galaxy 0902+34 at $z \approx 3.39$. This source was first identified by Lilly (1988) (from hereafter L88). The study of high redshift radio galaxies is interesting to analyze the physical conditions of the early universe and the galaxy evolution at cosmological redshifts. It has been claimed that some of these systems may be protogalaxies in the process of formation. Indications for this are the flat spectrum and the absence of the 4000 Å break, features which have already been observed in many cases. In particular, observations in the spectral range from V to K suggest that 0902+34 is a young galaxy (Eisenhardt and Dickinson 1992). Recent radio observations of the 21 cm line of neutral hydrogen have detected (Uson et al. 1991) an absorption against the radio continuum source. This absorption could also leave a track in the optical, redwards the Lyα line. Our observations were carried out with the ISIS spectrograph at the 4.2 m William Herschel Telescope (seeing $\approx$ 1.2–1.6 arcsec). A spectral dispersion of 0.78 Å/pixel (blue arm) and 1.38 Å/pixel (red arm) was obtained. A long slit of width 3" was used providing a spectral resolution of $\approx$ 5.4 Å in the blue arm and of $\approx$ 9.5 Å in the red one. Both resolutions are a clear improvement over that achieved by L88 of 20 Å, allowing us to resolve the Lyα line (and its possible structure) and any other possible strong features appearing in the spectral range observed (e. g., C IV λ1549, He II λ1640, ...). Six different observations of 2700 s of the radio galaxy 0902+34 were carried out. The slit was rotated to coincide with the parallactic angle at the beginning of each exposure. This will allow us to map spectroscopically

M. Kafatos and Y. Kondo (eds.), Examining the Big Bang and Diffuse Background Radiations, 499–500.

different regions of the galaxy (for more details see Martín-Mirones et al. 1994).

We have made two separate analysis of the data: individual spectra and summed spectrum. The observed properties of the lines detected in the summed spectrum appear in Table 1. We have arrived to the following conclusions:

– The Lyα and C IV lines have been resolved for the first time, whereas the He II line has been first detected and resolved in the present observation.

– The analysis of the line ratios in exposures with different orientations of the slit shows the existence of strong ionization and/or dust density gradients. Moreover, the ratios obtained are typical of radio galaxies with $z > 1.8$.

– We have also detected the optical continuum of the radio galaxy showing that it is almost flat in agreement with the recent conclusion, based on a much wider wavelength range, that 0902+34 is a young galaxy observed during its initial burst of star formation.

– Finally, the possible Lyα absorption corresponding to the H I cloud observed in 21 cm has not been detected by us. This result implies that either the absorbing cloud is in between the Lyα emitting region and the radio core or it is placed nearer us than the Lyα emitting region but with a maximum angular size in the range $\approx 0.00063 - 0.0063$" (of the order of tens of parsecs) for spin temperatures in the range $\approx 10^4 - 10^2$ °K.

TABLE 1: PROPERTIES OF THE LINES DETECTED

Property	Lyα	C IV	He II
Redshift	$3.3909^{+0.0004}_{-0.0003}$	$3.3969^{+0.0074}_{-0.0019}$	$3.3909^{+0.0015}_{-0.0012}$
FWHM (km s^{-1})	662^{+66}_{-66}	882^{+882}_{-209}	540^{+245}_{-147}
Surf. brig. (10^{-20} W m^{-2} arcsec^{-2})	$2.85^{+0.56}_{-0.49}$	$0.45^{+0.57}_{-0.22}$	$0.30^{+0.25}_{-0.18}$
Line surf. brig./Lyα surf. brig.	1.00	0.16	0.11
Equivalent width (Å)	182^{+36}_{-32}	55^{+70}_{-28}	43^{+36}_{-26}
Surf. lumin. (10^{35} W arcsec^{-2})[a]	$2.59^{+0.03}_{-0.03}$	$0.41^{+0.43}_{-0.10}$	$0.27^{+0.13}_{-0.08}$

[a]$H_0 = 50$ km s^{-1} Mpc^{-1}, $\Omega = 1$.

References

Eisenhardt, P.R.M., and Dickinson, M. (1992), *The Astrophysical Journal*, **Vol. no. 399**, L47.

Lilly, S.J. (1988), *The Astrophysical Journal*, **Vol. no. 333**, 161.

Martín-Mirones, J.M., Martínez-González, E., González-Serrano, J.I. and Sanz, J.L. (1994), *The Astrophysical Journal*, in press

Uson, J.M., Bagri, D.S., and Cornwell, T.J. (1991), *Phys. Rev. Letters*, **Vol. no. 67**, 3328.

FOREGROUND GALAXIES AROUND LUMINOUS QUASARS

J. VON LINDE, U. BORGEEST AND S. REFSDAL
Hamburger Sternwarte
Gojenbergsweg 112, D-21029 Hamburg, Germany

AND

K.-J. SCHRAMM AND E. VAN DROM
Université de Liège, Inst. d'Astrophysique
5, Avenue de Cointe, B-4000 Liège, Belgium

We compare galaxy counts in deep R-band exposures of the fields of 36 highly luminous, high redshift QSOs to those in control fields at a distance of 1 deg. We find indication for a weak overdensity of galaxies in the foreground of QSOs on scales of arcminutes on a low significance level. Counts inside rings around the quasars, stars in the quasar fields and stars in the control fields show evidence for an excess of galaxies on scales of several arcseconds around the quasars as well as for a stronger clustering of galaxies in the QSO fields than in the control fields. We interpret this in terms of an amplification bias by gravitational lensing.

Observations were done at the DSAZ 3.5 and 2.2 m telescopes at Calar Alto, Spain, the ESO NTT, Chile, and the *Nordic Optical Telescope* (NOT) at La Palma. Quasars were selected with respect to optical luminosity (12 QSOs, $z \geq 1$ $M_V \leq -29.0$; $H_0 = 50$ km s^{-1} Mpc^{-1}, $q_0 = 0$), and to both optical luminosity *and* radio flux (24 QSOs,, $z \geq 1.5$, $M_V \leq -27.0$ mag, S(6 cm) ≥ 0.8 Jy). All fields were observed in the Johnson R band.

Figure 1a shows the galaxy counts in the quasar vs. control fields for 32 objects down to a limiting magnitude of 21.5 rmag (the 4 fields observed at the NOT are not included here because of the small field or a significantly lower brightness limit, respectively). Assuming Poisson statistics, and deriving the variance from the galaxy counts in the respective control fields, we find 6 fields with an excess of galaxies of more than 3 σ vs. only one field with a significant underdensity. The confidence level, however, is very low, as can be seen by comparison with the distribution of stars in the fields (fig. 1b). The total overdensity of galaxies in the quasar fields is 5%.

M. Kafatos and Y. Kondo (eds.), Examining the Big Bang and Diffuse Background Radiations, 501–502.
© 1996 *IAU.*

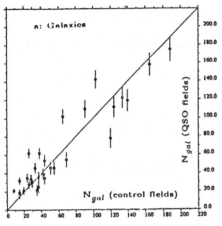
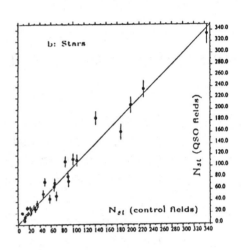

Fig. 1

The results of the 19 fields around radioloud quasars with $S(6\text{cm}) \geq 1\text{Jy}$ show no significant difference to those of the entire sample.

The table below shows the numbers of galaxies in rings of $4'' \leq r \leq 15''$ around the qso (N_{qso}^{qf}), one star in the qso field (N_{star}^{qf}), and two stars in the control fields at respective positions (N_{s1}^{cf}, N_{s2}^{cf}).

Sample (N)	N_{qso}^{qf}	N_{star}^{qf}	N_{s1}^{cf}	N_{s2}^{cf}	$<N_{stars}>$	$N_{qso}/<N_{stars}>$
all (36)	45	20	31	33	24.7	1.61 ± 0.22
S >1 Jy (19)	24	12	20	20	17.33	1.38 ± 0.34
opt (8)	8	0	3	4	2.43	3.42 ± 1.8

Mind that the enhancement factor is even larger, when only the counts inside the QSO fields are taken into account: $N_{qso}^{qf}/N_{star}^{qf} = 2.25$ and 2.0 for the entire sample and the 1 Jy-subsample, respectivly. If this is not due to statistical fluctuation it gives indication for clustering of galaxies in the QSO vicinity, since the average number density of galaxies is nearly equal in QSO and control fields.

The results do not give evidence for a magnification bias by clusters of galaxies. However, they show indication for a stronger clustering of galaxies around the quasars than in random fields at the sky. The overdensity of galaxies is weaker in the subsample of extremly radioloud objects. If this is not due to statistical fluctuations, it contradicts the hypothesis of the "multiple waveband bias". A possible explanation would be, that the luminosity function of flat spectrum quasars is not as steep as that of radio quiet objects.

EVOLUTION OF GALAXY LUMINOSITY IN THE CDM MODEL

Mirta Mosconi[1], Patricia Tissera[1,2] and Diego García Lambas[1,2]
[1]- Observatorio Astronómico de Córdoba, Argentina.
[2]- Consejo de Investigaciones Científicas y Técnicas (CONICET).

Abstract: we analyze the evolution of the luminosity function of galaxies using the CDM model in numerical simulations. There is an observational excess in the number counts of galaxies per square degree in the blur band N_b (i.e. APM, Maddox *et al.* 1990). Several authors have tried to reproduce it trying to fit the observations. Our model assumes an instantaneous star formation rate (SFR) proportional to a power of the local density. A 'single star burst' is produced each time step and we follow the evolution of the luminosity and colour of each 'stellar group'. The galaxies are identified with a density criterium. We compute U,V,B,K colours and N_b and our results agree quite well with observations.

Model and simulations

We have used a PM code with a 128^3 grid in the simulations. Each particle has dark matter and gas that can make stars ($\Omega_* \sim 0.015$). Its total mass is of the order of 7.6×10^9 $M_\odot$. We have studied the evolution of 200000 particles in a cube of 28 Mpc. The initial fluctuation spectra corresponds to a CDM universe (Bond & Efstathiou, 1986) with density parameter $\Omega=1$, baryon fraction $\Omega_b =0.1$, and present Hubble constant $H_o= 50$km s^{-1}Mpc^{-1}. Time unit: 1.09×10^9 years.

We have analyzed several simulations with different initial conditions. We identify 'galactic haloes' in each time step according to a criteria of maximal density (see Mosconi *et al.*, 1994). We calculate mass, luminosity in different bands and circular velocities at different redshifts. For each time step we approximate the luminosity function varying the parameters M_B^* and α of the Schechter function (1976):

$$\varphi(x)dx = \varphi^* x^\alpha \exp(-x)dx$$

where

$$x \equiv 10^{0.4(M_B^*-M)} \_and\_M_B^* = -0.72z - 21.18$$

and

$$\alpha = -1.35z - 1.5$$

.. We have calculated the number of galaxies per square degree per apparent magnitude interval, integring Schechter function with the obtained parameters for each z for a fixed absolute magnitude B_j:

M. Kafatos and Y. Kondo (eds.), Examining the Big Bang and Diffuse Background Radiations, 503–504.
© 1996 IAU.

$$\frac{dN}{d\Omega} = \mu \iint \varphi(B_j, z) f^2(z) g(z) dz dB_j$$

with $q_o = 0.5$. We obtain apparent magnitudes from:

$$b_j = B_j - 25 - \frac{5\log(r)}{2.3026} - K(z)$$

We have adopted a linear aproximation for the K(z) correction (from Pence data, 1970), and we have obtained:

$$K(z) = 3.2z$$

averaging over the different morphological types.

Results

We show our results superposed to observations (Figure 1). The results of the simulations show a strong evolution in the luminosity of galaxies at low redshifts. This is a very simple model and a better treatment should include energy injection due to supernovae, a better numerical resolution, and a different consideration of the gas and dark matter; we are working on the inclusion of star formation in a SPH-AP3M code (Abadi *et al.*, this volume).

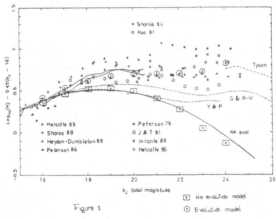

Figure 1

References

Bond. J. R. and Efstathiou, G. 1984, ApJ (Letters), 285, L45.

Maddox, S.J., Sutherland, W. J., Efstathiou, G., Loveday, J. and Peterson, B. A. 1990, MNRAS, 247, 1.

Mosconi, M., Tissera, P. and Abadi, M. 1994, *Numerical Simulations in Astrophysics,* ed. by J. Franco, S. Lizano, L. Aguilar and E. Daltabuit (Cambridge Univ. Press), 63.

Pence, W. 1976, ApJ., 203, 39.

COSMOLOGICAL PARAMETERS DETERMINATIONS FROM DEEP SKY REDSHIFT SURVEYS

Doru Marian Suran

Astronomical Institute of the Romanian Academy,

75212 Bucharest 28, ROMANIA

1. Observations

A new cosmological method and new calculations are presented in order to derive the cosmological fundamental parameters (H_0, Ω_0, Λ_0) using the *observed correlation function*.

The form of the *correlation function*:

$$\xi(r) = \begin{cases} (r/r_0)^\gamma & \begin{array}{l} \gamma = -1.77, \\ r_0 = 19 - 21h^{-1}\text{Mpc} \end{array} & \text{for r} \leq \text{ r}_a = 10h^{-1}\text{Mpc} \\ f_P(r) & \text{(periodic function)} & \text{for } r_a \leq r \leq r_b = 1000h^{-1}\text{Mpc} \\ f_G(r) & \text{(Gaussian fluctuations)} & \text{for r} \geq \text{ r}_b. \end{cases}$$

is derived from different 3D catalogs (Suran 1993):galaxies ($r \leq 750h^{-1}\text{Mp} \leq c$)- YALE catalog; clusters ($r \leq 500h^{-1}\text{Mpc}$) - North Galactic Cone Catalog; super-clusters ($r \leq 500h^{-1}\text{Mpc}$) .

2. Equations

For our model and calculations we used the following *input parameters:*

(H_0, Ω_0, Λ_0) , (a_0) , $\xi^{obs}(r)$ where a_0 - the scale parameter and $\xi^{obs}(r)$ the *observed* correlation function (see last section).

Using these parameters we could determine:

for z$\simeq$ 0

o the power spectrum: $P(k) = \frac{1}{2\pi^2} \int\limits_{0}^{\infty} \xi(r) \frac{\sin(kr)}{kr} r^2 dr$, $k = \frac{2\pi}{\lambda} = \frac{2\pi}{a_0 r} \Longrightarrow$

M. Kafatos and Y. Kondo (eds.), Examining the Big Bang and Diffuse Background Radiations, 505–507.

$$P(k) = \begin{cases} Ak^n = F_G(k) & k \le k_b \\ A\frac{(k/k_t)^{n_s}}{1+(k/k_t)^{n_s-n_l}} = F_P(k) & k_b \le k \le k_a \\ F_L(k) & k \ge k_a \end{cases}$$

which match with COBE spectrum at $F_P \to F_H$ (horizont limit).

o density perturbations: $|\delta_k|^2 = 8\pi^3 |P(k)|$;

o the transfer function: $T_f^2(k) = \frac{|\delta_k|^2}{Ak}$;

o *peculiar* velocity field: $v^2(r_b) = \left[\frac{a_0\Omega^{1.2}}{\pi^3}\right] H_0^2 \sum_k k^2 |P(k)| \, dr V_s(k, r_b)$;

where $V_s(k, r_b) = \frac{\sin(kr_b)}{kr_b}$ or $e^{-k^2 r_b^2}$;

o *biasing* parameters (second determination): $\sigma^2(r_c) = 2a_0 \sum_k k^4 |P(k)| \, dr W_s^2(k, r_c)$

where $W_s^2(k, r_c) = \left[\frac{3\sin(kr_c)-3kr_c\cos(kr_c)}{(kr_c)^3}\right]$, $b = \frac{1}{\sigma}$, $\xi_g = b^2\xi_m$, $b_m = \frac{1}{\sigma_{8,m}}$;

o mass fluctuations: $[(\delta M/M)(r_c)]^2 = \int_0^\infty k^2 \delta^2(k) W_s^2(k, r_c)$;

o the *local* topology (3D) of the Universe:

$C_\nu = \frac{1}{\pi} \left[\frac{\sigma_1}{\sqrt{3}\sigma_0}\right]^3 (1-\nu^2)e^{-\nu^2/2}$, where $\sigma_1 = \frac{1}{2\pi^2} \int_0^\infty P(k)k^4 dk$ and $\sigma_0 = \frac{1}{2\pi^2} \int_0^\infty P(k)k^2 dk$

for z=z_H

o temperature fluctuations:

▷ total: $(\delta T/T)_{rms}^2 = \frac{1}{4\pi} \sum_{l=2}^{l\max} (2l+1)C_l e^{-\frac{l^2\theta_c^2}{2}}$; quadrupole: $(\delta T/T)_Q^2 = \frac{5}{4\pi} C_2 e^{-2\theta_c^2}$

where $C_l^2 = 32\pi a_0 \Omega_0^{1.54} \left[\frac{H_0}{2c}\right]^4 \sum_k |P(k)| \, dr J_l^2(k, r_m)$;

o the topology (2D) of the Universe: $C_\nu = -\frac{1}{\pi} \left[-\frac{C''(0)}{C(0)}\right]^{3/2} (1 - \nu^2)e^{-\nu^2/2}$.

3. Results

We made the cosmological calculations using the following set of input parameters:

$n_s = -1.6, n_l = 2.4, \lambda_t = 175, \lambda_l = 10$ (power spectrum form);

$A = 24^4, \epsilon_0 = 3.10^{-5}, \theta_c = 0.051299$ (transfer function normalisation, COBE);

$a_0 = 1, b = 1. - 1.5$ (biasing parameters);

$r_a = 25, r_b = 50, r_c = 8$ (different scales); $l_{\max} = 20$ (nr.of harmonics).

The obtained results are presented in Table 1., where we denoted T≡topology, Sigma≡ σ.

Table 1: Results of cosmological calculations

H0	Omega	(DT/T)Q	(DT/T)rms	V(50)	Sigma	ampl(T)
100	0.1	8.37e-7	1.9e-6	74.	2.36	1.21
	0.2	1.42e-6	3.3e-6	112.	1.91	2.42
	0.5	2.88e-6	6.71e-6	194.	1.45	6.05
	0.7	3.37e-6	8.69e-6	237.	1.31	8.47
	1.0	4.91e-6	1.14e-5	294.	1.40	12.11
50	0.1	8.35e-7	1.94e-6	93.	2.36	0.94
	0.2	1.42e-6	3.31e-6	141.	1.91	1.89
	0.5	2.88e-6	6.71e-6	244.	1.45	4.73
	0.7	3.73e-6	8.69e-6	299.	1.31	6.63
	1.0	4.91e-6	1.14e-5	371.	1.18	9.46
50;b=1.5	0.1	6.82e-7	1.58e-6	76.	1.92	0.95
	0.2	1.16e-6	2.70e-6	115.	1.56	1.80
	0.5	2.35e-6	5.48e-6	199.	1.19	4.73
	0.7	3.05e-6	7.10e-6	244.	1.07	6.62
	1.0	4.01e-6	9.34e-6	392.	0.97	9.46

References

[1] Suran,M.D.,Popescu,N.A. (1992),in Symp. Observ. Cosmology, ed.G. Chin-carini,pg.154;

[2] Suran,M.D.,Popescu,N.A. (1993),Romanian Astron.J.,**3**,1,pg.1;

A STATISTICAL STUDY OF ENVIROMENT EFFECTS
ON GALAXY PROPERTIES

Carlos A. Valotto - Diego Garcia Lambas
IATE - Observatorio Astronomico de Cordoba
Laprida 854 - 5000 - Cordoba - ARGENTINA

We perform statistical analyses in the CfA and SRC 2 catalogs regarding the dependences of galaxy morphology and galaxy circular velocity on enviroment.

The morphological index t is found a continuous decreasing function of the galaxy density measured in shells (radial width 2 Mpc) at different radii in the range 1Mpc < r < 10 Mpc. Figure 1 shows the average normalized inner density (r < 1.5 Mpc), Di, versus the morphological index t, where the galaxies with index t<0 have been added in the bin t = -1.

At a given value of the inner galaxy density we find that the morphological index of the central object is more negative when larger is the galaxy density in the external shells (5Mpc < r < 10Mpc). This effect is present only if both early and late types are considered (positive and negative values of t).
We find that the galaxy density at external shells is not related to variations of morphology neither within early types nor within late types. These results may be explained in terms of a higher merger rate of spirals in global galaxy density enhancements originating early type objects. On the other hand, disk-bulge ratios of late types and ellipticities of early types are found to depend only locally on density.
All galaxies are binned according to Di. For each bin we compute the average difference delta t = t(high) - t(low) where high (low) referes to 1/3 the galaxies with the highest (lowest) external densities (3-5 Mpc). The histogram of the differences delta t is shown in figure 2-a (all types) and figure 2-b (t > 0). If the morphology-density relationship is of local character it would be expected this histogram to be centered in zero. As seen in these figures this ocurrs in case 2-b but not in case 2-a. We conclude that the abundance of elliptical galaxies requires of a global high density environment.

We find that the mean circular velocity of galaxies has an approxmately linear increase with shell density as expected in the linear regime of a hierarchical clustering scenario like the CDM model. This effect is shown in figure 3 which displays Di as a function of the circular velocity Vc.

Applying the previously described method we find that at a given Di the circular velocities of galaxies are ~10% smaller than the average for the sub sample with the largest external densities. This may be regarded as evidence for the stripping of halo material driven by encounters.
Figure 4-a displays the average difference in Vc between high and low external densities (3Mpc < r < 5Mpc) for galaxies with the same value of Di. In igure 4-b is shown the histogram corresponding to figure 4-a.

REFERENCES
Mo H.J. and Lahav O., 1993, in ASP Conferences Series, Vol. 51, 150.
Nicotra M., Abadi M., Garcia Lambas D., 1993, in ASP Conferences Series, Vol. 51, 152.

M. Kafatos and Y. Kondo (eds.), Examining the Big Bang and Diffuse Background Radiations, 509–510.

510

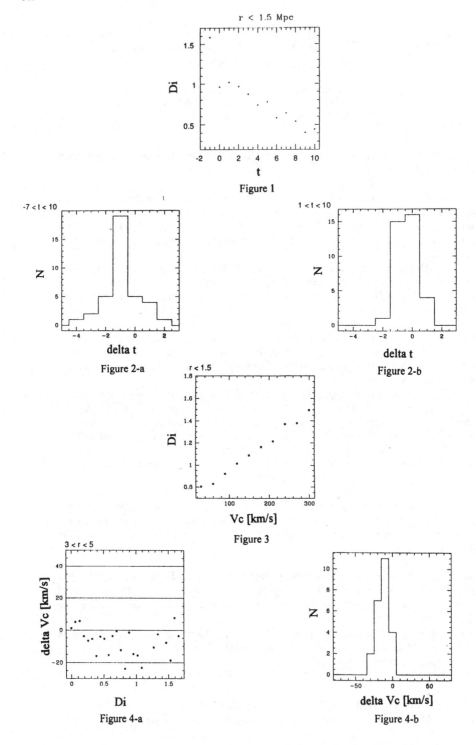

Figure 1

Figure 2-a

Figure 2-b

Figure 3

Figure 4-a

Figure 4-b

CHANCE AND NON-CHANCE CLUSTERING IN THE UNIVERSE AND PROBLEM OF HIGH REDSHIFT GALAXIES IN COMPACT GROUPS.

JOANNA ANOSOVA[1], LUDMILA KISELEVA[2]
[1] *National Astronomical Observatory, Tokyo 181, Japan.*
[2] *Institute of Astronomy, University of Cambridge,*
Madingley Road, Cambridge CB3 0HA, England.

1. Introduction

In previous work (Anosova, Kiseleva 1993) we have studied the statistical properties of Hickson's (1982) compact galaxy groups. We have shown that most of these groups are confident and probable physical systems of galaxies. Only about 13% of galaxies can be non-members of their groups due to the effects of projection.

In this work we apply a new classification technique to the galaxies which have concordant and discordant redshifts in compact groups. We define the probability that these galaxies are not chance objects inside their groups. Usually authors consider independently the angular separations ρ and differences of radial velocities dV of galaxies under study. Then they use various critical values of dV_{cr}. In our method ρ and dV are used simultaneously. We define the probability P and expectation EX of a number of chance groups with similar relative observed data for the components on the sky sphere (a random uniform distribution is assumed for field galaxies). We have classified the galaxies from compact groups into four classes:

I. confident non-chance groups,

II. probable non-chance groups,

III. probable chance groups,

IV. confident chance groups.

If the redshifts indicate the distances of the galaxies from the Sun, close non-chance groups of galaxies are systems with physically connected components; the discordant members of wide non-chance groups are con-

M. Kafatos and Y. Kondo (eds.), Examining the Big Bang and Diffuse Background Radiations, 511–515.
© 1996 *IAU.*

fident single galaxies belonging to voids (between members of wide non-chance groups there are almost empty regions). Chance groups are fortuitous groupings of field galaxies and may be called 'optical' ones.

2. Classification and basic characteristics of chance and non-chance pairs

In this work, we develop a new method for the determination of structures of fields of astronomical objects and apply this method to the observed distribution of galaxies in the Universe. Using the theory of probability, we carry out a quantative statistical analysis and find parameters which characterize differences in the distributions of galaxies in the sky and objects in a model uniform random field. In the case of galaxies, we consider the cut phase space taking into account simultaneously the angular separations ρ between galaxies, and their radial velocities V. We assume that the radial velocities of these objects indicate their Hubble distances, and we do not consider the peculiar velocities of galaxies.

The following classification of the objects is used:

1. members of clusters with different multiplicities;

2. uniform random distributed single objects in the galaxy field;

3. objects separated by voids.

We suggest a definition of the probability P_{nch} for objects to belong to these classes, and the critical values of P_{nch} for this classification. The relative uncertainties in the observations can be take into account.

Our statistical analysis involves the number $N = 4.5 \times 10^7$ of objects in the celestial sphere with radius corresponding to the maximum velocity $V_{max} = 45000$ km/s of galaxies from CfA-Catalogue. We take the mean number density of galaxies in the field $\nu = 0.05 gal/Mpc^3$ and the Hubble constant $H = 75 km/s/Mpc$. For every object under consideration, we calculate the probability P_{nch} and identify the confident and probable members of galaxy groups and clusters.

The formula for probability P_n of a chance realization of a group with multiplicity n, and for expectation EX_n of the number of such groups inside the whole sky sphere, were given by Anosova (1987) and Anosova & Kiseleva (1993), and include two important terms: (a) B^{n-1} and (b) $(1 - B)^{N-n}$. $B = V(\varsigma)/V(\sum)$, $V(\sum)$ and $V(\varsigma)$ are the volumes of $\sum$ and ς, ς being the phase space occupied by a group , while $\sum$ is the phase space of the whole sample of N objects.

The term B^{n-1} means that the given n objects occur in the region ς by chance, and the term $(1 - B)^{N-n}$ indicates that there are no other objects $N - n$ in the same region. Therefore, the proposed method can be used for the identification of clusters of objects as well as of voids in the Metagalactic field.

One can expect that with an increase of the phase volume occupied by the group, the values of $P(n)$ and $EX(n)$ also increase initially, reach their maxima, and then decrease till $P(n) \approx 0$, $EX(n) < 1$.

For every phase space volume $\Sigma(R)$ with a radius R and the assumed average number density ν of the objects, we can find the maximal number

$$EX_{max}(n) = EX_{ch} \tag{1}$$

of chance groups with a given multiplicity n. In order to obtain the typical basic characteristics of such chance groups the following modeling procedure was performed.

For the definition of $EX_{max}(n)$ we consider a model of the galaxy field which is the phase space sphere $\Sigma(R)$ of radius R corresponding to the largest radial velocity V of galaxies in this field. We examine the galaxy fields with various V corresponding to the observed data for the CFA galaxies: $1000 \leq V \leq 45000 \, \text{km/s}$. Using the expression $N = 4/3\pi\nu R^3$, for every V we find the total number N of galaxies inside this field. We construct model pairs $(n = 2)$ of galaxies with various values of angular separations ρ and differences of radial velocities dV: we scan dV and ρ in the following intervals:

(a) we choose dV such that dV/V lies in the interval $(0.0, 1.0)$

(b) in order to obtain typical values of the parameters for galaxy pairs of all classes I-IV we examined a large range of angular separations between galaxies from $0°$ to $50°$.

Table I shows the results for pairs of galaxies of the model fields: for a given V, this Table contains the number N of galaxies in the phase space $\Sigma(R)$, the values $EX_{ch} = EX_{max}(2)(n = 2)$, as well as the average values dV_{ch} and ρ_{ch} and their r.m.s. deviations for confident chance pairs of galaxies. The last column in Table I gives the maximal possible angular separation ρ between two physicaly connected galaxies for given V. A detailed description of a new method for classification of galaxies from any field into eight classes (I-VIII) between confident clusters and confident objects separated by voids was given by Anosova et al. (1994).

3. Results for compact groups

As an extention of our previous work (Anosova & Kiseleva 1993), in the present work we apply our new algorithm for the identification of chance and non-chance clusters and voids to Hickson's compact galaxy groups. These groups contain galaxies with concordant and discordant redshifts (Table 2).

TABLE 1. Characteristics of typical chance pairs

V	N	EX_{ch}	dV_{ch}	$\delta_d V_{ch}$	ρ_{ch}	$\delta\rho_{ch}$	ρ_{cl}
1000	5.0×10^2	91	98	85	898	278	374
2000	4.0×10^3	730	106	88	319	101	133
3000	1.3×10^4	2.5×10^3	162	138	174	55	72
5000	6.2×10^4	1.1×10^4	226	190	80	25	34
6000*	1.1×10^5	2.0×10^4	262	222	61	19	25
8000*	2.5×10^5	4.7×10^4	326	276	40	12	16
10000	5.0×10^5	9.1×10^4	387	327	29	9	12
20000*	4.0×10^6	7.3×10^5	666	562	10	2	4.1
35000*	2.1×10^7	3.9×10^6	1050	886	4.5	1.5	1.9
40000	3.2×10^7	5.8×10^6	1172	988	3.6	1.2	1.5
45000	4.5×10^7	8.3×10^6	1289	1087	3.0	1.2	1.3

TABLE 2. Classification of galaxies from compact groups

I. Concordant redshifts ($dV < 10^3$ km/s):
1. confident non-chance members of clusters, $N_{gal}=40$;
$10^{-5} < P_{ch} < 10^{-3}$, $dV < 500 km/s$;
2. probable non-chance members of clusters, $N_{gal}=244$;
$10^{-3} < P_{ch} < 10^{-2}$, $500 < dV < 10^3 km/s$;

II. Discordant redshifts ($dV > 10^3$ km/s):
1. 'bright discordants', $N_{gal}=11$;
$10^{-3} < P_{ch} < 10^{-2}$, $10^3 < dV < 10^4 km/s$;
2. 'double chance groups', $N_{gr}=6$, $N_{gal}=34$;
$10^{-3} < P_{ch} < 10^{-1}$, $6 \times 10^3 < dV < 11 \times 10^3 km/s$;
3. probable chance 'groups', $N_{gal}=24$;
$10^{-1} < P_{ch} < 0.20$, $\times 10^3 < dV < 18 \times 10^3 km/s$;
4. confident chance 'groups', $N_{gal}=10$;
$0.26 < P_{ch} < 0.98$, $8 \times 10^3 < dV < 19 \times 10^3 km/s$;

We find that 97% of galaxies are not by chance in the phase space of their groups with probability $P_{nch} > 0.80$. Another 3% of galaxies can be objects, from a field with random uniform distribution of galaxies, which are projected on to the groups. Most of the discordant galaxies had been identified as confident discordants in our previous work, but other types of discordant galaxies appear to have non-random connection with their groups; possibly these objects are separated by voids.

We can now redetermine multiplicities of compact groups, and we find that 7 of them are binary and 13 are triple galaxies (Hickson defined for his groups the multiplicity $n \geq 4$); the number of quartets is reduced from 59 to 55, the number of quintets from 26 to 22, of sextets from 7 to 4, and the number of groups with $N = 7$ from 6 to 4.

Thus we can conclude that the majority of the compact groups examined are confident non-chance systems of galaxies. We can suggest some possible explanations for confident non-chance members of compact groups with discordant redshifts:

- objects separated by voids;
- products of strong dynamical interaction between galaxies within groups;
- non-Hubble distances or other physical reasons.

Therefore, components of compact groups of galaxies with discordant redshifts require a special study. An application of redshift-independent methods of distance determination may be especially important in these cases.

Our results agree with the conclusions of Hickson and Sulentic (Sulentic 1987, 1988; Hickson 1990) that only about 3% of the compact groups are confident chance ones.

4. References

Anosova J.P., 1987, Astrofizika, **27**, 535.

Anosova, J.P., Kiseleva, L.G., 1993, Astrophys. Space Sc., 209, 181.

Anosova, J.P., Iyer, S., Varma, R.K., 1994, Ap.J., submitted.

Hickson, P.,1982, Astrophys. J., **255**, 382.

Hickson, P., 1990, Proc. IAU Coll. No. 124, 77.

Sulentic, J.W., 1987, Astrophys. J., **322**, 605.

Sulentic, J.W., 1988, in New ideas in Astronomy, ed. F. Bertola, J. Sulentic, B. Madore, 123.

ANALYSIS OF THE CFA "GREAT WALL"
USING THE MINIMAL SPANNING TREE

S.P. BHAVSAR AND D.A. LAUER
University of Kentucky
Lexington, KY 40506-0055, USA

Abstract. MST analysis shows the "Great Wall" as a statistically significant linear feature in the galaxy distribution. It is not as long as visual
impressions may suggest. Another "Wall" extends radially, perpendicular
to the GW. These two walls intersect at the Coma cluster.

The Minimal Spanning Tree or MST (Barrow, Bhavsar and Sonoda
1985) has proved to be a valuable tool for the identification and analysis of
filamentary structures (Bhavsar and Ling 1988). The operations of pruning
and separating the MST, called the reduced MST, allow one to extract
the prominent linear features from any point distribution. We have used
these constructs to extract the "backbone" of the linear features from the
CfA redshift survey extension (Geller and Huchra 1989). This makes for
an objective selection of these features, which can then be studied in a
quantitative way. Figure 1 shows the data and its reduced MST.
It is interesting to compare the visual impressions in the point data

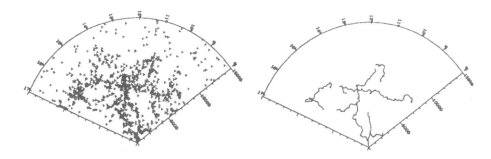

Figure 1. The CfA galaxy survey and the reduced MST

517

M. Kafatos and Y. Kondo (eds.), Examining the Big Bang and Diffuse Background Radiations, 517–519.
© 1996 IAU.

518

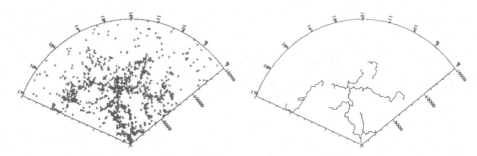

Figure 2. Same as figure 1 with "fingers of God" removed

set with features picked out by the MST. The MST does quite well as a filament finding algorithm. Its objectivity is an important bonus, since the same quantitative criteria are used on different data sets, a distinct advantage when dealing with visual bias (Barrow and Bhavsar 1987).

The Great Wall [GW] (Ramella, Geller and Huchra 1992) is one of the most prominent features, both visually and in the reduced MST. Notice that the traditional GW has a break in it as picked out by the reduced MST. The break may not initially be noticed in the point data set because of visual inertia (see Barrow and Bhavsar 1987). Another prominent feature, picked out by the MST analysis which one notices in the data in retrospect, is the long radial linear feature, starting from the vertex, going somewhat left of the vertical and then toward the right. This feature has more galaxies in it than any continuous portion of the GW. At the intersection of this feature and the GW is the Coma cluster. Figure 2 shows the same data with the "fingers of God" removed. The same prominent features that were picked out before are again picked out.

Bhavsar and Ling (1988) have described the statistical "shuffling" technique which keeps small scale correlations intact but breaks large scale coherence. This is used to determine the physical significance of the linear features. If the features are just visual artifacts as a result of small scale clumping, they will persist in shuffled distributions in a statistical way. If they are real then at that level of pruning and separating the filamentary structure is unique to the data and can be interpreted as (in fact to define) real filaments. Here the shuffling was done only among thin radial wedges. As a consequence, correlations in the radial direction are not disturbed. Thus in this analysis only the reality of the non-radial features, like the GW have been tested.

Figure 3 shows one such shuffled data set and its reduced MST. The GW feature is no longer seen. This is the case in repeated shufflings, indicating that the GW is a real feature in the data, not a visual artifact of chance

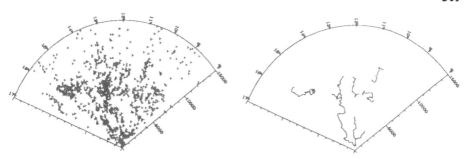

Figure 3. A shuffled data set and its reduced MST

superpositions and visual bias. The vertical feature still persists, but this analysis cannot make any statement about its reality.

References

Barrow, J.D. and Bhavsar, S.P. (1987), *QJRAS*, **28**, pp. 109–128.
Barrow, J.D. , Bhavsar, S.P and Sonoda, D.H. (1985), *MNRAS*, **216**, pp. 17–35.
Bhavsar, S.P and Ling, E.N. (1988), *Ap. J. Lett*, **331**, pp. L63–L68.
Geller, M.J. and Huchra, J.P. (1989), *Science*, **246**, pp. 897–903.
Ramella, M. , Geller, M.J. and Huchra, J.P. (1992), *Ap. J.*, **384**, pp. 396–402.

SUPERCLUSTERS AND SUPERVOIDS

I. Kuneva and M. Kalinkov
Institute of Astronomy
Bulgarian Academy of Sciences
72 Tsarigradsko Chausse blvd
1784 Sofia
e-mail: markal@bgearn

1 Introduction

A new catalog of superclusters from the entire sky cluster catalog of Abell *et al.* (1989) is compiled. In fact the last version of *Reference Catalog of ACO Clusters of Galaxies* (Kalinkov and Kuneva 1994) is used. The RC contains information of optical, radio and X-ray observations gathered from catalogs, lists and papers.

There are more than 1000 ACO clusters with measured redshift. We have applied multiple regression analysis to obtain redshift estimate for the rest clusters. Estimators in the last version of the RC are almost the same as in Kalinkov *et al.* (1994).

We use measured as well as estimated redshift for luminosity distance to ACO clusters assuming $q_0 = 1/2$ and $H_0 = 100$ km s^{-1}Mpc^{-1}.

2 Catalog of Superclusters of Galaxies

Our searching procedure for systems of galaxies is described by Kalinkov and Kuneva (1986) and Kuneva and Kalinkov (1990).

We use density contrast $f = 10$, 20, 40, 100, 200, and 400. But we define $f = D_\nu/D_{100}$, where D_ν is the density in a supercluster with multiplicity ν and D_{100} is the density in a sphere with 100 Mpc radius around clusters. Evidently D [cl Mpc^{-3}]. Thus f is a local density enhancement. After a supercluster is found f refers to a mean local density enhancement over all supercluster members.

There are 893 superclusters with $\nu \geq 2$ in our catalog altogether. We list all superclusters for $|b| \geq 30°$ and $R \leq 627$ Mpc ($z = 0.2$) independently whether redshift is measured or estimated. A few superclusters with $|b| < 30°$ and $R > 627$ Mpc are added – in all cases its cluster members have measured redshifts.

For $R \leq 627$ Mpc there are 302 superclusters with $b \geq 40°$ and 394 with $b \leq -40°$.

There are 4 superclusters with $R < 60$ Mpc: Nr 173 (A262, 347, 426), Nr 246 (A3163, 3226), Nr 721 (A3656, 3698, 3706, 3742). Hydra-Centaurus supercluster is really remarkable – Nr 542 (A1060, 3526, 3537, 3656, 3574) – for $f = 10 - 100$ the distance is 40 Mpc, while for $f = 200 - 400$ is 33 Mpc (A3526, 3565) and for $f = 200$ is 49 Mpc (A3537, 3574).

M. Kafatos and Y. Kondo (eds.), Examining the Big Bang and Diffuse Background Radiations, 521–522.
© 1996 IAU.

The main Table contains data for superclusters with $\nu \geq 3$, having at least one cluster with measured redshift:

ν – multiplicity

f – density contrast level

ACO – clusters

$RA(1950)Dec$ and b – galactic latitude

R – luminosity distance ($q_0 = 1/2, H_0 = 100$ km s^{-1} Mpc^{-1}), Mpc

$\triangle RA, \triangle Dec, \triangle R$ – corresponding extent, Mpc

r_ν – distance from the supercluster centre to the most distant member (radius; $r_\nu = \triangle r/2$ for $\nu = 2$), Mpc

$r_{\nu+1}$ – distance from the supercluster centre to the nearest cluster outside the configuration, Mpc

$D = D_\nu / D_{100}$ – ratio of the supercluster density to the mean local density, defined within a sphere of 100 Mpc radius

$\mathcal{M}$ – supercluster mass in units $10^{14} \mathcal{M}_\odot$, according to the relation of Bahcall and Cen (1993)

3 Catalog of Voids

Operational deffinitions of minivoids (without any ACO-cluster in volume $V < 10^6$ Mpc^3) and of supervoids, together with searching procedures are given by Kuneva and Kalinkov (1991).

Catalogs of superclusters, minivoids and supervoids could be received from authors.

This work was supported in part by the National Scientific Research Funds (Contracts F107/1991 and F260/1992).

References

Abell, G. O., Corwin, H. G. and Olowin, R. P. 1989, *ApJ* **70**, 1

Bahcall, N. A. and Cen, R. 1993, *ApJ* **407**, L49

Kalinkov, M. and Kuneva, I. 1986, *MNRAS* **218**, 49p

Kalinkov, M. and Kuneva, I. 1994, *Reference Catalog of ACO Clusters of Galaxies (on magnetic tape), Institute of Astronomy, Bulg. Acad. Sci., Sofia*

Kalinkov, M., Kuneva I. and Valtchanov I. 1994, *ADASS III*, eds. D. R. Crabtree, R. J. Hannisch & J. Barnes. ASP Conference Series (San Francisko), 61

Kuneva, I. and Kalinkov, M. 1990, *Paired and Interacting Galaxies*, eds. J. W. Sulentic, W. C. Keel & C. M. Telesco (NASA), 143

Kuneva, I. and Kalinkov, M. 1991, *Galaxy Environments and the Large Scale Structure of the Universe*, eds. G. Giuricin, F. Mardirossian and M. Mezzetti (Trieste)

CLUSTERS AND VOIDS IN GENERAL GALAXY FIELD

J.ANOSOVA[1], S.IYER[2], R.K.VARMA[2]
1.National Astronomical Observatory, Tokyo 181,Japan
2.Physical Research Laboratory, Navrangpura,
Ahmedabad 380009, India

In our previous works (Anosova 1987, Anosova,Iyer,Varma 1994), we developed a new method for determination of the structure of a clustered distribution. In this work, we apply this method studying the observed distribution of galaxies in the Universe.

Using the theory of probability, we carry out a quantative statistical analysis and find parameters which characterize differences in the distributions of galaxies in the sky and objects in an artificial uniform random field.

We are able to consider the phase space for objects in the field (for example, for stars), but in the case of galaxies, we consider the cut phase space taking into account observational data simultaneously - angular separations ρ between galaxies, and their radial velocities V.

We assume that the radial velocities of these objects indicate their Hubble distances and do not consider the peculiar velocities of galaxies.

We use the following classification of the objects:

1. members of clusters with different multiplicities;
2. uniform random distributed single objects in the galaxy field;
3. single objects inside voids.

We suggest a definition of the probability P_{nch} for objects to belong to these classes, and the critical values of P_{nch} for this classification. The relative uncertainties in the observations can be take into account.

In this paper, we consider only an area of the sky with coordinates approximately given by $0° < \alpha < 1°$ and $-90° < \delta < +90°$, and study a distribution of galaxies in the CfA-Catalogue. The number of galaxies with known redshifts inside this area is 116.

Our statistical analysis involves the number $N = 4.5 \times 10^7$ of objects in the celestial sphere with the radius corresponding to the maximum value $V_{max} = 45\,000$ km/s of velocities of galaxies from CfA-Catalogue.

M. Kafatos and Y. Kondo (eds.), Examining the Big Bang and Diffuse Background Radiations, 523–524.

We take the mean number density of galaxies in the field
$\nu = 0.05gal/Mpc^3$ and the Hubble constant $H = 75km/s/Mpc$.

For every object under consideration , we calculate the probability P_{nch} and classify its. We identify the confident and probable members of galaxy groups and clusters.

As a result, we have found 19 new confident and probable galaxy groups with multiplicity $n < 7$ and one new 'supercluster' with $n = 31$ containing a few small clusters.

The members of the galaxy clusters are determined, on the average, with the probability P_{nch}=0.932 . For every cluster and different samples of galaxies we define the average values of basic characteristics - a angular separation ρ and a difference dV of radial velocities V of objects.

Between clusters and single galaxies often are confidently determined empty regions (voids) with the mean value of $P_{nch} = 0.995$.

Our calculations produce a highly skewed distributions of P_{nch} for galaxies which have above mean values of P_{nch}, and about 95% of galaxies have $P_{nch} = 0.85$ or more .

We conclude that 75.0% of the galaxies under study are members of clusters. Other galaxies in the field are either confident single ones (15 galaxies- 12.5%) distributed randomly in the field or belong (14 galaxies- also almost 12.5%) to voids.

We have here the cluster-void structure; sometimes in the field we observe few close clusters, connected with each other; between clusters and single galaxies we have often confident voids.

We have found also the number of chance pairs of galaxies in an artificial 'galaxy' field with an uniform random distribution of objects and the mean values of their characteristics ρ and dV: the number of confident chance pairs of 'galaxies' in this field with $V_{max} = 45\ 000$ km/s and $N = 4.5 \times 10^7$ is 8.3×10^6; for such pairs with $V = V_{max}$ the mean values $< \rho_{ch} >= (3.0 \pm 1.2)'$ and $< dV_{ch} >= (1289 \pm 1087)$ km/s .

We also have found two confident real compact galaxy groups with large differences dV for their members: the values dV are equal to 2090 and 2076 km/s.

References

Anosova,J.P.1987,*Astrofizika*, **27**,535.

Anosova J.P.,Iyer S.,Varma R.K.1994.*Astrophys.J.* (submitted).

TYPICAL CHARACTERISTICS OF CHANCE AND NON-CHANCE COMPACT GROUPS OF GALAXIES

LUDMILA KISELEVA[1], JOANNA ANOSOVA[2]
1.*Institute of Astronomy, University of Cambridge, Madingley Road, Cambridge CB3 0HA, England.*
2.*National Astronomical Observatory, Tokyo 181, Japan.*

In our previous works (Anosova 1987, Anosova and Kiseleva 1993) we developed a new objective statistical method for an identification of members of star and galaxy clusters as chance or non-chance ones. In the case of galaxies this method uses simultaneously their radial velocities V and angular separations ρ.

In this work, we examine the galaxy fields with various V corresponding to the observed data for the CFA galaxies: $1000 \leq V \leq 45000\,\mathrm{km/s}$.

We construct model pairs of galaxies with various values of angular separations ρ and differences of radial velocities dV. Using our new method, we find the typical values for relative quantities dV_{ch} and ρ_{ch} for confident chance pairs of galaxies .

We shown that for small V, and correspondently, small dV, for typical chance pairs the value of $< \rho_{ch} >$ is large. With increasing V and dV these values decrease quickly. For the largest V ($45000\,\mathrm{km/s}$) $< \rho_{ch} >= (3.0 \pm 1.2)'$. For this value of V, $< dV_{ch} >= (1289 \pm 1087)\,\mathrm{km/s}$. Therefore, we can see that if the group of galaxies is very far from the Sun, the two dimensional projection may be compact, but velocity differences dV may be more then $1000\,\mathrm{km/s}$ for chance members. If dV is much less than this, then this group is a physically connected one; if dV is much more than $1000\,\mathrm{km/s}$ then it is a confident non-chance phenomenon. It may be an effect of projection or the radial velocities of the galaxies are not indicative of their Hubble distances.

References

Anosova J.P.,1987,Astrofizika,**27**,535.
Anosova,J.P.,Kiseleva,L.G.1993,Astrophys.Space Sc.,209,181.

M. Kafatos and Y. Kondo (eds.), Examining the Big Bang and Diffuse Background Radiations, 525.

A DEDICATED QUASAR MONITORING TELESCOPE

U. BORGEEST, K.-J. SCHRAMM AND J. VON LINDE
Megaphot, c/o Hamburger Sternwarte, Gojenbergsweg 112,
D-21029 Hamburg-Bergedorf, Germany;
uborgeest@hs.uni-hamburg.de, schramm@astra.astro.ulg.ac.be

1. The Megaphot association

Under the auspices of Sjur Refsdal, 25 astrophysicists and engineers from Germany and Scandinavia have founded a non-profit association, aiming at the use of an intelligent telescope for quasar monitoring in the optical (Borgeest et al. 1993). Beyond a better understanding of the physics in quasars, the scientific goals are determining the cosmic distance scale at large redshifts and constraining the nature of Dark Matter, both using the gravitational lens effect. Thus, targets of special interest are the multiply lensed quasars and some well-known violently variable blazars. The optical photometry will in part be carried out simultaneously to observations with, e.g., ISO, ROSAT, CGRO and various radio telescopes. For the first time, a complete quasar sample will be monitored continuously, namely a sub-sample of the all-sky 1 Jy catalogue (5 GHz). Since we will collect about 10^6 photometric data points during the programme, Megaphot has been chosen as name for the association. Members from Hamburg and Bochum intend to test the 1.5 m Hexapod Telescope (HPT) astronomically in the very near future. The HPT hardware was developed and built by *Vertex Antennentechnik, Duisburg* together with the *Ruhr-Universität, Bochum* and *Carl Zeiss, Jena*; the intelligent software and weather control requires still some work. When working well, the system will be placed at a site with excellent astronomical conditions. After a few years of exclusive quasar monitoring, it will be used as a German photometry telescope.

2. Hamburg Quasar Monitoring and Joint Monitoring

Some Megaphot members are actually involved in quasar monitoring programmes using existing multi-purpose telescopes; e.g., the *Hamburg Quasar*

M. Kafatos and Y. Kondo (eds.), Examining the Big Bang and Diffuse Background Radiations, 527–528.
© 1996 IAU.

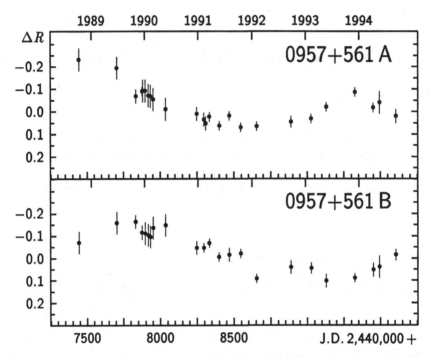

Figure 1. Results from automatic CCD photometry of the Double Quasar images, A and B, relative to stars in the field. Plotted is the mean for each observing campaign of typically 5 nights. Rough error estimates are indicated. Our CCD frames form a data set allowing a determination of the time delay Δt between A and B for a time interval fully independent of the measurements of Vanderriest et al. (see Pelt et al., this Symposium). For the plotted lightcurves, a correlation by eye suggests $\Delta t \approx 425$ days rather than values above 500 days. However, a reliable error calculation for Δt will only be possible after a more critical inspection of the single CCD frames.

Monitoring (HQM) programme at Calar Alto (since autumn 1988, Borgeest & Schramm 1994) and an international *Joint Monitoring Programme* (since summer 1994, Schramm et al. 1994). For both, we use an automatic routine reducing the CCD frames immediately after the exposure. As an example, we show in Fig. 1 the HQM lightcurve of the gravitationally lensed Double Quasar 0957+561. The time interval is independent from other, previously published data. A correlation by eye yields a time delay of ≈ 425 days.

References

Borgeest, U., Schramm K.-J. (1994) The Hamburg Quasar Monitoring Program (HQM) at Calar Alto: I. Low amplitude variability in quasars, *A&A* 284, 764

Borgeest, U., Schramm K.-J., von Linde, J. (eds.) (1993) *The Need for a Dedicated Quasar Monitoring Telescope; Proc. 1ˢᵗ Megaphot Workshop.* Megaphot, Hamburg

Schramm, K.-J., Bian, Y., Borgeest, U., Swings, J.P. (1994) The Joint Monitoring Programme of Quasars in the Optical, submitted to *Chinese Astronomy and Astrophysics*

ACCURATE MEASUREMENTS OF THE LOCAL DEUTERIUM ABUNDANCE FROM HST SPECTRA

JEFFREY L. LINSKY[1]
Joint Institute for Laboratory Astrophysics
University of Colorado and N.I.S.T.
Boulder CO 80309-0440

Abstract. An accurate measurement of the primordial value of D/H would provide a critical test of nucleosynthesis models for the early universe and the baryon density. I briefly summarize the ongoing HST observations of the interstellar H and D Lyman-α absorption for lines of sight to nearby stars and comment on recent reports of extragalactic D/H measurements.

1. The Importance of an Accurate Measurement of D/H

The Hubble expansion, microwave background, and light-element abundances are the main observational pillars upon which the standard Big Bang cosmology now rests. Of these three tests, the light-element abundances provide the main constraint on the total baryon density (luminous and dark matter). The D/H ratio provides the tightest constraint because of (1) the absence of any known significant sources of deuterium after about 10^3 s in the early universe, (2) the subsequent destruction by nuclear reactions in the cores of stars where D is the most fragile species, and (3) the steep monotonic slope between the primordial D/H ratio and the baryonic density in contemporary Big Bang nucleosynthesis models (e.g., Walker et al. 1991). Since none of the other light elements (^{3}He, ^{4}He, ^{6}Li, ^{7}Li, Be, or B) share these properties, their abundances provide more uncertain estimates of the baryon density of the universe. The importance of D/H has led to studies of D and deuterated molecules in many environments. See reviews by Boesgaard & Steigman (1985) and by Wilson & Rood (1994).

---

[1]Staff Member, Quantum Physics Div., National Institute of Standards and Technology.

M. Kafatos and Y. Kondo (eds.), Examining the Big Bang and Diffuse Background Radiations, 529–532.
© 1996 *IAU.*

The ratio of D to H column densities in warm interstellar gas ($T \approx 7000$ K) as inferred from absorption in the Lyman series lines is now thought to provide the most accurate D/H ratios in the Galaxy. Although this gas has been chemically processed and the D/H ratio must be lower than primordial, the relative ionization fraction, molecular association fraction, and degree of condensation onto dust grains should be the same for D and H in this environment. For Galactic lines of sight, only the Lyman-α line can be studied by IUE and HST, but the overlap of the H and D lines (-0.33 Å from the H line) limits the use of the Lyman-α line to nearby stars where $\log N_{HI} < 18.7$. A reanalysis of the best available Copernicus and IUE data led McCullough (1992) to estimate that the mean value of D/H by number in the local interstellar medium (LISM) is $(D/H)_{LISM} = 1.5 \pm 0.2 \times 10^{-5}$. Since these data have rather low S/N, the H line is very saturated, and the intrinsic shapes of the D and H lines are unresolved at the spectral resolutions of IUE and Copernicus, we initiated an observing program with the Goddard High Resolution Spectrograph (GHRS) on HST to obtain more accurate values of $(D/H)_{LISM}$.

2. HST Observations of D/H in the Local ISM

On 1991 April 15 we obtained echelle spectra of the resonance lines of H and D (Lyman-α at 1216 Å), the FeII multiplet UV1 (at 2599 Å), and the MgII h and k lines (at 2796 Å and 2803 Å) of the Capella binary system at orbital phase 0.26, very close to maximum radial velocity separation. A careful analysis of these Capella spectra by Linsky et al. (1993) showed that the neutral H column density is $N_{HI} = (1.7\text{--}2.1) \times 10^{18}$ cm^{-2} and $(D/H)_{LISM} = 1.65$ ($+0.07$, -0.18) $\times 10^{-5}$ for this line of sight.

A major systematic error in our analysis of the Capella phase 0.26 observations is the uncertain intrinsic Lyman-α emission-line profiles of the two stars in the Capella system, especially those portions of the emission lines that form the "continuum" against which the observed profile is compared to determine the interstellar column densities and broadening parameters for H and D. We therefore reobserved Capella on 1993 September 19 at orbital phase 0.80, close to the opposite orbital quadrature, to analyze the (assumed constant) interstellar absorption against the background of a somewhat different intrinsic emission line from the Capella system. Analysis by Linsky et al. (1994) of both Capella data sets, together with a more accurate representation of the instrumental point spread function, led to essentially the same D/H ratio, $(D/H)_{LISM} = 1.60 \pm 0.08 \times 10^{-5}$.

Our second target was Procyon, an F5 IV-V star located 3.5 pc along a line of sight about 54° from Capella. We observed this star on 1992 December 21 in the same way as we observed Capella at phase 0.26, except that the Lyman-α line was observed with the G160M grating through the small science aperture (SSA). The spectral resolution at Lyman-α was only 20,000 (15 km s^{-1}) instead of 84,000 (3.57 km s^{-1}) when we used the echelle-A grating. These observations and their analysis will be described in more detail by Linsky et al. (1994). Using a broadened solar profile for Procyon's intrinsic Lyman-α emission line, they concluded that Procyon data are consistent with but do not prove that $(D/H)_{LISM} = 1.60 \times 10^{-5}$.

The GHRS has been used by Lemoine to study interstellar H and D absorption toward the hot white dwarf G191-B2B (50 pc) and by Alexander for the line of sight toward λ And (24 pc), but the analyses of these data are not yet published. We have requested GHRS spectra to observe α Cen A and B (1.3 pc) and the binary system HR 1099 (33 pc). Observations at both quadratures in the orbit of HR 1099 should help remove the uncertainty of the intrinsic stellar Lyman-α emission line. Other lines of sight should be explored through the use of the GHRS echelle-A grating. HST programs to obtain D/H ratios for extragalactic lines of sight have been approved, but as yet there have been no reports of results from these difficult observations.

3. The Range of Ω_B Implied by D/H Measurements

An accurate determination of the primordial number ratio, $(D/H)_p$, should tell us the number density of baryons during the period 100–1000 s after the Big Bang when the temperature became low enough for the light nuclei to form. This conclusion follows from the density sensitivity of nuclear reaction rates that yield a higher abundance of ^{4}He and lower abundance of D for larger densities at that time. Since the Hubble expansion relates the baryon densities then and now, $(D/H)_p$ also determines the mean baryon density in the universe today and the ratio Ω_B of the baryon density to the critical density needed to eventually halt the expansion. Thus $(D/H)_p$ is a critical parameter for experimental cosmology.

Although our data do not allow us to measure $(D/H)_p$ directly, we can infer its value from our measurement of $(D/H)_{LISM}$ and chemical evolution calculations for the Galaxy. Steigman & Tosi (1992) and others have calculated the survival fraction of D as the primordial D is converted to heavier elements in the cores of stars and this deuterium-depleted gas is dispersed into the interstellar medium from which later generations of stars are formed. Their calculations indicate that $(D/H)_p = (1.5–3.0) \times (D/H)_{LISM}$, so that $(D/H)_p = (2.2–5.2) \times 10^{-5}$. Comparison of the Capella value for $(D/H)_p$ with recent Big Bang nucleosynthesis calculations (Walker et al. 1991) indicates that $\eta_{10} = 3.8–6.0$, where η_{10} is 10^{10} times the ratio of nucleons to photons by number. This range in η_{10} leads to the very important result that $0.06 \leq \Omega_B h_{50}^2 \leq 0.08$, where h_{50}^2 is the Hubble constant in units of 50 km s^{-1} Mpc^{-1}. Thus, no matter what value one assumes for the Hubble constant, $\Omega_B \ll 1$, and the universe must be open if the cosmological constant is zero and if only baryons are present. Tremaine (1992) and others, however, have argued that $\sim 90\%$ of the universe consists of dark nonbaryonic matter. Thus whether $\Omega = 1.0$ remains an open question.

4. Comments on Recent Reports of Extragalactic D/H

One way to avoid the uncertainties in Galactic chemical evolution models is to measure D/H in warm gas with very low metallicity. The measured D/H ratio should therefore be close to zero metal abundance, the primordial value. Songaila et al. (1994, and, independently, Carswell et al. 1994) has reported on observations of the line of sight toward the $z_{em} = 3.42$ quasar Q0014+813, which has a metallicity of $<10^{-3.5}$ solar. The proposed D feature has a column density of 2×10^{12} cm^{-2}, a factor of 2.5×10^{-4} times that of the most opaque H cloud. Is this feature D or H? If it is D, then $(D/H)_p \approx 2.5 \times 10^{-4}$, a factor of 5–10 times larger than $(D/H)_{LISM}$, whereas Galactic chemical evolution calculations (e.g., Steigman & Tosi 1992) indicate that this factor should lie in the range 1.5–3.

The major uncertainty in the estimates of D/H in absorbing clouds toward Q0014+813, and by implication other distant lines of sight, is the possibility that a low-column-density H-absorbing cloud at the predicted D velocity is masquerading as D. This possibility was recognized by Songaila et al. (1994) and Carswell et al. (1994). I will discuss how to test this hypothesis elsewhere, but the possibility of H masquerading as D is high if an increasing number of clouds are present in the line of sight with decreasing values of N_{HI}, and the velocity centroid of the cloud distribution is centered on the mean velocity of the observed clouds.

This work is supported by Interagency Transfer S-56460-D from NASA to the National Institute of Standards and Technology.

References

Boesgaard, A.M. & Steigman, G. (1985): ARAA **23** 319
Carswell, R.F., et al. (1994): MNRAS **268** L1
Linsky, J.L., et al. (1993): ApJ **402** 694
Linsky, J.L., Diplas, A., Wood, B., Brown, A., & Savage, B.D. (1994), in preparation
McCullough, P.R. (1992): ApJ **390** 213
Songaila, A., Cowie, L.L., Hogan, C.J., & Rugers, M. (1994): Nature **368** 599
Steigman, G. & Tosi, M. (1992): ApJ **401** 150
Tremaine, S. (1992): *Physics Today* **45** 28
Walker, T.P., et al. (1991): ApJ **376** 51
Wilson, T.L. & Rood, R.T. (1994): ARAA **32** 191

MUTUAL INTERFERENCE AND STRUCTURAL PROPERTIES OF OBJECT IMAGES IN THE VICINITY OF THE GRAVITATIONAL LENS CUSP POINT

A.V. MANDZHOS

Astronomical Observatory of Kiev University
Observatorna St., 3 Kiev 254053 Ukraine
e-mail: aoku@gluk.apc.org

Gravitational lens system could form multiple images of the same radiating surface of a cosmic object. The radiation fluxes of these images interfere to one another (Refsdal, 1964). The practical importance of this effect is seems to be as follows: (i) it is the only direct test of microlensing (Schneider et al. 1985); (ii) this effect may say about fine structure of a lensed object; (iii) it must be allowed for simulations of microlensing process. The maximum effect is expected in the case when a cusp point of the microlensing caustic is projected on the lensed object. To investigate the interference effect in this case, the exact solution of the lens equation in the vicinity of the cusp point was obtained (Mandzhos,1993):

$$x_{1,2} = 2|A| \cdot \cos\{\Phi_0 + [1 \pm \text{sign}(\xi)]\frac{\pi}{3}\} \cdot \text{sign}(\xi),$$

$$x_3 = 2|A| \cdot \cos\{\Phi_0 + \frac{4}{3}\pi\} \cdot \text{sign}(\xi), \quad y = \frac{1}{2}\eta - \frac{1}{4}vx^2 \qquad (1)$$

were ξ, η are dimensionless coordinates of a radiating element on the object plane; x, y are coordinates of the point at which the lens plane is intersected by light beam; furthermore:

$$\Phi_0 = \frac{1}{3}\text{Arctan}\sqrt{h\frac{\eta^3}{\xi^2} - 1}, \quad |A| = k\eta^{1/2}$$

$$h = \frac{2}{9}\frac{v^3}{3v^2 - 2u}, \quad k = \sqrt{\left|\frac{2v}{3v^2 - 2u}\right|}. \qquad (2)$$

where $u = \partial^4\psi / \partial x^4$; $v = \partial^3\psi / \partial x^2 \partial y$, ψ is the lens scalar potential.

M. Kafatos and Y. Kondo (eds.), Examining the Big Bang and Diffuse Background Radiations, 533–534.
© 1996 IAU.

534

It is well known, that there are three images of the internal cusp point region of the object. The mutual coherence degree γ_{ij} of the ij-th image pair is the main theoretical quantity in this problem; its module can vary from 0 to 1. The mutual coherence degree was calculated analytically for a homogeneous disk with the center at the cusp point (Mandzhos, 1993):

$$\gamma_{13} = P \cdot \left(\frac{r_g}{c}(1+z_d) \right)^{-1/3} \cdot R^{-2/3} \left(1 - hR \right)^{5/2} \omega^{-1/3} \qquad (3)$$

where P, h are the dimensionless lens parameters of order unity , r_g is the gravitational radius of the microlens, z_d is the redshift of the lens, R is the dimensionless radius of the object and ω is the circular frequency of radiation. The result for the optical range is presented in Fig. 1. Analyzing the real situation in astrophysics, it may be inferred that the actual manifestation of the mutual interference effect takes place the following cases: (i) microlensing of a quasar core in radio region; (ii) microlensing of emission clouds from BLR of a quasar in optics; (iii) microlensing of a star from the Magellan Clouds by both stars or MACHO.

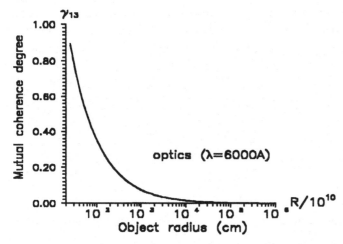

Fig. 1. *The mutual coherence degree of the radiation from the 1th and the 3rd microlens images of the internal cusp region of the object. The lens geometry corresponds to the Einstein Cross (Q2237+030)*

References

Mandzhos A.V. 1993 Mutual interference and structural properties of object images in the vicinity of the gravitational lens cusp point. SAO RAS prepr. No.98

Refsdal S. 1964 *Mon. Not. Roy. Astron. Soc.* **128** 295

Schneider P. and Schmid-Burgk J. 1985 *Astron. Astrophys.* **2** 369

COMPLEX THEORY OF GRAVITATIONAL LENSES PART II

Cluster-Lensing: Statistics of the Arclet Distribution

T. SCHRAMM AND R. KAYSER
Hamburger Sternwarte
D-21029 Hamburg-Bergedorf, Germany

Applying the complex formalism of Part I we investigate the transformation of the ellipticity and orientation distribution of elliptical background sources by local lensing. One important result is that we can apply the techniques from Part I directly in most parts of the cluster field from which we can also reconstruct the distribution of the background sources, without further information. Knowing the distribution we can apply the techniques from Part I everywhere.

Observed arclets are in general not images of circular sources. Additionally, the sources are distributed over a redshift range. Although the redshift distribution can be accounted for in the Beltrami equation we here assume for simplicity that all sources are at the same redshift.

However, we can also describe the situation including intrinsic ellipticities using the Beltrami equation introduced in Part I if a composed mapping $h(z) = \omega(\mathrm{w}(z))$ is introduced. h is a composition of two mappings w and ω. The mapping w maps the observed elliptical arclets to the unknown *real* elliptical source and ω maps this source to a circle. The corresponding Beltrami parameters are μ_h, μ_w and μ_ω, respectively.

For simplicity we use now indices in parentheses to denote derivatives. For the Beltrami parameter of the composed mapping we find after some calculation

$$\mu_h = \frac{\mathrm{w}_{(\bar{z})} + \mu_\omega \overline{\mathrm{w}}_{(\bar{z})}}{\mathrm{w}_{(z)} + \mu_\omega \overline{\mathrm{w}}_{(z)}} \quad . \tag{1}$$

If we suppose now that we are in the linear regime of the mapping, we can always write

$$\omega = \mathrm{w} + \mu_\omega \overline{\mathrm{w}}, \quad 0 \le |\mu_\omega| \le 1 \quad and \quad \mathrm{w} = (1-\kappa)z + \gamma\bar{z} \quad . \tag{2}$$

Note that we do not have to restrict the lens equation to the case of real γ. However, the only difference would be a constant phase in the appropriate Beltrami parameter, which could also be attached afterwards.

M. Kafatos and Y. Kondo (eds.), Examining the Big Bang and Diffuse Background Radiations, 535–536.
© 1996 *IAU.*

Applying Eq.(1) we find

$$\mu_w = \frac{\gamma}{1-\kappa}, \quad \text{real !} \tag{3}$$

$$\mu_h = \frac{\gamma + \mu_\omega(1-\kappa)}{1 - \kappa + \gamma\mu_\omega} = \frac{\mu_w + \mu_\omega}{1 + \mu_w \mu_\omega} \quad \text{and the inverse} \tag{4}$$

$$\mu_\omega = \frac{\mu_w - \mu_h}{\mu_w \mu_h - 1} \quad . \tag{5}$$

These equations describe the transformation of the Beltrami parameter under local lensing, given by μ_w. We are now interested to see how the mean of a local distribution Φ of μ_ω transforms under local lensing. This means how the statistical properties of a field of elliptical sources alter if mapped by a lens locally described by constant density and shear. For isotropical distributions we find

$$\langle \mu_h \rangle = \frac{1}{2\pi} \int_0^{\mu_\omega^{max}} \left[\int_0^{2\pi} \frac{\mu_w + \mu_\omega}{\mu_w \mu_\omega + 1} d\varphi_\omega \right] \Phi(|\mu_\omega|)|\mu_\omega| \, d|\mu_\omega|. \tag{6}$$

We therefore find for $\langle \mu_h \rangle$ using residue calculus for the integral in square brackets

$$\langle \mu_h \rangle = \mu_w \int_0^{\frac{1}{\mu_w}} |\mu_\omega| \Phi(|\mu_\omega|) \, d|\mu_\omega| + \tag{7}$$

$$\frac{1}{\mu_w} \int_{\frac{1}{\mu_w}}^\infty |\mu_\omega| \Phi(|\mu_\omega|) \, d|\mu_\omega| \quad , \tag{8}$$

$$= \left(\mu_w - \frac{1}{\mu_w} \right) \int_0^{\frac{1}{\mu_w}} |\mu_\omega| \Phi(|\mu_\omega|) \, d|\mu_\omega| + \frac{1}{\mu_w} \quad , \tag{9}$$

$$= \mu_w \quad \text{if } \Phi(|\mu_\omega|) = 0 \text{ for } |\mu_\omega| > \frac{1}{\mu_w} \quad . \tag{10}$$

Since $|\mu_\omega| \leq 1$ the last condition is fulfilled for $|\mu_w| \leq 1$, which is valid for a region where the lens mapping is orientation preserving (quasi conformal). This important results states that outside the critical curves (subcritical regions) the mean of the measured Beltrami parameter is only determined by the local lens. **The bad news is that if we know that $|\mu_w| > 1$ from other (global) information, the mean axail ratio of the observed arclets is influenced by the distribution of the intrinsic axial ratios. The good news is that we can conclude the intrinsic distribution from regions where the mapping is surely quasi conformal.** For details see Schramm & Kayser 1994c in the reference list of Part I.

FIELD SPECTROSCOPY OBSERVATIONS OF THE GRAVITATIONAL LENSES H1413+117 AND Q2237+030. PRELIMINARY RESULTS

V.L.AFANASIEV*, S.BALAYAN*, S.N.DODONOV*,
S.V.KHMIL**, A.V.MANDZHOS**, O.I.SPIRIDONOVA*,
V.V.VLASSIUK*, and V.I.ZHDANOV**

*Special Astrophysical Observatory
Nizhnij Archys, 357147 Russia
**Astronomical Observatory of Kiev University
Observatorna St., 3 Kiev 254053 Ukraine
e-mail: aoku@gluk.apc.org

The long-term observational programme for investigation of quasar microlensing spectral effects was devised and now is implemented at the Special Astrophysics Observatory (Russia, N.Arkhys) and at the Astronomical Observatory of Kiev University (Ukraine, Kiev). The main goal of this programme is to study the fine quasar structure. For this purpose the observations of the gravitational lenses H1413+117 and Q2237+030 are carried out with the 6m SAO telescope in the integral field spectroscopy mode. The suitable equipment for such observations - the Multi-Pupil Field Spectrograph - was designed and built in SAO. From February 1993 several observations of each of these objects were performed. Here we present the preliminary results of the Clover Leaf observations on April 23/24 , 1993. The quasar images are separated by angles nearly 1 arcsec. At seeing of the order of 1.5 arcsec the separation of the individual quasar image spectra was a problem. For this purpose a special procedure is developed. This procedure is based on the maximum likelihood hypothesis to restore a simple 4-point source picture plus the background. The effect of light scattering for a point source was simulated by means of observations of a reference star. Statistical properties of the radiation flux in question were studied by using the data of photon registration each 40 msec. The individual spectra of quasar images were obtained. The internal relative errors of the spectra separation procedure vary from 7 to 12% depending on wavelength. The spectra of quasar images 1, 2, 4 are not actually distorted . On the contrary, in the spectrum of the third image the profile of the emission line CIV differs considerably from the profile of the same line in the "undisturbed" quasar spectrum. Fig. 1a,b show the spectra of the 1st and the 3rd images.

This research was made possible in part by Grant No.U4P000 from the International Science Foundation and by Grant A-01-055 within the ESO C&EE Programme.

M. Kafatos and Y. Kondo (eds.), Examining the Big Bang and Diffuse Background Radiations, 537–538.
© 1996 IAU.

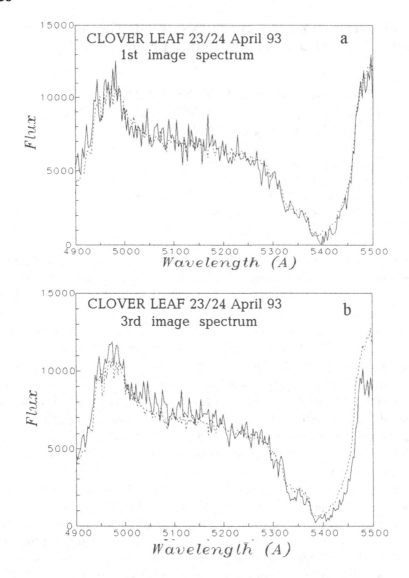

Fig.1. *The normalized spectra of the 1st and the 3rd quasar images as it is obtained by use of the separation algorithm. Dashed lines depicts an "undisturbed" quasar spectrum that is given by integration over total observation field minus background.*

THE TIME DELAY BETWEEN QSO 0957+561 A,B

J. PELT
Tartu Astrophysical Observatory
EE-2444 Toravere, Estonia

R. KAYSER
Hamburger Sternwarte
D-21029 Hamburg-Bergedorf, Germany

R. SCHILD
Harvard Smithsonian Center for Astrophysics
Cambdridge, MA 02138, U.S.A.

AND

D.J. THOMSON
AT&T Bell Laboratories
Murray Hill, NJ 07974, U.S.A.

We here present a short report about the application of a simple time delay estimator to the extensive data set compiled and partly observed by R. Schild, containing 707 A and B observations with corresponding error estimates from JD 244194 to JD 249169.

For our analyses we used a cross-sum type dispersion estimation statistic (for details see Pelt et al. 1994) to measure the dispersion around the combined data set $C_k, k = 1, \ldots, K$, where the B data are shifted by τ and corrected against the magnification b:

$$D^2 = \min_b \frac{\sum_{i=1}^{K-1} \sum_{j=i+1}^{K} S_{i,j} G_{i,j} (C_i - C_j)^2}{2 \sum_{i=1}^{K-1} \sum_{j=i+1}^{K} S_{i,j} G_{i,j}}$$

Here the *Smoothing Window* is defined as $S_{i,j} = 1$ for $|t_i - t_j| \leq \theta$ and otherwise $S_{i,j} = 0$. The statistic D^2 contains a free parameter θ. An absolutely parameter-free (high resolution) statistic can be obtained when $S_{i,j} = 1$ only for the pairs of (C_i, C_j) which occur neighboring in the combined data set. The *Selection Window* $G_{i,j} = 1$ is equal to one only for measurement pairs C_i, C_j for which C_i and C_j belong to different original series and zero otherwise. Consequently we take into account only relevant pairs.

M. Kafatos and Y. Kondo (eds.), Examining the Big Bang and Diffuse Background Radiations, 539–540.

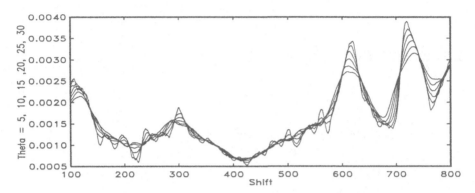

Figure 1. Smoothed dispersion spectra

An important statistic to describe the data is the *Data Window*, which counts the number of neighboring A,B pairs in the combined data set for different time delays τ. The minimum number for relevant pairs for the Vanderriest (1989) data occurred just for time delays near the 535 days (43 pairs for $\tau = 528, 534, 537$). For the extended data set there is a minimum of 155 pairs for time delays of 514 and 515 days. In the high resolution dispersion spectrum for the extended data set (with all pairs included) the strongest minimum occured at a delay of 515 days, which (again) coincides with the minimum of the *Data Window*. However, the spectrum with only A,B pairs included shows a minimum around 425 days.

We finally computed a series of spectra using smoothed estimators with different values of the resolution parameter $\theta = 5, 10, 15, 20, 25, 30$. For $\theta = 5, 10$ the absolute minimum occurred still around 425 days, but starting from $\theta = 15$ it moved to smaller values (414,416,412 and 413 days accordingly). From this follows that small shifts from 415 to 425 days must be attributed to the high frequency components of the A and B lightcurves. The convergence of the spectra is well illustrated on Fig. 1.

The application of the simple nonparametric delay estimation method to the extended data set shows clearly that time delay value around 415 days can not be ruled out. Contrarily to that, the feature around 535 days in the dispersion spectra is less pronounced than for the subsets analyzed earlier. Additional evidence of its spurious nature stems from the fact that the minimum in the *Data Window* for extended data set (515 days) shows itself also in the spectrum for which all the pairs are taken into account. However, this evidence for the delay of 415 days against the delay of 525 days can not be treated as final proof of its reality. Further analysis with use of refined statistical techniques is needed to solve the QSO 0957+561 controversy.

References
Vanderriest, C. et al. (1989) *A&A* 215, 1
Pelt J. et al. (1994) *A&A* 286, 775

GRAVITATIONAL MACROLENSING AND QUASAR SPECTRA

S.V.KHMIL
Kiev University Astronomical Observatory
Observatorna st 3, Kiev 254053, Ukraine
e-mail: aoku@gluk.apc.org

Gravitational microlensing, i.e. lensing by stars in a foreground galaxy, can affect not only the radiation flux observed from quasars (Chang and Refsdal, 1979) but also the spectral line profiles (Nemiroff, 1988; Schneider and Wambsganss, 1990). As a rule, in contrast to it, gravitational macrolensing, i.e. lensing by a whole galaxy or a cluster of galaxies, affects only the radiation flux; differences in image spectra arise mainly from source variability and travel time delays between images. However, if the source, the lens, and the observer are in motion relative to one another, then an additional shift in the source's image spectrum occurs due to the aberration of light (Birkinshaw and Gull, 1983; Khmil, 1988 and references therein). Here we consider the special case of intrinsic motions in the source and study their influence on spectra of macroimages.

Let δ be the small angular displacement of an image due to lensing, $\mathbf{v}$ be the (intrinsic) velocity of the source with respect to the cosmological reference frame, and z be the source spectral shift. Then from the general formula for the additional spectral shift Δz_S due to the aberration (Khmil, 1988) we have

$$\left(1+z_S\right)^{-1}\Delta z_S = -\left(1-v^2\right)^{-1/2}\delta_{\text{eff}}\cdot\mathbf{v}. \tag{1}$$

Here δ_{eff} is the "effective" angular displacement which is related to δ through

$$\delta_{\text{eff}} = \left(1+z_L\right)\left(1+z_S\right)^{-1}D_L D_S^{-1}\delta, \tag{2}$$

where z_L is the lens redshift and D_L, D_{LS} are the angular diameter distances to the lens and from the lens to the source, respectively. Here and below the speed of light is taken to be unity. Combining eq. (1) and an expression for the Doppler shift, one obtains for the relation between the wavelength λ as measured by the observer and the wavelength λ_S in the source rest frame:

$$\lambda/\lambda_S = \left(1+z_{S0}\right)\left(1-v^2\right)^{-1/2}\left[1+\left(\mathbf{n}-\delta_{\text{eff}}\right)\cdot\mathbf{v}\right], \tag{3}$$

where z_{S0} is the source cosmological redshift, $\mathbf{n}$ is the unit vector along the line of sight. Note two features that follow from eqs. (1) - (3). First, the aberration shift

M. Kafatos and Y. Kondo (eds.), Examining the Big Bang and Diffuse Background Radiations, 541–542.
© 1996 IAU.

has, generally speaking, different values for different images. Second, although δ is small and does not exceed 10 arcsec for the known lenses, the effective displacement δ_{eff} may be much greater than δ due to distance dependence.

Considering the influence of the aberration shift on emission spectra of quasars, we assume that the line emitting gas is described by a velocity field $\mathbf{v}(\mathbf{r})$ corresponding to Keplerian rotation model and by an isotropic emissivity $\varepsilon(\mathbf{r}, \lambda_s)$ as measured in the gas rest frame (see Schneider and Wambsganss, 1990). We assume also that no absorption takes place. Then within a constant factor the radiation flux is given by

$$F(\lambda) = \int [1 + \mathbf{n} \cdot \mathbf{v}(\mathbf{r})]^{-2} [1 - \delta_{eff} \cdot \mathbf{v}(\mathbf{r})]^{-3} \varepsilon(\mathbf{r}, \lambda_s) dV, \tag{4}$$

where λ and λ_s are related by (3) and the integration is performed over the volume in the cosmological rest frame.

The computations with eq. (4) within the framework of Keplerian rotation model show that the aberration shift does not distort line profiles but only causes their broadening. This broadening depends on relative orientation of the rotation axis, the line of sight and the image displacement. Unfortunately, in typical cases of macrolensing the broadening is too small to be of practical interest and has a detectable value only for $\delta_{eff} > 10^3$ arcsec. It should be noted, however, that in macrolensig of objects with relativistic motion (e.g., superluminal sources, radio jets) the aberration effect studied here may probably have an appreciable value.

References

Birkinshaw, M., and Gull, S.F.: 1983, *Nature*, **302**, 315.

Chang, K., and Refsdal, S.: 1979, *Nature*, **282**, 446.

Khmil, S.V.: 1988, *Sov. Astron. Letters*, **14**, 461.

Nemiroff, R.J.: 1988, *Astrophys. J.*, **335**, 593.

Schneider, P., and Wambsganss, J.: 1990, *Astron. Astrophys.*, **237**, 42.

HALO/THICK DISK CVS AND THE COSMIC X-RAY BACKGROUND

JONATHAN E. GRINDLAY AND EYAL MAOZ

Harvard-Smithsonian Center for Astrophysics,
60 Garden Street, Cambridge, MA 02138

Abstract. In a recent study (Maoz and Grindlay 1995) we have found that a number of previously recognized anomalies in the diffuse x-ray background at soft energies (~0.5-2 keV) can be understood if about 20-30% of the diffuse flux arises from a population of low luminosity sources in a thick disk or flattened halo distribution in the Galaxy. Here we summarize our results and review the arguements that these objects are not accreting neutron stars or black holes but rather white dwarfs (i.e. CVs) which may have been produced in a primordial population of disrupted globular clusters.

1. Introduction

The COBE results (Mather et al 1990) provided final confirmation that the cosmic x-ray background (CXB) radiation is not due to hot gas but, as long suspected (Giacconi et al 1979, Soltan 1991) due to the superposition of discrete sources. It is the nature of these sources that has been the major question. Although many arguements (e.g. the fluctuation analyses of Hamilton and Helfand (1987) and Barcons and Fabian (1988)) point to the dominance of relatively low luminosity active galactic nuclei (AGN), the lack of clustering strongly suggests that in fact no more than about 70% of the CXB can arise from known classes of AGNs. A new population of weak sources is probably required. Thus it was extremely interesting when in an epochal study of the CXB and source counts with the Deep Survey observations undertaken with ROSAT, Hasinger et al (1993; hereafter H93) concluded that in fact about 60% of the CXB at fluxes $\geq 2.5 \times 10^{-15}$ cgs (hereafter cgs = erg cm^{-2} s^{-1}) in the 0.5-2 keV band could indeed be resolved into discrete sources but that the source counts are inconsistent with models for the x-ray luminosity functions (XLF) of AGNs (Boyle et al

543

M. Kafatos and Y. Kondo (eds.), Examining the Big Bang and Diffuse Background Radiations, 543–548.
© 1996 IAU.

1993). H93 concluded, therefore, that either more complicated models for the evolution of the XLF or – again– that a new population of sources is needed at the faintest flux levels.

It is this possibility that we have explored (Maoz and Grindlay 1995, hereafter MG). In this brief addendum to that work, we summarize our analysis and results for constraints on the possible luminosity and density of a *galactic* population of faint x-ray sources (halo/thick disk CVs) which could satisfy all the new constraints from ROSAT as well as the earlier CXB flux and fluctuation constraints.

2. Generalized Halo/Thick Disk Models

We have considered whether either a spherical halo ($\propto 1/R^2$) or a thick disk (exponential in both R and Z) distribution of discrete sources of fixed (standard candle) luminosity L and spatial density n_0 (in the solar neighborhood) can satisfy all the constraints imposed by the source counts in the 27 deep fields surveyed by H93 as well as the fluctuation and flux constraints imposed by the CXB. Our models are normalized to a possible galactic contribution to the source counts of 120 deg^{-2}, at the flux limits of H93, which allows for the measured contributions of stars and all known classes of extragalactic sources (galaxies, clusters, BL Lacs and QSOs with the XLF evolution of Boyle et al 1993). Details of the calculation are given in MG and our results are shown below in Figure 1, which presents all the constraints on the allowed values of L and n_0 for two spherical halo models with small and large core radii, and for two thick disk models with modest and large scale heights. For each model, the solid line represents the combinations of L and n_0 that would exactly resolve the discrepancy found in the faint source counts; the shaded area corresponds to combinations which would produce a number density of unidentified bright sources which is consistent with observations; the two inclined dashed lines confine the parameter space which would not violate constraints on the contribution to the CXB intensity; the horizontal dash-dot lines confine the region which would be consistent with the results of fluctuation analyses of the unresolved background, and the region below the dotted line corresponds to models which contribute $\leq 10^{40}$ erg/s to the Galaxy's total x-ray luminosity.

The striking result is that for either halo or thick disk models a relatively well defined region of parameter space is defined by all the constraints. In particular, the typical luminosity of the sources is $10^{31\pm0.5}$ erg/s (0.5-2 keV), and is not very sensitive to changes in the scale of the spatial distribution. Thus although Kshyap *et al.* (1994) have recently suggested that MACHOs may have x-ray coronae with typical luminosities of order $\sim 10^{27}$ erg/s, and may produce a small fraction ($\sim 10\%$) of the soft CXB *flux*, their luminosity is far too low to satisfy the constraints imposed by number counts and the excess of faint sources detected by H93.

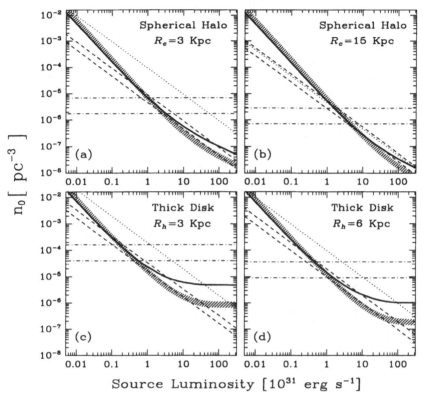

Figure 1. Combined observational constraints on the typical x-ray luminosity of the sources, L, and their number density in the Solar vicinity, n_0, for two different halo models (a,b) and thick disk models (c,d).

All our models have an approximately Euclidean logN-logS relation at fluxes above $\sim 10^{-14}$ cgs, and which flattens considerably below a flux of a few times 10^{-16} cgs. In the flux range of 10^{-15}-10^{-14} cgs, the logN-logS relation is steeper than that expected for an evolving population of QSOs (Boyle *et al.* 1993), thus providing an increasing relative contribution of the Galactic sources with decreasing flux.

We have also calculated the constraints imposed by the spectrum and anisotropy (limits) of the CXB on our proposed source population. H93 found that the average spectrum of the resolved sources becomes harder with decreasing flux. Since the extragalactic sources have a steeper average spectrum in the 0.5 - 2 keV band, we conclude that the average spectral index of the proposed population of sources must be < 1 in this band. Allowing for the Galactic emission components of the CXB which already indicated thin *and* thick disk components at higher flux levels, we conclude that the source population proposed here most probably has a flattened distribution and thus should have a significant contribution to the detected anisotropy. The fractional contribution of the proposed population to the CXB in the 2-60 keV band, averaged over the entire sky, is $\sim 8\%$ but can be 20-30% near the galactic center depending on the model. Anisotropy constraints would allow a slightly higher source temperature in more spherical models than in flattened models.

3. Identification of the Sources as CVs

Our primary requirements are sources with typical $L_x \sim 10^{31}$ erg/s and a relatively hard spectrum. These are not consistent with dM stars (which are only this bright during brief flares, and are typically soft), with LMXBs in quiescence (the ROSAT studies of Verbunt et al 1994 have shown these sources to be typically very soft, with $kT \sim 0.3$ keV) or with isolated neutron stars (e.g. old pulsars, for which thermal emission is also expected to be very soft – cf. Ögelman et al 1993). However pulsars also appear to have harder emission components (Ögelman 1994), but with much lower L_x than we require. Magnetized neutron stars (NSs) accreting from the ISM could produce hard cyclotron emission (cf. Nelson et al 1994) with $L_x \sim 0.05 L_{acc} \sim 3 \times 10^{25} n_{-3}/v_{100}^3$ erg/s for a typical pulsar at 100 km/s moving through a halo interstellar medium of density $n_H \sim 10^{-3}$ cm^{-3}, which is again far too low. Thus although (old) pulsars would nicely satisfy the requirement for some $10^{8.5}$ total sources in a thick disk or halo, it seems they could not satisfy the luminosity constraint.

Similarly, we now consider $\sim 10^8$ galactic black holes (BHs) with mass $\lesssim 10^4 M_\odot$ (so as not to exceed the halo mass) and moving on halo orbits with $v \sim 200$ km/s. Note these are probably not the $\sim 10^6 M_\odot$ halo BHs invoked by Lacey and Ostriker (1985) to stabilize the disk although it is conceivable that our lower mass could be accomodated with changes in the primordial fluctuation spectrum (Ostriker, private communication). To obtain the required L_x by accretion from the halo ISM would imply a radiation efficiency $\epsilon \sim 10^{-3}$ (v_{200}^3/n_{-3}. However at the low Bondi accretion rates implied ($\sim 10^{-9} n_{-3}/v_{200}^3$ of Eddington), the recent results of Narayan and Yi (1994) for the effective accretion efficiencies onto BHs when advection is included would suggest much lower efficiencies (typically $\lesssim 10^{-5}$, but dependent on the assumed viscosity parameter α). Thus BHs are unlikely but perhaps cannot be totally excluded.

Cataclysmic variables (hereafter CVs) can most naturally fit our requirements for luminosity and local space density. Hertz *et al.* (1990), from optical identifications of the x-ray selected Einstein galactic survey (Hertz and Grindlay 1984), find a local space density for low-$\dot{M}$ CVs which is consistent with our prediction. Howell and Szkody (1990) discuss the halo CV population and report that their significantly lower mean orbital period implies that these systems are relatively old. Magnetic CVs (*e.g.*, AM-Her type) are also predominantly below the period gap, and are also optically fainter than nonmagnetic CVs (larger L_x/L_v ratio) since they do not have well developed accretion disks. Thus, it is possible that the proposed population of x-ray sources are some kind of magnetic CVs. Their low-$\dot{M}$ could push their characteristically soft x-ray spectral components into the XUV band leaving just their hard components.

Globular clusters contain dim sources with typical soft x-ray luminosity of $\sim 10^{31.5-32.5}$ erg/s in the ROSAT band, and the bulk of these sources are fully consistent with being CVs (Grindlay 1994a, and references therein). X-ray and optical observations indicate a presence of $\gtrsim 10$ CVs per cluster (at least for (near-) core collapsed globulars) with $L_x \sim 10^{30.5-31.5}$ erg/s. These may be optically fainter than disk CVs for a given x-ray luminosity (Grindlay 1994a), deficient in dwarf novae, and contain a sifnificant fraction of magnetic CVs (Grindlay 1994a,b).

Thus the thick disk or halo CV population proposed in this paper resembles the globular cluster CV population, suggesting they may have been ejected from globular clusters. Indeed, Grindlay (1994b) shows that the dim sources in globulars appear to have an extended radial distribution which extends to radial offsets well beyond what purely relaxed objects (CVs) should obtain and may thus imply ejection. Although this process may contribute to the proposed Galactic population, it obviously cannot provide the required number of $\sim 10^{8.5}$ CVs in the halo from the present ~ 150 globular clusters. Thus the required halo CVs, if from globulars, must imply a population of some 10^{6-7} primordial globulars which have been disrupted and which had a flatter radial distribution than the $\sim 1/R^3$ distribution of the survivors. In particular, a thick disk may form from globular clusters on low-inclination orbits which have been disrupted by encounters with giant molecular clouds, as suggested by Grindlay (1984) for the origin of the field LMXBs.

4. Conclusions

The CXB may partially originate in a large population of intrinsically faint (and presumably old) CVs which are distributed within a thick Galactic disk (or an oblate halo) with a scale height of a few kpc. The inferred $\sim 3 \times 10^8$ CVs would have luminosity of $\approx 10^{30-31}$ erg/s and a number density of $\sim 10^{-5}$-10^{-4} pc^{-3} sources in the Solar vicinity. Our model can be tested by searching for a large-scale variation in the surface *number density* of resolved x-ray sources. We predict that the density of sources with flux below 10^{-14} cgs will decrease either with an increasing galactic latitude ($|b|$) or with an increasing angular distance to the Galactic center (θ). The data presented by H93 are insufficient for performing this test since they do not represent a broad range in either b or θ and the number of sources is limited. Deeper HRI surveys (for more precise source positions) at several θ and $|b|$ values are needed.

The most direct test of our CV contribution to the CXB is to optically identify the predicted $\sim20\%$ of the faintest ROSAT survey sources. Regardless of whether these are non-magnetic CVs or magnetic ones (*e.g.*, AM-Her CVs), they should appear almost as "pure" emission line sources and could thus be found by very deep images in broad band (R) vs. Hα. We emphasize that *all* faint x-ray sources in the ROSAT fields should be examined, including ones already "identified" as AGNs since it is probably that some fraction of these have been misidentified given an average AGN angular separation of only $\sim4'$.

We thank Jerry Ostriker, Rosanne Di Stefano, and Elihu Boldt for interesting discussions. This work was supported by the U.S. National Science Foundation, grant PHY-91-06678, and NASA, grant NAGW-3280.

References

Barcons, X., & Fabian, A.C. 1988, MNRAS, 230, 189

Boyle, B.J., Griffiths, R.E., Shanks, T., Stewart, G.C., & Georgantopoulos, I. 1993, MN-RAS, 260, 49

Finley, J.P. 1994, to appear in Proceedings of the ROSAT Science Symposium

Giacconi, R. et al 1979, ApJ, 234, L1

Grindlay, J.E., 1984, Adv.Space.Res., 3, 19

Grindlay, J.E. 1994a, in *Evolution of X-Ray Binaries*, eds. S. Hott & C. Day, AIP Conf.Proc., 308, 339

Grindlay, J.E. 1994b, in *Millisecond Pulsars: Decade of Surprise*, ASP Conf. Proc., Aspen Workshop, eds. A Fruchter, M. Tavani, D. Backer, in press

Hamilton, T.T., & Helfand, D.J. 1987, ApJ, 318, 93

Hasinger, G., Burg, R., Giacconi, R., Hartner, G., Schmidt, M., Trümper, J., & Zamorani G. 1993, A&A, 275, 1 (H93)

Hertz, P., & Grindlay, J.E. 1984, ApJ, 278, 137

Hertz, P., Bailyn, C.D., Grindlay, J.E., Garcia, M.R., Cohn, H., & Lugger, P.M. 1990, ApJ, 364, 251

Howell, S.B., & Szkody, P. 1990, ApJ, 356, 623

Kashyap, V., Rosner, R., Schramm, D. & Truran, J. 1994, Submitted to ApJ.Letters

Lacey, C.G. and Ostriker, J.P. 1985, ApJ, 299, 633

Maoz, E. and Grindlay, J.E. 1995, ApJ, in press (May 1, 1995)

Mather, J. *et al.* 1990, ApJ, 354, L37

Narayan, R. and Yi, I. 1994, preprint

Ögelman, H., Finley, J.P., & Zimmerman, H.U. 1993, Nature, 361, 136

Ögelman, H. 1994, to be published in *Lives of Neutron stars*, NATO ASI, Kemer, Turkey

Soltan, A.M., 1991, MNRAS, 250, 241

Verbunt, F. *et al.* 1994, A&A, in press

ARCS IN X-RAY SELECTED CLUSTERS

I.M. GIOIA[1,2], G.A. LUPPINO[1], J. ANNIS[1,3], F. HAMMER[4,5]
AND O. LE FÈVRE[4,5]

[1] Insitute for Astronomy, University of Hawaii, USA;

[2] Istituto di Radioastronomia del CNR, Bologna, ITALY;

[3] Fermilab, Batavia, IL, USA;

[4] DAEC, Observatoire de Paris, Meudon, FRANCE

AND

[5] CFHT Corporation, Kamuela, Hawaii, USA

Abstract

We present the results of an imaging survey to search for gravitationally lensed arcs in a sample of 40 EMSS clusters of galaxies. The lensing frequency obtained suggests that the X-ray selection is the preferred method for finding true, massive clusters which are more likely to exhibit the lensing phenomenon.

1. Introduction

Giant arcs are images of distant galaxies, gravitationally distorted by massive foreground clusters. Arc redshifts are systematically measured to be greater than the cluster redshifts (see among others Soucail et al. 1988; Pelló et al. 1992) and very similar to those of galaxies found in deep redshift surveys. This fact, combined with the symmetry and locations of the arcs, confirm the gravitational lensing idea. As the physics of the lensing is understood, we may use the properties of the arcs to probe the mass distribution in rich clusters. Another promising field of investigation is the study of the properties of very distant normal galaxies which are the "source" of the lensing. A large number of clusters needs to be observed to derive global properties of either the lens or the sources and large amounts of telescope time need to be allocated for both imaging and spectroscopy. A high efficiency procedure to search for arcs is then required.

M. Kafatos and Y. Kondo (eds.), Examining the Big Bang and Diffuse Background Radiations, 549–552.

We have been involved in an observational program to image a sample of 40 of the highest X-ray luminosity Extended Medium Sensitivity Survey (EMSS; Gioia et al. 1990) clusters with $L_x > 2 \times 10^{44}$ ergs/s and with redshifts in the range $0.15 \leq z \leq 0.83$, in order to search for gravitational lensing (Gioia and Luppino, 1994; Luppino et al. 1994). We assume that the hot, X-ray emitting gas is a tracer of the total gravitational potential of the cluster, both visible and dark matter. High X-ray luminosity is therefore a sign of a deep potential well and thus is an indicator of a true massive cluster, likely to exhibit the lensing phenomenon.

2. Observations

CCD images of the clusters were obtained during the period from May of 1992 through January of 1994 using the University of Hawaii 2.2m telescope equipped with a Tektronix 2048 × 2048 CCD. R band images of all 40 clusters, and B band images of some of the clusters were acquired during 8 observing runs in photometric conditions and in excellent seeing (median seeing value of $0''.8$ FWHM in R and $0''.9$ FWHM in B). Additional V and I images of a few clusters using both the UH 2.2 m and the CFHT were taken. The image scale was $0''.22$/pixel and the field of view was $7'.5 \times 7'.5$ ($1.54h_{50}^{-1}$ Mpc × $1.54h_{50}^{-1}$ Mpc at z=0.15 and $3.74h_{50}^{-1}$ Mpc × $3.74h_{50}^{-1}$ Mpc at z=0.82). Although gravitationally lensed arcs are known to be relatively blue, they are still more easily detected in the R band since their average $B - R$ color is roughly 1.3 (Soucail 1992). The additional B or V band images were taken to allow us to recognize any gravitational arcs by their relative blue color compared to the red cluster galaxies.

3. Results and Discussion

There is evidence of strong gravitational lensing in the form of giant arcs (length $l \geq 8''$, axis ratio $A \geq 10$) in 8 of the 40 clusters. Two additional clusters contain shorter arclets, and 6 more clusters contain candidate arcs that require follow-up observations to confirm their lensing origin. Luppino et al. (1994) present a fully description of the Hawaii arc survey and discuss the results obtained. We give here a brief summary of the resulting lensing frequency which is 20% for giant arcs, 25% for "secure" cases of lensing, and may be as high as 40% if all the candidate arcs prove to be real. There is a trend for giant arcs to be contained in the higher X-ray luminosity clusters ($L_x > 10^{45}$ erg s^{-1}), thereby confirming our hypothesis that high X-ray luminosity does identify the most massive systems more likely to exhibit lensing. Figure 1 is a mosaic of eight EMSS clusters, seven of which contain giant arcs. The top two are MS0302.7+1658 (arc first reported in Girauud, 1991; see Mathez et al. 1992 for a complete study) and

MS0451.6−0305 (discovered by Luppino et al. 1994). The following clusters are (from left in clockwise order) MS0440+0204 (Luppino et al. 1993); MS1006.0+1202 Le Fèvre et al. 1994); MS1455.0+2232 (arclet candidate, Smail et al. 1994a,b); MS2137.3−2353 (Fort et al. 1992); MS2053.7-0449 (Luppino and Gioia, 1992); MS1621.5+2640 (Luppino and Gioia, 1992).

The results from our survey are consistent with the predictions by several authors that compact lenses produce more arcs (Hammer, 1991; Wu and Hammer, 1993). The high rate of occurrence of arcs (eight out of forty clusters contain giant arcs corresponding to a frequency of 20%), combined with the observed geometry of the arcs, indicate that the cluster central mass density profiles must be compact (Hammer, 1991, Hammer et al. 1993, Grossman and Saha 1994). There is also evidence of mass substructure in the central cluster cores as indicated by the presence of several arcs with large radii of curvature and from the observed orientation of the lensed arcs. We refer the reader to Luppino et al. 1994 for a comprehensive discussion of the results presented here.

We are presently obtaining spectra of the newly discovered arcs with the CFH and Keck telescopes to confirm the lensing hypothesis. Deep imaging observations with HST and X-ray observations with ROSAT and ASCA are also planned. The multiwavelength approach will allow to probe the structures of the sources and the state of the cluster X-ray gas.

This work received partial financial support from NSF Grant AST-91199216 and NASA Grants NAG5-1752, NAG5-1880, and NAG5-2587. The UH CCD cameras were constructed using NSF Grant AST-9020680.

References

Fort, B., Le Fèvre, O., Hammer, F., Cailloux, M. 1992, *ApJL*, **399**, L125.

Gioia, I. M., Maccacaro, T., Schild, R.E., Wolter, A., Stocke, J.T., Morris, S. L., Henry, J. P. 1990a, *ApJS*, **72**, 567.

Gioia, I.M., and Luppino, G. A. 1994, *ApJS*, in press.

Giraud, E., 1991, *ESO Messenger*, **66**, 65.

Grossman, S. & Saha, P. 1994, *ApJ*, in press.

Hammer, F. 1991, *ApJ*, **383**, 66.

Hammer, F., Angonin-Willaime, M.C., Le Fèvre , O., Wu, X. P., Luppino, G. A., and Gioia, I. 1993, in *Gravitational Lenses in the Universe*, 31st Liege Int. Astroph. Colloquium 1993.

Luppino, G.A., and Gioia, I.M. 1992, *A&A*, **265**, L9.

Le Fèvre , O., Hammer, F., Angonin, M.C., Gioia, I., and Luppino, G. A., 1994, *ApJL*, **422**, L5.

Luppino, G.A., Gioia, I.M., Annis, J., Le Fèvre , O., and Hammer, F. 1993, *ApJ*, **416**, 444.

Luppino, G.A., Gioia, I.M., Annis, J., Le Fèvre , O., and Hammer, F. 1994, *ApJ*, submitted.

Mathez, G., Fort, B., Mellier, Y., Picat, J.-P., Soucail, G. 1992, *A&A*, **256**, 343.

Pelló, R., Le Borgne, J.F., Sanahuja, B., Mathez, G., Fort, B., 1992, *A&A*, **266**, 6.

Smail, I., Ellis, R., and Fitchett, M. 1994a, *MNRAS*, in press.

552

Smail, I., Ellis, R., Fitchett, M., and Edge, A. 1994b, *MNRAS*, in press.
Soucail, G., Mellier, Y., Fort, B., Mathez, G., and Cailloux, M., 1988, *A&A*, **191**, L19.
Soucail, G. 1992, in *Clusters and Superclusters of Galaxies*, ed. Fabian, NATO ASI Series Vol. **336**, 199.
Wu, X. P., and Hammer, F. 1993, *MNRAS*, **262**, 187.

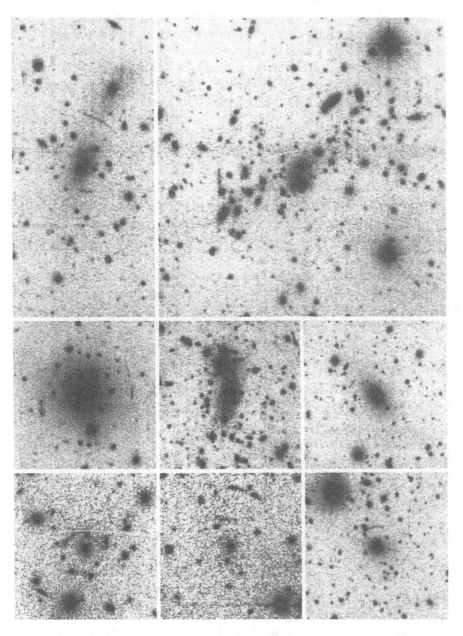

Figure 1 - CCD mosaic of 8 EMSS clusters containing arcs.

TAUVEX AND THE NATURE OF THE COSMOLOGICAL UV BACKGROUND

NOAH BROSCH
The School of Physics and Astronomy, Raymond and Beverly Sackler Faculty of Exact Sciences, Tel Aviv University, Tel Aviv 69978, Israel

TAUVEX is a three-telescope array intended to image wide sky areas in the UV. It is being constructed in Israel for flying on-board the Spectrum X-γ (SRG) international high-energy observatory. SRG will be orbited by Russia in early 1996 for a three-year+ mission. TAUVEX will operate in parallel with X-ray imaging telescopes on board SRG to provide time-resolved photometry and deep UV imaging.

Observations in the UV region longward of Lyman α up to the atmospheric limit at $\sim$3000Å take advantage of reduced sky background because of a fortuitous combination of zodiacal light decreasing shortward of $\sim$3000Å and other backgrounds remaining low up to near the geocoronal Lyman α. In this spectral region it is therefore possible to observe faint astronomical sources with high signal-to-noise ratio, even with a modest telescope.

The present design of TAUVEX includes three co-aligned 20 cm diameter telescopes in a linear array on the same mounting surface. Each telescope images a field of 0°.9 with 10" resolution onto photon-counting position-sensitive detectors with wedge-and-strip anodes, produced by DEP of Roden, Holland. The payload is designed and assembled by El-Op Electro-Optics Industries, Ltd., of Rehovot, the top electro-optical manufacturer of Israel, under close supervision of Tel Aviv University astronomers.

The TAUVEX telescopes are equipped with a set of six filters, which sample the spectral segment 1400–2800Å. Almost all sub-assemblies and parts are at least doubly-redundant, and in most cases the redundancy is invoked automatically. Significant image processing is done on-board, to reduce the volume of the downlinked data.

The combination of long observing periods of SRG (three days out of every four), a high orbit (200,000 km apogee) with low radiation and solar

M. Kafatos and Y. Kondo (eds.), Examining the Big Bang and Diffuse Background Radiations, 553–554.
© 1996 IAU.

scattered background, and long staring sessions to every target (typically 4-6 hours), implies that TAUVEX will be able to detect and measure star-like objects of 20.5 mag with a S/N of 10. This corresponds to V$\simeq$22.5 mag QSOs, given typical [UV-V] colors of QSOs, and at least 10 such objects are expected in every TAUVEX field-of-view. During the guaranteed SRG mission, at least 50,000 QSOs will be observed, if the targets will be different and at high galactic latitude. This is $\sim$an order of magnitude more QSOs than cataloged now.

Diffuse objects, such as nearby large galaxies, will be measured to a surface brightness of about 20 UV mag/square arcsec for each resolution element. Combining the information for larger areas will reach much fainter surface magnitudes. The UV sky background has been reviewed by Bowyer (1991) and Henry (1991). The background is correlated with the Galactic HI column density (Bowyer 1990), and the lowest acceptable value is about 100 photons/sec/cm^2/Å /steradian, as derived from the extrapolation to zero HI column density.

The main limitation of TAUVEX in detecting the diffuse UV back-ground is straylight; this becomes problematic for very faint values of brightness whenever the Sun is in the forward direction of SRG. In cases whenever straylight will not be a problem the detection will be limited by photon statistics, as the TAUVEX detectors are virtually noiseless. In a typical SRG pointing of six hours, using the broad-band filter with HPBW$\simeq$1000Å , TAUVEX will detect the lowest UV background value quoted above with a S/N$\geq$5 in each 10" resolution element. Other filters may be used, with a corresponding reduction of S/N, or this may be boosted by area binning, yielding the wavelength distribution of the UV sky back-ground.

The results have important implications in understanding the nature of the UV background and its astrophysical sources. For example, is it indeed only the summed UV emission from galaxies that produces the extragalactic component ? What is the contribution of resolved galaxies at intermediate redshifts, and how do Butcher-Oemler galaxies fit in this picture ?

References

Bowyer, S. (1991) The cosmic far ultraviolet background, *Ann. Rev. Astr. Astrophys.* **Vol. no. 29**, pp. 59–88.

Bowyer, S. (1990) The galactic far ultraviolet background, in *The Galactic and Extra-galactic Background Radiation* (S. Bowyer and Ch. Leinert, eds.), Kluwer Academic Press, pp. 171–18.

Henry, R.C. (1991) Ultraviolet background radiation, *Ann. Rev. Astr. Astro-phys.* **Vol. no. 29**, pp. 89–127.

PEAKS AND PERIODICITIES IN THE REDSHIFT DISTRIBUTION
OF QUASI-STELLAR OBJECT

D. DUARI, P. DASGUPTA AND J. V. NARLIKAR
Inter-University Centre for Astronomy and Astrophysics
PB-4, Pune-411 007, India.

1. Introduction

There have been claims from time to time that there are periodicities in the redshift distribution of QSOs. These claims are examined from various statistical angles for a sample of 2164 QSO redshifts, with redshifts lying in the range 0.025 − 4.43 taken from the latest compilation (1993) by Hewitt & Burbidge. We have not included QSOs whose redshifts were obtained by 'grism' technique, because of a possible selection effect (Scott 1991). For our analysis we use several staistical techniques, including the Burbidge - O'Dell type Power spectrum analysis (1972), the Rayleigh test, the Kolmogorov-Smirnov test, and the so called 'Comb-tooth' template test. the tests are not inter related and have the advantage of checking the claim for nonuniformity of the distribution from various independent angles. We have also carried out Monte Carlo simulations of the redshift distribution of QSOs to check for any systematic effect that can give rise to a spurious periodic effect.

2. Tests and Results

The Power spectrum analysis gives two peaks of the spectral power function that exceeds 90 % confidence levels. The two peaks corresponds to a periodicity value of $\xi = 0.0565$ and 0.0129 with confidence level of 90.21% and 98.28% respectively.

The generalized Rayleigh test involves the analysisi of phases in the signal (i.e., the number counts of the QSOs against their redshifts) and helps to fine tune the value of the periods obtained by power spectrum analysis. The result of this test showed the presence of the periodicity of 0.0565 with confidence level exceeding 99%. The test showed two other

M. Kafatos and Y. Kondo (eds.), Examining the Big Bang and Diffuse Background Radiations, 555–557.
© 1996 *IAU.*

periods, that of 0.0121 and 0.0127 for two different degrees of freedom for which the test was carried out.

The Kolmogorov-Smirnov test yielded a result in favour of the claim of presence of a periodicity in the data and showed a strong rejection of the null hypothesis of an expected uniform distribution of QSO redshifts.

The Comb-Template test was constructed keeping in mind our requirement of checking the presence if any, of a periodic signal in the observed redshift distribution. It is especially suited to detect periodicity. We consider a comblike template with "teeth" at regular intervals, which is made to slide across the redshift histogram. If the period of the comb matches the underlying periodicity of the histogram, then we expect to see peaks arising when and only when there is a tooth for tooth matching between the two distributions. The test showed the presence of a periodicity of $\xi = 0.0565$ with a 4.1σ level of significance.

The question of redshift periodicity, its validity and cause can be put to tougher challenge if one takes into account all the factors that can creep into the observational data due to the solar system's position and motion in the Milky way galaxy. Using the value of the circular speed of Local Standard of Rest (LSR) with respect to the Galactic Centre(220 ± 19 km s^{-1}; Fich et al, 1989) and the solar system velocity with respect to the LSR (20 km s^{-1} towards $\alpha = 18^h$ and $\delta = 30°$; Kerr and Lynden-Bell, 1986), redshift reduction to the Galactocentric frame was carried out for all the QSOs under investigation. It should be emphasized that with the reduction to the Galactocentric frame, the statistical significance of $\xi = 0.0565$ actually increases. These results lend a greater confidence to the reality of the effect. Moreover, even after considering the error in determining the solar velocity, the effect persists, and in all the cases the confidence level for both $\xi = 0.0565$ and 0.0129 increases.

We next considered the hypothesis of periodicity in $\ln(1 + z)$, especially to test the claim that a periodicity of $\Delta\ln(1 + z) = 0.206$ is present in the QSO redshift distribution (Arp et al, 1990), but failed to detect any significant periodic signal in the $ln(1 + z)$ distribution.

To check whether the periodicity obtained is an artifact due to the statistical tests and to look for any systematic bias in the chosen sample which can produce a periodic effect, we took help of large number of Monte Carlo simulations. From these simulations it was found that there seems to be no systematic effect in the sample we have chosen that can give rise to a spurious periodic signal not present in the actual data; and moreover, our statistical tests are not biased towards picking up periodic signals around the $\xi = 0.0565$ value.

3. Conclusion

Our various statistical tests and Monte Carlo simulations confirm an underlying spiky nature of the redshift distribution of QSOs. There is considerable evidence to support the claim of periodicity of $\xi = 0.0565$ and also perhaps of a periodicity in the range 0.0121–0.0129. The former periodicity, which survives with greater confidence level when the redshifts are transformed into Galactocentric frame, is close to that observed by Burbidge (1968) in a much smaller sample. It is extremely difficult to draw any deeper conclusion from the results, beyond stating the fact that the peaks and periodicities have withstood the challenge of various researchers, and have remained for more than two decades despite the enormous increase of the data.

References

Arp, H. , Bi, H. G., Chu, Y., Zhu, X. 1990, *A & A*, **239**, 33
Burbidge, G. R. 1967, *ApJ*, **147**, 851
Burbidge, G. R. 1968, *ApJ. Lett.***154**, L41
Burbidge, G. R., and O'Dell, S. L. 1972, *ApJ*, **178**, 583
Fang, L. Z., Chu, Y., Liu, Y. and Cao, C. 1982, *A& A*,**106**, 287
Fich, M., Blitz, L., and Stark, A., 1989, *ApJ*, **342**, 272
Hewitt, A. and Burbidge, G.R. 1993, *ApJS*, **87**, 451
Kerr, F.J., and Lynden-Bell, D., 1986, *M.N.R.A.S.*, **221**, 1023
Scott, D. 1991 *A&A*, **242**, 1

INFLATIONARY COSMOGONY, COPERNICAN RELEVELLING AND EXTENDED REALITY

YUVAL NE'EMAN
Wolfson Distinguished Chair of Theoretical Physics
Raymond and Beverly Sackler Faculty of Exact Sciences
Tel-Aviv University, Tel-Aviv, Israel 69978

and

Center for Particle Theory, Physics Department
University of Texas, Austin, Texas 78712

ABSTRACT. "Eternal" Inflation has relevelled the creation of universes, making it a "routine" physical occurence. The mechanism of the Big Bang, from the conditions triggering it, to the eventual creation of the entire matter content of the resulting universe, involves no singular physical processes. However, causal horizons, due to General Relativity, separate the newborn universe from the parent universe in which it was seeded as a localized vacuum energy. The new universe's expansion only occurs "after" infinite time, i.e. "never", in the parents frame. This forces a reassessment of "reality". The two universes are connected by the world line of the initial localized vacuum energy, originating in the parent universe. Assuming that the parent universe itself was generated in a similar fashion, etc., an infinite sequence of previous universes is thus connected by one world-line, like a string of beads.

1. Copernican relevellings in Inflationary Cosmogony

Inflation [1-3] was conceived as the solution to paradoxes within conventional Friedmann cosmology and in the interface with Standard-Model inspired unified gauge (or string) theories. As spin-off, it yields in addition a mechanism for the Big-Bang, in the form of a "de Sitter model" exponential expansion, locally triggered by a large vacuum-energy in a microscopic region. This is the quantum field theory version of the "cosmological constant", in which the latter represents a quantum vacuum energy density (localized), e.g. a fluctuation of the Higgs field responsible for spontaneous symmetry breakdown of the GUT (at $10^{16} GeV$, in the presently experimentally favoured "minimal supersymmetric" version), i.e. of an *inflaton* field. Another unexpected bonus consists in the energy-conserving features of that same mechanism, in creating the full particle content of the universe (a null total energy throughout the process, with the gravitational binding energy cancelling the mass).

Several aspects of Inflation represent a Copernican relevelling. First, the Big-Bang-originated universe is very much larger (beyond the observational horizon at 1.5×10^{10} light-years) than the observable "village". Secondly, it is *eternal* [2]: (a) our Big-Bang was born in an existing universe and (b) flatness ensures a quasi-eternal expansion. Thirdly, the "Big Bangs" are "normal" phenomena and occur stochastically, provided the conditions

559

M. Kafatos and Y. Kondo (eds.), Examining the Big Bang and Diffuse Background Radiations, 559–562.
© 1996 *IAU.*

we described (large vacuum energies) happen to materialize. There are thus infinitely many "universes" – using the term to imply something like our observable universe plus its unobservable embedding (resulting from the same Big Bang). Very roughly, this is a return to the Bondi-Gold-Hoyle (1948) Steady-State universe, yet on an infinitely grander scale – and with continuous creation replaced by creation of spacetime and matter in discrete "bursts".

2. Classical General Relativity Horizons Stretch the Twin Paradox

In this presentation, however, I shall discuss yet another revolutionary feature, which I recently pointed out [4]. This is *a new conceptual watershed in our understanding of time, and even more so of "reality"*. We shall see that even though the "parent" and "offspring" universes are not disconnected (the offspring being the outcome of a "vacuum fluctuation" in a tiny region of the parent), *the newborn will never develop and never 'exist' – within the eternal time frame of the parent!* And yet it will exist in its own time frame, an existence with a time-stretch spreading over billions of years – years that will *never* "come" for the parent's clocks. This leads to surprising metaphysical conclusions about our idea of *reality*, which has to be replaced by a new *"surreality"*. We develop these points in the next sections. Before this, however, we review in this section the seeds of this conceptual revolution, as they already appear in the simplest problems in classical GR.

Horizons, as produced by GR, have been thoroughly studied by Penrose and described in the literature [5,6]. The conceptual issue we discuss here is present *classically* in the simplest Schwarzschild horizons. After the discovery of the quasars, Hoyle et al.[7] suggested that their energy originate in the gravitational collapse of very massive stars. It was then realized that in the formation of a black hole, the collapsing matter never really reaches its Schwarzschild radius, in the reference frame of a distant outside observer A. For a quasar this is of the order of $10^{16} cm$, thus yielding a density of $10^{-4} g/cm^3$ with very little chance for nuclear reactions to be initiated. This led to the suggestion that quasars are (extremely dense)*white holes*, rather than such rarefied *black holes* [8]. Returning to the collapsing star case, its matter accumulates as a shell close to the Schwarzschild radius, gradually becoming infinitely red-shifted, with time-dilation causing it to emit less and less all the time. The whole of A's 'eternity' then corresponds to one hour, in the reference frame of an observer B in the collapsing star, falling into the black hole. This is a common GR extension of the *twin paradox* of Special Relativity. In B's frame, however, things happen very fast: the Schwarzschild radius is reached and crossed within that hour and after another comparable stretch of B's time, he (or she) disappears in the $r = 0$ (classical) singularity. A metaphysical problem then arises – namely 'when' does this last half-hour of the collapsing star occur? Clearly, *half an hour after the end of (our) time!* The issue disappeared, however, when, as a result of the work of J. Bekenstein [9] and of S. Hawking [10], it was realized that quantum black holes, unlike the classical ones, evaporate away through quantum tunneling and through pair creation at the microscopic level. This causes a gradual shrinkage of the Schwarzschild radius and the vanishing of the horizon. Thus, the issue is only marginally present in black hole physics.

The constraints fixing the size of this contribution do not allow us to go into the actual formalism and we refer the reader to the publications listed in ref. [4]. We also recommend gaining insights through the study of the Schwarzschild solution in Kruskal-Szekeres coordinates [11] and through the latter's adaptation to de Sitter geometry by Gibbons and Hawking [12].

3. Non-overlapping Time-Extensions in "Eternal" Inflation

We now come to the related conceptual revolution with respect to time, in the context of Eternal Inflationary Cosmology [2]. The model assumes that the first stage (lasting some 10^{-35} sec) of a Big Bang follows a de Sitter Model (i.e. an exponential expansion), triggered by a large quantum vacuum energy-density $\lambda = < 0|V(\Phi)|0 >$; V is the potential of the *inflaton*, e.g. the 'upper' Higgs at $E_{GUT} = 10^{16}GeV$. The scale function $S(t)$ is then given by $S(t) = exp(Ht)$, with Hubble constant $H = (8\pi G\lambda/3c^2)^{1/2}$. For this stage to last for a brief instant only and then to transit into the Friedmann model we observe, the vacuum energy 'trigger' has to correspond to a 'false' vacuum (e.g. the symmetric $\Phi = 0, V = 0$ solution for the quartic potential of the Higgs field), reached through a supercooling-like unstable procedure and easily replaced (through tunneling) by the true vacuum and Friedmann's slowed expansion. The 'falseness' of that vacuum is a necessary but not a sufficient condition as a 'gracious exit' from the inflationary regime proceeds through the merger of 'bubbles' of 'true' vacuum, forming inside the prevailing 'false' vacuum – a merger which has to overcome the exponential growth of the interbubble intervals. In the latest version [13] this is achieved by assuming Einsteinian gravity to represent the low-energy (long-range) regime of an Affine [14,15] or Conformal quantum gravity. Newton's "constant" is then $G = (16\pi < 0|\sigma|0 >)^{-1/2}$, $\sigma(t)$ the (Brans-Dicke like) dilaton field. In the Planck-energy regime of the de Sitter stage, G isn't yet 'frozen' at this present value; the increasing σ and decreasing G then decrease $S(t)$, letting the bubble-merger process catch up.

Let us follow the birth of a new universe [16,17]. A vacuum fluctuation occurs (e.g. as the energy concentration in a topological defect, such as a cosmic string). The dimensions of this trigger could be as small as $10^3 - 10^9$ Planck lengths. At the end of the inflationary stage it will have reached the size of an orange – and 10^{10} years later (in its own frame B) it will look like our observable universe.

Outside observers A *will just note the creation of a tiny black-hole like object*, with only the very beginnings of an expansion, lasting in this state "forever", i.e. while $t \to \infty$ – very much like the case of the Schwarzschild horizon above. Our entire universe is an A frame and will *never* see the transformation of that tiny false vacuum region into anything else. However, for an inside frame of reference B, we have the birth of a de Sitter universe, a Big Bang, followed by the exit phase, then evolving into a new Friedmann (flat) universe – and perhaps, some 10^{10} years later, astronomers discussing horizons and concepts of reality. Note that classically, the new universe would have involved a singularity (a time-like half-line) due to the Penrose theorem – except that quantum tunneling makes it now possible for that budding 'world' to escape the theorem. In one such solution [16], the new universe starts with a configuration which, classically, would make it recollapse without inflation, a true black hole; *would thus have reached its ordained singularity in the future*. Instead, however, it quantum-tunnels into an exponentially inflating solution *whose classical singularity would have lain in the past*, thus avoiding the singularities altogether. It then goes on to make a universe, with the latter carrying no singularity 'blemish' and being in no way different from its parent, "our" present universe. Presumably, this is also how the universe we live in came into being, with an *eternal* lifetime and with no singularities. We should thus extend the Principle of Covariance to all such universes. They are all eternal - except that this is meaningless within our present conceptual framework: the new universe *will never exist, in our frame A, in all our time;* and yet it is as good as our own universe, will have (in its B frame) galaxies, suns, astronomers and physicists. So where and when does it exist? Note that the time variables in the two frames overlap before the "happy event" which triggers the birth of a universe. They then separate, B going it by itself, observing A fading away, flashing out its eternity in the infinitely red-shifted environment of the new Big Bang..

4. Transcendant time and Surreality

There is, presumably a countable infinity of such "eternities", branching out from each other, then separating, with the offspring, eternally "incubating" -- without ever being born -- in the parent universe's *reality*. [Semantically, *real* as against *abstract* implies *existence in spacetime as perceived in the user's frame*, which justifies our discussion of the effects of the above de Sitter horizons on *reality*.] And yet, beyond the parent's eternity, there is another full-fledged universe, the offspring, flourishing and "realizing itself".

This new picture calls for our conceptual framework to admit "surrealism", *i.e.* "*existence*" *beyond our subjective space and time [4]*. Note that in the direction of the past, there is one world-line tying together all past eternities. (This construction misses 'brother' or 'cousin' universes, selecting only the line of direct descendance). We may use a "transcendant- time variable" τ (A refers here to a frame in the $n-th$ universe, distant from the point where the $n+1-th$ universe will be born), $\tau = \sum_{A^n, n \in Z(n)^n}^{\oplus} arctan[tanh(t_{A^n})]$ for a linear sequence. $Z(n)$" denotes all integer values between 0 and n. This time-resembling variable spans surreality; the genealogy of universes can be represented by $(\tau, \sum^{\oplus} t_n)$.

References.

1. A.H. Guth, *Phys. Rev.* **D23** (1981), 347; A. Linde, *Phys. Lett.* **B108** (1982) 389; A. Albrecht, and P.J. Steinhardt, *Phys. Rev. Lett.*, **48** (1982) 1220.
2. A. Linde, *Phys. Lett.***B175** (1986) 395.
3. Recent reviews of Inflationary Cosmology: A.H. Guth, *Proc. Nat. Acad. Sci. USA*, **90** (1993) 4871; A. Linde, in *Gravitation and Modern Cosmology* 1991, A. Zichichi, ed., (New York: Plenum Press); P.J. Steinhardt P.J., *Class. Quantum Grav.*, **10** (1993) S33.
4. Y. Ne'eman, *Found. of Phys. Lett.*, to be pub. ; also "Inflation, the Top and all that" (Erice lectures, May 1994) to be pub. ; "Time after Time in Eternal Inflation" PASCOS 94 contribution, to be pub.
5. R. Penrose, *Phys. Rev. Lett.* **14** (1965) 57.
6. S.W. Hawking and G.F.R. Ellis, *The Large-Scale Structure of Spacetime*, Cambridge Un. Press, Cambridge (1973).
7. F. Hoyle, W.A. Fowler, G.R. Burbidge and E.M. Burbidge E.M., *Ap. J.* **139** (1964) 909.
8. I.D. Novikov I.D. *Astr. Zh.* **41** (1964) 1075; Y. Ne'eman *Ap. J.***141** (1965) 1303; Y. Ne'eman and G. Tauber *Ap. J.* **150** (1967) 755.
9. J.D. Bekenstein, *Lett. Nuovo Cim.* **4** (1972) 737; *Phys. Rev.* **D7** (1973) 2333.
10. S.W. Hawking, *Comm. Math. Phys.***4** (1975) 199; *Phys. Rev.* **D14** (1976) 2460.
11. M.D. Kruskal M.D., *Phys. Rev.* **119** (1960) 1743; G. Szekeres, *Pub. Math. Debrecen* **7** (1960) 285.
12. G.W. Gibbons and S.W. Hawking, *Phys. Rev¿***D15** (1977) 2738.
13. D. La and P.J. Steinhardt *Phys. Rev. Lett.* (1989) **62** 376.
14. Y. Ne'eman and Dj. Sijacki, *Phys. Lett.***B200** (1988) 286.
15. F.W. Hehl, J. Dermott McCrea, E. Mielke and Y. Ne'eman, "Metric Affine Gravity", to be pub. in *Physics Reports*.
16. E. Farhi, A.H. Guth and J. Guven, *Nucl. Phys.***B339** (1990) 417.
17. W. Fischler, D. Morgan and J. Polchinski, *Phys. Rev.* **D42** (1990) 4042.

DO MASSIVE NEUTRINOS IONIZE INTERGALACTIC HI ?

M. ROOS
High Energy Physics Laboratory
POB 9, FIN–00014 University of Helsinki, Finland

S. BOWYER AND M. LAMPTON
Center for EUV Astrophysics
2150 Kittredge, University of California, Berkeley, CA 94720

AND

J. T. PELTONIEMI
International School for Advanced Studies
Via Beirut 2–4, 34013 Trieste, Italy

Abstract. The radiative decay of massive relic 30eV neutrinos could explain several observational puzzles including the missing dark matter in the universe and the anomalous degree of ionization of interstellar matter in the Galaxy. We note that various non-standard particle physics models with extended scalar sector or minimal supersymmetry have sufficient freedom to accommodate such neutrinos. We discuss observational constraints in the immediate Solar neighborhood, in nearby regions of low interstellar absorption, in the Galactic halo, in clusters of galaxies, and in extragalactic space. Although some observations have been interpreted as ruling out this picture, we note that this is true only for models in which extreme concentrations of neutrinos occur in clusters of galaxies. An instrument is under development to measure the cosmic diffuse EUV background in the local Solar neighborhood, for flight on the Spanish Minisat satellite platform. This instrument will have the capability of providing a definitive test of the radiative neutrino decay hypothesis.

1. Introduction

A relic neutrino ν could decay radiatively into a lighter neutrino ν' and a photon if it has a nonvanishing mass m_ν and an electromagnetic coupling.

563

M. Kafatos and Y. Kondo (eds.), Examining the Big Bang and Diffuse Background Radiations, 563–567.
© 1996 IAU.

The decay photons would be monochromatic, with energy

$$E_\gamma = \frac{m_\nu^2 - m_{\nu'}^2}{2m_\nu} \approx \frac{1}{2}m_\nu \ .$$

The wavelength λ of a photon radiated by a relic dark matter neutrino in the mass range of 10–30 eV would be in the far ultraviolet (FUV) or the extreme ultraviolet (EUV) bands. Recall that the dividing line between FUV and EUV is the hydrogen ionizing Lyman limit at energy $E_\gamma = 13.6$ eV corresponding to the wavelength $\lambda = 91.1$ nm.

The hypothesis of a radiatively decaying relic neutrino is attractive in that it establishes a plausible identity for dark matter manifested in the large scale motions of galaxies and clusters and for the high degree of ionization of intergalactic gas as evidenced by the Gunn-Peterson effect [1]. If these neutrinos have an appropriate mean life of about 10^{24} s, the photons produced can furnish the ionizing energy needed to maintain the high ionization state of intergalactic hydrogen [2, 3, 4]. Moreover, Sciama has shown [5] that the decay radiation is capable of explaining the anomalously high degree of hydrogen ionization within our Milky Way Galaxy if the radiatively decaying neutrino has a mass and mean life in the narrow window $m_\nu = (29.21 \pm 0.15)$eV, $t_\nu = (2 \pm 1) \cdot 10^{23}$s, provided that the neutrinos can coalesce within galaxies. This massive neutrino could be the τ or μ neutrino because their masses are only bound to ≤ 31 MeV and ≤ 0.22 MeV, respectively [6].

2. Non-standard interactions

While physical theories admitting massive neutrinos can easily be formulated, the electromagnetic coupling of these neutral particles requires more elaboration. In the minimal standard model of electroweak interactions the neutrinos are massless and left-handed (negative helicity states, spin and momentum vectors antiparallel). Enlarging the model to include also right-handed states, all neutrinos would be naturally massive four-component Dirac particles like the electrons, but there would be no explanation for the smallness of their mass. Radiative decay is then possible, but only in second-order processes where the rate depends on the small external neutrino mass. This makes the decay extremely slow, yielding a mean lifetime six orders of magnitude larger than that required by Sciama. Also, the standard model predicts a magnetic transition moment of only $3 \cdot 10^{-19} \mu_B (m_\nu/\text{eV})$ for a massive Dirac neutrino, whereas the required moment is 0.5–$1 \cdot 10^{-14} \mu_B$.

One can obtain a larger electromagnetic coupling by introducing new non-standard interactions allowing the chirality flip to occur due to the mass of the virtual fermion present in the second order radiative loop correction.

The minimal scenarios involve one new charged scalar boson coupling to neutrinos and charged leptons [8, 9], and generating a magnetic moment up to $10^{-11}\mu_B$. The simplest scenario generating consistently both the appropriate mass matrix and the required electromagnetic coupling is then a hybrid model, involving both the Zee model [10] and the see-saw mechanism. Other alternative remedies are new exotic fermions, a more complicated radiative scenario involving two loop diagrams or supersymmetric models. Suffice it to say that models can be constructed [11] to realize the Sciama parameters.

All these models can have other observable consequences for low energy phenomenology. In the mass matrix the neutrino flavors naturally mix, as will be tested in on-going or planned experiments. Actually, present experiments already constrain the parameters of the simplest scenarios to being marginally allowed. Also, the resulting Majorana masses imply neutrinoless double-beta decay close to present search limits. Of course, all these models have variants, so that a negative result would only imply that the simplest versions are ruled out.

3. Observations

The confrontation of models of decaying neutrino populations with observations demands a careful treatment of combined attenuation and emission [11]. The absorption cross section of neutral hydrogen becomes enormously smaller for wavelengths greater than the Lyman edge and the highest terms of the Lyman series lines 91.1–91.5 nm.

Because of the very different distance scales involved, the observations divide into two kinds: EUV ($\lambda < 91$ nm) observations that constrain the emissivity from neutrinos within our Galaxy, and FUV ($\lambda > 91$ nm) observations that constrain extragalactic emissivity at redshifts ≥ 0.1.

3.1. EXTREME ULTRAVIOLET OBSERVATIONS

• *Component 1:* Emission from material in the local Solar neighborhood, assuming a uniformly intermixed emitter and absorber of density 0.1 cm^{-3}, predicts a local diffuse EUV line intensity of 60 photons cm^{-2} s^{-1} sr^{-1}.

• *Component 2:* The hot interstellar medium provides no absorption for 80 nm photons. Attenuation would only occur in localized wisps of cooler matter containing neutral or partly ionized atoms. Thus corridors of high EUV transmission define long lines of sight along which neutrino decay emissivity could integrate to significant values. The predicted flux in the direction of the hot ISM corridors is 300 photons cm^{-2} s^{-1} sr^{-1} which is considerably greater than the flux in the case of the local medium alone, in spite of attenuation by the local solar-neighborhood cloud, because of the 100 times greater mean free path in the hot ISM corridors.

• *Component 3:* A Galactic halo of massive decaying neutrinos has a decay luminosity which is not self-absorbed but radiated efficiently into intergalactic space. Unfortunately, the Galactic hydrogen column density constitutes an optical depth which eliminates the possibility of directly observing decay photons in the 80–91 nm band. Thus, regardless of the extent of the halo of the Galaxy, no direct observational limit on the halo neutrino decay luminosity can be set by EUV observations.

• *Component 4:* An 80–91 nm emission process from galaxies and clusters of galaxies at $z <$ 0.1 is again unobservable by direct EUV photometry owing to our Galactic hydrogen, even in the most favorable scenario of the source neutrino halo of the cluster being fully exterior to the hydrogen content of its member galaxies.

The most sensitive upper limits that have been placed on diffuse EUV line emission in the 80–90 nm band are those from the Voyager UVS [12] (re-evaluated): 6000 photons cm^{-2} s^{-1} sr^{-1}.

3.2. FAR ULTRAVIOLET OBSERVATIONS

• *Component 5:* A uniform extragalactic emission from decaying neutrino populations beyond $z = 0.1$ benefit from the transparency of the ISM at wavelengths longer than the Lyman edge. Consequently, a diffuse decay emission line is extended into a diffuse continuum. The flux would appear spatially uniform, spectrally smooth, steadily rising toward shorter wavelengths. One can show that the flux intensity at 140 nm should be 4000 photons cm^{-2} s^{-1} sr^{-1} nm^{-1}. Detecting such an emission source is a considerable challenge.

An observational upper limit to diffuse extragalactic light is at the level of 3600 photons cm^{-2} s^{-1} sr^{-1} nm^{-1} [13], marginally consistent with the expected flux. The most sensitive available FUV diffuse spectroscopy detected no positive extragalactic flux and yielded an upper limit of 2800 photons cm^{-2} s^{-1} sr^{-1} nm^{-1} at 140 nm [14].

• *Component 6:* The amount of non-uniform extragalactic emission at $z >$ 0.1 depends on the degree of clumping. A rich cluster of galaxies with mass 10^{15} $M_\odot$ at redshift $z = 0.2$ would be expected to yield an observable flux of intensity 2 photons cm^{-2} s^{-1} at an observed wavelength $\lambda = \lambda_e(1 + z)$.

Observations of the rich cluster Abell 665 at $z = 0.181$ with the HUT instrument revealed no emission at a local wavelength of 99 nm. This set a mean life limit for cluster-binding neutrinos of $\geq 3 \times 10^{24}$ s, assuming that all the dark matter is neutrinos. Observations of IUE spectra of the cluster of galaxies surrounding the quasar 3C263 at $z = 0.646$ also revealed no emission line, setting a limit of $\geq 2 \times 10^{23}$ s if the neutrinos dominate the cluster mass. Both the IUE and the HUT workers point out that these limits are weakened if absorbing material intervenes. Neutral hydrogen absorbs 15 eV photons strongly, and dust absorbs hydrogen recombination radiation. Consequently the presence of small amounts of neutral gas and dust could destroy the emission signature. Thus the IUE and HUT observations rule out only models in which extreme concentrations occur in clusters of galaxies.

4. Conclusions

Theoretically, there exist many possibilities for massive neutrinos with non-standard interactions [11] which could fit the Sciama scenario [3, 5]. Observationally, the search for photons from decaying neutrinos divides into two parts: neutrinos local to our own Galaxy, whose decay photons appear in the EUV band, and neutrinos at cosmologically significant distances, whose photons appear in the FUV band. Neither population has been detected, but existing data rule out only highly clumped neutrino populations [11]. We conclude that 30 eV neutrinos can be responsible for most of the mass density of the universe, but large galaxies and clusters are bound by dark baryons. This reduces the total number of decaying neutrinos in clusters, and partially attenuates their decay radiation. An instrument to measure the diffuse EUV background, including the neutrino decay line predicted by Sciama, will be flown in the near future on the Spanish Minisat satellite.

References

1. J. E. Gunn and B. A. Peterson, Astrophys. J. 142 (1965) 1633.
2. Y. Rephaeli and A. S. Szalay, Phys. Lett. B 106B (1981) 73.
3. D. W. Sciama, Mon. Not. R. Astron. Soc. 198 (1982) 1P.
4. A. L. Melott et al., Astrophys. J. 324 (1988) L43, *ibid* 421 (1994) 16.
5. D. W. Sciama, Nature 346 (1990) 40; Astrophys. J. 364 (1990) 549; Comments Astrophys. Space Phys. 15 (1990) 71; Phys. Rev. Lett. 65 (1990) 2839.
6. L. Montanet et al., (Particle Data Group), Phys. Rev. D 50 Part II (1994) I.1.
7. J. Schechter and J. W. F. Valle, Phys. Rev. D 24 (1981) 1883.
8. M. Fukugita and T. Yanagida, Phys. Rev. Lett. 58 (1987) 1807.
9. K. S. Babu and V. S. Mathur, Phys. Lett. B 196 (1987) 218.
10. A. Zee, Phys. Lett. 93B (1980) 389.
11. S. Bowyer, M. Lampton, J. T. Peltoniemi, and M. Roos (in preparation, 1994).
12. J. Holberg, and H. B. Barber, Astrophys. J. 292 (1985) 16.
13. R. Kimble, S. Bowyer, and P. Jakobsen, Phys. Rev. Lett. 46 (1981) 80.
14. C. Martin, M. Hurwitz, and S. Bowyer, Astrophys. J. 379 (1991) 549.

PHASE TRANSITION AT THE METRIC ELASTIC UNIVERSE

ALEXANDER GUSEV
Department of General Relativity and Astronomy,
Kazan State University
18, Lenin str., Kazan, 420008, Russia

At the last time the concept of the curved space-time as the some medium with stress tensor $\sigma_{\alpha\beta}$ on the right part of Einstein equation is extensively studied in the frame of the Sakharov - Wheeler metric elasticity(Sakharov (1967), Wheeler (1970)). The physical cosmology pre- dicts a different phase transitions (Linde (1990), Guth (1991)). In the frame of Relativistic Theory of Finite Deformations (RTFD) (Gusev (1986)) the transition from the initial state $\mathring{g}^{o}_{\alpha\beta}$ of the Universe (Minkowskian's vacuum, quasi-vacuum(Gliner (1965), Zel'dovich (1968)) to the final state $\mathring{g}_{\alpha\beta}$ of the Universe(Friedmann space, de Sitter space) has the form of phase transition(Gusev (1989) which is connected with different space-time symmetry of the initial and final states of Universe(from the point of view of isometric group G_n of space). In the RTFD (Gusev (1983), Gusev (1989)) the space-time is described by deformation tensor $\epsilon_{\alpha\beta} = \mathring{g}_{\alpha\beta} - \mathring{g}^{o}_{\alpha\beta}$ of the three-dimensional surfaces, and the Einstein's equations are viewed as the constitutive relations between the deformations $\epsilon_{\alpha\beta}$ and stresses $\sigma_{\alpha\beta}$. The vacuum state of Universe have the visible zero physical characteristics and one is unsteady relatively quantum and topological deformations (Gunzig & Nardone (1989), Guth (1991)). Deformations of vacuum state, identifying with empty Mikowskian's space are described the deformations tensor $\epsilon_{\alpha\beta}$, where $\mathring{g}_{\alpha\beta} = g_{\alpha\beta} - U_{\alpha}U_{\beta}$ the metrical tensor of deformation state of 3-geometry on the hypersurface, which is ortogonaled to the four-velocity U^{α}, $\mathring{g}^{o}_{\alpha\beta} = \delta^{\gamma}_{\alpha}\delta^{\zeta}_{\beta}g_{\gamma\zeta}$ is the 3 -geometry of initial state, $\delta^{\gamma}_{\alpha} = \delta^{\gamma}_{\alpha} - U_{\alpha}U^{\gamma}$ is a projection tensor.

The phenomenological description of thermodynamics systems " creating matter - elastic deformations of gravitational field " can be make on the basis of general theory phase transitions of 2nd kind Landau(Landau & Lifshich (1976)). The characteristic of the phase transitions theory of 2nd kind is to vanish a some element symmetry of thermodynamics system which are

569

M. Kafatos and Y. Kondo (eds.), Examining the Big Bang and Diffuse Background Radiations, 569–570.
© 1996 IAU.

connected with appearance a order parameter η and to creating at the non-symmetrical phase the some new macroscopical physical values, which are vanished at the symmetrical phase. For a description of macroscopical properties media (Prigogin (1988)) a depende- nce of free energy $F = 1/2 \, \sigma^{\alpha\beta} \epsilon_{\alpha\beta} + \varrho$, entropy S of the Universe on the temperature T, deformation tensor $\epsilon_{\alpha\beta}$, order parameter $\eta = \sqrt{T - T_c} = \sqrt{1 - l_o^2/R^2}$ is being found out. In the initial symmetrical phase (from the point of view of isometry group G_{10} – Poincare's group) for $t < t_o$ $R = l_o$, $\varrho = P = F = T = S = 0$. In the final nonsymmetrical phase (from the point of view of isometry group G_7 of space) for $t > t_o$ evolution of the Universe is developing by the following law: $R \rightarrow \cosh \hat{t}$, $\rho \rightarrow \tanh^2 \hat{t}$,

$$P = \sigma_\alpha^\alpha/3 \propto \sum_{n=1}^{6} C_n/R^n, \tag{1}$$

$$S = -\frac{\partial F}{\partial T} = S_0 - a_1 T + a_2 T^2 - a_3 T^3, \tag{2}$$

$$F = -2a_1 - \frac{1}{2}a_2\eta^2 + \frac{3}{4}a_1\eta^4 + 2a_1(1 - \eta^2)^{3/2} - \frac{3}{2}a_3(1 - \eta^2)ln(1 - \eta^2) \tag{3}$$

where $\hat{t} = \sqrt{8\pi a_{is}/3}(t - t_o)$, $a_{ad} = \alpha(T - T_c)$, S_o is an initial entropy of Universe, can be equal a zero, α is a thermal expansion coefficient, $T_c \geq 0$ is a critical temperature, a_1, a_2, a_3 are a parameters of RTFD. This solution has the asymptotic of de Sitter Universe (Hoyle, Barbidge and Narlikar (1993)) The analyses of the phase portraits of one-, two- , three-parameter DS and of the bifurcation space of parameters (Gusev (1989)) confirms the character of chaotic inflation(Linde (1990)) at the early times.

References

Gliner, E.B., (1965), Zhurnal eksper. i teoretich. fiziki., **49**, p.542

Gunzig, E., Nardone, P., (1989), Int. J. Theor. Phys., **28**, p.943

Gusev, A.V., (1983), Proc. Kazan Astr. Observatory, **47**, p.126

Gusev, A.V., (1986), Proc. Kazan Astr. Observatory, **50**, p.33

Gusev, A.V., (1989), Proc. Kazan Astr. Observatory, **52**, p.114

Guth, A.H., (1991), Physica Scripta, **36**, p.237

Hoyle, F., Barbidge, G., Narlikar, J.V. (1993), Ap. J, **410**, p.437

Landau, L.D., Lifshitz, E.M., (1976), Statisticheskay fizika, Moscow, Nauka

Linde, A.D., (1990), Elementary Particle Physics and Inflationary Cosmology, Moscow, Nauka.

Misner C.W., Thorne K.S. and Wheeler J. A., (1973), Gravitation, San Francisco, Freeman

Prigogine, I. et al, (1988), Proc. Nat. Acad. Sci. USA, **85**, p.7428

Sakharov, A.D., (1967), Dokl. Akad. Nauk USSR, **177**, p.70

Zel'dovich, Ya.B., (1968), Usp. Phys. Nauk, **95**, p.209

NONSINGULAR METRIC ELASTIC UNIVERSE

ALEXANDER GUSEV
Department of General Relativity and Astronomy,
Kazan State University
18, Lenin str., Kazan, 420008, Russia

In the RTFD(Gusev (1986)) the conception of a Sakharov - Wheeler Metric Elasticity(SWME)(Sakharov (1967), Wheeler (1970)) had been worked out. On the basis of the exact solutions of Einstein equations and qualitative analysis RTFD the global evolution have been studied and the phase portraits of the early Universe is being constructed. An analysis of phase portraits show on the possibility description of spontaneous creation of Universe from an initial Minkowskian's vacuum to an inflationary de Sitter space-time in the frame of phenomenological non-quantum theory (Guth (1991)). During the past decade, a radically new picture of cosmology has emerged. The present homogeneous expanding Universe would have stated out with a de Sitter phase. The purpose of this paper is to shown that the geometry-dynamical approach to the Einstein's gravitation theory in the frame RTFD also is leaded to the nonsingular cosmological models (Brandenberger (1993)). Let us to propose that before the some moment of time the Universe is at the vacuum state and is described the geometry of Minkowskian's space. Deformations of vacuum state, identifying with empty Mikowskian's space are described by the deformations tensor $\epsilon_{\alpha\beta} = \mathring{g}_{\alpha\beta} - \mathring{g}^{o}_{\alpha\beta}$, An arising of deformation $\epsilon_{\alpha\beta}$ is leaded to appearance of the stress tensor $\sigma_{\alpha\beta}$ and the energy-momentum $T_{\alpha\beta}(\epsilon_{\gamma\delta})$ which is connected with "creating" particles in the Universe. Here we are considered the deformations of Minkowskian's space (the initial vacuum state with $\mathring{g}_{\alpha\beta} = 0$) at the linear theory ($\sim \epsilon$) of finite deformations . The final deformation state $g_{\alpha\beta}$ are searched in the metric class of Friedmann's cosmological spaces. In the comoving reference system $U^{\alpha}(0,0,0,1)$ the Friedmann's equations have form (Narlikar & Padmanabhan (1983), and Gusev (1989)):

$$\frac{\dot{R}^2}{R^2} + \frac{k}{R^2} = \frac{8\pi\varrho}{3}, \qquad (1)$$

571

M. Kafatos and Y. Kondo (eds.), Examining the Big Bang and Diffuse Background Radiations, 571–572.
© 1996 *IAU.*

$$-\frac{2\ddot{R}}{R} - \frac{\dot{R}^2}{R^2} - \frac{k}{R^2} = 8\pi P = \frac{8\pi\sigma_k^k}{3} = k_1(1 + \frac{k_2}{R})(\frac{l_o^2}{R^2} - 1), \qquad (2)$$

where $R(t)$ is so called the expansion factor at the Robertson - Walker line element, k is the curvature parameter with the possible values -1, 0, +1, P is pressure, k_1, k_2 are the some combination from a Lame coefficients, l_o^2 is a "initial radius" Universe, a free parameter model. The phase space of this model is the two-dimensional $(R, \dot{R})$ plane. We note that there is only two singular points $(\dot{R} = 0, \ddot{R} = 0)$ in the phase plane. The one of those points is $R = l_o, \dot{R} = 0$ and corresponds to Minkowski space - time. There are two classes trajectories which are asymptotically de Sitter. Those starting at large positive values of $\dot{R}$ go off to $\dot{R} = +\infty$, reaching their asymptotic value of H from above. Those starting with large negative values of $\dot{R}$ tend to $R = +\infty$ with $\dot{R} > 0$. For small values of $\dot{R}$ and R we can see that there are periodic solutions about Minkowski space. The corresponding solutions oscillate with frequency given by H_o (which is possible equal planck scale) about Minkowski space. Based on the preceding discussion of asymptotic solutions we see that there is a separatrix (Gusev, (1989)) in phase space dividing solutions which tend to $R = +\infty$ from those which oscillate or tend to $R = l_o$. The above analyses of the phase portraits is an indication that in our theory Minkowski space may be unstable toward homogeneous deformations. We stress that all the general features of the phase portrait analyses are true for quadratic deformations of gravitational vacuum. Our model incorporates a very important feature: in the asymptotic de Sitter region, the quadratic deformations and temperature effects does not have an important effect on the geometry.The effective gravitational constant of coupling goes to zero as space - time approaches de Sitter space. In this sense the model is asymptotically free(gravitational confinement Linde, (1990)). At the late times the solutions are described a evolution of the de Sitter Universe $R{\sim}expHt$ (Hoyle et al. (1993)).

References

Brandenberger, R., Mukhanov, V. and Sornborger, A., (1993), Phys. Rev. **D48**, p.1629

Gerlach, U.H., Scott, J.F., (1986), Phys.Rev., **D34**, p.3638

Gusev, A.V., (1986), Proc. Kazan Astr. Observatory, **50**, p.33

Gusev, A.V., (1989), Proc. Kazan Astr. Observatory, **52**, p.114

Guth, A.H., (1991), Physica Scripta, **36**, p.237

Hoyle, F., Barbidge, G., Narlikar, J.V. (1993), Ap. J, **410**, p.437

Linde, A.D., (1990), Elementary Particle Physics and Inflationary Cosmology, Moscow, Nauka.

Misner, C.W., Thorne, K.S. and Wheeler, J. A., (1973), Gravitation, San Francisco, Freeman

Narlikar, J. and Padmanabhan, T., (1983), Ann. Phys. (N.Y.), **150**, 289

Sakharov, A.D., (1967), Dokl. Akad. Nauk USSR, **177**, p.70

Stability Of Cosmological Models

M.I.Wanas[1] and M.A.Bakry[2]

(1. Astronomy and Meteorology Department, Faculty of Science, Cairo University, , Giza, Egypt. E.mail wanas@frcu.eun.eg)
(2. Department of Mathematics, Faculty of Education, Ain Shams University, Cairo, Egypt.)

Abstract

The stability problem in cosmology is studied using the equations of geodesic deviation. General conditions for stability are obtained. The method is applied to a number of cosmological models, resulting from different field theories. The results are compared with those obtained from FRW -models.

Most of the cosmological models assume the validity of the cosmological principle. So, if the universe is isotropic and homogeneous, then what is the origine of large scale structures in the universe?. These structures cannot be formed in stable cosmological models. Therefore, it is of importance to study the stability of models as a first step to answer the previous question.

Generalizing the classical scheme, for studying the stability of a dynamical system, the authors found that stability can be studied [1] by examining the limit $(x^1 = r, x^2 = \theta, x^3 = \phi, x^4 = t)$

$$\lim_{t \to b} \xi^\alpha, \tag{1}$$

where $[a, b]$ is the interval in which the functions ξ^α behave monotonically. The functions ξ^α are the components of the deviation vector resulting as a solution of the geodesic deviation equation [2],

$$\frac{d^2 \xi^\alpha}{ds^2} + 2\{^{\alpha}_{\beta\gamma}\}U^\beta \frac{d\xi^\gamma}{ds} + \{^{\alpha}_{\beta\gamma}\}_{,\lambda} U^\beta U^\gamma \xi^\lambda = 0, \tag{2}$$

M. Kafatos and Y. Kondo (eds.), Examining the Big Bang and Diffuse Background Radiations, 573–574.

where $\{^{\alpha}_{\beta\gamma}\}$ is Christoffel symbol, U^{β} is the unit vector resulting from the solution of the geodesic equation:

$$\frac{dU^{\alpha}}{ds} + \{^{\alpha}_{\beta\gamma}\}U^{\beta}U^{\gamma} = 0. \tag{3}$$

If the limit (1) tends to infinity as t tends to b then the system will be unstable, otherwise the system will be stable. The following scheme is suggested to study the stability of cosmological models, assuming the validity of the cosmological principle. This scheme can be applied as follows: 1- Using the values of Christoffel symboles, of a certain model, to solve (3) we get U^{α}. 2- Using $\{^{\alpha}_{\beta\gamma}\}$ and U^{β} to solve (2) we get ξ^{α}. 3- Examining the limit (1) to see whether the model is stable or not.

The study of stability of cosmological models resulting from the following different field theories: 1- the General Theory of Relativity, 2- the Relativistic Theory of Gravitation [3], 3- the Generalized Field Theory [4], 4- Moller's Tetrad Theory of Gravitation [5], shows [1] that the suggested scheme represents an easy and fast method for exploring the stability of any cosmological model.

References

[1]M.I.Wanas and M.A.Bakry (1994) Astrophys. Space Sci. (to appear).

[2]S.Weinberg (1972) "Gravitation and Cosmology", John Wiley Sons.

[3]A.A.Logonov and M.A.Mestvirishvilli (1989) "The Relativistic Theory of Gravitation" Mir Pub.

[4]F.I.Mikhail and M.I.Wanas (1977) Proc. Roy. Soc. Lond.A, 356, 471.

[5]C.Moller (1978) Mat. Fys. Skr. Dan. Vid. Selsk 39, 13, 1.

Cosmological Models in AP-Spaces

M.I.Wanas

(Astronomy and Meteorology Department, Faculty of Science, Cairo University, , Giza, Egypt. E.mail wanas@frcu.eun.eg)

Abstract

To solve some of the problems of standard cosmology, the Absolute Parallelism (AP) space is used to construct alternative field theories other than GR. Four different field theories are being reviewed. The consequences of their cosmological applications are being compared with those of GR. The results are promising and show that more investigations, concerning this space, are needed, especially generalization of the singularity theorems.

Most of the cosmological models are built using the General Theory of Relativity (GR). The most successful and famous is the standard Big Bang model. This model gained its success from its predictions of the CMBR, confirmed observationally, and the abundance of light elements in our Universe. Although the model is successful, it suffers from some problems (e.g. singularity, horizon, flatness,...). In order to go around these problems, some authors (cf. [1]) suggested the modifications of the model. Others (cf. [2]) proposed the construction of field theories other than GR. They suggested either to extend the field equations of GR (cf. [3]), or to extend its geometric structure (cf. [4]).

We believe that the Absolute Paralellism (AP) geometry (cf. [5]) is a good candidate for this purpose. Some authors started constructing field theories (cf. [6]) using this geometry. The cosmological applications of such theories are promising. Some of the problems of the standard cosmology are removed. For example the problem of creation of matter [7], the horizon and flatness problems [8], were removed using these models.

We still hope that other problems of standard cosmology might find a solution within the framework of theories constructed in the AP-geometry.

M. Kafatos and Y. Kondo (eds.), Examining the Big Bang and Diffuse Background Radiations, 575–576.

For instance the singularity problem which needs some efforts to generalize the singularity theorems (cf. [9]), in the AP-geometry. This work is in progress now.

References

[1]A.D. Linde (1990) "Inflation and Quantum Cosmology" Acad. Press.

[2]F.I. Mikhail (1964) IL Nuovo Cimento, series x 32, 886.
F.W. Hehl, J. Nitsch and P. von der Heyde (1980) in the G.R.G.
Einstein Commemorative Volume, Ed. A. Held, Plenum, New York.

[3]F. Hoyle (1948) Mon. Not. Roy. Astr. Soc. 108, 372.

[4]M.I. Wanas (1990) Astrn. Nachr. 11, 253.

[5]T. Levi Civita (1950) "A Simlified Presentation of Einstein's Unified
Field Theory", Blackie.
F.I. Mikhail (1950) Ph.D. Thesis, London University.

[6]F.I. Mikhail and M.I. Wanas (1977) Proc. Roy. Soc. Lond. A 356,
471.
C. Moller (1978) Mat. Fys. Medd. Dan. Vid. Selsk. 39, 13.
K. Hayashi and T. Shirafuji (1979) Phys. Rev. D 19, 12.

[7]W.H. McCrea and F.I. Mikhail (1956) Proc. Roy. Soc. Lond. A 235,
11.

[8]M.I. Wanas (1989) Astrophys. Space Sci. 154, 165.

[9]S.W. Hawking and G.F. Ellise (1973) "Large Scale Structure of Space
-Time" Cambridge Univ. Press.

COSMOLOGICAL SIMULATIONS WITH SMOOTHED PARTICLE HYDRODYNAMICS

Mario G. Abadi , Diego G. Lambas and Patricia B. Tissera

IATE-Observatorio Astronomico Cordoba, Argentina.
CONICET, Argentina.

We have developed and tested a code that computes the evolution of a mixed system of gas and dark matter in expanding world models. The gravitational forces are calculated with the Adaptative P3M algorithms developed by H. Couchmann , 1993. The calculation of gas forces follow the standard SPH formalism (Monaghan, 1989).
The system expands according to a cosmological solution and consist of a cube with periodic boundary conditions. Our simulations run typically with 5000 gas particles and 32000 collisionless particles.
We have tested the code using the Lazer-Irvine equation (cosmic energy conservation) modified to include the presence of gas particles.

We have also analyzed the ability of the code to reproduce the self-similar solution for the acretion to a spherically symmetric perturbation in a homogeneous expanding background of a collisional gas. Figure 1 show the dimensionless density (D), pressure (P), mass (M) and velocity (V) profiles vs Lambda (radius/turn-around radius) and the corresponding analytical solutions found by Bertschinger, 1985. The vertical bar in the figures indicates the shock radius.

The inclusion of energy dissipation allows the segregation of the dark and the baryonic components and the formation of disk systems. We follow the colapse of a rotating spherical pertubation of 512 gas particles with initial dimensionless spin parameter 0.04. In figure 2a,b are shown the final particle distributions for the gas projected into the xy and xz plane respectively.

Future research will be centered in several cosmological applications of this code which will take into account star formation.

We have benefited with useful discussions with Dr. Simon White and Dr. Julio Navarro.
This work has been supported by grants from CONICET, CONICOR and The British Council, Argentina.

M. Kafatos and Y. Kondo (eds.), Examining the Big Bang and Diffuse Background Radiations, 577–578.

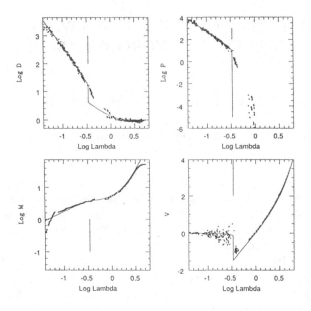

Figure 1

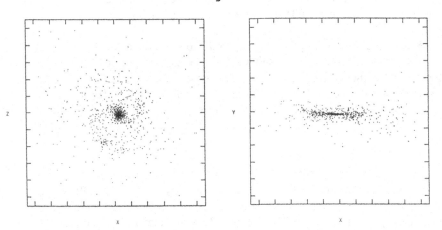

Figure 2a   Figure 2b

Author Index

Abadi, M.G.	577	Elson, A.W.	219
Afanasiev, V.L.	537	Forbes, D.A.	219
Aghanim, N.	447	Ford, V.L.	175
Annis, J.	549	Freudling, W.	183
Anosova, J.	511, 523, 525	Frieman, J.A.	321
Arp, H.	369, 401	Gilmore, G.F.	219
Bakry, M.A.	573	Gioia, I.M.	549
Balayan, S.	537	Giovanelli, R.	183
Bechtold, J.	237	Gispert, R.	447
Benitez, N.	493	Glazebrook, K.	219
Benn, C.R.	481	Gonzalez-serrano, J.I.	
Bhavsar, S.P.	517		493, 497, 499
Borgeest, U.	501, 527	Green, R.F.	219
Borovsky, R.	45	Griffiths, R.E.	219
Bouchet, F.R.	447	Grindlay, J.E.	543
Bowyer, S.	563	Gurvits, L.I.	477
Brjukhanov, A.	45	Gusev, A.	569, 571
Brosch, N.	553	Gutierrez, C.M.	453
Browne, I.W.A.	95	Hammer, F.	549
Budilovich, N.	45	Hancock, S.	453
Burbidge, G.	329, 407	Hasinger, G.	245
Cai, Z.D.	491	Hauser, M.	99
Casertano, S.	219	Haynes, M.P.	183
Cayon, L.	493	Henstock, D.R.	95
Cesarsky, C.J.	109	Hetem, A.	465
Chamaraux, P.	183	Hivon, E.	447
Comastri, A.	263	Horvath, J.E.	465
Da Costa, L.N.	183	Hoyle, F.	329
Dal Pino, E.M.G.	465	Huchra, J.P.	143, 219
Daly, R.A.	469	Illingworth, G.D.	219
Dasgupta, P.	555	Im, M.	219
Davies, R.D.	453	Iyer, S.	523
Dodgson, M.	489	Jackson, J.C.	489
Dodonov, S.N.	537	Jakobsen, P.	229
Driessen, P.	467	Jonas, J.L.	487
Driver, S.P.	219	Kafatos, M.	431
Duari, D.	555	Kalinkov, M.	521
Dunlop, J.S.	79	Kayser, R.	539
Dutta, S.N.	441	Khmil, S.V.	537, 541
Efstathiou, G.	441	Kiseleva, L.	511, 525
Einasto, J.	193	Kniffen, D.A.	279
Elbaz, D.	109	Kondo, Y.	399
Ellis, R.S.	219	Koo, D.C.	201, 219

580

Korogod, V.	45
Kosov, A.	45
Kuneva, I.	521
Lambas, D.G.	503, 509, 577
Lampton, M.	563
Lasenby, A.N.	453
Lauer, D.A.	517
Laurent, P.	271
Lefevre, O.	549
Linsky, J.	529
Loan, A.J.	481
Lubin, P.M.	125
Luppino, G.A.	549
Mandzhos, A.V.	533, 537
Maoz, E.	543
Martin-Mirones, J.M.	497, 499
Martinez-Gonzalez, E.	
	445, 493, 497, 499
Mather, J.C.	419
Mathewson, D.S.	175
Matter, J.	17
Melek, M.	461
Morrison, P.	423
Mosconi, M.	503
Mutz, S.B.	219
Nan, R.D.	491
Narlikar, J.V.	329, 555
Nechaev, V.	45
Neeman, Y.	559
Nemlikher, Y.	45
Neuschaefer, L.W.	219
Novikov, I.	289
Okuda, H.	117
Ostrander, E.J.	219
Partridge, R.B.	59, 427
Pearson, T.J.	95
Pelt, J.	539
Peltoniemi, J.T.	563
Phillips, A.C.	219
Puget, J.L.	447
Ratnatunga, K.U.	219
Readhead, A.C.S.	95
Rebolo, R.	453
Rees, M.J.	389
Refsdal, S.	501
Roos, M.	563
Rottgering, H.	89
Rukavicin, A.	45
Salzer, J.J.	183
Sandage, A.	163
Santiago, B.	219
Sanz, J.L.	445, 497, 499
Schild, R.	539
Schneider, P.	209
Schramm, K.J.	527
Schramm, T.	535
Sciama, D.W.	381
Setti, G.	263
Singal, A.K.	71
Skulachev, D.	45
Smoot, G.F.	31
Spiridonova, O.I.	537
Sreekumar, P.	279
Strukov, I.	45
Suran, D.M.	505
Tammann, G.A.	163
Taylor, G.B.	95
Thomson, D.J.	539
Tissera, P.B.	503, 577
Trimble, V.	9
Turner, M.S.	301
Tyson, A.J.	219
Valotto, C.A.	509
Van den Bergh, S.	157
Van Drom, E.	501
Varma, R.K.	523
Vavilova, I.B.	473
Vermeulen, R.C.	95
Villela, T.	465
Vlassiuk, V.V.	537
Von Linde, J.	501, 527
Wall, J.V.	481
Wanas, M.I.	573, 575
Watson, R.A.	453
Wegner, G.	183
Wilkinson, P.N.	95
Windhorst, R.A.	219
Woltjer, L.	3
Xu, W.	95
Zhdanov, V.I.	537

Subject Index

γ-Ray Background   271
π^0 annihilation   274, 275
$1 < z < 2$ radiogalaxies   493
Abell clusters   147, 157, 256
Active galactic nuclei (AGN)
    3, 110, 234, 238, 246, 247,
    260, 264, 275, 280, 416, 449,
    470, 477, 493, 543
Advanced Cosmic Microwave
    Explorer (ACME)
          40, 129, 132
Alternative cosmological models
          32
Anisotropy   125, 293, 321
ARCADE (Absolute Radiometer
    for Cosmology, Astrophysics
    and Diffuse Emission)   139
ARGO   40
ASCA   252, 253
Baryon/photon ratio   389
Baryonic
    Dark matter   96, 213
    Density   13
    Matter   230, 292, 293,
          331, 349, 361
    Model   292
Big bang
    31, 289, 332, 347, 399, 424,
    431, 529, 559, 575
BLAST (Balloon Absolute
    Spectrometer)   138
C-field   333, 334
Chance and Non-chance compact
    groups   529, 525
CJ1, CJ2   97
Clusters of galaxies
    115, 193, 213, 291, 301, 399,
    435, 448, 473, 523
    EMSS clusters of galaxies 549
COBE Differential Microwave
    Radiometer
          17, 33, 34, 35, 37, 42, 256

COBE (Cosmic Background
    Explorer)   6, 17, 32, 42, 45,
    59, 99, 119, 120, 121, 125,
    198, 199, 200, 302, 308, 310,
    311, 326, 389, 419, 423, 447,
    453, 461, 465, 482, 506
Coma redshift   157
Compton Gamma Ray Observatory
    (CGRO)   280
Cosmic background radiation
    33, 34, 45, 56, 65, 59, 117,
    121, 125, 143, 245, 441, 468
Cosmic Infrared Background
    17, 421, 99, 100, 101, 105
Cosmic Microwave Background
    6, 17, 31, 175, 184, 293, 349,
    419, 427, 432
Cosmic string   308, 449
Cosmological constant
    13, 211, 292, 301, 321, 330,
    332, 346, 420, 489, 531, 559
Cosmological distance   123
Cosmological model
    126, 289, 292, 298, 400, 453,
    469, 477, 571, 573, 575
Cosmological parameters
          42, 209, 238, 447, 505
Cosmological principle
          3, 72, 431, 432, 466, 573
Cosmological Ultraviolet
    Background   553
D/H   354, 529
Dark matter
    13, 34, 211, 289, 304, 305,
    308, 381, 390, 399, 424, 447,
    489, 531, 550, 564
    Cold Dark Matter
          195, 199, 202, 292, 298,
          301, 393, 432
    Cosmic dark matter   215, 291
    Missing dark matter   563
    Halo dark matter   291

582

Deceleration   13, 316, 396, 480
Decoupling   419
Density fluctuations
    22, 197, 321, 445, 465
Density parameter
    195, 321, 391, 482, 503
Deuterium   302
Diffuse background radiation   117
Diffuse Gamma Rays   279
Diffuse Infrared Background
    Radiation (DIBRE)   17, 99
Domain walls   34, 466
Doppler effects   295, 448
Doppler shifts   449, 541
Energetic Gamma Ray Experiment
    Telescope   279
EUV   10, 237
Far Infrared Absolute Spectro-
    photometer   17
Far-Ultraviolet Background   229
Flatness and horizon problems 431
Fluctuations   59, 62
Friedmann cosmologies   201
    Robertson-Walker
    38, 330, 335, 394, 431, 572
Galactic Black Holes (BHs)   546
Galaxies   301, 399,
    435, 501, 525
    High redshift   511
Galaxy rotation curves   321
Gamma Rays   10
General relativity
    31, 339, 346, 431, 433, 560
Grand Unified Theories   431
Gravitational redshift   407
Gravitational Macrolensing   541
Gravitational Microlensing   391
Gravitational lenses
    72, 95, 160, 209, 291, 305,
    315, 322, 392, 429, 494, 496,
    501, 527, 533, 535, 537, 549
Great Wall
    149, 175, 301, 400, 517
Great Attractor
    146, 147, 149, 152, 175, 400

GRO/OSSE   276
HEAO-1   252, 256, 264, 276
Helium
    ^{3}He   302, 354
    ^{4}He   302
    Helium abundance   389
HEMT amplifiers   40
Hot dark matter   292, 301
Hot big bang   301
Hubble constant   61, 144, 146,
    163, 193, 195, 210, 301, 348,
    391, 394, 442, 503, 531, 561
Hubble diagram   79, 349
Hubble flow   14, 159, 165,
    176, 393, 424, 435, 468
Hubble's law   302, 311
Hubble parameter   49, 157, 159,
    160, 322, 399, 461
Hubble time   220, 224, 394
IGM   232
Inflation   32, 309, 315, 321,
    332, 333, 425, 433, 559, 561
Infrared   10
Interstellar Medium (ISM)
    105, 279, 282
IRAS galaxies   364, 494
IRAS   110
IRIS, IRTS   117
ISAS   121
ISM   238
ISOCAM   109
ISOPHOT   110, 112, 113
Lens   95
LIBRIS   54
Local Supercluster   165
Local group
    144, 165, 176, 181, 184, 188
logN-logS   287, 350, 545
Lyman forest clouds (Lyα forest)
    230, 231, 235, 237, 529
Mach's principle   3
Magellanic Clouds   408, 534
Mass-density fluctuations   175
Mass-to-light ratio
    211, 321, 361, 362

Massive neutrinos   229, 292
MAX   132
Medium Scale Anisotropy
    Measurement (MSAM)   40
Metric Elastic Universe
                            569, 571
Microlensing
    210, 290, 316, 442, 533, 537,
    541
Milky Way system   157
MIRS (Middle IR Spectrometer)
                                118
Missing mass   362
Molonglo Quasar Sample   73, 77
Multipole   135
Neutrinos
    10, 306, 307, 315, 382, 563
NII towards βCMa   386
Non-Baryonic matter
                    293, 360, 392, 393
Non-zero cosmological constant
                    49, 204, 393
Nucleosynthesis   273, 389
Olber's paradox   3
Omega
    96, 312, 321, 323, 341, 344,
    345
    $\Omega_{b, cl, vis}$ 290, 291
Optical   10
Owens Valley Radio Observatory
    (OVRO)   60
Phase transition   396, 569
Planck time   303
Planck particles
                    342, 344, 345, 346
Primeval density fluctuations
                                303
Primeval galaxies and quasars
                                113
Primordial nucleosynthesis
                    263, 308, 309
Primordial abundances   498
Primordial density fluctuations
                                175
Quantum gravity   31, 437

Quasi-Steady State
                    329, 363, 425
Radio   10
Redshifts
    79, 89, 97, 201, 407, 424,
    497, 555
RELICT-1   45
RELICT-2   45
ROSAT   251, 252, 253, 258
Sa galaxy   370
SAMBA   139
South Pole   60
Spontaneous symmetry breaking
                                 66
Standard big-bang cosmology
                            331, 350
Standard model of cosmology   31
Standard Big Bang   12, 14
Standard cosmology   302
Superclusters 301, 393, 435, 521
    Supercluster-Void   193
Supernovae   158, 274
Supersymmetry   302
Supervoids   193, 521
Supnernovae Ia (B)   169
Sy 1 galaxies   264, 267
Sy 2 galaxies   266
Sy2/Sy1 ratio   269
Tired light   13, 74, 407
Total density   13
Tully-Fisher   159, 183
Type Ia   158
ULISSE   40
Vacuum Energy Density
                    310, 323, 559, 561
VLBA   97
VLBI   96, 491
Voids   32, 301, 435, 512, 523
X-ray background (XRB)
                            245, 267
X-ray luminosity function   246
X-rays   10
XRB   267
Zwicky clusters   494

Object Index

0740+768	97
0902+34 (z=3.4)	91, 499, 501
0943-242	91
0956+47	495
1217+36	495
3C232	376, 380, 381
3C256	495
3C263	569
3C274	404
3C275.1	373, 379, 381
3C279	404
3C294	495
3C309.1	378, 380
3C324	83, 495
3C368	83
4C41.17	79, 83, 84
B2 0902+34	79, 84, 85, 86
Cloverleaf quasar	92
Cluster Abell 665	568
Cluster Abell 2218	67, 394
Cluster of galaxies A665	385
Clusters A370 and MS2137-23	214
Coma cluster	289, 321, 363, 365, 392, 396, 411, 519
Double Quasar 0957+561	530
E galaxies	157, 165, 167
Fornax cluster	157
Galaxy, Milk Way	61, 283, 364, 365, 383
Galaxies Arp 220 and NGC 6240	364
H1413+117	539
Halo star HD 93521	383
Hercules cluster	365
HI Cloud 1225+01	384
Hydra-Centaurus Supercluster	146, 164, 175, 183, 523
IRAS galaxy F10214+4724 at z = 2.3	92
Large Magellanic cloud	282, 291, 364, 444
M101	158
M31	12, 157, 167
M33 with ScII-III galaxies	158
M81	366, 411, 470
M87	158, 366, 470
Mark 205	377, 380
Mark 474	372, 380, 381
MK 205	416, 470
MS 0302.7+1658	552
MS 0451.6-0305	552
MS 0440+0204	552
MS 1006.0+1202	552
MS 1455.0+2232	552
MS 2137.3-2353	552
MS 2053.7-0449	552
MS 1621.5+2640	552
NGC 1097	469
NGC 3067	376, 380, 381
NGC 4147	496
NGC 4291	377
NGC 4319	377, 416, 469
NGC 4472	167
NGC 4651	373
NGC 5682	372
NGC 5689	372, 380
NGC 5832	378
NGC 7603	413
NGC 891	388
Nr 173 (A262, 347, 426)	523
Nr 246 (A3163, 3226)	523
Nr 721 (A3656, 3698, 3706, 3742)	523
Nr 542 (A1060, 3526, 3537, 3656, 3574)	523
Pavo	178, 182
Perseus-Pisces Supercluster	176, 181, 183
Procyon, star	532

PSR   B0531+21,   B0833-45,
    0630+178, B1706-44, B1055-
    52   280
Q2237+030   539
QSO 0957+561   210, 541
QSO 0414+0534   212
QSO 2237+0305   209
QSO with z=4.90   302
QSO UM673   212
QSO pair 1343+266   212
QSO pair UM 680/681   212
QSOs   63,   209,   254,
    258, 291, 416, 418, 469, 495,
    503, 546, 556, 557
Quasar PKS 2000-33 (Z=3.78)239
Quasar Q0014+813   533
Quasars   76, 234, 503
Radio galaxy 0211-122 (z=2.34)
                              91
Radio galaxy NGC 1275   365
Sc I galaxies NGC309 and M100
                             158
Seyfert galaxy NGC4258   403
Seygert galaxies and quasars
                 372, 380, 470
Shapley 8   182
Shapley Supercluster   149, 184
Small Magellanic Clouds 249, 282
Ursa Major clusters   144
Virgo cluster
    143, 157, 163, 366, 404, 470